33rd Air Infiltration and Ventilation Centre Conference (AIVC 2012) and 2nd TightVent Conference 2012

Optimising Ventilative Cooling and Airtightness for [Nearly] Zero-Energy Buildings, IAQ and Comfort

Copenhagen, Denmark
10 – 11 October 2012

ISBN: 978-1-5108-3761-4

Printed from e-media with permission by:

Curran Associates, Inc.
57 Morehouse Lane
Red Hook, NY 12571

proceedings.com

Some format issues inherent in the e-media version may also appear in this print version.

Printed by Curran Associates, Inc. (2017)

International Network for Information on Ventilation and Energy Performance (INIVE)
Lozenberg 7
B-1932 Sint-Stevens-Woluwe
Belgium

Fax: +32 (0)2 529 81 10

info@inive.org

Additional copies of this publication are available from:

Curran Associates, Inc.
57 Morehouse Lane
Red Hook, NY 12571 USA
Phone: 845-758-0400
Fax: 845-758-2633
Email: curran@proceedings.com
Web: www.proceedings.com

TightVent Europe
BUILDING AND DUCTWORK AIRTIGHTNESS PLATFORM

AIVC

Joint Conference
33rd AIVC Conference and 2nd TightVent Conference

Optimising Ventilative Cooling and Airtightness for [Nearly] Zero-Energy Buildings, IAQ and Comfort

List of available papers (status September 25, 2012)

TightVent Europe

BUILDING AND DUCTWORK AIRTIGHTNESS PLATFORM

AIVC

Copenhagen, Denmark
Axelborg Building
10–11 October 2012

Joint Conference
33rd AIVC Conference and 2nd TightVent Conference

Optimising Ventilative Cooling and Airtightness for [Nearly] Zero-Energy Buildings, IAQ and Comfort

PROCEEDINGS

In cooperation with :

International Energy Agency
Energy Conservation in Buildings and Community Systems Programme

International BUILDAIR Symposium

REHVA
Federation of European Heating, Ventilation and Air-conditioning Associations

Acknowledgments

This conference is possible thanks to the support from:

AIVC with its member countries : Belgium, Czech Republic, Denmark, France, Germany, Greece, Italy, Japan, the Netherlands, New Zealand, Norway, Portugal, Republic of Korea, Sweden and USA.

Since 1980, the annual AIVC conferences have been the meeting point for presenting and discussing major developments and results regarding infiltration and ventilation in buildings. AIVC contributes to the programme development, selection of speakers and dissemination of the results. pdf files of the papers of older conferences can be found in AIRBASE. See www.aivc.org.

TightVent Europe

The TightVent Europe 'Building and Ductwork Airtightness Platform' was launched in January 2011. It aims at facilitating exchanges and progress on building and ductwork airtightness issues, including the production and dissemination of policy oriented reference documents and the organization of conferences, workshops, webinars, etc. The platform receives active support from the organisations listed below. More information can be found on www.tightvent.eu.

VELUX

VELUX has given substantial support during the planning process of the conference as well as logistic and financial support (www.velux.com)

ICIEE, DTU Civil Engineering

International Centre for Indoor Environment and Energy, DTU Civil Engineering has given substantial support in the planning process of the conference as well as logistic support (www.dtu.dk)

INIVE EEIG (International Network for Information on Ventilation and Energy performance)

INIVE was founded in 2001.

INIVE is a registered European Economic Interest Grouping (EEIG), whereby from a legal viewpoint its full members act together as a single organisation and bring together the best available knowledge from its member organisations. The present full members are all leading organisations in the building sector, with expertise in building technology, human sciences and dissemination/publishing of information. They also actively conduct research in this field - the development of new knowledge will always be important for INIVE members.

INIVE has multiple aims, including the collection and efficient storage of relevant information, providing guidance and identifying major trends, developing intelligent systems to provide the world of construction with useful knowledge in the area of energy efficiency, indoor climate and ventilation. Building energy-performance regulations are another major area of interest for the INIVE members, especially the implementation of the European Energy Performance of Buildings Directive.

With respect to the dissemination of information, INIVE EEIG aims for the widest possible distribution of information.

The following organisations are members of INIVE EEIG (www.inive.org) :

BBRI - Belgian Building Research Institute - Belgium
CETIAT - Centre Technique des Industries Aérauliques et Thermiques - France
CSTB - Centre Scientifique et Technique du Bâtiment - France
IBP - Fraunhofer Institute for Building Physics - Germany
SINTEF - SINTEF Building and Infrastructure - Norway
NKUA - National & Kapodistrian University of Athens - Greece
TNO - TNO Built Environment and Geosciences, business unit Building and Construction - Netherlands

The following organisations are associated members.

CIMNE - International Center for Numerical Methods in Engineering, Barcelona, Spain
eERG - End-use Efficiency Research Group, Politecnico di Milano, Italy
ENTPE - Ecole Nationale des Travaux Publics de l'Etat, Vaulx en Velin, France
TMT US - Grupo Termotecnia, Universidad de Sevilla, Spain

INIVE

HEALTH AND INDOOR AIR QUALITY CHALLENGES

Marie-Eve Héroux[*1] and Matthias Braubach[1]

[1] WHO European Centre for Environment and Health
Hermann-Ehlers-Str. 10
53113 Bonn, Germany
**herouxm@ecehbonn.euro.who.int*

EXTENDED ABSTRACT

The quality of the air that we breathe inside our homes, offices, schools and other public or private buildings is an important determinant of healthy life and people's well being. Indoor air pollutants, either infiltrating from the outside or produced by indoor sources, are associated with a wide range of both acute and chronic health outcomes. These include respiratory symptoms, asthma, effects on the nervous system, cardio-vascular diseases and lung cancer [1].

However, the relationship between conditions in indoor environments in general, and indoor air exposure in particular, and health is still poorly understood. Moreover, the range in indoor exposures of individuals is large, and therefore the assessment of the population exposure to indoor air pollutants occuring in indoor settings is difficult. Relevant data for the quantification of the burden of disease from exposure to the various indoor air pollutants are scarce. While first attempts to quantify the impact of built and indoor environments indicate that there is a significant health burden, quantitative relationships remain uncertain to a large extent [2]. Intensive research on emerging health risks of exposures in indoor environments continues to provide better basis for policy making and sound source management. Meanwhile, the public health significance of "well known" indoor air exposures, such as carbon monoxide and environmental tobacco smoke, remains. The education of the general public on the sources of air pollution in their indoor environments and the situations under which important exposures may occur is an ongoing challenge.

Effective housing policies and risk management for indoor air quality is multifaceted and requires integrated, concerted action by public authorities, industry and individuals at the national, regional and in some cases even international levels. A few successful interventions aiming at improving housing conditions and/or reducing exposure to indoor pollutants have been published in the literature. Of relevance for the reduction of chemical and biological agents, the few interventions with sufficient evidence to demonstrate their effectiveness are smoke free policies, integrated pest management, lead hazard control, moisture intrusion elimination, radon gas mitigation through active soil depressurization and multi-faceted tailored asthma interventions [2, 3, 4, 5, 6, 7]. However, in general, data on the effectiveness of interventions to reduce exposure, and subsequently show measured improvements in health, are rare. Furthermore, the attribution of benefits to a particular intervention is challenging, especially since the health effects of air pollution are typically related to long-term exposure to several co-pollutants, most likely coming from multiple sources. In addition, rehabilitation projects often tend to focus on one or a limited number of housing factors and may, if not carried out with adequate consideration of the building context, even have negative side effects on health and wellbeing. One example is the conflict arising from energy efficiency measures that can increase the risk of indoor air pollution by reducing ventilation rates [8].

There is a complex relationship between indoor environmental quality, energy efficiency, environmental sustainability, and human health [9]. The residential sector has one of the highest potentials for energy efficiency [10]. In Europe, buildings account for roughly 40 to 45% of total energy consumption [11]. They also contribute a great proportion to ambient air pollution, mainly through emissions from combustion sources such as domestic wood stoves [12]. The building sector also globally emitted close to 18% of direct carbon dioxide from energy combustion in 2008, with 11% due to household use of grid lectricity and district heating, and the remainder due to emissions at household level (e.g. cooking and heating with gas, coal, oil, etc.) [13]. There are therefore co-benefits to addressing healthy housing through good building practices that go beyond the sole positive, direct impact on health.

In the absence of clear definitions and guidelines on healthy buildings, professional judgment of the experts and consensus-based decision making therefore plays an important role in establishing guidance that can be used to indicate acceptable levels of population exposure. Based on expert advice, WHO started developing guidelines for indoor air quality in 2006, addressing three distinct groups of issues. WHO guidelines for biological indoor air pollutants (dampness and mould) were published in 2009 [14]. As well, WHO published pollutant-specific guidelines for nine selected pollutants in 2010 [1]. Finally, indoor air guidelines for household combustion of fuels for cooking, heating and lighting are currently being developed. They will also provide recommendations for household fuels and technology use that will enable a move towards the air quality guidelines. If these guidelines are sensibly applied as part of policy development, indoor exposures to air pollutants should decline and a significant reduction in adverse health effects should follow.

Further work by WHO addresses the dimension of inequalities in relation to housing and built environments, assessing the differences in exposure associated with sociodemographic determinants such as age, sex, income or education. Unfortunately, there is a scarcity of data for such inequality assessments although the available data on housing-related inequalities (e.g. damp homes, thermal comfort, noise, exposure to environmental tobacco smoke or sanitary equipment) indicate that large exposure differentials exist across the WHO European Region. Giving two examples, the data suggest that within the EU, the lowest-income population reports having no bath or shower at home 13 times more often than the richest while 16 million people in relative poverty cannot afford to heat their homes in winter. It is only realistic to expect that similar inequalities will exist for indoor exposures related to chemical substances, but currently there is little data on such indoor exposure in general and basically no data that allow stratification by socioeconomic or demographic factors [15].

In conclusion, indoor air pollution is an important contributor to the burden of disease of the population, but there are many limitations in the quantitative assessment of its impact on human health. There is a need for a stronger scientific evidence base to convince policy makers of the importance of the issue and the need for preventive action to benefit health. Policies for healthy housing and best practices for source control are necessary to reduce the health risks from indoor air pollution to a minimum.

KEYWORDS
Indoor air quality; health; exposure; housing interventions; guidelines

REFERENCES

[1] WHO. 2010. *WHO guidelines for indoor air quality: selected pollutants*. Copenhagen, WHO Regional Office for Europe. (http://www.euro.who.int/en/what-we-publish/abstracts/who-guidelines-for-indoor-air-quality-selected-pollutants.)

[2] WHO. 2011. *Environmental burden of disease associated with inadequate housing*. Copenhagen, WHO Regional Office for Europe. (http://www.euro.who.int/en/what-we-do/health-topics/environment-and-health/Housing-and-health/publications/2011/environmental-burden-of-disease-associated-with-inadequate-housing.-full-version).

[3] Thomson, H., Petticrew, M., and Morrison, D. 2001. *Health effects of housing improvement: systematic review of intervention studies*. BMJ. 323:187-90.

[4] National Center for Healthy Housing. 2009. *Housing Interventions and Health: A Review of the Evidence*. National Center for Healthy Housing. (http://www.nchh.org/Research/Archived-Research-Projects/Housing-Interventions-and-Health-Outcomes.aspx).

[5] Jacobs, D.E., Brown, M.J., Baeder, A., Sucosky, M.S., Margolis, S., Hershovitz, J., Kolb, L., and Morley, R.L. 2010. *A Systematic Review of Housing Interventions and Health: Introduction, Methods, and Summary Findings*. J Public Health Management Practice. 16(5) E-Supp, S5-S10.

[6] Sandel, M., Baeder, A., Bradman, A., Hughes, J., Mitchell, C., Shaughnessy, R., Takaro, T., and Jacobs, D.E. 2010. *Housing Interventions and Control of Health-Related Chemical Agents: A Review of the Evidence*. J Public Health Management Practice. 16(5) E-Supp, S24-S33.

[7] Krieger, J., Jacobs, D.E., Ashley, P.J., Baeder, A., Chew, G.L., Dearborn, D., Hynes, H.P., Miller, J.D., Morley, R., Rabito, F., and Zeldin, D.C. 2010. *Housing Interventions and Control of Asthma-Related Indoor Biologic Agents: A Review of the Evidence*. J Public Health Management Practice. 16(5) E-Supp, S11-S20.

[8] WHO. 2008. *Preliminary results of the WHO Frankfurt housing intervention project*. Copenhagen, WHO Regional Office for Europe. *(*www.euro.who.int/__data/assets/pdf_file/0011/98696/E91699.pdf*)*.

[9] Schenck, P., Ahmed, A.K., Bracker, A., and DeBernardo, R. 2010. *Climate Change, Indoor Air Quality and Health*. University of Connecticut Health Center. *(*www.epa.gov/iaq/pdfs/uconn_climate_health.pdf*)*.

[10] European Environment Agency. 2008. *Energy and environment report 2008*. Copenhagen, EEA. (http://www.eea.europa.eu/publications/eea_report_2008_6).

[11] United Nations Economic Commission for Europe. 2009. *UNECE Discussion Papers. UNECE climate change activities*. Geneva, UNECE. (http://www.unece.org/oes/disc_papers/climat_change.html).

[12] WHO. 2006. *Health risks of particulate matter from long-range transboundary air pollution*. Copenhagen, WHO Regional Office for Europe. (www.euro.who.int/document/e88189.pdf).

[13] WHO. 2011. *Health in the green economy: health co-benefits of climate change mitigation – housing sector*. Copenhagen, WHO Regional Office for Europe. (*www.who.int/hia/hgehousing.pdf*)

[14] WHO. 2009. *WHO guidelines for indoor air quality: dampness and mould*. Copenhagen, WHO Regional Office for Europe. (http://www.euro.who.int/en/what-we-publish/abstracts/who-guidelines-for-indoor-air-quality-dampness-and-mould).

[15] WHO. 2012. *Environmental health inequalities in Europe*. Copenhagen, WHO Regional Office for Europe. (http://www.euro.who.int/en/what-we-publish/abstracts/environmental-health-inequalities-in-europe.-assessment-report).

LONG TERM MONITORING OF RESIDENTIAL HEAT RECOVERY VENTILATION WITH GROUND HEAT EXCHANGE

Bart Cremers
Zehnder Group Nederland
Lingenstraat 2
8028 PM Zwolle, The Netherlands
bart.cremers@zehndergroup.com

ABSTRACT
The monitoring of a demand controlled heat recovery ventilation system with ground heat exchange in a zero-energy building in Groenlo, The Netherlands, revealed interesting practical insights.

A healthy indoor climate can be obtained with a high comfort in terms of CO_2 levels and supply air temperatures. The CO_2 level stays well within the comfortable range in the living room and three bed rooms (parents, child, and guests), thanks to the demand controlled ventilation. Supply air temperatures are in the comfortable range thanks to the heat recovery in combination with the ground heat exchange by an earth pipe. This is shown for ambient temperatures between -8°C and +33°C.

The energy efficient behaviour is proven by the avoided heating load of 3465 kWh and free cooling of 1052 kWh during a full year. The observed seasonal performance factor SPF is 17 for the avoided heating and 8 for the free cooling. The thermal efficiency based on the supply air temperature is observed to be as high as 91% for a slightly unbalanced air flow. When this is mathematically corrected, the thermal efficiency would have been 97% for a perfect balance between supply and return air flow.

KEYWORDS
Indoor air quality (IAQ), residential ventilation, heat recovery, ground heat exchange, passive cooling, monitoring, CO_2, recovery efficiency, demand control

INTRODUCTION
This article reports the results of a full year monitoring of a zero-energy residential building in Groenlo, the Netherlands. The ventilation system in this building is a demand controlled heat recovery ventilation system in combination with ground heat exchange in the form of an earth pipe. The results of the monitoring show that the ventilation system is highly energy efficient and provides a healthy and comfortable indoor climate.

THE BUILDING
The monitored building displayed in fig. 1 has been built according to the passive house standards. In general terms, the house has a compact, well insulated envelope and south oriented windows with triple glazing. Photovoltaic panels and solar thermal collectors on the roof provide electricity and hot tap water during sunny weather. A heat pump coupled to a vertical bore hole is providing heating and cooling via a floor distribution system. Details of the house can be found in [1] and [2].

Figure 1. The monitored building in Groenlo, The Netherlands.

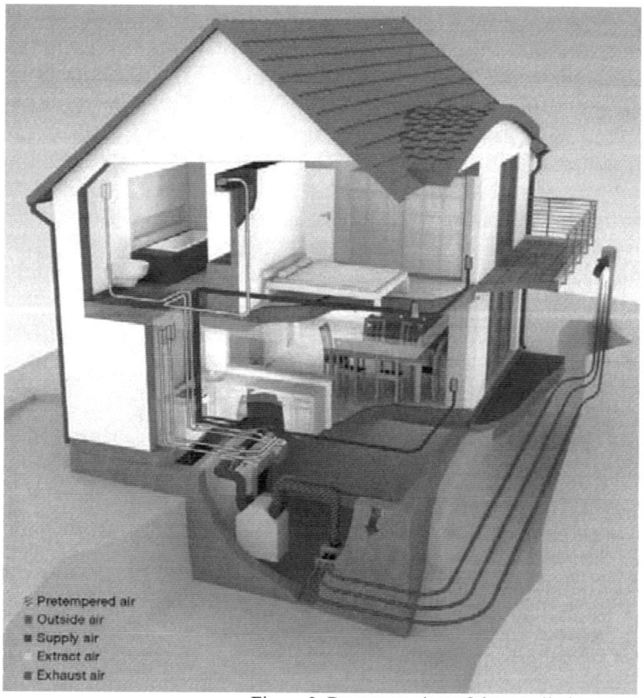

Figure 2. Representation of the ventilation system.

THE VENTILATION SYSTEM

A heat recovery unit is bringing fresh outdoor air into the building and is removing stale air from the building (see fig. 2). The heat from the extract air is recovered and returned back into the fresh air for energy efficient ventilation. The standard ventilation volume is 160 m^3/h. In a house with a volume of 840 m^3 this corresponds to a ventilation rate of around 0.2 h^{-1}.

The outdoor air is supplied to the individual rooms by 7 individual flexible circular ducts. Four of them lead to low induction grilles near the floor of the bedrooms (parents, child and guests) and an office room, all situated on the ground floor. The rest leads to the first floor to the living room. Extract air is extracted from the living room, the loft, the bathroom and the toilets via 7 return ducts. Supply air as well as return air are distributed and collected respectively via sound attenuators, one in the extract air stream and two in the supply air stream. The kitchen is ventilated by a separate HRU which is not subject of the monitoring project.

The manual setting of the ventilation volume (standard position 1; 160 m^3/h) is increased automatically by a demand control based on 4 individual CO_2 sensors in the living room and the bedrooms (parents, child and guests). If one of the CO_2 levels is above a pre-set threshold level, a signal is brought to the HRU to increase the air volume.

Ground heat is provided by an earth pipe. The earth pipe is 50 m long with a diameter of 200 mm and a mean depth of 2.5 m. It is buried into the earth at a slope to remove any possible condensation in the pipe. An air damper is installed in the fresh air duct upstream of the HRU. The HRU controls the damper to decide whether outdoor air is brought into the building directly from outside (north façade) or via the earth pipe (inlet see foreground in fig. 1). Note: fig. 2 shows a slightly different version with 3 parallel and shorter earth pipes without a damper installed.

THE MONITORING

The relevant parameters of the ventilation system have been collected at an interval of 1 minute by a laptop connected to the HRU. The collected data is sent weekly by the resident accompanied by any relevant feedback. The data is transformed into hourly values and analysed in the form of so-called carpet plots, duration graphs, correlation diagrams or bar charts. This report gives the results of a full year starting in February 2011 until February 2012. An intermediate report for a half year period can be found in [3].

COMFORTABLE CO_2 LEVELS

The comfort in the house is assessed by the CO_2 levels in the living room, the master bedroom (2 parents), the child's bedroom and the guest bedroom. The threshold level for the living room was set at 800 ppm and for the bedrooms at 1000 ppm.

As expected, the hourly CO_2 values showed an increased CO_2 level when the rooms were occupied. As an example, fig. 3 shows the CO_2 levels in the child's bedroom for the period February to May 2011. During the day, CO_2 levels are close to the natural background level of 400 ppm while during the night they were in the range 800 – 1000 ppm. When the CO_2 level exceeded the threshold level of 1000 ppm, the ventilation was increased automatically by the HRU to maintain the CO_2 level within a healthy and comfortable range.

From mid-April on, there is a generally lower CO_2 level in all of the rooms resulting from window ventilation used in this period of higher solar irradiation on the south façade.

Another observation is that the CO_2 level during the night was generally higher in the child's bedroom (occupied by one child) than in the master bedroom (occupied by two adults). The reason for this is that the child sleeps with the door closed and the parents sleep with the door open. An open bedroom door results in an exchange of the air in the bedroom (with CO_2

source) with the air in the hallway (without CO_2 source). This pattern was confirmed by CO_2 levels above normal when the master bedroom door was closed occasionally.

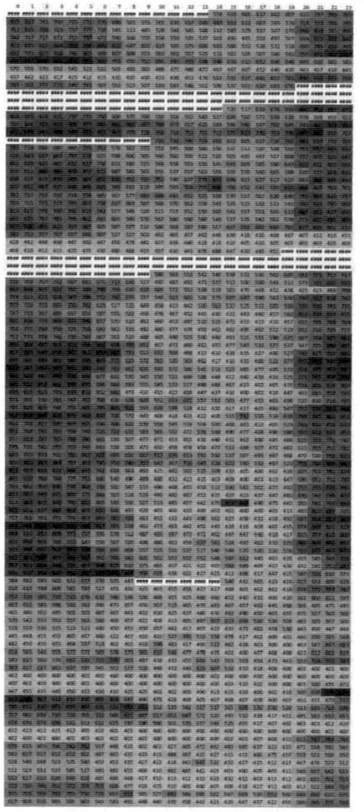

Figure 3. Carpet plot of CO_2 level in the child's bedroom. Rows indicate days from Feb 5th (top row) to May 16th (bottom row). Columns indicate hour of day from 0:00 (left column) to 24:00 (right column). Grey shades indicate CO_2 level ranging from 400 ppm (grey) to 1200 ppm (black), and missing data (white). Note: black represents a CO_2 value that is still below the Dutch guideline values!

The observation of lower CO_2 levels with the door open is confirmed by theoretical calculations. The natural exchange of air by temperature differences between bedroom and hallway can be calculated as 370 m^3/h for a door of 1 m wide and 2 m high with a temperature difference of 1 °C! This is roughly 6 times more than the amount of fresh air of 58 m^3/h provided by the HRU on the maximal level. This means that an open door leads to 6 times faster dilution of CO_2 when compared to a closed door. Note that in case of an open door the CO_2 loaded air is replaced by air from the hallway with unknown air quality, while the HRU ensures the necessary amount of fresh air from outside.

A duration graph of CO_2 levels is given in fig. 4. The uncomfortable level of 1200 ppm is exceeded extremely rare (densely occupied bedroom with closed door). Again, one can see higher CO_2 levels in the child's room than the master bed room during the nights. The CO_2 level of 1200 ppm has never been exceeded in the living room and the child's bedroom. In the master bedroom and the guest bedroom, 1200 ppm has been exceeded only 0.1 % (8 hrs) and 0.2% (16 hrs) of the time, respectively.

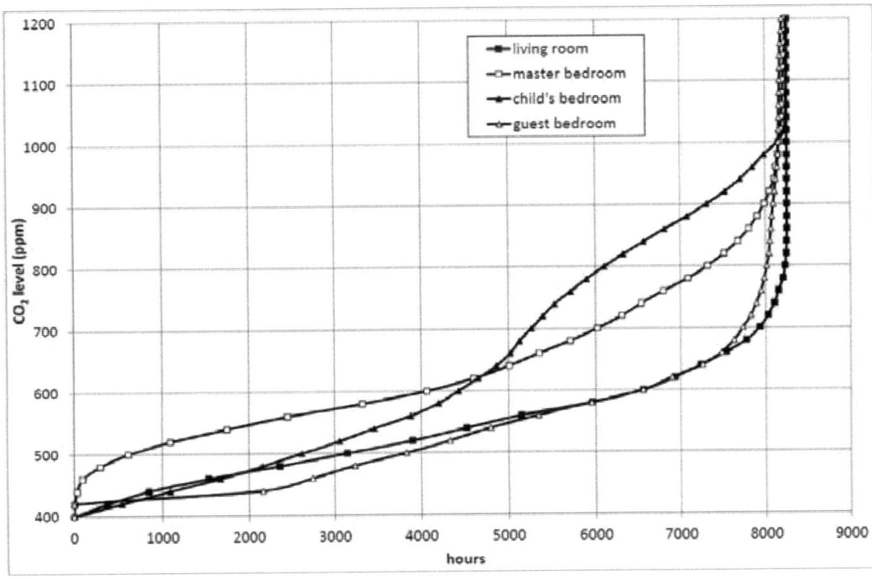

Figure 4. Duration graph of CO_2 levels in living room, master bedroom, child's bedroom and guest bedroom.

COMFORTABLE TEMPERATURES

Throughout the year, the temperature of the earth at 2.5 m depth is much less varying than the outside air temperature. Therefore, the earth can be used for preheating incoming outdoor air in winter and precooling it in summer. In winter, the temperature of the earth is generally higher than the outside air temperature. Fig. 5 shows that the preheated air (at the exit of the earth pipe) is between 8 and 12°C for outside temperatures between -5 and 10°C. In summer, the temperature of the earth is generally lower than the outside air. Fig. 5 shows that the precooled air (at the exit of the earth pipe) is between 12 and 17°C for outside temperatures between 16 and 33°C. For mild outside temperatures between 10 and 16°C the ground heat exchange is switched off by controlling an air valve; outdoor air is taken into the house directly from the north façade (not via earth pipe).

The advantages of the ground heat exchange are the following. In winter, it ensures frost-free operation of the heat exchanger in the HRU, without the need for an electrical anti-freeze heating element. In summer, it decreases the temperature of the outdoor air to a level below the inside temperature, so that free cooling is used for the whole summer period, and not only during cool nights. The extra fan power to draw the air through the earth pipe is negligible (approximately 3 W).

In winter, the (preheated) outdoor air is entering the HRU where it is efficiently heated by the return air in the heat exchanger. Fig. 6 shows that ventilation air is supplied to living room and bedrooms with a comfortable temperature of 18°C in winter even at very low outside temperatures[1]. Without heat recovery, ventilation air would enter the rooms with a temperature equal to outside which would result in uncomfortable draughts.

[1] some hours with supply temperature below 18°C are situations with the central heating switched off during absence

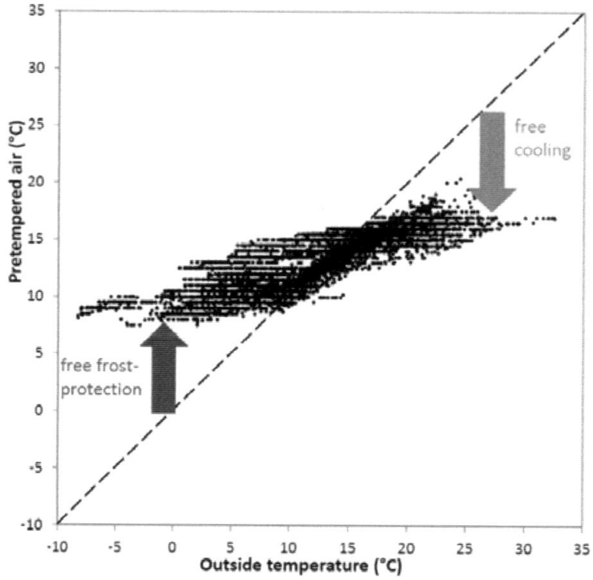

Figure 5. Hourly values of outdoor air entering the house. Ground heat exchange ensures frost protection of the HRU in winter and free cooling for the house in summer.

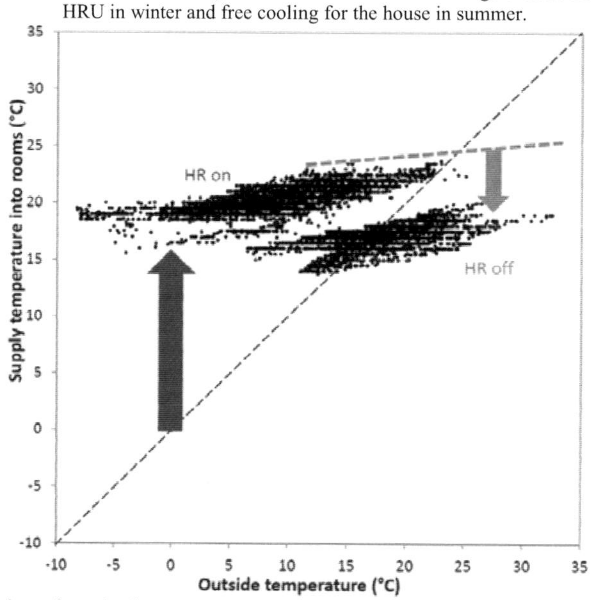

Figure 6. Hourly values of supply air temperature entering the rooms with and without heat recovery (HR on, and HR off). Upward arrow indicates avoided heating and downward arrow indicates free cooling compared to indoor temperature.

For outdoor temperatures above 13°C, the heat recovery is switched off when cooling is both requested and available. This occurs when both of the following conditions are true:

- Actual indoor temperature is above the setting of the comfort temperature (here: 21°C)
- Actual pretempered air temperature is lower than actual indoor temperature.

The heat recovery is switched off by bypassing the heat exchanger in the HRU. Outdoor air is transported directly (without heat recovery) to the rooms. This results in free cooling of the house as the supply temperature is always below the actual indoor temperature. The monitoring shows that the supply temperature is always below 20°C. As the ventilation air flow rate is not large (160 m^3/h), the free cooling cannot be compared with air conditioning equipment, but it raises the comfort and reduces the cooling load of the building.

ENERGY EFFICIENT VENTILATION

The benefits of heat recovery ventilation with ground heat exchange are expressed in terms of avoided heating and free cooling.

With heat recovery switched on, the avoided heating load (or recovered heat) reflects the fact that, thanks to the HRU, the central heating system does not have to heat cold outdoor air to the desired indoor temperature (see upward arrow in fig. 6). The exact amount can be calculated using the actual ventilation flow rate and the actual difference between supply air temperature and outdoor air temperature.

With heat recovery switched off, the free cooling reflects the fact that the indoor air is cooled by the incoming (lower) supply air temperature (see downward arrow in fig. 6). The exact amount can be calculated using the actual ventilation flow rate and the actual difference between indoor air temperature and supply air temperature.

Fig. 7 shows cumulative avoided heating load and free cooling per week during the monitoring period. The energy benefits have been obtained at the expense of the electrical consumption of the fans in the HRU, which consume only 33 W at 160 m^3/h thanks to the low resistance of the flexible air distribution system.

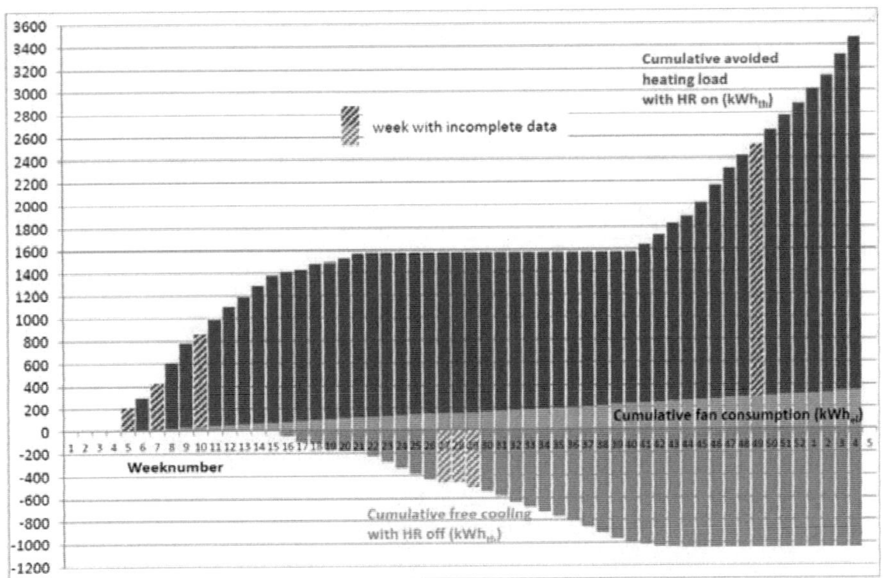

Figure 7. Cumulative sums for avoided heating load (above horizontal axis), free cooling load (below horizontal axis) and fan electricity consumption (above horizontal axis) from week 4 in 2011 to week 4 in 2012.

Table 1 shows a summary of values for the reported period February 2011 until February 2012. The seasonal performance factor SPF for avoided heating (or free cooling) is given by the ratio between avoided heating load (or free cooling load) and the electricity consumption of the fans during hours when the bypass was closed (or open). The observed SPF for avoided heating corresponds reasonably well with an SPF of 22 for the expected gain of a heat recovery system using comparable climate data of Milan, Italy from [4].

	Full year	Electrical consumption of fans during season	Seasonal Performance Factor SPF
Avoided heating load	3465 kWh	199 kWh	17
Free cooling load	1052 kWh	137 kWh	8

Table 1. Annual energy benefit of heat recovery ventilation and seasonal performance factors.

THERMAL EFFICIENCY OF HEAT RECOVERY IN PRACTICE

The thermal efficiency is defined as the ratio between outdoor air temperature increase and maximal temperature increase $(T_{supply\ air}-T_{fresh\ air})/(T_{return\ air}-T_{fresh\ air})$. When ground heat exchange is used, the outdoor air temperature in this formula is the preheated or precooled outdoor air temperature.

The thermal efficiency of an HRU is dependent on a lot of variables, among which ventilation flow rate and mass balance between supply air and extract air are dominant. Fig. 8 shows practical efficiency as a function of fan percentage. The HRU is most frequently in position 1 (fan percentage 35%). Fan positions 2, 3 and absent can also be discerned. Intermediate fan percentages occur when the CO_2 demand control increases fan percentage gradually.

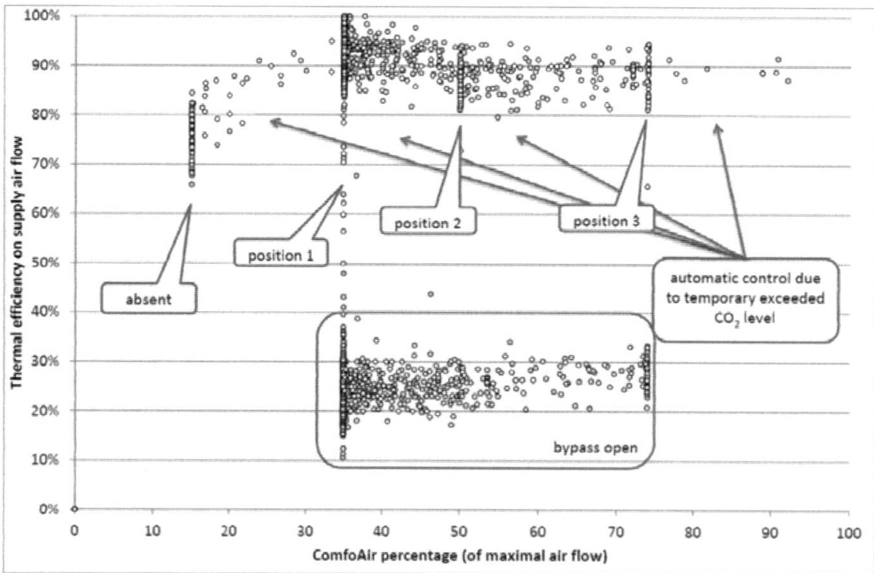

Figure 8. Thermal efficiency of HRU as a function of fan percentage (of maximal rotational speed).

With the bypass open (heat recovery switched off), the average undesired thermal efficiency is still 24%. Optimally, this efficiency would be 0%, but the fans add a small amount of heat (approximately 2°C) to the outdoor air, in spite of the use of efficient EC fans. If AC fans had been used, the thermal efficiency would be even higher.

With the bypass closed (heat recovery switched on), the optimal efficiency is obtained for the most frequently used fan position 1 (160 m^3/h). For the position 'absent' efficiency is decreasing, probably coming from imbalance in mass flows for very low flow regions. For higher fan speeds, the thermal efficiency is slowly decreasing because air is moving fast in the heat exchanger so that the limited exchanger surface becomes noticeable.

The observed average thermal efficiency with bypass closed is as high as 91%. This is a high number considering the fact that the supply air flow and the extract air flow are not perfectly balanced. The resident of the house has commissioned the HRU with a lower extract air flow than supply air flow rate. Detailed flow rate measurements revealed a 6% imbalance in volume flows. Mathematically, one can correct for this imbalance to obtain 91%/(100%-6%) = 97%. This means that, if the HRU system was commissioned in balanced flow, a thermal efficiency of 97% would be obtained, which corresponds perfectly with the thermal efficiency as measured in laboratory.

CONCLUSION

The monitoring of a demand controlled heat recovery ventilation system with ground heat exchange in a zero-energy building in Groenlo, The Netherlands, revealed interesting practical insights. A healthy indoor climate can be obtained with a high comfort in terms of CO_2 levels and supply air temperatures. The energy efficient behaviour is proven by the avoided heating load of 3465 kWh and free cooling of 1052 kWh during a full year. The observed seasonal performance factor SPF is 17 for the avoided heating and 8 for the free cooling.

REFERENCES

[1] Vollebregt, R. 2011. *Koelen zonder energiegebruik door zomernachtventilatie (in Dutch)*, Verwarming & Ventilatie, May 2011, 268–271

[2] Mooi R. 2011. *Huis vol technische installatiesnufjes (in Dutch)*, Installatie & Sanitair, 3, April 2011, 12–14

[3] Cremers B.E. 2012. *Vraaggestuurde ventilatie met wtw en grondwarmte (in Dutch)*, TVVL Magazine, 3, March 2012, 16–19

[4] Kriesi R. 2011. *Comfort ventilation - a key factor of the comfortable, energy-efficient building*, REHVA Journal, May 2011, 21–26

HUMAN PREFERENCE AND ACCEPTANCE OF INCREASED AIR VELOCITY TO OFFSET WARM SENSATION AT INCREASED ROOM TEMPERATURES

Giulio Cattarin[1,2], Angela Simone[*2] and Bjarne Olesen[2]

1 DFT – Dipartimento di Fisica Tecnica, Università degli Studi di Padova, Via Venezia 1, 35131 Padova, Italy

2 DTU – Department of Indoor Climate, Technical University of Denmark, Nils Koppels Allé, 2800 Kgs. Lyngby, Denmark
**Corresponding author: asi@byg.dtu.dk*

ABSTRACT

Previous studies have demonstrated that in summertime increased air velocities can compensate for higher room temperatures to achieve comfortable conditions. In order to increase air movement, windows opening, ceiling or desk fans can be used at the expense of relatively low energy consumption.
The present climatic chamber study examined energy performance and achievable thermal comfort of traditional and bladeless desk fans. Different effects of mechanical and simulated-natural airflow patterns were also investigated. 32 Scandinavians, performing office activities and wearing light clothes , were exposed to a increased air movement generated by a personal desk fan. The subjects could continuously regulate the fans under three fixed environmental conditions (operative temperatures equal to 26 °C, 28 °C, or 30 °C, and same absolute humidity 12.2 Kg/m^3).
The experimental study showed that increased air velocity under personal control make the indoor environment acceptable at higher air temperatures. This will during summer season and in warmer countries improve thermal comfort without too high energy costs. There was significant individual difference in the preferred air velocities, which indicate that personal control is important. The accepted air velocities depended on the type and source of the increased velocity. The Scandinavian subjects did not accept so high velocities as found in studies with Chinese subjects.

KEYWORDS
Thermal comfort, air velocity, personal control, desk fan

INTRODUCTION

Buildings' construction and operation are considered of central importance on the path of a sustainable development. Passive techniques for heating and cooling have gained more and more audience for their feasibility, their efficacy and the positive effects on human health when compared to traditional air-conditioned systems. Previous studies have broadly demonstrated that in summertime increased air velocities can compensate for higher room temperatures to achieve comfortable conditions (from [1] to [4]). In order to increase air movement, windows opening, ceiling or desk fans can be used at the expense of relatively low energy consumption (from [5] to [7]).

METHOD

The present climatic chamber study examined energy performance and achievable thermal comfort of traditional and bladeless desk fans. Different effects of mechanical and simulated-natural airflow patterns were also investigated. 32 Scandinavians, performing office activities (1.2 met) and wearing light clothes (I_{cl} equal to 0.5-0.6 clo), were exposed to a direct air movement generated by a personal desk fan in continuous regulation under three fixed environmental conditions (operative temperatures (t_o) equal to 26 °C, 28 °C, or 30 °C, and relative humidity (*RH*) varying in the range of 40%-50% (at constant dew point of 14.8 °C)).

After an adaptation time, the subjects were invited to adjust the air movement for achieving their preferred thermal comfort. Is the preferred equal to the predicted thermal comfort? Will the personal control or different type of fans affect the results?
The individual preferred air velocities were recorded, and the relative energy consumptions were collected in order to estimate the potential energy savings when comparing to AC systems.

Experimental SET UP

The experiment was carried out in an office-like climatic chamber with dimensions 5.9*5.8*3.2 m^3 at the International Centre for Indoor Environment and Energy of Technical University of Denmark (ICIEE-DTU). The chamber reproduces a typical office room, providing occupants with a view on the outdoors garden. Internal and external blinds can be operated in order to let diffuse sunlight enter the room and meanwhile shade the direct sunlight. Eight workplaces, 4 on the right side and 4 on the left, were arranged with desk, office chair, desk lamp, and desk fan. A partition between the right and left side was located in the middle of the room in order to avoid any possible influence of air moviment due to others occupants (see Figure 1).

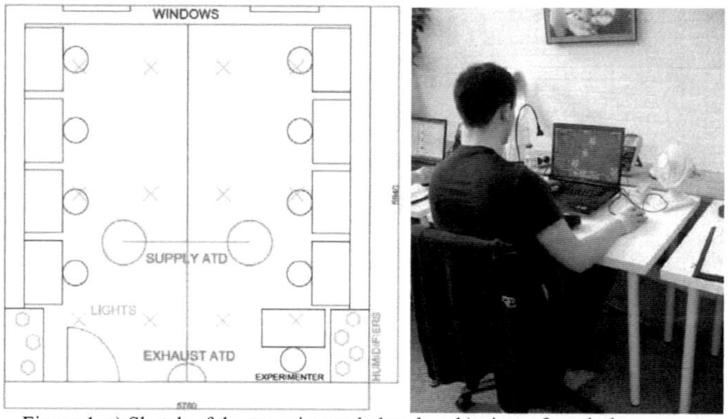

Figure 1. a) Sketch of the experimental chamber, b) view of workplace set up

The air and globe temperature sensors, developed at the ICIEE-DTU [8], have been calibrated for a temperature range of 20-35 °C with an accuracy of ±0.3°C and used during the experimental measurements. Omnidirectional anemometers were used for mesuring air velocity. Temperatures and air velocity at 0.1, 0.6, 1.1 m and 1.7 m heigh above the floor level were recorded in the center of the room.The skin temperatures of both hands and forehead of the subjects were continuously recorded. For the skin temperatures measures iButtons sensors were used as suggested by van Marken Lichtenbelt et al. [9] and Smith et al. [10]. During the experiment three different types of desk fans were used:

- 8 common two-steps desk fans which were regulated with a dimmer switch in order to obtain a continuous variation of the air speed (**CF** = Continuous regulation Fans)
- 2 bladeless fans (**BL**)
- 2 fans which simulate natural wind (**SN**)

The voltage applied to the CF fan has been correlated with the air speed generated at a distance of 60-70 cm from the fan, distance that intercor between the occupant and the fan, so that the air speed profile for each exposed participant was recovered by the recorded voltage data. The SN fans used on the present experiment were built at Tsinghua University, Beijing, China, where they have been used in a previous experiment with Chinese subjects [11].

Along the experiment the subjects were asked to fill in twelve surveys, 4 *long* and 8 *short* (containing respectively 19 and 10 questions) regarding: 1) thermal environment, namely thermal comfort, thermal acceptability, thermal preference, air movement preference, local thermal sensation and air movement sensation; 2) air quality, namely acceptability of air quality, perception of air humidity, preference on air humidity; 3) satisfaction with light and noise level; 4) experience of symptoms such as headache, dry eyes, irritated throat and nose irritation.
A total of 32 Scandinavian voluntiers with good health participated in the experiments, most of them being university students. Their anthropometric data are reported in Table 1.

Sex	No. of subjects	Age (years)	Height (cm)	Weight (kg)	Du Bois area (m^2)	Body Mass Index (BMI)
females	16	23 ± 2	170 ± 6	66 ± 10	1.76 ± 0.14	22.9 ± 3.6
males	16	25 ± 4	180 ± 9	78 ± 22	1.97 ± 0.26	23.8 ± 5.0
females + males	32	24 ± 4	175 ± 9	72 ± 18	1.86 ± 0.23	23.3 ± 4.3

Table 1. Anthropometric data of the subjects.

Each subject was esposed at three different conditions of 4-hours experiments in different days. The subjects were asked to wear a typical summer clothing ensemble, consisting in: panties/briefs, bra (if female), T-shirt, jeans or normal trousers, light socks, trainers or normal shoes. No garments that would protect the subjects from the air movement were allowed. The overall clothing insulation, considering the chair insulation of 0.1 clo (EN ISO 7730), resulted of about 0.5-0.6 clo.

Experimental procedure

The conditions investigated in the experiment are reported in Table 2.

Condition	t_o [°C]	RH %
A	26	50
B	28	45
C	30	40

Table 2. Physical parameters set during the experiment.

The relative humidity was set in order to keep the dew point constant at 14.8 °C. All experiments were carried out in afternoon sessions, in order to exclude confounding factors related to the circadian rhythms of the participants. Eight participants were esposed at the same time and at the same condition. They were allowed to work at their laptop, read a book or perform similar sedentary activities estimated equal to 1.2 met. The exposure time of each condition last in total 4 hours and only comsumption of provided mineral water was allowed. Each condition consisted in 4 periods, as reported in table 2, with the adaptation time followed by three rounds of exposure to different air velocities or type of fans.
During the adaptation time (AD) the occupants were exposed for 90 minutes at the room environment having 0.5 m/s of air movement at the upper body part generated by the CF desk fan; it was extlabished during the pre-test analyses. The three round text consisted on 30 minutes of exposure where the occupants were encouraged to freely adjust the air speed level of their desk fan (with fixed orientation) and 15 minutes of exposure to their preferred air velocity. At the end of each round the subjects were assigned to another desk and different type of fan. Each round consisted of same principle and exposure time.
CF fan was connected to a multimeter, which allowed to record the voltages correspondent to the preferred air velocities. The experiment was conceived to that in each session four

subjects would be exposed to all three different types of fans, whereas the other four subjects would experience twice the exposure to the CF fan.

	Adaptation	1st round		2nd round		3rd round	
air movement	0.5 m/s	free adjust.	fixed setting	free adjust.	fixed setting	free adjust.	fixed setting
abbreviation	**AD**	**1 R**	**1 R***	**2 R**	**2 R***	**3 R**	**3R***
duration [min]	90′	30′	15′	30′	15′	30′	15′

Table 3. Time schedule of the experimental condition.

RESULTS

Table 4 reports the usage of fans at the different environmental conditions (data from R1*, R2* and R3* are pooled). The three types of fans presented peculiar differences of usage. In condition A the CF fans were used by 71% of subjects, whereas only 26% and 35% used the BL and the SN fans, respectively. In condition C a vast majority of subjects kept the CF on. The usage of BL fans increases by 32% and 33% from condition A to B and from B to C, reaching 91% of usage in condition C. While the SN fans had a large increase of use from condition A to B (65%) and slightly decrease from B to C (4%).

% USAGE of FAN	CF		BL		SN	
condition	**% on (n)**	**% off (n)**	**% on (n)**	**% off (n)**	**% on (n)**	**% off (n)**
A	71 (22)	29 (9)	26 (6)	74 (17)	35 (8)	65 (15)
B	89 (29)	11 (3)	58 (14)	42 (10)	100 (18)	0 (0)
C	97 (30)	3 (1)	91 (21)	9 (2)	96 (22)	4 (1)

(n) represent the number of exposed participants

Table 4. Usage of fans at the different environmental conditions

Resulted analyses when CF fans were used

The mean preferred air velocity (±standard deviation (SD)) of the three investigated conditions when CF fan was used are reported in Figure 2. It is increasing from condition A to B to C (from 0.56 m/s to 0.69 m/s to 0.85 m/s).
No statistically significant difference was found between male and female subjects.
The number of subjects choosing a certain air velocity is reported in Figure 3. In condition A nine subjects chose to keep the fan off, and 17 subjects chose an air speed between 0.3 and 0.7 m/s. In condition B the highest preferred air velocity was 0.5-0.6 m/s,while 0.9 m/s was chosen in condition C. A tendency towards higher air speeds at increasing air temperatures is clearly visible, large individual can still be observed.

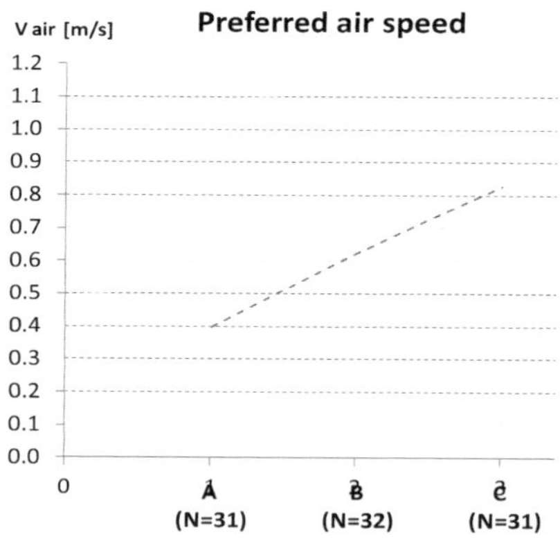

Figure 2. Preferred air velocity at the three investigated conditions.

The mean skin temperature (± (SD)) are reported in table 5 for the occupants exposed to the use of CF fan and for subjects not using their desk fan. The difference was found to be significant in condition A (p <0.0001) and condition B (p <0.002) for the forehead skin temperature, confirming the cooling effect of the CF fans. A slight decrease of forehead skin temperature was noticed with the increasing air velocities. No significant difference was observed between male and female.

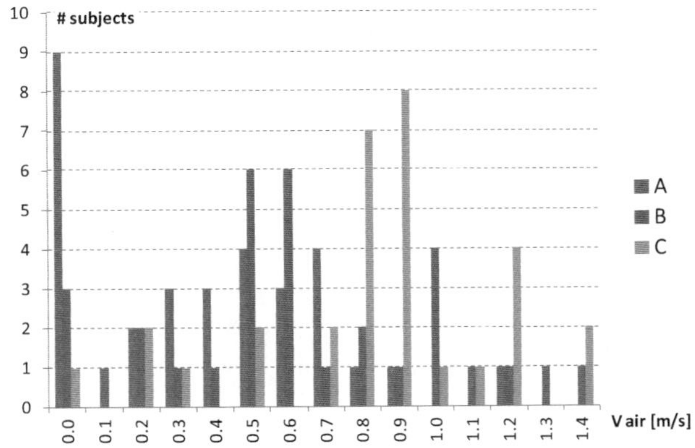

Figure 3. Number of subjects choosing a certain air velocity.

TEMPERATURE ±SD	Forehead (°C)		Right hand (°C)		Left hand (°C)	
Condition	CF-ON	Fan-OFF	CF-ON	Fan-OFF	CF-ON	Fan-OFF
A	33.0±0.4	34.2±0.8	31.6±1.0	32.4±1.1	31.6±1.0	32.4±1.0
B	33.8±0.6	34.5±0.5	33.1±0.8	33.5±0.7	33.3±0.9	33.2±0.7
C	34.2±0.6	34.8±0.7	33.9±0.7	34.0±0.8	34.1±0.6	34.0±0.8

Table 5. Mean skin temperature of forehead, right and left hands of subjects using the CF fans and not using local air moviment.

Responses on the thermal environment

Table 6 shows the mean thermal sensation votes (TSV) for females and males at the three investigated conditions. Females felt slightly warmer than males in condition C, but the difference was not statistically significant. The mean TSV increased with increasing room temperatures, passing from neutrality (0) in condition A to "slightly warm" (+1) in condition C. A comparison of mean TSV between subjects using the CF fan and not using any local air movement is shown on the right side in Table 6. In conditions B and C a slight difference can be noticed, as expected the occupants using the fans tend to have a slightly cooler thermal sensation; butthat is not statistically significant.

The thermal environment for subjects using CF fans was generally considered acceptable in conditions A and B, while it became critical in condition C, with a percentage of dissatisfied rising up to 55%.

TSV ±SD				
condition	**females**	**males**	**Fan On**	**Fan Off**
A	0.0±0.6	-0.1±0.7	0.0±0.6	0.0±0.6
B	0.5±0.6	0.7±0.9	0.5±0.7	0.8±1.0
C	1.5±0.6	1.1±0.7	1.3±0.7	1.7±1.1

Table 6. TSV organized by sex and by the use or not of the CF fan

Figure 4a reports the actual TSV and the Predicted Mean Votes (PMV) of subjects using the CF fans. As shown in figure 4b, the PMV model confirmed to be a fairly good predictive tool for Danish people. In the warm side the percentage of dissatisfied rapidly increases from "neutral (0) to "slightly warm" (+1) to "warm" (+2) as for the PMV/PPD model; while a large discrepancy was found in the "slightly cold" (-1) condition where 26% of people dissatisfied were expected.

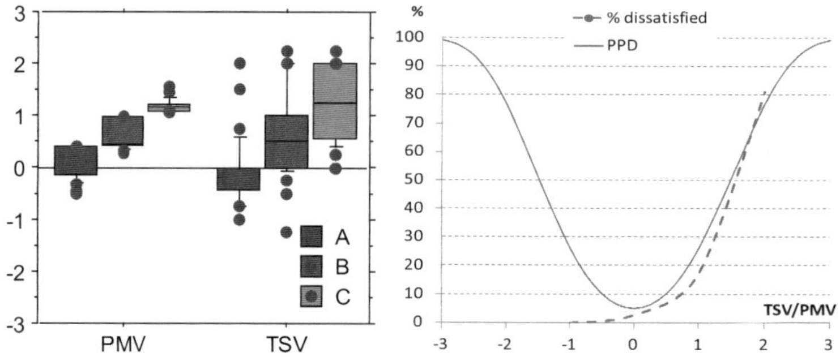

Figure 4. a) Comparison of PMV and TSV; b) Comparison of PMV/PPD model with present results.

The thermal preferences at the three conditions are reported in Table 7. Despite the subjects could increase the air movement, they were not able to reach a satisfying thermal comfort. Infact, in condition C about 85% of the subjects operating CF fans were asking for a cooler environment.

	Thermal preference		
condition	warmer	no change	cooler
A	4	13	5
B	0	12	16
C	0	4	25

Table 7. Thermal preference at the three investigated conditions.

No correlation was found between TSV and preferred air velocity or TSV and forehead skin temperature.

Responses on air movement

The vast majority of the subjects using their desk fan could feel air movement. The body parts where the air movement was most commonly felt were the face (more than 95%) and the right arm (60% to 85% depending on the fan type). It was noted that 20 % of subjects keeping their fan off could anyway feel air movement. The mostly recurrent body part where the air movement was felt was the right arm, possibly due to the chosen setting (fan on the right side of the desk, thus blowing partly towards the right arm of the desk neighbour, see Figure 1b.
In Table 8 are reported the acceptancy of the use of different type of fan for the three investigated indoor thermel conditions. Reminding the limits conditions of the draught (DR) model (reported in ISO7730), DR were calculated at different air velocities for an environment having 26°C and 30% of turbolence (Tu) and reported in Table 9.

	CF		BL		SN		OFF	
	acc.	not acc.	acc.	not acc.	acc.	not acc.	acc.	not acc.
A	21	1 (5%)	2	4 (67%)	6	2 (25%)	43	3 (7%)
B	27	2 (7%)	10	4 (29%)	14	4 (22%)	10	6 (38%)
C	22	8 (27%)	11	10 (48%)	15	7 (32%)	2	2 (50%)

Table 8. Air movement acceptability for the different types of exposure.

DR %	5	9	13	17	22	27	32	37	42
V air [m/s]	0.1	0.15	0.2	0.25	0.3	0.35	0.4	0.45	0.5

Table 9. Draught risk at 26 °C and 30% Tu for increasing air velocities

DISCUSSIONS

Preferred air velocity

When analysing the preferred air velocity (v_{air}), a preliminary question pops up: "Preferred by who?". In order to give a complete view of the human behaviour and response in the present experiment, at least three different groups of subjects can be analysed.

The first group is constituted by all subjects sitting at a desk equipped with a CF fan. This group can be seen as the most representative, however about 30% of the subjects kept their CF fans off in condition A, resulting in low values of mean air velocity of the whole group.

A second group can be identified in the subjects actually using the CF fans. In this way the mean air velocity will not be affected by the null data of the subjects keeping their fan off. In this case, it is important to keep in mind what was the actual usage of the CF fans during the three conditions.

A third group can finally be constituted by those "comfortable subjects" expressing acceptability for both the thermal environment and the air movement, and voting between "slightly cold" (-1) and "slightly warm" (+1) in TSV scale. This group is not reflecting the actual mean level of satisfaction of the experimental sample, but it can be useful to point out which environmental parameters may potentially result in an acceptable thermal balance. It should be noticed that in condition C only 11 subjects (35%) are considered "comfortable".

The TSV and the preferred air velocity of the three mentioned groups is reported in Table 10, where also the effective number of subjects in the three conditions is indicated.

Condition / Group	A		B		C		Nr. of subjects		
	TSV±SD	v_{air} ±SD	TSV±SD	v_{air} ±SD	TSV±SD	v_{air} ±SD	A	B	C
	[-]	[m/s]	[-]	[m/s]	[-]	[m/s]	[-]	[-]	[-]
CF on&off	0.0 ±0.6	0.40 ±0.32	0.6 ±0.8	0.62 ±0.36	1.3 ±0.7	0.83 ±0.34	31	32	31
CF on	0.0 ±0.6	0.56 ±0.23	0.5 ±0.7	0.69 ±0.31	1.3 ±0.7	0.85 ±0.30	22	29	30
comfortable	-0.2 ±0.4	0.41 ±0.29	0.4 ±0.5	0.64 ±0.33	0.6 ±0.3	0.91 ±0.25	26	22	11

Table 10. TSV and preferred air velocity in the three investigated conditions.

PMV and TSV

The PMV resulted to be a good indicator of TSV. The PMV index was actually devised on the basis of tests conducted mainly on subjects from temperate climates. Studies conducted in tropical climates ([13], [14], and [15]) show that the PMV overestimates the thermal sensation of people, and it is today recognized that it is necessary to take into account the adaptation of people to their local climate.

A study by Humphreys & Hancock [16] questioned "what is the actual preferred thermal sensation of subjects expressing their TSV on an ASHRAE scale?". They found that the most common personal desire was "neutral" followed by "slightly warm" and that it varied with the TSV currently experienced. The data were collected in dwellings and lecture rooms for temperatures between 16°C and 24 °C. In the present work the same question resulted in "neutral" tending to "slightly cool" desire for condition A and B, and "slightly cool" in condition C. The different result suggests that culture and climate may affect people's thermal preferences.

On SN fans use

The SN fans provided an average air velocity of 0.7 ÷ 0.8 m/s, with a dynamic profile of air velocities and gusted up to 1.4 m/s. The comparison between TSV of the present study with the previous one performed by Hua (et al. [11]) is shown inFigure 5. In the Chinese experiment the subjects felt on average cooler than the Danish ones, with a difference in TSV of 0.6-0.7 units on the ASHRAE TSV scale at 28°C and 30 °C.

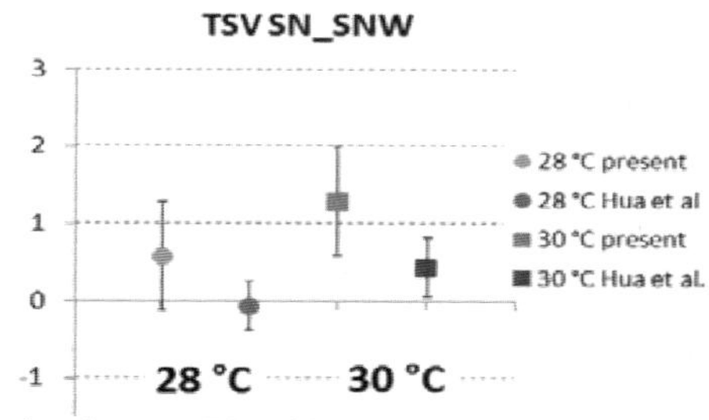

Figure 5. Comparison between TSV of the present study and Hua's similar experiment.

Sixty percent of the subjects using the SN fans were asking for less air movement in condition A, while 40% of them requested more air movement in condition C. This may imply that the dynamic airflow profile should be adjusted to the room temperature in order to guarantee a proper cooling effect.

On BL fans use

BL fans could not be operated at low air velocities (lower than 1.3 m/s), it implied that a consistent number of subjects were asking for less air movement and when possible preferred to turn it off. The BL fans had a strong cooling effect in condition A (see figure 6), causing a mean drop in the forehead skin temperature of about 1.4 °C more than the one obtained by the use of CF fans that was 2.6 °C lower than when no local air movement was used. The difference is less evident in the room conditions where temperatures are higher

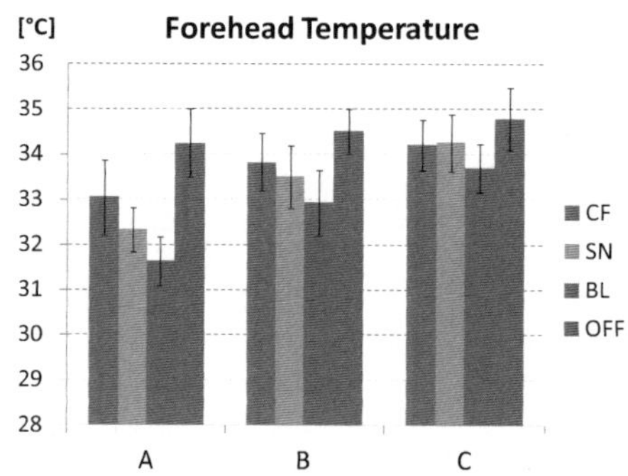

Figure 6. Forehead mean skin temperatures at different room conditions and fans exposure

The bladeless fans present an attractive design style, and the different airflow pattern (a rather constant, non-buffeting airflow) that could result in a high level of comfort. However, new prototypes with also a wider range of air velocities, starting from 0.2 m/scirca, is suggested.

Impact of the operability of CF, BL, and SN fans

The characteristics of the desk fans imposed several limits of use:

- The CF fans had an upper air velocity setting of 1.2-1.4 m/s. It implied that the subject could not further increase the air velocity. However, none of the subjects choosing the highest air velocity asked for more air movement.
- The experimenter was setting the simulated-natural mode of the SN fans. It appeared having an influence on the behaviour of the occupant that repeatedly asked for changes of the fan setting.
- The operability of BL fans operating only at high range of air velocities, reduced the flexibility of use in a sort of on/off use.
- A minor role could have been played by the **aesthetics** of the three types of desk fans, both caused by their "design pleasantness" and by the "perceived familiarity".
- Noise appeared to be one of the reasons why the occupants preferred not to increase the air velocity of the provided fans having instead a warmer thermal sensation. That was mainly related to the BL fan use that increased from 45 dBA to 54 dBA in the middle of the room when it was switched on and it could reach 72 dBA at the occupant place.

Several subjects indicated that their preferred air velocity was chosen as a trade off between the cooling effect and the drying effect on the eyes.
The present study confirmed the findings of Fang et al.[12], who observed that the perceived air quality (PIAQ) decreases with increasing air temperature and humidity. The use of desk fans has to be carefully considered also as regards the eventuality of cross infection.

CONCLUSIONS

The experimental measurements showed that higher air velocity and personal control make the indoor environment acceptable at higher air velocity with a benefit on energy consumption applicable during the summer seasons and in warmer countries. There was significant individual difference in the preferred air velocities, which indicate that personal control is important. The accepted air velocities depended on the type and source of the increased velocity.
The PMV resulted to be a good indicator of TSV, however the PPD curve overestimated the percent of people dissatisfied. The "slightly cool" sensation was actually chosen by 45% of subjects as preferred TSV in condition C, suggesting that culture and climate may affect people's thermal preferences.
The responses from subjects exposed to a simulated natural airflow suggest that the dynamic airflow profile should be adjusted to the room temperature in order to guarantee a proper cooling effect.
Although the bladeless fans had a consistent cooling effect, their low flexibility of use resulted in a large number of subjects dissatisfied. New prototypes with a wider range of air speeds should be designed for a deeper investigation on the potentialities of this technology.
The fan usage did not show correlation with perceived indoor air quality, while it was observed that with the increasing of air temperatures the indoor air quality was negatively perceived.

ACKNOWLEDGEMENTS

Financial support for this study from the Technical University of Denmark is gratefully acknowledged.

REFERENCES

[1] Scheatzle, D.G., Wu, H. and Yellott J. 1989. *Extending the summer comfort envelope with ceiling fans in hot, arid climates*. ASHRAE Trans. 95, 269-279.

[2] Rohles, F.H., Konz, S.A. and Jones B.W. 1983. *Ceiling fans as extenders of the summer comfort envelope*, ASHRAE Trans. 89 (1), 245-263.
[3] Fountain, M., Arens, E., de Dear, R., Bauman F. and Miura, K. 1994. *Locally controlled air movement preferred in warm isothermal environments*. ASHRAE Trans. Symp. 14 (1), 937-949.
[4] Cândido C, de Dear R.J., Lamberts, R. and Bittencourt L. 2010. *Air movement acceptability limits and thermal comfort in Brazil's hot humid climate zone.* Build. and Env. 45, 222-229.
[5] Koranteg, C. and Mahdavi, A. 2011. *An investigation into the thermal performance of office buildings in Ghana.* Energy & Build. 43, 555-563.
[6] Aynsley, R. 2007, (November). *Circulating fans for summer and winter comfort and indoor energy efficiency*. Environment Design Guide.
[7] Yamtraipat, N., Khedari, J., Hirunlabh, J. and Kunchornrat, J. 2006. *Assessment of Thailand indoor set-point impact on energy consumption and environment.* Energy policy 34, 765-770.
[8] Simone, A., Olesen, B.W., Babiak, J., Bullo, M., Langkilde, G. 2007. *Operative temperature control of radiant surface heating and cooling systems*. In: *Proc. Clima 2007,* Helsinki.
[9] van Marken Lichtenbelt, W.D., Daanen, H.A.M., Wouters, L., Fronczek, R., Raymann, R.J.E.M., Severens, N.M.V. and Van Somenen, E.J.W. 2006. *Evaluation on wireless determination of skin temperature using iButtons*. Physiol. and Behaviour 88, 489-497.
[10] Smith, A.D.H., Crabtree, D.R., Bilzon, J.L.J. and Walsh, N.P. 2010. *The validity of wireless iButtons and thermistors for human skin temperature measurement*. Physiol. Meas. 31, 95-114.
[11] Hua, J., Ouyang, Q., Wang, Y., Li, H. and Zhu, Y. 2012. A *dynamic air supply device used to produce simulated natural wind in an indoor environment*. Build. and Env. 47, 349-356.
[12] Fang, L., Clausen, G. and Fanger, P.O. 1998. *Impact of temperature and humidity on the perception of Indoor Air Quality*, Indoor Air 8, 80-90.
[13] Xia, Y.Z., Niu, J.L., Zhao, R.Y. and Burnett, J. 2000. *Effects of turbulent air on human thermal sensations in a warm isothermal environment*, Indoor Air 10, 289-296.
[14] Wong, N.H. and Khoo, S.S. 2003. *Thermal comfort in classrooms in the tropics*, Energy and Build. 35, 337-351.
[15] Feriadi, H. and Wong, N.H. 2004. *Thermal comfort for naturally ventilated houses in Indonesia*, Energy and Build. 36, 614-626.
[16] Humphreys, M.A. and Hancock, M. 2007. *Do people like to feel "neutral"? Exploring the variation of the desired thermal sensation on the ASHRAE scale*, Energy and Build. 39, 867-874.

INTELLIGENT ENERGY CONSUMPTION IN LOW ENERGY HOUSING

Ditte Marie Jørgensen[1],
1 Danish Technological Institute
Gregersensvej
Taastrup, Denmark

Anders Høj Christensen[*2]
2 Danish Technological Institute
Gregersensvej
Taastrup, Denmark
**Corresponding author: anhc@teknologisk.dk*

ABSTRACT

BR10 requires that all new residential constructions should be built as low energy housing. In order to meet these requirements residential buildings must be equipped with far more complex technology, than conventional housing. This, for example, could be a combination of mechanical balanced ventilation, natural ventilation, heat pumps, solar heating, solar cells or automatic sunscreens.

The users' needs and wishes for the energy consumption (extent and quality), and thereby the interaction between consumer and building, installations and electrical consuming equipment, is vital for reaching the energy- and emission goals.
The project's focus is energy efficiency in residential housing by controlling the interaction between technology and behaviour. The project is looking into all of the significant energy consuming applications, such as heating, ventilation, lighting, household, entertainment, IT etc.

The main part of the project has been the interaction with the users of the house. Different scenarios have been tested by the family living in the house. The family has through its daily living tested if the building, the installations and the utilities have been functioning according to plan – what have been good and what have been bad?

KEYWORDS

Energy performance, indoor environment, comfort, hybrid ventilation, intelligent controlling

INTRODUCTION

EnergyFlexHouse in Taastrup, is a high-tech laboratory for development, testing and demonstrating total innovative energysolutions for the building sector. The construction consists of two almost identical buildings, from which the one – EnergyFlexHouse Family (EFHfamily) – is Denmark's first energy neutral residential home. The home is supplied with energyefficient systems and sustainable energy facilities, which produce the energy, used both by the family who lives inside the house but also for an electrical car used for transportation by the family.

The two buildings are equipped with extensive data logging equipment (more than 500 measuring points all together), which gives a unique opportunity to monitor and study, cause and effect. In October 2009 the first test family moved in to EFHFamily and in December 2011, the fifth family moved out. The families live in the house for 3 to 6 months at a time depending on different practicalities. The families take part in the activities, focusing on the interaction between technology, users, controls, energy services and energy consumption.

The energy neutral home is supplied with the technology used in an average residential home. Although everything is new – putting the newest technology to a test – constructions, installations, kitchen appliances, IT etc.

EFHfamily gives a rare opportunity to develop, test and document the energy- and user-related advantages of simple or more complex control concepts and systems for energy services, primarily focusing on indoor climate.

The global challenges regarding energy supply and significant reduction of the CO2 level, must be accomplished by energy efficiency and usage of sustainable energy. 35-40% of the energy consumption is used in the buildings for indoor climate, hot water and other appliances, which demands energy in accordance with the buildings usage.

Experience from newly constructed low energy housing has so far shown that the interaction between the user and the technology has not been working properly. Mainly because the users are not taught how to use the technology in the houses correct. As a consequence the houses use more energy than calculated, but also the indoor climate is challenged by overheating during the summer and low temperatures during winter.

The technological challenges are therefore to:

- develop cost- and energy-efficient technology
- mastering the dynamic interaction between consumption and supply
- finally, mastering the interaction between user and the energy consuming technologies – building, installations and equipment.

This project focuses on the last and important part of the challenges which has "intelligent" controlling of the energy services as its focus.

The energy- and user-related advantages of simple or more complex controlling-_concepts and systems for energy services have been tested and documented during the project. .
This has been done with focus on, that the controlling solution can manage the comfort level in the house as efficient and energy optimized as possible. All the while the users demand and need for information and control, of the comfort level and energy consumption has to be accommodated.

A central part of the project has therefore been the interaction with the users. Different scenarios and processes have been tested in practice. The family has through its daily living tested how the building, the installations and the equipment have been working. What have been good – what have been bad?

The questions have been many, for instance:
How is the building being used, and what is the resulting energy consumption? Can the family master the technical installations to get a good indoor climate? Is intelligent control an advantage – or is it just an annoyance? Which information concerning energy consumption and energy supply is the family interested in? Can we develop a manageable interface to make it simpler and interesting to monitor energy consumption? Can an interactive interface affect the family's behaviour and energy consumption?

For this project there has been developed automatic controlling of the mechanical ventilation, natural ventilation, heating, and sunscreens, so that these subsystems could be connected in random order. With this platform as a starting point, there has been developed 4 control strategies:

a. A **basic control** – A simple control, which the three following controls, is based on.
b. An **indoor optimized control**, which insures that the indoor climate fulfil category I of DS/EN 15251
c. An **energy optimized control** which ensures that the house is managed energy efficient according to the guidelines in Bygningsreglementet 2010
d. A **user-defined control**, which makes the users able to make manually individual changes to ventilation, sunscreens and heating.

The difference between the **basic control** and the more advanced control is the following:

- Dividing the house into zones
  In the more advanced controls, the house has 5 zones instead of just 1 zone.

- Temperature set points (heating and cooling) changes during the day and according to the season. The temperature is registered in each zone, so each zone has the possibility of having different temperature set points and settings for natural ventilation and sunscreens.

- Variation in opening area of automatic windows.
  In the **basic control** it is only possible to open the windows 0 % or 100%. In the advanced controls any opening area is possible.

- Temperature regulation with hysteresis
  The hysteresis (range) prevents the windows, sunscreens and heat pumps from opening/going on and off all the time. The value changes during the day and according to the season.

- Automatic control of the bypass.
  The automatic control can contribute to a better indoor climate, so it does not get too hot, along with an energy saving in transition periods in which the bypass is not yet activated.

- Use of motion detectors to determine if the family is home or not.

The four controlling strategies are fitted to EnergyFlexHouse (EFH) at Danish Technology Institute (DTI) and the strategies were tested by a family living in the house.

For the **user defined control** and for choosing strategies the family was given a tablet, with a related application

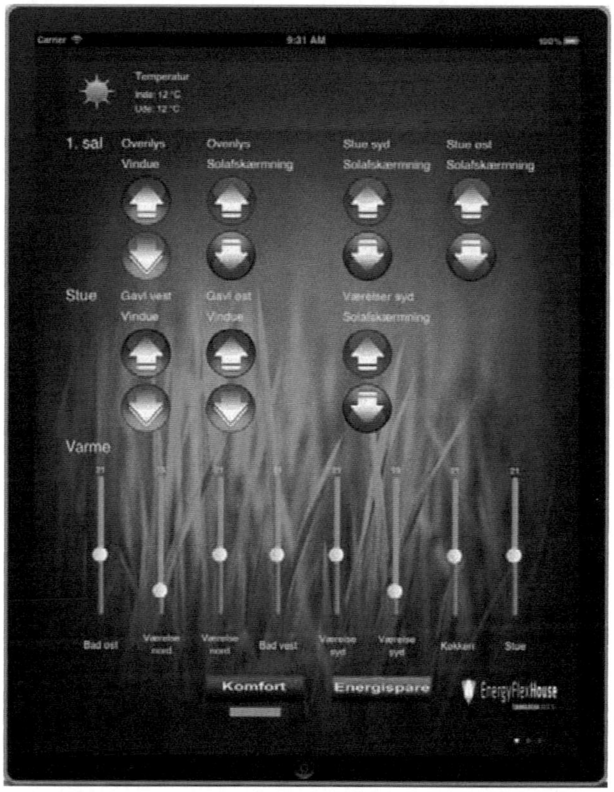

Figure 1 - Tablet for controlling strategies

CONCLUSION

The tests show that it is one thing to design and implement energy optimized solutions and systems and it is an entirely different thing to get the users to use them correctly in their daily living. If the users cannot handle the systems, e.g. because they are too complicated or too time demanding, creative ideas and thoughts behind the energy efficient solutions and systems will not do any good. This also concerns the effect of promoting the awareness of the consequences of the users' actions and subconscious behavioural patterns.

The tests with the 4 controlling strategies in cooperation with the family in EFH, show that it is not possible to develop a satisfying controlling strategy without involving the users. The users will not be satisfied with the indoor climate, if they do not have influence on the controls. This should be taken into context with that energy efficient controlling requires as little user influence as possible, in order to realize the expected energy efficiency.

The optimal control is characterized by its ability to adapt user demands along with resulting in a change in user behaviour adapting to the control.

The project also shows that there is a substantial barrier in getting products from different manufactures to communicate. Problems are partly technical, partly responsibilities related, in that it can be difficult to place the responsibility in any operational problems, manufactures are also hesitant to provide technical specifications.

Based on the concept, which is developed through cooperation between the residents, the involved companies and DTI, a prototype for the advanced controls has been developed and tested by the family living in EnergyFlexHouse.

Experiments with the 3 advanced controlling strategies, "optimized indoor climate", "optimized energy consumption" and "user controlled", have been performed over a period of 4 months.

If you only focus on the potential energy saving on the ventilation, the energy consumption has been reduced by 83 %, comparing a situation of constant mechanical ventilation to a situation with similar outdoor conditions, but with hybrid ventilation, controlled by the developed strategies.

The family expressed general satisfaction with the indoor climate no matter which strategy was chosen.

The family was surprised that their heating consumption was nearly 30% higher using the user controlled strategy and admitted that the energy optimized strategy definitely was acceptable, especially considering the consequences.

The project has shown that it is of great significance for the energy consumption in low energy housing, that the indoor climate is precisely controlled along with demands for the indoor climate is fulfilled. It is also of high importance that the installations are engaged in the right order according to weather conditions. The project shows that, an intelligent controlling, giving the users certain degrees of freedom, has a decisive significance on the energy saving potential in low energy housing.

This has to be seen in perspective of that the significance of the user behaviour, is relatively larger for low energy housing than for conventional construction.

The test family in EnergyFlexHouse has given unique opportunities for mapping the effect of the changes in both systems and products, among others the subjective user evaluations of the physical and controlling actions have given a valuable knowledge on the subject. The fact to include the end user this way – in the real environment on an average day – can ensure that the solutions developed is getting impact on the market and delivering the wanted energy savings.

ACKNOWLEDGEMENTS
The project was supported by Dansk Energi through PSO-project PSO-342-058 "Intelligente energiydelser i lavenergi-boliger baseret på brugerdreven innovation"

We would also like to acknowledge the participants in the project:
Asger Bay Christensen – Asger BC lys
Claus R. Johnsen – Region Sjælland, Sygehus Nord Roskilde, Medicinsk afdeling
Ejvind Degn Larsen – Fischer
Daniel Lux – Seluxit
Lis Jacobsen – Nilan

REFERENCES

[1] DS/EN 15251 1. Edition 2007. Indoor environmental input parameters for design and assesment of energy performance of buildings addressing indoor quality, thermal environment, lighting and acoustics.

COMPARISON OF DISPLACEMENT VENTILATION AND MIXING VENTILATION SYSTEMS WITH REGARD TO VENTILATION EFFECTIVENESS IN OFFICES

Michele De Carli*[1], Roberta Tomasi[1], Roberto Zecchin[1], Giacomo Villi[1]

1 Department of Industrial Engineering
Via Venezia 1, 35131
Padova, Italy

**Corresponding author: michele.decarli@unipd.it*

ABSTRACT

Air quality in offices depends on the ventilation system ability to remove contaminants from the occupied zone. In a low polluted building air quality mainly depends on the human presence and carbon dioxide is normally used as indicator of human bioeffluents.
The aim of this paper is to investigate, by using computational fluid dynamics (CFD) methods, the effect of the supply and exhaust positions on the contaminant distributions in an office equipped with a ceiling cooling system. Mixing ventilation was compared with different displacement ventilation solutions, adopting either floor or wall mounted displacement diffusers. Exhaust vents in the mixing ventilation cases were placed directly under the supply or on the opposite wall; in the displacement ventilation cases, exhaust vents were placed on the upper part of a sidewall or on the ceiling in different positions. Besides the percentage of dissatisfied PD [%] and the contaminant removal effectiveness, a discomfort index for the whole office was introduced and calculated. It resulted that, for the considered scenario, displacement ventilation performance was very sensitive to the position of exhaust grilles; moreover, it was found that displacement ventilation does not result always in better air quality than mixing ventilation.

KEYWORDS

Computational fluid dynamics, contaminants, ventilation effectiveness, mixing ventilation, displacement ventilation, office.

INTRODUCTION

Computational fluid dynamics represents an appropriate instrument for the prediction of contaminant distribution in offices, where the indoor air quality is important for both the health and productivity. Wargocki et al. [1] estimated that performance of office work increases on average by 1.5 % for every 10 % decrease in the percentage of persons dissatisfied with the air quality; they also studied the effects of pollution loads and outdoor air supply rates on Sick Building Syndrome (SBS) symptoms in office [2].
Within an office air quality is affected particularly by human presence. The occupants emit bioeffluents like carbon dioxide, carbon monoxide and water vapor in quantities variable with metabolic activity and age, as reported in [3] and [4]. Carbon dioxide is considered a good indicator of air quality in rooms with human presence even if it does not represent a serious

health problem at concentrations that generally occur indoor. Furthermore it was found that carbon dioxide concentration is connected to the acceptability of the space in terms of body odor (Persily, [5]). European Standard EN 15251 [6] recommends ventilation rates values in non residential buildings should be based on occupant density (to take into account the pollution emitted from persons) and floor area (to take into account the pollution emitted from building materials) basing on three indoor climate categories. In order to analyze the effectiveness of a ventilation system, the contaminant removal effectiveness index (ε^c) defined by REHVA [7] is meant to compare the concentration at the exhaust with the concentration in the room. CFD made it possible to calculate the contaminant removal effectiveness index in rooms equipped with different ventilation systems and to make comparisons between them. In the prediction of room airflow, the turbulence model plays an important role with respect to the accuracy of prediction and the required simulation time. It was demonstrated by Zhai et al.[8] that, among different turbulence models, the LES model was able to capture very detailed flow features for natural convection, forced and mixed convection, although computational time could take very long. The effect of the air supply location on the performance of a displacement ventilation system in a large office was investigated by Lin et al. [9], while Novoselac et al.[10] studied the vertical concentration of active pollutant sources (CO_2) and VOC from carpet in a conference room with a cooled ceiling combined with displacement ventilation.

In this work the carbon dioxide concentration was calculated in an office for three persons equipped with a ceiling cooling system and mechanical ventilation. Mixing ventilation was compared with different displacement ventilation strategies and the effects of both supply and exhaust locations were investigated.

METHOD

An office room for three persons, equipped with mechanical ventilation combined to ceiling cooling was analyzed under typical summer conditions by means of CFD in order to analyze the effect of different ventilation systems on the resulting ventilation effectiveness which was quantified by the contaminant removal effectiveness index. Numeric simulations were performed be means of the FDS (Fire Dynamics Simulator) model which solves a modified form of the Navier-Stokes equations appropriate for low-speed thermally-driven flow; all spatial derivatives in the conservation equations are discretized by second order finite difference scheme and all the thermodynamic variables are updated in time by means of an explicit second order predictor-corrector scheme (McGrattan et al. [11]). Sub-grid modelling is performed by means of the Smagorinsky model which is based on the eddy viscosity assumption. FDS is characterized by fast computational speed and relatively modest requirements in terms of computational resources but is restricted to regular geometries. Although it has been developed for addressing fire related problems, the low Mach number assumption is also appropriate to describe building ventilation scenarios which do not include fires (McGrattan et al.[11]). Different examples of how the model can be used to investigate indoor airflow scenarios can be found in (Musser et al.[12]), (Lin et al.[13]), (Cho and Liu.[14]) and (Farnham et al.[15]).

The investigated office has dimensions 7 m (x dimension) × 4 m (z dimension) × 3 m (y dimension,(Figure 1) and the total internal heat gain considered is 680 W (3 persons and 3 computers). The ceiling surface temperature is supposed to be at 22°C while the floor and all the vertical walls are at 26°C. As occupants were considered sitting at their desk, the maximum height of the modeled person is 1.1 m corresponding to the head level according to ISO 7726 [16]. It was supposed that each person emits 0.02 m^3/h of carbon dioxide (which is the recommended value for an adult in sitting position [4]) and the contaminant source was

described as a rectangular patch (0.2 m × 0.1 m) positioned on the upper part of block representing the occupant body.
The office is ventilated with an outdoor air change rate of 1.5 h^{-1} at 20°C of temperature. The ventilation rate was calculated assuming a 7 l/s ventilation rate per occupant and a 0.5 l/sm^2 ventilation rate for emissions from building materials. Different ventilation strategies were considered:

- Wall mounted diffusers (mixing ventilation: case 1 and 2);
- Floor mounted displacement ventilation inlets with radial, horizontal jet (case 3, 4 and 6);
- Floor mounted displacement ventilation diffusers with inclined discharge (case 5);
- A wall mounted displacement ventilation diffuser (case 7, 8, 9 and 10).

As to the exhaust, different (in size and location) outlets were adopted. Exhausts were placed on one of the two short side walls, at different heights, or on the ceiling, along the room's longitudinal axis. The ten cases considered in this work can be categorized as a function of the type, the number and the position of inlets and outlets as reported in Table 1 and shown in Figure 1.
Carbon dioxide concentration was calculated in 23 points considered relevant (taking into account the possible position of a generic occupant) for comfort (listed in Table 2), as shown in Figure 2. They are:

- In front of the occupants, at 1.1 m height, 20 cm from the table; this zone is reference to as "Table" and includes 3 points;
- Next to the desks, 80 cm far away and at 1.7 m height. This zone is referenced to as "Perimeter" and includes 4 points;
- In room's corners at 1.7 m height. This zone is referenced to as "Corner" and includes 4 points;
- Close to the desks, at 1.1m and 1.7 m. This zone is referenced to as "Desk1", Desk 2" and "Desk3" and 4 points are considered for each desk.

	Unit	Case 1	Case 2	Case 3	Case 4	Case 5	Case 6	Case 7	Case 8	Case 9	Case10
Code		MIX-1	MIX-2	UFAD-1	UFAD-2	UFAD-3	UFAD-4	DV-1	DV-2	DV-3	DV-4
Total Internal heat loads	W	680	680	680	680	680	680	680	680	680	680
Wall and floor surface temperature	°C	26.0	26.0	26.0	26.0	26.0	26.0	26.0	26.0	26.0	26.0
Ceiling surface temperature	°C	22.0	22.0	22.0	22.0	22.0	22.0	22.0	22.0	22.0	22.0
Primary air inlet temperature	°C	20.0	20.0	20.0	20.0	20.0	20.0	20.0	20.0	20.0	20.0
Inlet air velocity	m/s	0.29	0.29	0.22	0.22	0.22	0.22	0.18	0.18	0.18	0.18
Primary air volume flow	h^{-1}	1.5	1.5	1.5	1.5	1.5	1.5	1.5	1.5	1.5	1.5

Table 1. Simulated configurations scheme.

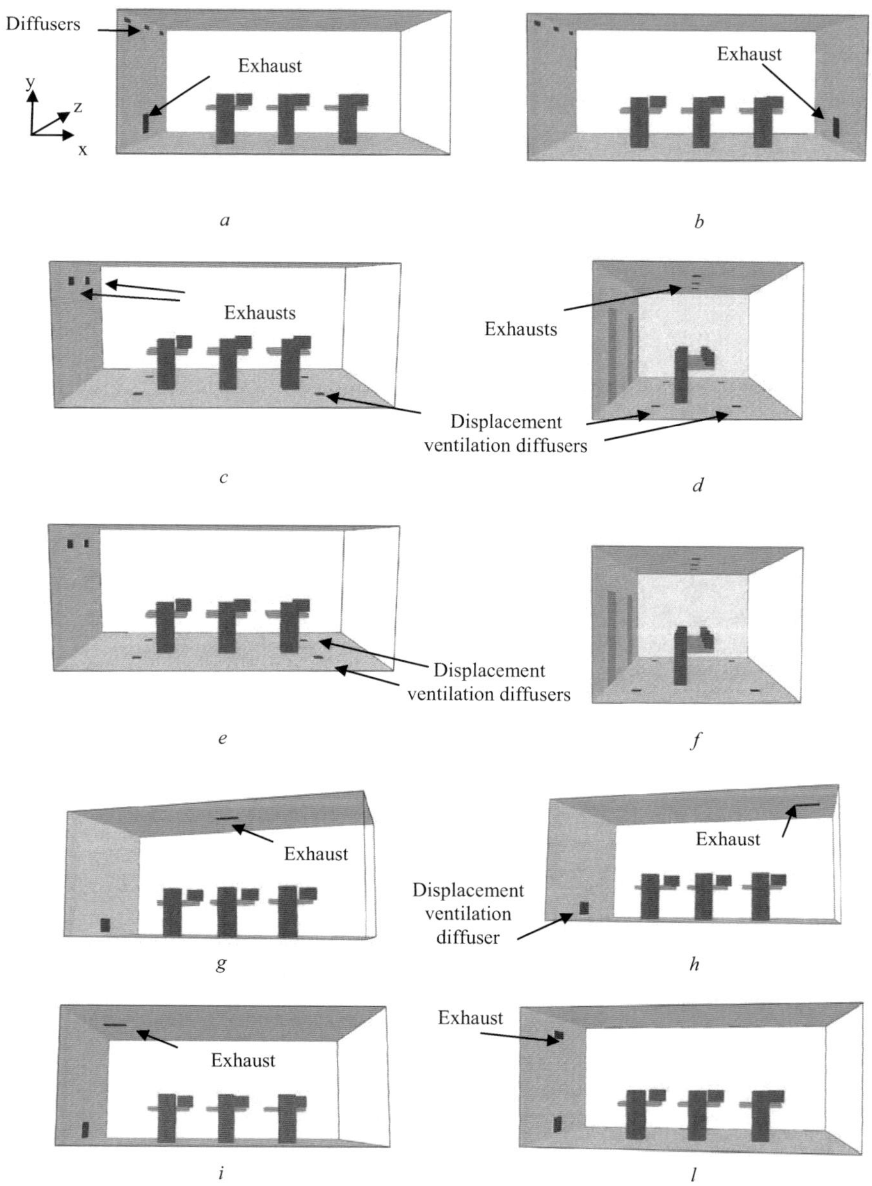

Figure 1 Layout of the simulated office for the different investigated cases: inlet air vent (blue) and outlet air vents (violet) for case 1 (a), case 2 (b), case 3 (c), case 4 (d), case 5(e), case 6 (f), case 7 (g), case 8 (h), case 9 (i) and case 10 (l).

point n°	Zone	x [m]	z [m]	y [m]
1	Table 1	2	2	1.1
2	Table 2	3.5	2	1.1
3	Table 3	5	2	1.1
4	Perimeter	2.8	0.8	1.7
5		4.3	0.8	1.7
6		2.8	3.2	1.7
7		4.3	3.2	1.7
8	Corners	1	0.5	1.7
9		1	3.5	1.7
10		6	0.5	1.7
11		6	3.5	1.7
12	Desk 1	1.7	1.3	1.1
13		1.7	1.3	1.7
14		1.7	2.5	1.1
15		1.7	2.5	1.7
16	Desk 2	3.2	1.3	1.1
17		3.2	1.3	1.7
18		3.2	2.5	1.1
19		3.2	2.5	1.7
20	Desk 3	4.7	1.3	1.1
21		4.7	1.3	1.7
22		4.7	2.5	1.1
23		4.7	2.5	1.7

Table 2. Cartesian Coordinates of the 23 points in which carbon dioxide was calculated.

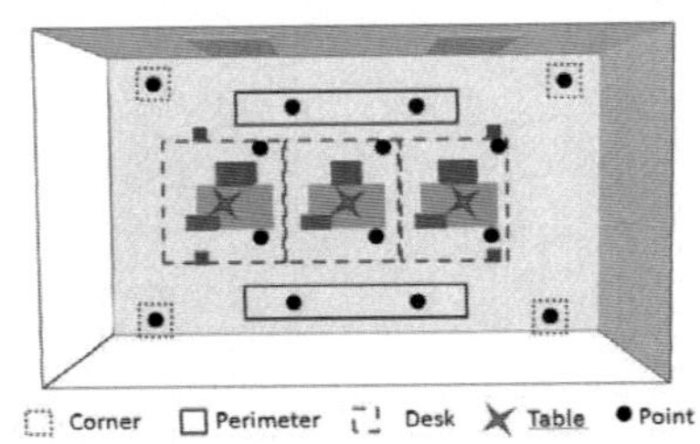

Figure 2. Assumed zones in the office, layout of case 4.

Zone	N° points	Weighting coefficient (c_i)
Table	3	10
Perimeter	4	5
Corner	4	3
Desk 1,Desk 2, Desk 3	4	10

Table 3. Weighting coefficient c_i for the different sampling points as a function of the macro zones the analyzed office was divided in.

Carbon dioxide is an indicator of human bioeffluents and, if occupants are the exclusive pollution source in a space, it is possible to determine the percentage of persons dissatisfied PD [%] with the level of perceived air quality by using the ASTM D6245:98 [17] equation:

$$PD\% = 392 \cdot e^{(-15.5 \cdot dCO_2^{(-0.25)})} \tag{1}$$

where dCO_2 is the difference between the outdoor and the indoor concentration. In the present analysis, it was assumed that the outdoor CO_2 concentration is zero, so that the calculated indoor concentration was used for the dCO_2 term.
In order to better compare the ten cases, a discomfort index ID [%] was introduced as the weighted value of the predicted percentages of persons dissatisfied PD [%] in a representative number of points in the room; the weighting coefficient (Equation (2)) is based on the relevance of the zone with respect to occupants.

$$ID\% = \frac{\sum_{23} (c_i \cdot PD\%)}{\sum c_i} \tag{2}$$

In the present case, the proposed parameter was applied to 23 sampling points which were conveniently distributed in the room, varying the weighting coefficients from 3 to 10 (e.g. the highest value, 10, is for the sampling points closest to the tables (Table 3) occupants are seated at). Finally, the contaminant removal effectiveness was calculated according to the REHVA method [7] where the adopted parameter (ε^c) is defined (Equation (3)) as the ratio between the contaminant concentration at the exhaust (c_e) and the contaminant concentration in the room $<c>$ to express how quickly an airborne contaminant is removed from the room. For all the investigated cases, $<c>$ was assumed to be the average of the CO_2 concentrations in the different zones of the office was divided in (i.e. "Perimeter", "Desk1"...) while the concentration at the exhaust (i.e. c_e) was assumed to be the steady state concentration under the well mixed hypothesis and calculated according to Standard ASTM D6245-98 [17].

$$\varepsilon^c = \frac{c_e}{<c>} \tag{3}$$

RESULTS

Results show that CO_2 equilibrium concentrations in the room were quite different depending on the adopted ventilation strategy. Figure 3 reports the calculated carbon dioxide concentrations at 1.1 m height for three different cases (i.e. mixing ventilation and two strategies for displacement ventilation). It is evident that mixing ventilation resulted in concentrations in all the volume (calculated concentrations are about 800 ppm for the case 1 simulation (Figure 3 a); as to the "case 2" scenario results were similar, (see Table 4)).
When displacement ventilation from floor is adopted (i.e. "UFAD" cases, Table 4) with exhaust on a wall (Figure 3 b), CO_2 concentrations resulted locally much higher than those resulting from the two mixing ventilation cases; as reported in Table 4, CO_2 concentration at Desk 1 (which is the closest to the exhaust) is higher than 1000 ppm in case 3 and case 5 and it decreased as the centre of the room is considered. In particular, it was found that case 3 and 5 presented the highest percentages of dissatisfied persons PD [%] and discomfort index ID [%], as shown in Tables 5 and 6. A significant improvement on air quality resulted from placing the exhaust on the ceiling with the same supply location: in the case three exhaust

outlets were placed on the ceiling, the average CO_2 calculated concentration is about 550 ppm (Table 4, Case 4) and the resulting discomfort index ID [%] decreased to 17.1% (Table 6). If the floor diffusers were placed far away from the desks (i.e. case 6) the average CO_2 was about 800 ppm and the discomfort index ID [%] was 22.5% even in the case the exhaust is placed on the ceiling.

Finally if displacement ventilation from wall mounted diffusers is considered (i.e. "DV" cases) the discomfort index ID [%] decreased to about 20%, regardless of the exhaust position (Table 6). Case 7, with the exhaust in central position, resulted in the lower CO_2 concentrations and Case 10 results, where the exhaust was placed just above the wall mounted displacement diffusers, are not so different from the previous DV cases: the discomfort index ID [%] is 20.1% (Table 6).

Similar considerations apply for the contaminant removal effectiveness: it was found that, excluding the table zone, the highest values resulted for Case 4 ($\varepsilon^c > 0.9$ for all the room's volume) followed by cases 7, 8, 9 and 10 ($\varepsilon^c > 0.7$), as shown in Figure 4 and Table 7.

N° case	Case	Zone					
		Table	Perimeter	Corner	Desk1	Desk2	Desk3
1	Mix 1	895.9	792.4	784.7	798.9	789.6	784.6
2	Mix 2	905.9	842.7	818.3	821.7	831.5	826.7
3	UFAD-1	1015.3	883.9	896.8	1018.9	911.6	801.7
4	UFAD-2	616.4	533.0	539.4	535.9	526.4	513.3
5	UFAD-3	1032.4	873.5	896.2	1024.8	907.4	791.8
6	UFAD-4	857.8	771.9	785.3	803.6	762.4	736.7
7	DV-1	804.3	673.9	663.3	656.2	670.8	656.7
8	DV-2	810.5	694.4	699.0	683.2	686.2	680.5
9	DV-3	850.3	715.7	726.6	729.0	720.3	682.3
10	DV-4	797.9	656.7	657.2	663.4	652.1	627.2

Table 4. Calculated average carbon dioxide concentration above outside air for each zone [ppm].

N° case	Case	Zone					
		Table	Perimeter	Corner	Desk1	Desk2	Desk3
1	Mix 1	24.6	22.7	22.6	22.9	22.7	22.6
2	Mix 2	24.9	23.7	23.2	23.3	23.5	23.4
3	UFAD-1	26.8	24.5	24.8	27.0	25.0	22.9
4	UFAD-2	18.7	16.9	17.0	16.9	16.7	16.4
5	UFAD-3	27.1	24.3	24.7	27.1	25.0	22.7
6	UFAD-4	23.9	22.3	22.6	22.9	22.1	21.5
7	DV-1	22.7	20.2	19.9	19.8	20.1	19.8
8	DV-2	22.9	20.6	20.7	20.4	20.5	20.3
9	DV-3	23.7	21.1	21.3	21.4	21.2	20.4
10	DV-4	22.6	19.8	19.8	20.0	19.7	19.1

Table 5. Percentage of persons dissatisfied PD [%] due to the calculated values of CO_2 concentration.

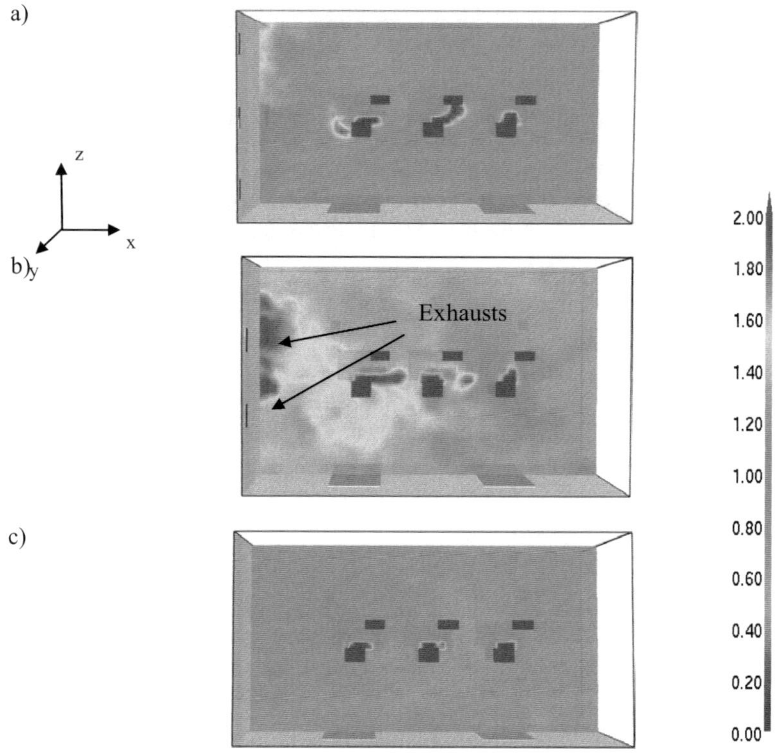

Figure 3. CO_2 distribution at 1.1 m height, at equilibrium, for case 1 with mixing air flow(a), cases 3(b) and case 4 (c) with displacement ventilation. Concentrations are expressed in [kg/kg $\times 10^{-3}$].

N° case	1	2	3	4	5	6	7	8	9	10
Case	Mix 1	Mix 2	UFAD-1	UFAD-2	UFAD-3	UFAD-4	DV-1	DV-2	DV-3	DV-4
Discomfort Index [%]	23	23.7	25.2	17.1	25.2	22.5	20.4	20.9	21.5	20.1

Table 6. Discomfort index ID [%] for each case.

N° case	Case	Zone					
		Table	Perimeter	Corner	Desk1	Desk2	Desk3
1	Mix 1	0.5	0.6	0.6	0.6	0.6	0.6
2	Mix 2	0.5	0.6	0.6	0.6	0.6	0.6
3	UFAD-1	0.5	0.5	0.5	0.5	0.5	0.6
4	UFAD-2	0.8	0.9	0.9	0.9	0.9	0.9
5	UFAD-3	0.5	0.5	0.5	0.5	0.5	0.6
6	UFAD-4	0.6	0.6	0.6	0.6	0.6	0.6
7	DV-1	0.6	0.7	0.7	0.7	0.7	0.7
8	DV-2	0.6	0.7	0.7	0.7	0.7	0.7
9	DV-3	0.6	0.7	0.7	0.7	0.7	0.7
10	DV-4	0.6	0.7	0.7	0.7	0.7	0.8

Table 7. Contaminant removal effectiveness ε^c [-] for all the analyzed cases and in different zones [-].

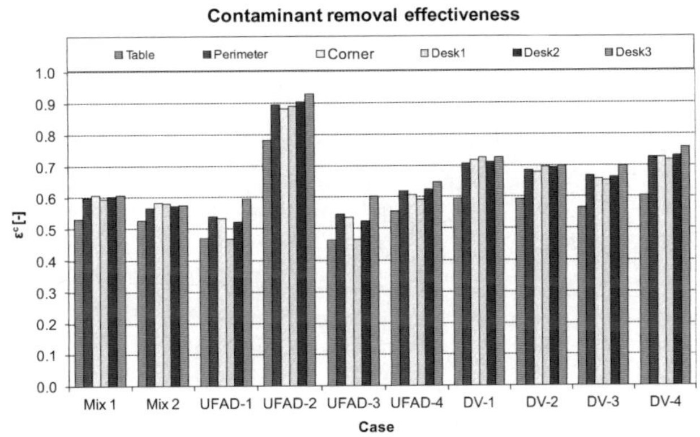

Figure 4. Contaminant removal effectiveness ε^c [-] for different zones and for all the analyzed cases.

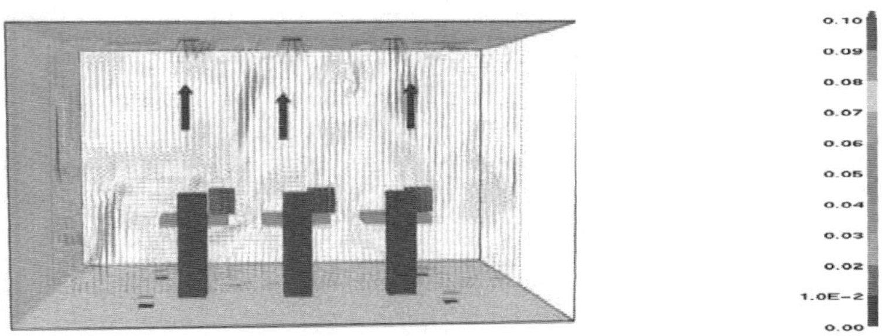

Figure 5: Air velocity distribution for case UFAD-2 [m/s]. The distinguishing airflow patterns are highlighted with grey arrows.

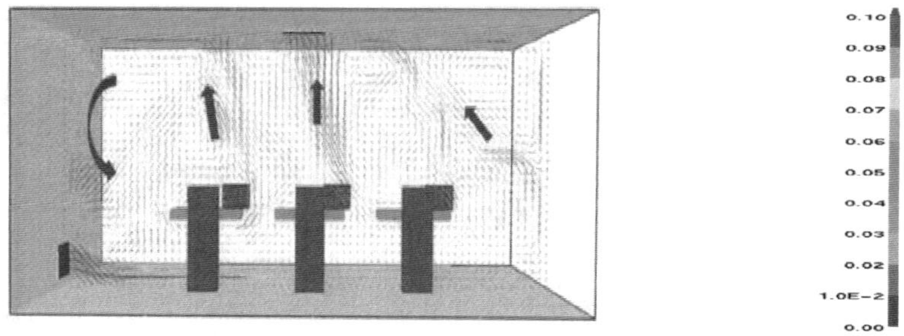

Figure 6: Air velocity distribution for case DV-1 [m/s]. The distinguishing airflow patterns are highlighted with grey arrows.

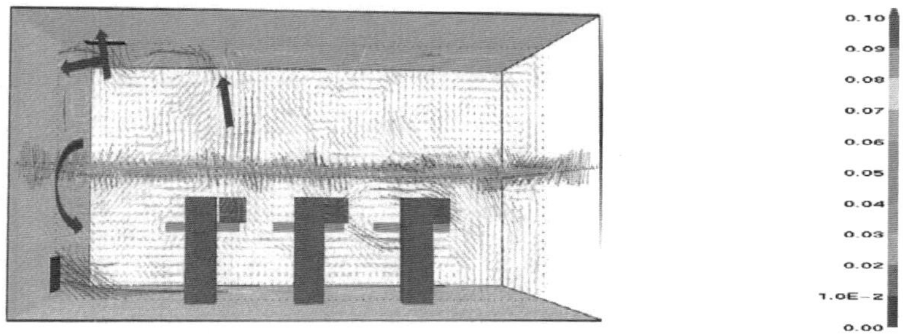

Figure 7: Air velocity distribution for case DV-3 [m/s]. The distinguishing airflow patterns are highlighted with grey arrows.

DISCUSSION

The presented contaminant removal efficiency analysis shows that displacement ventilation is not always the best alternative to traditional mixing systems. The combination of floor mounted diffusers and wall exhaust can result in an asymmetrical distribution of CO_2 in the analyzed office, penalizing the persons closest to the exhaust. The discharge direction does not affect the breathing zone (1.1 m height) because the velocities involved are low. As a matter of fact, case 3 (UFAD-1) and case 5 (UFAD-3), with tangential or inclined flow, were comparable in terms of contaminant removal effectiveness and discomfort index.
When floor mounted diffusers are combined to ceiling exhaust, contaminant removal effectiveness increased significantly, particularly in the case the exhaust is located directly above the persons (case 4). However, the improvement is less evident if floor diffusers are distant from the desk: by comparing case 4 (UFAD-2) and 6 (UFAD-4) (Table 6), it is evident that if diffusers are moved 40 cm towards the walls (so that the distance from the desks is about 0.8 m instead that 0.4 m) the discomfort index ID% increased from 17.1% to 22.5%.
In order to better understand the phenomenon, it is important to analyze the velocity distribution in the room: due to the thermal plume, contaminants are entrained upwards, reaching the cold ceiling surface. At this point, only in the case the exhaust is placed on the ceiling, CO_2 is directly removed; on the contrary (i.e. in the case the exhaust is located on the walls), it takes longer for being exhausted. If air is supplied at floor level, at low velocities and at a significant distance from the desks, the effect of the of thermal plume is reduced. Also in the case displacement ventilation from wall diffusers it is evident that contaminants

are removed more efficiently from the center of the ceiling because of the smaller distance from the sources, as shown in Figure 6; on the contrary Figure 7 shows that a potential risk exists when exhaust is lateral because of the resulting air velocity directions. As a matter of fact, Table 4 shows that case 7 (DV-1) results in lower CO_2 concentration than case 8 (DV-2) and 9 (DV-3). When the room air flow patterns are markedly influenced by the direction of the supply discharge, such as in the case of displacement ventilation from wall diffusers, the wall mounting of the exhaust (case 10) can be effective in determining a uniform CO_2 concentration in the occupied zone. The performed study is consistent with Kobayashi and Chen [18] findings about the influence of the exhaust position on air quality in the case of floor supply displacement ventilation in small offices: they demonstrated that contaminants can be removed faster when exhausts are placed above the contaminant sources rather than in lateral position on the ceiling. However, results from numerical simulations should be completed by experimental tests investigating the effect of supply air terminals and exhaust positions on the resulting air quality.

As regards the mixing ventilation cases, in the case the inlet air velocity is low, because of its temperature (20°C) air falls at the beginning of the occupied zone; then it generally starts a horizontal movement under the tables towards the opposite wall, it gets warmer due to the presence of humans and pcs and it eventually rises upwards removing CO_2 from the occupied zone. Nevertheless, if the exhaust vent is on the wall opposite to the supply location, it is possible that a fraction of air is exhausted before it reaches the occupied zone; this can explain why the discomfort index ID [%] of case 2 is higher than that calculated for case 1 (Table 6).

The occupants position in the room was a relevant parameter: in this study it was assumed that occupants were seated at their desks, but if this is not the case, it can be thought that air from the floor could reach earlier the source. However the aim of this work was to compare different systems with the same layout of the room, and therefore only relative differences between contaminant concentrations for each case have been determined.

CONCLUSION

A numerical investigation about the effects of the supply and exhaust locations on the CO_2 distribution in a typical office with cooled ceiling was carried out by means of computational fluid dynamics (CFD) methods. Two mixing ventilation configurations were compared with different displacement ventilation solutions with regards to the resulting CO_2 concentration measured in 23 positions which were recognized of being relevant for occupants' comfort. The percentage of dissatisfied due to the CO_2 concentration was calculated for all the sampling locations and a discomfort index for the whole office was introduced and determined. Finally, the contaminant removal effectiveness was calculated.

Results showed that mixing ventilation results in a higher contaminant removal effectiveness if exhaust and supply share the same wall. Displacement ventilation performance was very sensitive to the relative location of inlets and outlets. In the case of air supplied from the floor and exhausted from a sidewall, the corresponding contaminant removal effectiveness was lower than those resulting from the mixing ventilation scenarios and, in particular, the positions next to the exhaust were penalized. Nevertheless, if wall mounted displacement diffusers are adopted, the exhaust from a wall can be a good solution.

Finally it was found that the location of the exhaust on the ceiling, results in higher air quality levels regardless of the position of the inlet diffusers.

ACKNOWLEDGEMENTS

This work was a part of a research program on radiant ceiling panels of Rhoss S.p.A., Italy.

REFERENCES

[1] Wargocki, P., Wyon, D. P., Baik, Y. K Fanger, P O. 2000. *Productivity is affected by air quality in offices.* Proceedings of Healthy Buildings 2000, Vol.1, 635-640.

[2] Wargocki, P., Wyon, D. P., Sundell, J., Clausen, G., Fanger, P. O. 2000. *The Effects of Outdoor Air Supply Rate in a Office on Perceived Air Quality, Sick Building Syndrome (SBS) Symptoms and Productivity.* Indoor Air 2000 (10), 222-226.

[3] CEN report CR 1752-1998. Ventilation for buildings- design criteria for the indoor environment. CEN/TC 156/WG 6.

[4] De Gids, W.F., Wouters, P. 2010. *CO_2 as indicator for the indoor air quality - general principles.* AIVC.VIP N°33 July2010.

[5] Persily, A.K., 1997. *Evaluating building IAQ and ventilation with indoor carbon dioxide.* ASHRAE Transaction 1997 Vol. 103 Pt 2.

[6] EN 15251-2007. Indoor environmental input parameters for design and assessment of energy performance of buildings addressing indoor air quality, thermal environment, lighting and acoustics.

[7] Mundt E., Mathisen, H.M, Nielsen, P.V., Moser, A. RHEVA Guidebook n°2. 2004.Ventilation effectiveness.

[8] Zhai, Z. J., Zhang, Z., Zhang, W., Chen, Q.J. 2007. *Evaluation of various turbulence models in predicting airflow and turbulence in enclosed environment by CFD: Part 1- summary of prevalent turbulence models.* HVAC &R Resarrch (6), Vol 14.

[9] Lin, Z., Chow, T.T., Tzang. C.F., Fong. K.F., Chan. L.S.2005. *CFD study on effect of the air supply location on the performance of the displacement ventilation system.* Building and Environment 40 (2005), 1051-1067.

[10] Novoselac, A., Sbrebric, J. 2002. A critical review on the performance and design of combined cooled ceiling and displacement ventilation systems. Energy and Buildings 34 (2002), 497-509.

[11] McGrattan, K., Hostikka, S, Floyd, J., McDermott, R. 2010. *Fire Dynamics Simulator (Version 5) Technical Reference Guide. Volume 3: Validation.* National Institute of Standards and Technology, NIST Special Publication 1018-5.

[12] Musser, A., McGrattan, K.B., Palmer, J. 2001. *Evaluation of a fast, simplified computational fluid dynamics model for solving room airflow problems.* National Institute of Standards and Technology (NIST), NISTIR 6760

[13] Lin, C.H., Wu, T.T, Horstman, R.H., Lebbin, P.A., Hosni, M.H., Jones, B.W, and Beck, B.T. 2006. *Comparison of large eddy simulation predictions with particle image velocimetry data for the airflow in a generic cabin model.* HVAC&Research, 12, 953 – 951.

[14] Cho, Y.H., Liu, M. 2010. *Correlation between minimum airflow and discharge air temperature.* Building and Environment, 45, 1601 – 1611.

[15] Farnham, C., Nakao, M., Nishioka, M., Nabeshima, M., Mizuno, T. 2011. *Study of mist-cooling for semi-enclosed spaces in Osaka, Japan.* Procedia Environmental Sciences, 4, 228 – 238.

[16] ISO 7726:1998. Ergonomics of the thermal environment - Instruments for measuring physical quantities, 1998.

[17] ASTM D6245-98 -2002. Standard guide for using indoor carbon dioxide concentrations to evaluate indoor air.

[18] Kobayashi, N. and Chen, Q. 2003. *Floor-supply displacement ventilation in a small office.* Indoor and Built Environment, 12(4), 281-292.

NATURAL VENTILATION STRATEGY POTENTIAL ANALYSIS IN AN EXISTING SCHOOL BUILDING

Laura Lion[1,2], Annamaria Belleri*[1,3], Roberto Lollini[1], Dino Zardi[2], Lorenzo Giovannini[2]

[1] *EURAC Research*
Viale Druso 1
39100, Bolzano, Italy
**annamaria.belleri@eurac.edu*

[2] *University of Trento*
Department of civil and environmental engineering
Via Mesiano 77
38123 Trento, Italy

[3] *Università degli Studi di Bergamo*
Salvecchio, 19
24129 Bergamo, Italy

ABSTRACT

Natural ventilation is increasingly considered a promising solution to improve thermal comfort in buildings, including schools. However in order to support its planning and implementation, quantitative analysis on airflow paths and heat-airflow building interactions are needed. This requires an adequate accounting of both internal effects, from building layout and structure, and external forcings from atmospheric factors.

The paper analyses the performances of natural ventilation strategies as retrofit solutions to improve thermal comfort in an existing school building in Lavis (Trento, Italy).

A climatic analysis is performed to define the potential of wind driven natural ventilation. Meteorological data collected on site are analysed to identify typical wind conditions during the cooling season. The resulting daily cycle of wind speed and direction in sunny days reflects the typical dynamics of a regular valley wind, but also displays the peculiar characteristic of being strongly affected by the outbreak of a lake breeze flowing from a nearby valley and originated from Lake Garda.

Based on these findings, three natural ventilation strategies are proposed (night cooling, wind driven cross ventilation and stack and wind driven cross ventilation), and their effectiveness on thermal comfort are compared by means of dynamic simulation tools.

The thermal comfort in classrooms is evaluated according to the standard UNI EN 15251. For a standard occupant behaviour, discomfort situations from overheating occur in 34% of occupational period hours in the spring-summer season. The proposed ventilation strategies allow to reduce this value by up to 4%. Natural ventilation turns out to be an interesting low cost solution to control indoor temperatures without mechanical cooling systems.

KEYWORDS

Natural ventilation, school building, passive cooling, thermal comfort, wind.

INTRODUCTION

Existing school buildings in Italygenerally have no cooling system. However nowadays an increasing number of overheating discomfort situations occurs, as school building facilities are used also during the summer season for extra-scholastic activities. These situations are likely to increase, as a consequence of observed modifications of local atmospheric regimes, connected with either urbanisation [1] or global warming effects [2].

Natural ventilation seems a promising technique to improve thermal comfort in classrooms[3]. Besides requiring higher energy performances, the Italian legislation (D. Leg.

311/2006) recommends the exploitation of natural ventilation to reduce cooling demand. A more recent regulation (D. Leg. 5/2012) promotes the modernization of school buildings, improving energy efficiency and reducing management costs.
Natural ventilation as passive cooling strategy is particularly suitable in school buildings, as they have a defined use pattern and a flexible indoor layout [4].The present paper analyses the performance of natural ventilation strategies as retrofit solutions to improve thermal comfort in an existing school buildinglocated in Lavis, a suburban zone near the city of Trento (Italy), in the Alpine Adige Valley.

BUILDING DESCRIPTION

The Lavis secondary school was built about seventy years ago, and renovated in the last years (Figure 1). There is no cooling systems or mechanical ventilation plants, except in the canteen underground. The four storey building has a very complex layout due to the recent extensions. The present analysis focuses on the southern building part, where the classrooms are located. Every floor has seven classrooms connected by a corridor to the rest of the building and to two stairwells. One of the stairwells is enclosed by fire resistive elements. Two classrooms are oriented eastward, two westward, while three classrooms face south. Each classroom has two windows with triple-casement and tilt and turn opening. No vents are installed. Teachers and students report discomfort situations during the middle season. The rest of the building is allotted to services, offices, library and laboratories for artistic and musical activities, and has not been modelled in detail for the purpose of the present analysis.

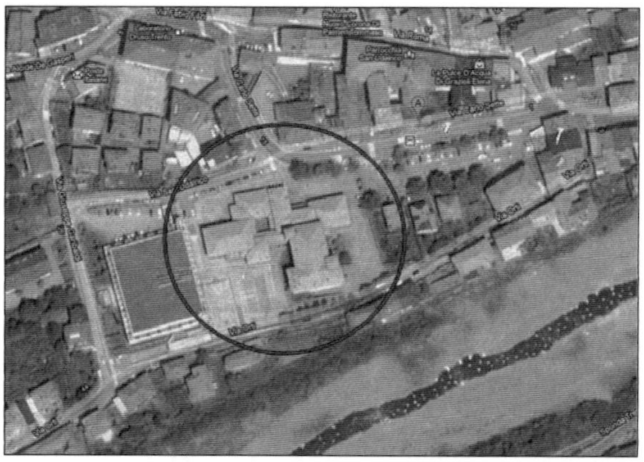

Figure 1. LavisSchool location. Source: © Google 2012

WIND POTENTIAL ANALYSIS

In the design of natural ventilation it is mandatory to understand how and when a wind induced internal flow is exploitable. To accomplish this goal, the building orientation and the internal space layout with respect to the main surrounding wind directions must be considered.
The Adige Valley in the Alps, where Lavis lies, is north-south orientated. Furthermore two tributary valleys join the Adige Valley near Lavis: the AvisioValley, east of the town, and the Lakes Valley, south-west of Lavis. Therefore wind speed and direction are expected to be strongly influenced by airflows occurring in this complex topography. During sunny days in the warm season valley winds generally blow in the above-mentioned valley. Valley winds,

which typically blow up-valley during the day and down-valley at night, develop as a consequence of the horizontal pressure gradients due to the temperature differences between different valley cross-sections or between the valley and the plain [5]. Moreover the local circulation blowing in the Lakes Valley is not a typical valley wind, but a combined "valley-lake" circulation, which starts blowing on the shores of Lake Garda, located in the southern part of this valley. This "valley-lake" breeze is generally rather strong and arrives into the Adige Valley from the Lakes Valley in the early afternoon.
In the present case typical daily cycles of wind speed and direction representative of the conditions occurring around the school building have been calculated from wind measurements acquired every 10 min by an amateur weather station located close to the school. A statistical analysis performed on wind speed and direction data for the whole 2011 shows that during the spring-summer season a typical daily cycle of wind speed and direction occurs(Figure 2). At night and in the early morning wind blows from east, following the down-valley wind flowing from the Avisio Valley, as the building is located close to the end of this tributary valley. Later in the morning and in the early afternoon wind blows from south-west, following the development of the up-valley wind and the outbreak of the "valley-lake" breeze from the Lakes Valley. Finally in the evening a transition can be seen from an up-valley wind, to a wind from east flowing from the Avisio Valley.

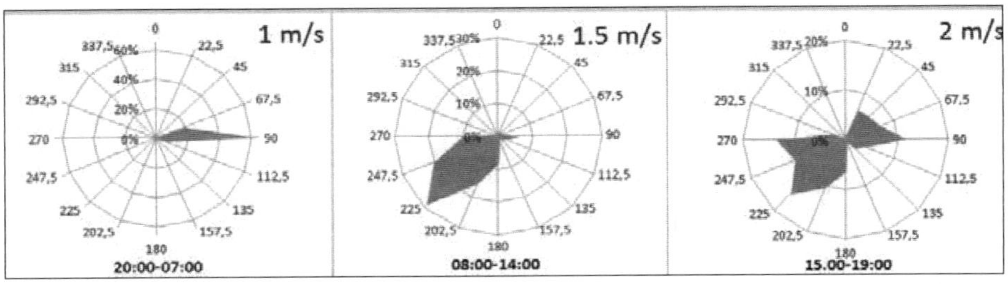

Figure 2. Average wind speed and directions in the typical day of spring-summer period.

NATURAL VENTILATION STRATEGIES

The three main strategiesproposed here are based on foundamental principles of natural ventilation and implemented to the case study, namely (i) night cooling, (ii) wind driven cross ventilation and (iii) stack driven cross ventilation.
Night ventilation (Figure 3) rejects excess heat cooling the building structure, taking advantage of the lower night external temperatures [6][7]. Opening actuators can be applied to the existing windows, controlled by external temperatures and humidity sensors. Night ventilation is activated if:

- outdoor temperature is lower than indoor temperatures;
- indoor temperature is higher than 24°C;
- outdoor temperature is higher than 14°C;
- it is not raining.

During the day a venting schedule based on the current school occupation pattern and use has been set.

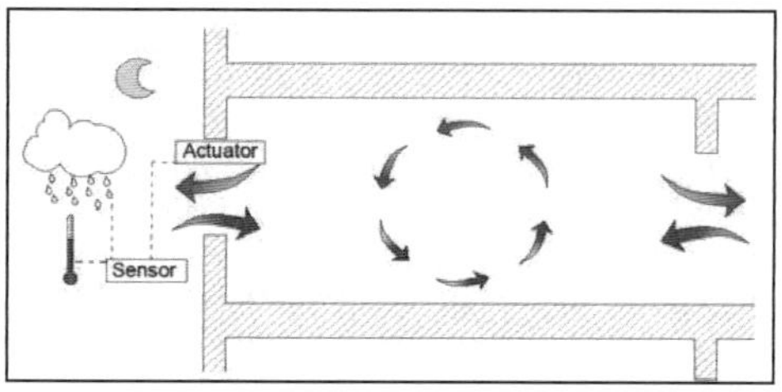

Figure 3. Night ventilation strategy.

During school time classroom doors stay closed, and single-sided ventilation occurswhen windows are opened manually by the occupants. A wind driven cross ventilation (Figure 4) can be easily implemented by installing vents above doors, allowing airflow from one side to the other of the building even if doors stay closed. Specific vents that combine acoustic attenuation, to fullfil acoustic standard requirements in classrooms, with very low airflow resistance (discharge coefficient $c_d = 0.71$) are available on the market.

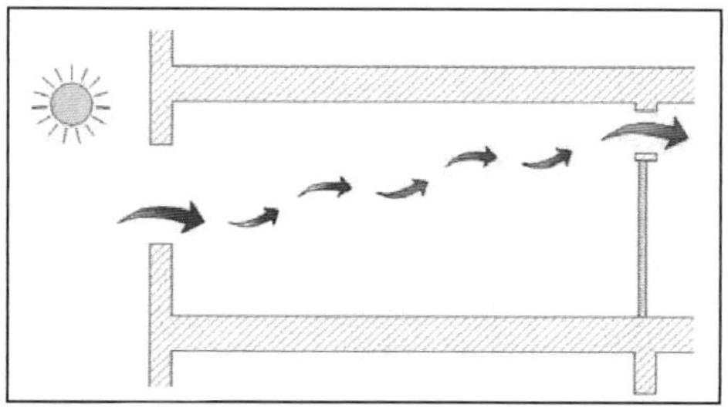

Figure 4. Wind driven cross ventilation strategy.

The third strategy proposed exploits the staircase in the South-East part of the building to increase the stack effect (Figure 5). A stack driven cross ventilation can be implemented by connecting corridors to the staircase through fire-resistant vents and adding a chimney on the top of the staircase. Classroom openings act as inlets allowing fresh air flowing through the building driven by the passive stack force of the exhaust air outlet at the top of the staircase. Some companies are specializing in the field of natural ventilation and provide specific products.

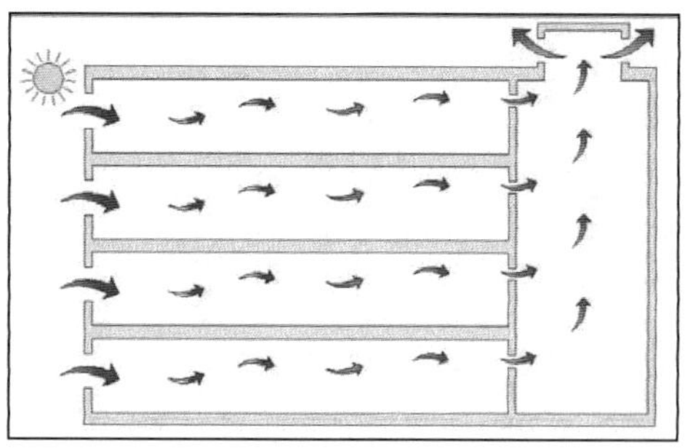

Figure 5. Stack driven cross ventilation strategy.

THE THERMAL – AIRFLOWNETWORK SIMULATION MODEL

The building model has been set up in Design Builder v.3, a graphical interface of the EnergyPlus building energy simulation engine[8][9]. The southern building part (grey volume in Figure 6) has been divided into thermal zones, according to occupation patterns, occupant activities, comfort needs, zone orientation and elevation [10]. The rest of the school building is modeled as an adiabatic block (red volume in Figure 6).
Standard year weather file data for the city of Trento are used [11]. Wind speed profile is modified by terrain roughness parameters forsuburbs.

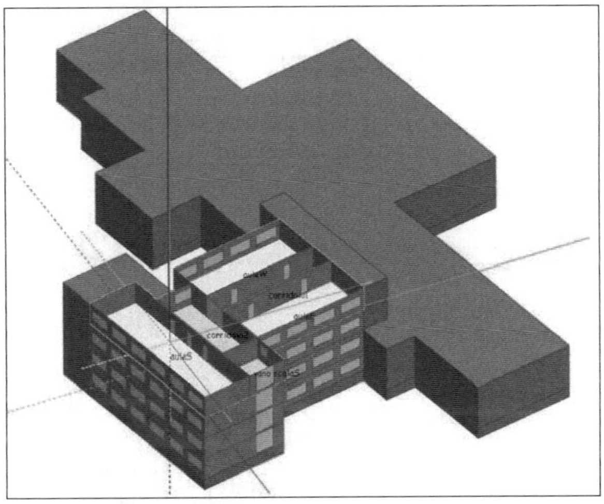

Figure 6. Lavis School model in Design Builder.

The proposed natural ventilation strategies were modelled by means of the AirflowNetwork object implemented in EnergyPlus [11] and compared.The airflow network object represents each thermal zone as a node of a network, characterized by a uniform temperature and a pressure varying hydrostatically (Figure 7). Windows, vents and cracks are represented as leakages, which link pressure differences due to wind or air density variations to power law equations. Each airflow path is calculated by Bernoulli equations and numerical solution of the problem gives the thermo-hygrometric conditions in the nodes. When temperatures and pressure are known, the program calculates airflow rates and latent and sensible heat exchanges. Thermal and airflow network model are coupled: they run in sequence and each uses the results of the other model (zone temperatures and pressures, airflow) in the previous time step [13].
The simulations are performed in free-running mode to analyse the passive behaviour of the building during the cooling season (15 April - 15 October). External openings are controlled by time schedules and by external temperature and humidity.Comfort temperatures are calculated from the CEN 15251 adaptive comfort model. Windows open if venting availability schedule allows venting and zone operative temperature is higher than the comfort one. Doors and vents are controlled by venting availability schedules according to a school standard occupation pattern. In general, doors must be closed during lesson periods and are a barrier to the natural airflows. Vents stay open everytime. AIVC wind pressure coefficient data sets[14]areused to calculate wind-induced pressure on each external node.They are defined for wind incidence angles in 45°increments for each surface in the model.

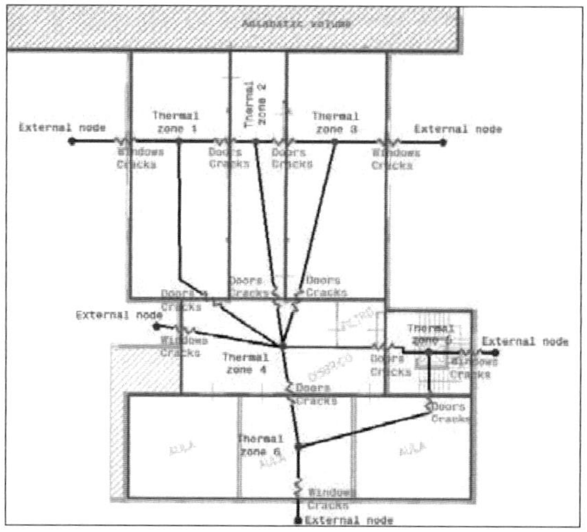

Figure 7. Schematic representation of AirflowNetwork in a floor type.

RESULTS

The three natural ventilation strategies proposed have been compared in terms of thermal comfort conditions. The european standard UNI EN 15251 defines three comfort categories limited by three temperature ranges. Thermal comfort is evaluated depending on the difference between the optimal operative temperature, according to the climate condition in the last week, and the simulated operative temperatures. The operative temperature outputs are compared with the comfort categories defined in the standard.

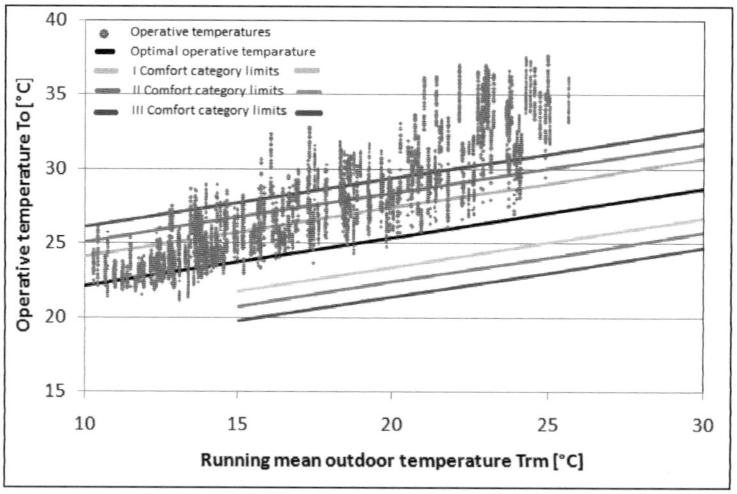

Figure 8. Base case simulation results. Operative temperatures of the most disadvantaged thermal zone (south oriented – third floor) are plotted against the outdoor running mean temperature. Green points above the superior limit of the third comfort category represent discomfort conditions due to overheating.

Simulations results of the most disadvantaged thermal zone, i. e. the one oriented to south at the third floor, were compared in the base case and in the natural ventilation strategies cases. In fact, the worst discomfort conditions are complained for that building part (Figure 8). Figure 9 shows the percentage of time during the occupation period when operative temperatures are within the comfort categories bounds.
Assuming a standard occupant behaviour, discomfort situations occur in the 34% of occupational period hours during the spring-summer seasondue to overheating.
Discomfort situations can be reduced to 5% by the natural night ventilation strategy, to 4% by the cross driven natural ventilation strategy and to 8% by the stack driven natural ventilation strategy. Furthermore, the percentage of hours when operative temperatures are within the first category boundaries increases significantly: from 43% in the base case, to 76% by natural night ventilation, to 69% by cross driven natural ventilation and to 59% by wind and stack driven cross ventilation. There is a 2% of time in which UNI EN 15251 standard comfort can not be applied as the Running Mean Outdoor Temperature is out of the required limits: upper limit temperature has to be between 10°C and 30°C and lower limit temperature has to be between 15°C and 30°C.
Simulation results show the global behaviour of the building in the assumed standard conditions. Mean values of solar and internal loads during the whole cooling season have been compared with the estimated heat rejected by airflow and the results are shown in Figure 10. It could be noted that the best results are reached by night ventilation, where the 90% of internal gains can be rejected.

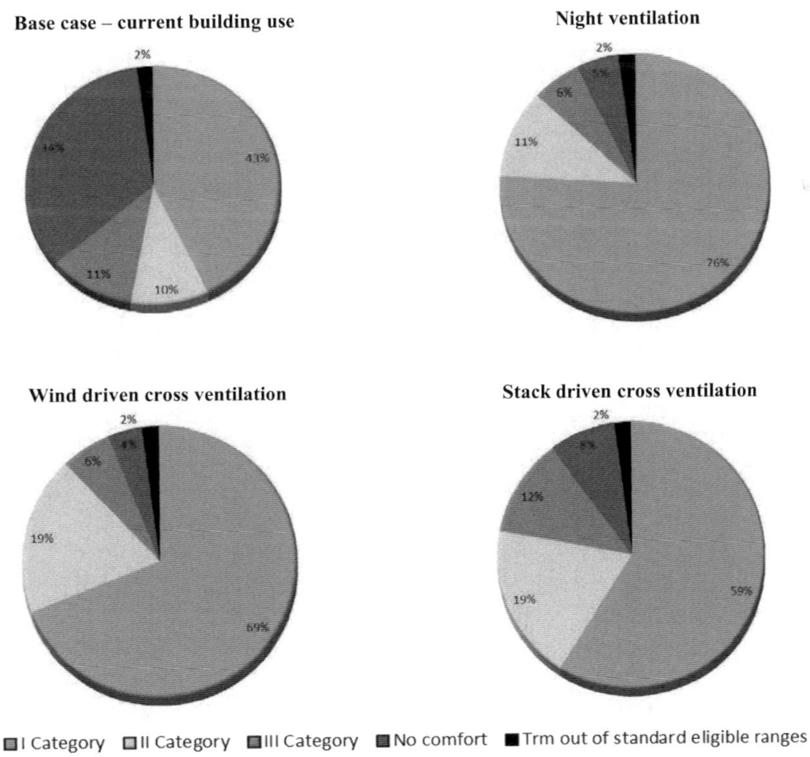

Figure 9. Comfort conditions in thermal zone considered according to UNI EN 15251 comfort standard.

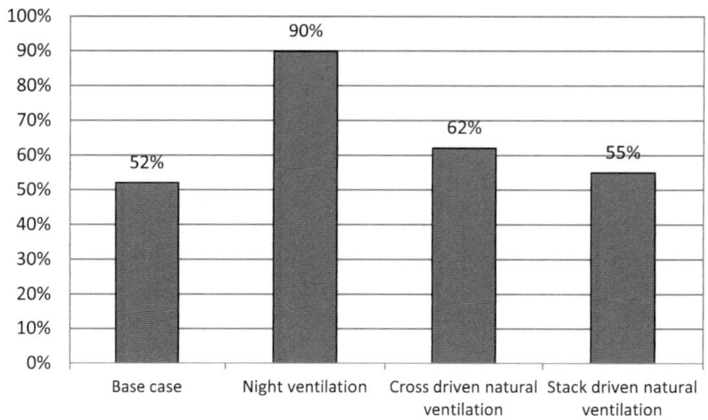

Figure 10. Mean percentage of heat (due to solar and internal loads) rejected by ventilation during the cooling season.

CONCLUSIONS

Natural ventilation seems to be a promising technique to improve thermal comfort in school buildings during the cooling season. In this work the potential of natural ventilation strategies in an existing school building has been studied.

The climate analysis showed how typical wind scenarios can be found by analysing data from a local weather station. The wind conditions at the building site are proven to rely on the dynamics of valley winds.

Three feasible natural ventilation strategies have been proposed on the basis of the climate analysis and the existing natural ventilation design guidelines: night ventilation, wind driven cross ventilation and stack driven cross ventilation.

The strategies effects on thermal comfort have been compared by means of coupled heat and airflow dynamic simulation models. The thermal comfort in classrooms has been evaluated according to the standard UNI EN 15251. Assuming a standard occupant behaviour, discomfort situations occur in the 34% of occupational period hours during the spring-summer seasondue to overheating. The night ventilation strategy proposed allowsreducing this value to 5% and also allows increasing the percentage of hours when operative temperatures are within the first comfort category boundaries.The study was completed by a survey on existing technologies, available on the market, to evaluate the practical fesibility of the proposed solutions.

It has been demonstrated that natural ventilation is an interesting low cost solution to control indoor temperatures and avoid mechanical cooling systems installation.

REFERENCES

[1] Giovannini, L., Zardi, D., de Franceschi, M., 2011.*Analysis of the urban thermal fingerprint of the city of Trento in the Alps*, J. Appl. Meteor. Climatol. 50, 1145-1162.

[2] Hansen, J., Ruedy, R., Sato, M., Lo, K., 2010.*Global surface temperature change*, Rev. Geophys., 48, RG4004,doi:10.1029/2010RG000345.

[3] ISBN. *Building Bulletin 101. Ventilation of School Buildings*. 2006.

[4] Santamouris M. et al.*Natural ventilation in buildings. A design handbook*. Allard F. 2002.

[5] Zardi, D., and C. D. Whiteman, 2012: Diurnal mountain wind systems. In: Chow, F.K., De Wekker, S.F.J., and B.J. Snyder (eds.): Mountain Weather Research and Forecasting: Recent Progress and Current Challenges, 37-122. Berlin: Springer..

[6] Artmann N., Manz H.,Heinselberger P.*Climatic potential for passive cooling of buildings by night ventilation in Europe*. Applied Energy, Vol. 84, Issue 2, Pages 187-201, February 2007.

[7] CISBE. Natural ventilation in non-domestic buildings. 2007.

[8] US Department of Energy. *The Encyclopedic Reference to EnergyPlus Input and Output*. 2011.

[9] US Department of Energy. *The Reference to EnergyPlus calculation*. 2011.

[10] O'Brien W, Athienitis A., Kesik T. *Thermal zoning and interzonal airflow in the design and simulation of solar houses: a sensitivity analysis*. Journal of Building Performance Simulation, Vol. 4, Issue 3, Pages 239-256, 2011.

[11] Typical climatic year of Trento (Italy), http://shop.cti2000.it/

[12] Lixing Gu, *Airflow network modelling in Energyplus*, Building simulations 2007

[13] Huang J., Winkelmann F., Buhl F., Curtis P., Fischer D., Liesen R., Taylor R., Strand R., Crawley D., Lawrie L., *Linking the COMIS multizone airflow model with the EnergyPlus building energy simulation program*, IBPSA, Kyoto, 1999

[14] Martin Liddament, *Air Infiltration Calculation Techniques, An Applications Guide*, AIVC.

*** Formatting Issue - Best Available Paper/Graphic ***

VENTILATED COURTYARD AS A PASSIVE COOLING STRATEGY IN THE HOT DESERT CLIMATE

Mohamed Hssan Hassan

Department of Architecture - Faculty of Engineering – Aswan University
AboElrish, 81542 Aswan, EGYPT
mamado_moh@yahoo.com

ABSTRACT

Traditional architecture gives ideas to enrich modern architecture. In traditional architecture, local materials and renewable energy resources have been used. The courtyard was one of the traditional architecture solutions as a climate modifier. The inclusion of an internal courtyard in buildings design is attributed to the optimization of natural ventilation in order to minimize indoor overheating conditions.

The paper investigates the potential of a ventilated courtyard for passive cooling in a small building in a hot desert climate. The analyzed model is one of the low-income housing models in New Aswan City - Egypt. Which was built and provided with main services by the government, these housing models were characterized by their improper design in many cases, especially, concerning with climatic design.

To evaluate the performance of a ventilated courtyard, building simulation software TRNSYS 16 (The coupling between TRNSYS and COMIS) was used. The courtyard parameters considered were the courtyard orientation and the courtyard geometry. To evaluate the performance of a ventilated courtyard, the average monthly indoor air temperature for the purposed building determined in the overheating summer season depended on the weather data for the building site. The results of the investigations of the courtyard parameters indicate that there are some important parameters and other are of less significance which affect the thermal performance of the courtyard building model.

KEYWORDS: Courtyard, hot desert climate, natural ventilation, low-income housing.

INTRODUCTION

Natural ventilation is one of the natural passive cooling strategies recommended for hot desert regions. In desert areas, this strategy is utilised to conserve energy while maintaining appropriate thermal comfort for inhabitants inside the building.

The aim of this study is to further the understanding and optimisation of natural ventilation cooling in buildings that have a ventilated courtyard. The performance of the courtyard was evaluated by using of TRNSYS 16 simulation tool, in a multi-story house located in New Aswan City, which is about 850 km South of Cairo, Egypt.

BRIEF LITERATURE REVIEW

Many researches such as Dunham, Fathy, Givony, Hinrichs, Konya, Lippsmeier, Olgyay and Saini reached the conclusion that a building with a courtyard is a desirable concept in the hot dry climate [1].

In a recent study, Al-Hemiddi describes an experiment to investigate the effect of a ventilated interior courtyard on the thermal performance of a single-family house in a hot–arid region.

*** Formatting Issue - Best Available Paper/Graphic ***

Statistical analysis of data recorded during the summer of 1997 was carried out. The results indicate that the courtyard gives high efficiency in providing cool indoor air through cross-ventilation [2].

In another study, a comparison between different geometries of courtyards in terms of wind flow characteristics and indoor air speed is performed based on the validation of Computational Fluid Dynamics (CFD) simulations with 2D published wind-tunnel experiments. In addition, assessment of thermal comfort is made inside a number of selected dwelling rooms facing different courtyard geometries. It is confirmed that rooms with cross ventilation have higher indoor air speed values and therefore a better thermal comfort than with single-side ventilation. The courtyard dimensions, the position of the room and the orientation are important aspects influencing the indoor air speed and thermal comfort [3].

In their research, Rajapakshaa, Nagaib and Okumiya investigate the potential of a courtyard for passive cooling in a single storey high mass building in a warm humid climate. From the results of thermal measurements, a significant correlation between wall surface temperatures and indoor air temperatures is evident. A reduction of indoor air temperature below the levels of ambient is seen as a function of heat exchange between the indoor air and high thermal mass of the building fabric. However, this behaviour is affected by indoor airflow patterns, which are controlled through the composition between envelope openings and the courtyard of the building. From a computational analysis, several airflow patterns are identified. A relatively better indoor thermal modification is seen when the courtyard acts as an air funnel discharging indoor air into the sky, than the courtyard acts as a suction zone inducing air from its sky opening. The earlier pattern is promoted when the courtyard is ventilated through openings found in the building envelope [4].

The passive cooling effects of a courtyard of a small building were determined numerically by safarzadeh and Bahadori in their paper, employing energy-analysis software developed for that purpose. The passive cooling features considered were the shading effects of courtyard walls and two large trees (of various shapes) planted immediately next to the south wall of the building, the presence of a pool, a lawn and flowers in the yard, and the wind shading effects of the walls and trees. It was found that these features alone cannot maintain thermal comfort during the hot summer hours in Tehran, but reduce the cooling energy requirements of the building to some extent. They have an adverse effect of increasing the heating energy requirements of the building slightly. The same savings in cooling energy needs of the building can be obtained through many features such as wall and roof insulation, double-glazed windows, and special sealing tapes to reduce infiltration. They all save on heating energy requirements as well [5].

It is clear from the foregoing review that the courtyard building form, although being described as a suitable solution in hot dry climate, has not investigated with regard to the multi-storey residential building and evaluation concerning the interaction between the courtyard geometry and the orientation. Consequently, the importance and significance of the present study are apparent.

ANALYSIS OF THE CASE STUDY

Climatic features of New Aswan City

New Aswan City is located on the west bank of the Nile, 10 Km northern of the present Aswan City and 850 km South of Cairo, so it is located in the south of Upper Egypt region which characteristic of a hot, dry climate, with a very wide difference between day and night temperatures. During the summer season, the day-by-day mean of maximum outdoor air temperature reaches 35° C in the north of Upper Egypt, and 41° C in the south. In winter season, the day-by-day mean of minimum outdoor air temperature reaches 6° C in the north of Upper Egypt, and 10° C in the south. In general, the Upper Egypt has a typical desert climate with large variations between seasons and between day and night temperatures [1].

The level of humidity in Upper Egypt is relatively low, especially during the summer months. The yearly mean of humidity in the Upper Egypt differs from 55% to 20%, north to south respectively. In New Aswan City relative humidity records the lowest value in May and June (12%), whereas reach to the highest value in December and January (36%, 34% respectively). On the other hand, The annual wind rose for New Aswan City indicates that most winds blow from north, northwest, and northeast (49.2%, 21%, and 12.9% respectively) [1].

Building description

The use of courtyards in residential buildings in Egypt, and other countries in the Middle East, is many centuries old. The courtyards provided security and privacy for the residents, and daylight for the rooms which were built around them. By building a pool, fountain and planting trees in the yard, the architects created a very pleasant space for the residents to spend a portion of their time during summer months in the yard.

In Egypt at the present time, the courtyard usually employed in buildings to collect sanitary pipes of the service's rooms (w.c, kitchen), which were built around it. Also, provide security, privacy and daylight for the residents in these rooms.

The original residential building was chosen for the study consists of three floors, and has not a courtyard as shown in Fig. 1. It was constructed in 2005 within one of national housing projects was built and provided with main services by successive governments. Such a policy was adopted and applied since the fifties and is up until now with the main care devoted to produce as many residential units as possible, with less care of units' quality. Consequently, these projects were characterized by their improper design in many cases, especially, concerning with climatic design.

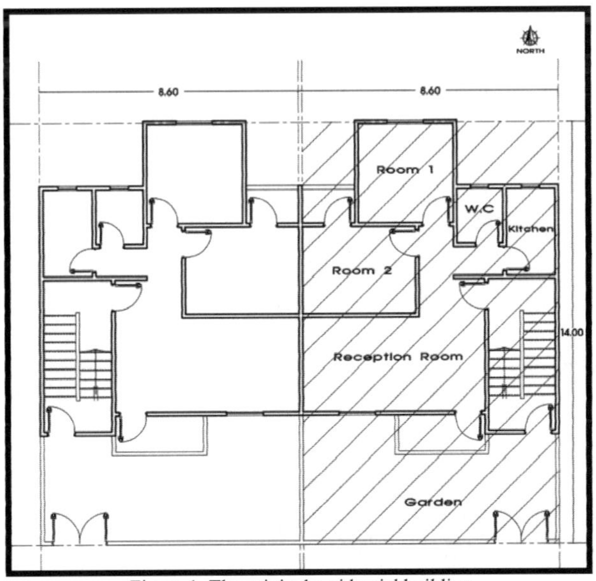

Figure 1. The original residential building

*** Formatting Issue - Best Available Paper/Graphic ***

The building does not have insulation on the walls and roof. It is constructed using standard local building materials. The external walls of the rooms are constructed with single bricks, 16 cm in thickness. The roof is flat and made of a concrete slab, 12 cm in thickness. External walls are painted light brown.

In a recent study TRNSYS 16 simulation tool was used to predict the hourly indoor air temperature which shows that the results indicates that the maximum temperature obtained is 43.6° C in the ground floor, while, in the first floor the maximum temperature obtained is 43.9° C, and in the last floor the maximum temperature obtained is 44.9° C.

In addition, the predicted average monthly indoor air temperature in the summer season (as shown in Table 1) indicates the overheating conditions which exceed the comfort limits to some extent [6].

Month		**Jun.**	**Jul.**	**Aug.**
Temperature (deg. C)	Ground floor	36.6	37.1	37.5
	First floor	36.5	37.0	37.4
	Last floor	37.1	37.7	38.0

Table 1. The predicted average monthly indoor air temperature in the summer season

The previous results declare the need for cooling system to exceed climatic difficulties, so, the results guide to the importance of employing solutions such as courtyard in buildings to act as a climate modifier.

The modified building has a courtyard in the central area of the house. The same area and dimensions were preserved; the courtyard opened to sky, and has ventilation open, and connected with the rooms around it with windows.

In order to determine the efficiency of the courtyard, the simulation process will be carried out for the last floor which considers the worst case. On the other hand, the courtyard parameters investigated will be in different four scenarios as follow.

Case 1: rectangular courtyard with ventilation open facing the north orientation (Fig. 2).
Case 2: rectangular courtyard with ventilation open facing the south orientation (Fig. 3).
Case 3: square courtyard with ventilation open facing the north orientation (Fig. 4).
Case 4: square courtyard with ventilation open facing the south orientation (Fig. 5).

Figure 2. Rectangular courtyard with north orientation

Figure 3. Rectangular courtyard with south orientation

*** Formatting Issue - Best Available Paper/Graphic ***

Figure 4. Square courtyard with north orientation

Figure 5. Square courtyard with south orientation

Fig. 6 shows the cross section of the purposed building.

Figure 6. Cross section through the building

TRNSYS program

The present research uses the computer simulation program TRNSYS 16 (The coupling between TRNSYS and COMIS).

TRNSYS (the TRaNsient SYstem Simulation program) is a flexible tool designed to simulate the transient performance of thermal energy systems. TRNSYS can be connected to COMIS (Conjunction Of Multizone Infiltration Specialists) through use of an add-on link component called Type157. This type recasts COMIS as a TRNSYS component. In this case, the COMIS input file is generated not using a separate graphical interface but using the TRNSYS simulation studio itself [7].

*** Formatting Issue - Best Available Paper/Graphic ***

RESULTS

Effect of the orientation

The comparing between the north and the south orientations clear that:

In the case of the courtyard which facing the north, the indoor air temperature recorded either in room 2 or in reception room values less than the indoor air temperature which recorded in the case of the courtyard facing the south (Fig 7). Also, the same results were recorded in the case of the square courtyard (Fig 8).

The square courtyard is more efficient than the rectangular courtyard.

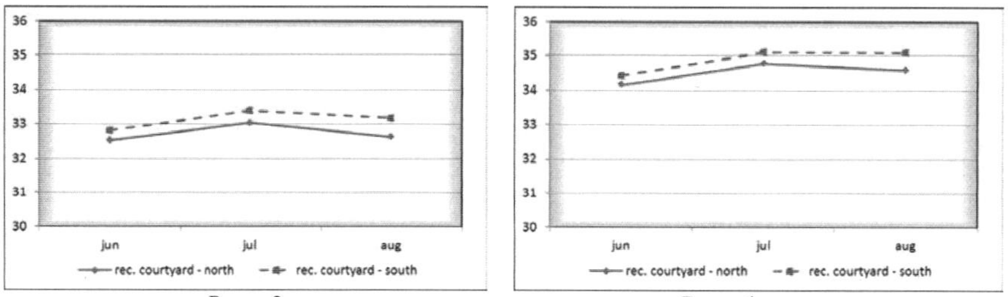

Room 2 Reception room

Figure 7. Results recorded in the case of rectangular courtyard

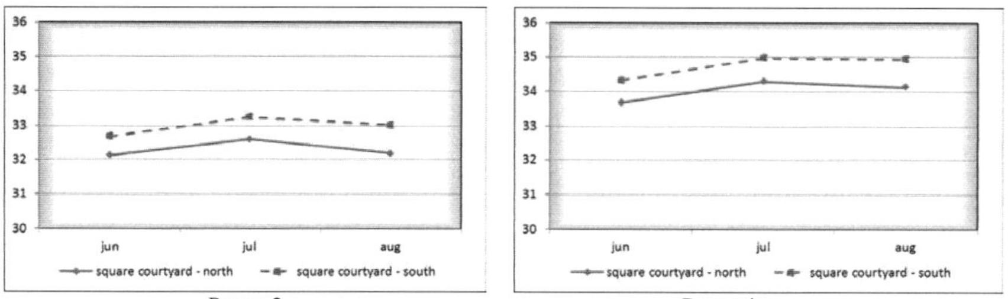

Room 2 Reception room

Figure 8. Results recorded in the case of square courtyard

Effect of the geometry

On the other hand, the comparing between the rectangular and the square courtyards indicates that the square courtyard is the best in all cases (Fig 9 – Fig 10).

In the same time, the north orientation is more efficient than the south orientation.

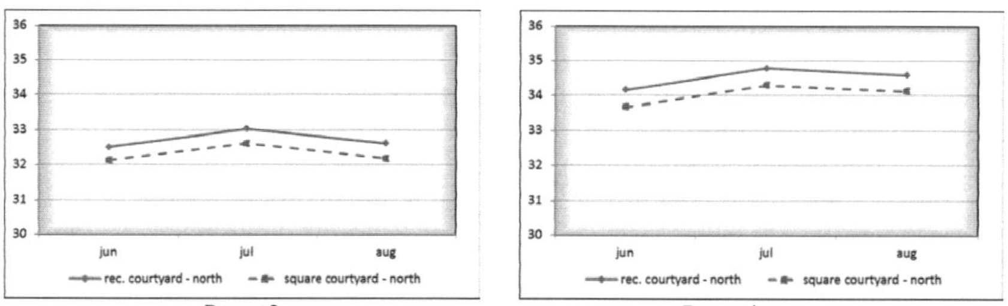

Room 2 Reception room

Figure 9. Results recorded in the case of courtyard facing the north

*** Formatting Issue - Best Available Paper/Graphic ***

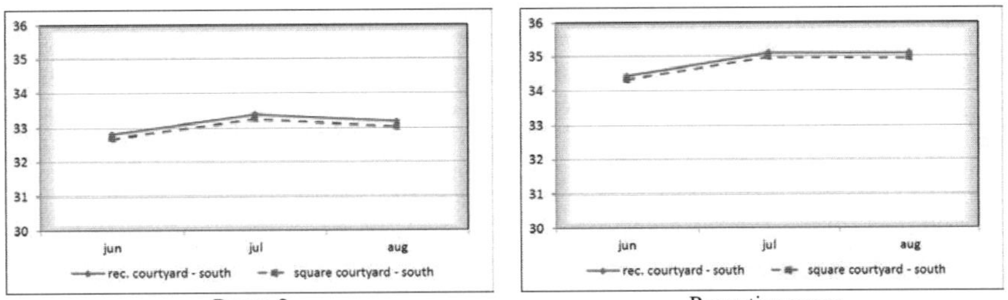

Room 2 Reception room

Figure 10. Results recorded in the case of courtyard facing the south

In general, it is clear to note that employing the courtyard in the building reduce the indoor air temperature in all cases by 3°C : 5°C (as shown in Table 2).

	Indoor air temperature in room 2			Indoor air temperature in reception room		
	jun	**jul**	**aug**	**jun**	**jul**	**aug**
Building without courtyard	37.1	37.7	38.0	37.1	37.7	38.0
Rectangular courtyard facing the north	32.5	33.0	32.6	34.2	34.8	34.6
Rectangular courtyard facing the south	32.8	33.4	33.2	34.4	35.1	35.1
square courtyard facing the north	32.1	32.6	32.2	33.7	34.3	34.1
square courtyard facing the south	32.7	33.3	33.0	34.3	35.0	34.9

Table 2. The comparison between the four scenarios

The reception room records more values of indoor air temperature than the room 2.

The square courtyard which facing the north orientation is the best case correspond to the rooms around the courtyard.

In all cases, it is noted that the reduction of the indoor air temperature is not enough to achieve the thermal comfort during the summer season.

CONCLUSION

The results showed that it is possible to get considerable reduction in the indoor air temperature with natural ventilation by using the ventilated courtyard in hot desert climate. However to ensure a good performance designers and other professionals should pay attention to some parameters like the orientation and the geometry of the courtyard.

The courtyard is an applicable design strategy for a building from the perspective of climatic and cost–benefit analyses. These strategies can be applied to single storey or multi-storey building, also, the use of the courtyard with a pond, fountain and trees could be an applicable strategy during the hottest periods.

*** Formatting Issue - Best Available Paper/Graphic ***

Finally, the development of the above formulae helps designers and architects to predict the thermal performance of courtyard buildings cooled by natural ventilation.

However, this approach only helps to predict the indoor air temperature of the purposed building, under given climatic conditions. Therefore this approach is not universally applicable, and values and judgements based upon it could change when we explore the effects of other designs and/or climatic conditions on courtyard performance.

ACKNOWLEDGEMENTS

The author would like to thank Aswan University for the funding of this research

REFERENCES

[1] Aly, A.M. 1994. *The effect of courtyard on the human thermal comfort inside residential building spaces in Upper Egypt*, Ph.D. thesis, Assiut University.

[2] Al-Hemiddi, N.A., Al-Saud, K.A. 2001. *The effect of a ventilated interior courtyard on the thermal performance of a house in a hot–arid region*, Renewable Energy 24, 581–595.

[3] Tablada, A., Blocken, B., Carmeliet, J., De Troyer, F., Verschure, H. 2005. *The influence of courtyard geometry on air flow and thermal comfort: CFD and thermal comfort simulations*, PLEA2005 - The 22nd Conference on Passive and Low Energy Architecture. Beirut, Lebanon.

[4] Rajapaksha, I., Nagai, H., Okumiya, M. 2003. *A ventilated courtyard as a passive cooling strategy in the warm humid tropics*, Renewable Energy 28, 1755–1778.

[5] Safarzadeh, H., Bahadori, M.N. 2005. *Passive cooling effects of courtyards*, Building and Environment 40, 89–104.

[6] Hassan, M.H., *Development of Building Simulation Model for Passive Cooling in Hot Desert Climate*, 2010. Ph.D. thesis, South Vally University (Egypt) on behalf of Dortmund University (Germany).

[7] http://www.trnsys.com/about.htm

TRANSFER OF ULTRAFINE PARTICLES AND AIR IN MULTI-STOREY BUILDINGS

Amalie Gunner[1], Siamak Rahimi Ardkapan[1], Alireza Afshari[1], Niels Christian Bergsøe[1]

[1]Department of Energy and Environment, Danish Building Research Institute, Aalborg University, Dr. Neergaardsvej 1, DK-2970, Hørsholm, Denmark

*Corresponding email: amg@sbi.aau.dk

ABSTRACT

An emerging issue in Denmark is passive smoking in residential buildings where non-smokers are exposed to harmful smoke from their neighbours. There are various ways that smoke infiltrates from one flat to another. The air infiltration rate between two flats in a multi-storey building depends on its construction, tightness and age.

This paper presents some of the results from of a study with transfer of ultrafine particles and tracer gas carried out in an older multi-storey building in Copenhagen. The aim of the study was to quantify the transfer of ultrafine particles and gases from one flat to another flat before and after sealing of the floor. The floor between the two flats was sealed using a new floor sealing method. The sealing method is developed by a specialist firm in sealing.

Indoor ultrafine particle concentrations and tracer gas were measured continuously in the two flats during the measuring periods. The gas source was N_2O and the particle source was burning cigarettes in the unoccupied flat. Reduction of the concentration of ultrafine particles and gases by sealing the floor with polyethylene and joint filler of bitumen was studied.

It was evaluated how the sealing performed with regard to decreasing the amount of transferred ultrafine particles and a tracer gas between two flats separated by a floor. When the floor between the flats was not sealed, the results showed that the transfer of ultrafine particles was about 4% and the transfer of tracer gas was 14%. After sealing, the transfer was reduced to 1.6% and 5% respectively.

Keywords

Particles, Air pollution, Ventilation, Full-scale experiments

INTRODUCTION

In multi-storey buildings, it can happen that residents are exposed to polluted air from the surrounding flats. The polluted air is usually associated with odour from activities as cooking and / or tobacco smoke. Tobacco smoke is harmful and therefore technical solutions needs to be developed to reduce the transmission of ultrafine particles in multi-storey buildings.

To what extent air pollution – airborne particles – is transferred from one flat to another depends on the construction of the building, the age of the building, density and the ventilation system. In an earlier study (1), the transfer of ultrafine particles from one flat (source flat) to another (exposed flat) was studied. The study showed, among other things, that approx. 9% of the particles from tobacco smoke were transferred when the exposed flat was directly above the source flat.

Previous research studied three technical solutions to reduce the transmission of ultrafine particles in multi-storey buildings (2). The first study examined sealing of the floor of the exposed flat. The other study examined the use of a novel air cleaning duct (Photochemical Air Purification) and the third examined a portable air cleaner (AC).

The first technical solution was tested in a flat in a multi-storey building from the 1930s. Cardboard and plastic foil of polyethylene were used for sealing the whole existing wooden floor in the exposed flat. After the sealing, the results of the study showed that the concentration in the exposed flat was independent of the generation of particles in the source flat.

The second solution was examined in a laboratory environment. The efficiency of the novel air circulating ductwork was examined by investigating the removal of ultrafine particles from a lit cigarette. The test showed that the removal efficiency ranged from 30% and 60% after 10 minutes, i.e. when the cigarette had burned out.

The third solution with a portable air purifier (AC) was investigated in a flat of 110 m2. Here, the efficiency of an air cleaner was studied in relation to the reduction of ultrafine particles from tobacco smoke. By use of an air purifier with a CADR (clean air delivery rate) of 240 m^3/h, the effectiveness of the air purifier ranged from 65% to 75% depending on its location in the flat in relation to the location of the particle generation (3).

Older multi-storey buildings often have leakages where particles can get through. There are leakages from piping in the floor and along the walls. Studies show that particles are not only transferred by the floor but also along the stairwell (1).

METHODS

The study was carried out in a block of flats from 1881. The exposed flat was directly above the source flat. The experiments were performed in January and February 2012. During measurements, no indoor activities took place - i.e. no cooking, cleaning or other activities that could generate particles.

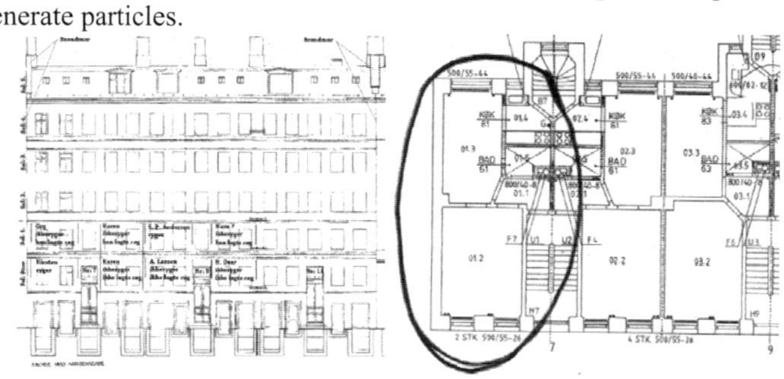

Figure 1: Facade and floor plan of the flats

Figure 1 shows the façade and floor plan of the two flats. The exposed flat and the source flat are laid out the same way. The volume of the flats is approximately 120 m^3. The flats consisted of a living room facing a busy road, a room and a kitchen facing a courtyard, a combined toilet/bath and an entrance in the middle of the flat. The floor of the flats was a lacquered wooden floor in the living room, the room and the hallway. The kitchen had vinyl flooring and the bathroom floor was covered with tiles. The kitchen had cabinets along the inner wall and a sink under the window. The living room and the room had skirting board along the walls and the ceiling had mouldings. In the middle of the ceiling in the living room and in the room was a rosette with a power outlet. Under each window in the living room and the room was a radiator with a heat pipe leading through the floor. There was ventilation in the toilet/bathroom. There was fresh air vents in all the windows.

With the purpose of creating an overpressure in the source flat compared to the exposed flat a higher temperature in the source flat was established. This was ensured by maintaining the temperature of the source flat at 5°C above the temperature in the exposed flat. During the examination the temperature in the exposed flat was approximately 20°C and the source flat was approx. 25°C.

The source for generation of particles consisted of four lit cigarettes that were placed in the lower flat, two in the living room and two in the room. All cigarettes were simultaneously burned out during a period of 10 minutes. The concentration of ultrafine particles were measured by using two particle counters *NanoTracer PNT 1000* from Philips and one condensation particle counter (*CPC*) model *3007* from TSI Incorporated. The particle counters measured concurrently. The two *NanoTracer PNT 1000* were placed in each of the flats and the *CPC 3007* was placed outside.

Besides particles, the air change rate and the transfer of air between the flats was measured by use of two Multi-Gas monitors, type 1302 from Brüel and Kjær, placed in each of the flats. Temperature and humidity were measured every 5 minutes with TinyTags dataloggers type TGU-4500 from Gemini. During particle generation, complete mixing was achieved by using small fans in the room. In the source flat, a top window in the facade facing the busy road was opened approx. 2 cm. All fresh air vents were also opened and the exhaust in the toilet/bath was closed. In the exposed flat all windows, exterior doors and fresh air vents were closed. In the toilet the exhaust was open.

The first part of the experiment was performed without sealing of the floor in the exposed flat. The second part of the experiment was carried out with sealing of the floor in the exposed flat. When the measurements were started the particle concentration in the flat was almost constant with a background concentration of about 10,000 particles per cm^3 in the source flat and 4000 particles per cm^3 in the exposed flat. The cigarettes were lit in the source flat, two in the room and two in the living room. The cigarettes were extinguished just before they burned out. The measurements continued until the ultrafine particle concentration in the two flats reached almost the initial concentration.

Before the last measurements, the floor in the exposed flat was sealed. The sealing method was developed by a specialist firm in sealing. The company chose to use a sealing of a vapour barrier, Icopal Blackline, made of polyethylene. The vapour barrier was put together by overlapping pieces bonded together with joint filler of bitumen. The vapour barrier was placed along the floor and up the walls, where it was sealed with building sealant approved for indoor use along the skirting boards. At the heating pipes, the vapour barrier was sealed with joint filler of bitumen. It is not included in the study to determine whether the products used contribute to the concentration of particles in the exposed flat.

CALCULATION METHODS

The transfer of ultrafine particles and gas was calculated with a calculation method [1] used in previous experiments with second-hand smoke (1).

$$c_{r(t)} = \frac{c_s\dot{V}}{\dot{V}+rV} + \frac{\dot{M}}{\dot{V}+rV} - \frac{\dot{V}}{\dot{V}+rV}\left[c_s + \frac{\dot{M}}{\dot{V}} - \frac{\dot{V}+rV}{\dot{V}}c_{r(0)}\right]e^{-\left[\frac{\dot{V}}{V}+r\right]\tau} \qquad [1]$$

Where
$\dot{V}$ = airflow rate [m^3/h]
$\dot{M}$ = particle transfer from flat 1 to flat 2 [(p/m^3) · (m^3/h)]
c_s = supply air concentration of UFP [p/m^3]
c_r = air concentration of UFP in the flat [p/m^3]
V = volume of the flat [m^3]
r = particle removal rate [h^{-1}].

The air change rate in the two flats was calculated by using the decay method [2].

With the decay method, the air change rate is determined by dosing a tracer gas in the flat. The tracer gas is distributed, so total mixing is guaranteed. After dosing of the tracer gas, the decay rate of the tracer gas is measured over time. The air change rate is calculated by the decay curve.

$$C(t) = C_0 e^{-nt} \quad [2]$$

Where C_0 is the start concentration in ppm
$C(t)$ is the concentration in ppm after t
t is the time in hours
n is the air change rate in h^{-1}

RESULTS

Before sealing

Four cigarettes were lit in the source flat, two in the living room and two in the room. The cigarettes were lit at the same time and extinguished just before they burned out. Figure 2 shows the measured concentrations of ultrafine particles in the source flat and the exposed flat. Before generation of ultrafine particles, the background concentration of ultrafine particles in the source flat was approx. 10,000 particles/cm^3 and in the exposed flat approx. 4000 particles/cm^3. The reason that the background concentration was higher in the source flat was that the resident in the source flat was a smoker and smoked indoors. The particles from the tobacco smoke deposited in materials such as furniture, walls and curtains (4). The maximum concentration of ultrafine particles in the source flat was measured to approx. 650,000 particles/cm^3. In the exposed flat, the maximum concentration of ultrafine particles was measured to approx. 11,000 particles/cm3. Figure 2 shows that the maximum concentration in the exposed flat took place 30 minutes after the cigarettes were extinguished.

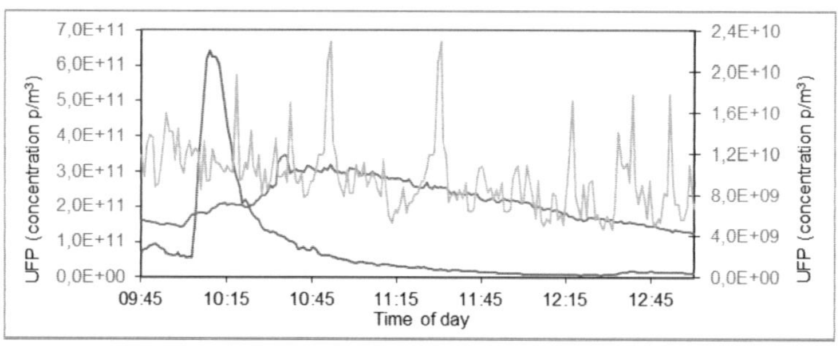

Figure 2: Particle concentration in the flats. Red: The source flat, Blue: The exposed flat, Grey: Outdoor

The transfer of ultrafine particles from the source flat to the exposed flat was found to approximately 4%.

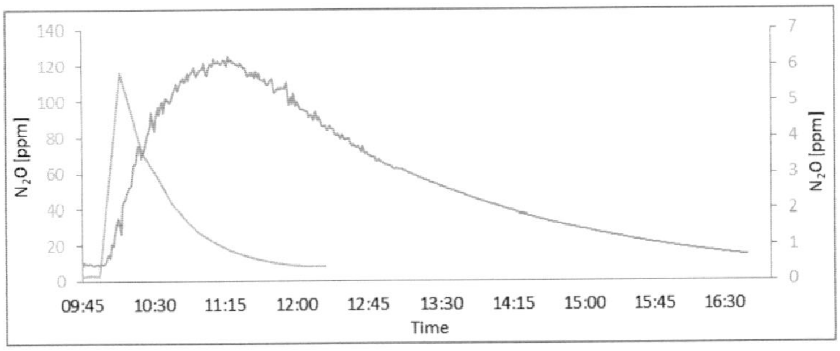

Figure 3: Tracer-gas concentration in the two flats. Yellow: The exposed flat, Green: The source flat

The air change rate in the source flat was calculated to 1.25 h^{-1} and to 0.18 h^{-1} in the exposed flat.

The transfer of tracer gas was calculated to approx. 14%.

After sealing

After the sealing, the particle transfer was reduced from 4% to 1,6% and the transfer of tracer gas was reduced from 14% to 5%. The results showed that sealing more than halves the transfer of particles and tracer gas.

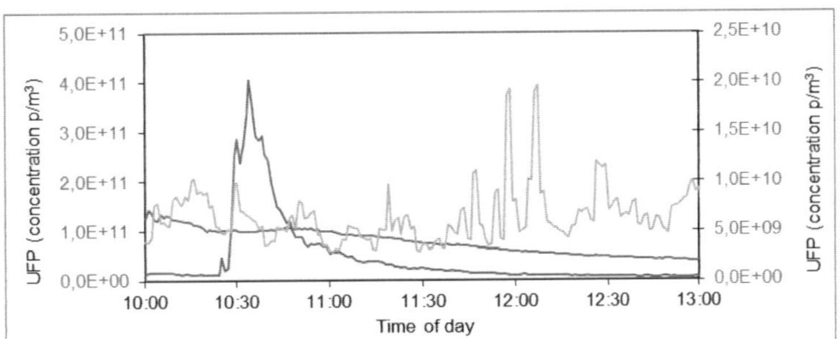

Figure 4: Concentration of ultrafine particles in the flats. Red: The source flat, Blue: The exposed flat, Grey: Outdoor

After the cigarettes were lit in the source flat, the concentration in the exposed flat was at the highest level after approx. 30 min, see Figure 4.

The initial concentration of ultrafine particles in the exposed flat was higher at the beginning of the experiment, than when the cigarettes in the source flat were ignited. This was because the residents were at home right up to the start of the study.

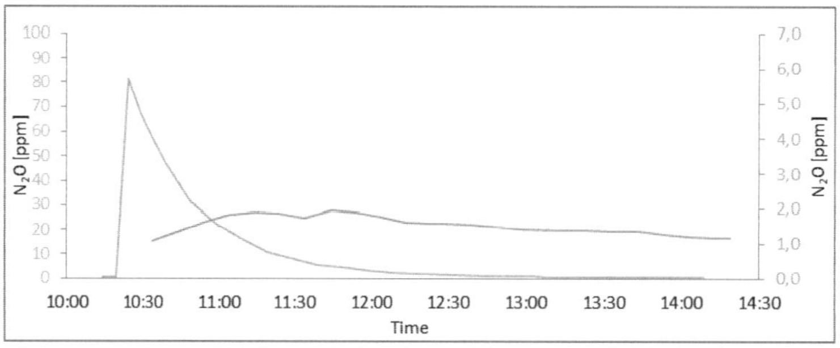

Figure 5: Tracer gas concentration in the two flats. Yellow: source flat, Green: Exposed flat

After sealing of the floor in the exposed flat, the air change rate in the source flat was calculated to be 1.46 h^{-1}. In the exposed flat, the air change rate was calculated to be 0.52 h^{-1}. The air change rate before sealing of the floor was 1.25 h^{-1} and 0.18 h^{-1} respectively. The difference in the air change rate may be due to the fact that when the floor was sealed, a hole was made in the outer wall for a balcony. During the experiment, the hole was sealed with a mattress, but no insulation. Due to the less insulated flat, the infiltration might have had an influence on the air change rate and thereby the air change rate is different. The wind speed for outdoor air was on average 6 m/s and 3 m/s for the two cases respectively (5). This indicated that the wind speed can not explain the increase in the air change rate.
After sealing of the floor of the exposed flat, the transfer of tracer gas from the source flat was calculated to be 5%. Before sealing the floor, the transfer was 14%.

DISCUSSION
In tobacco smoke, there are more than 30 different volatile compounds. The highly volatile compounds deposits on surfaces and in the environment where the pollution occurs (6). Tobacco smoke contains more than 4000 different chemical compounds during combustion (1). During combustion, both gases and particles are emitted. For most people, it is the odour that puts a limit on whether or not to accept tobacco smoke in the indoor environment. The concentration of the particles depends on their volatility (7). The concentration of particles also depends on temperature (increasing degassing with increasing temperature), moisture changes and ventilation (4). Some of the particles deposits on other materials, in particular furry textiles. If for example, the air change rate is low during a weekend, the emitting products deposit in the environment. They do not disappear just because the space is ventilated.

Gas
To investigate the transfer of gases, a tracer gas of N_2O was used. Since there are thousands of different gases consisting of several molecules in a lit cigarette, it is not possible to measure them all, and instead a measurable nitrogen tracer gas was used.
From the investigation, it was found that the transfer of tracer gas was reduced from 14% to 5% after sealing the floor in the exposed flat.
Because the tracer gas was N_2O, the measurements were certain to be only of the tracer gas. But it was uncertain whether it was transferred through the floor or through other leaks in the structure.
Since only a single gas of a specific chemical compound was measured, it is difficult to say how much of the gasses from tobacco smoke can be reduced, when compared with the large amount of different gasses in tobacco smoke.

Particles

The transfer of ultrafine particles was examined by measuring the concentration before and after sealing of the floor. The transfer of ultrafine particles was reduced from 4% to 1,6% after sealing the floor of the exposed flat.

This study showed that the transfer of ultrafine particles was less than the transfer of tracer gas. During the experiments without sealing of the floor, 14% of the tracer gas was transferred while only 4% of the ultrafine particles were transferred. This could be due to the fact that the ultrafine particles deposits on materials such as furniture and walls, and in cracks in the flat. Other factors, such as coagulation, sedimentation, condensation of water vapour on small particles, etc. also play a role.

CONCLUSION

After sealing of the floor in the exposed flat the transfer of ultrafine particle was reduced from 4% to 1,6%. This was more than half of what was transferred before the sealing. Transfer of gases was reduced from a transfer rate of 14% to 5%.

The investigation was performed in an older multi-storey building from the late 1800s. In Copenhagen, there is a great many of this kind of buildings but to verify the study, it is necessary that several different types of multi-storey buildings are tested.

The applied sealing method resulted in a reduction by more than half, of ultrafine particles and tracer gas from the source flat to the exposed flat.

New sealing materials are developed, and there is a need to test them in various types of buildings.

REFERENCES

(1) Afshari A, Shi B, Bergsøe NC, Ekberg L, Larsson T. Quantification of ultrafine particles from second-hand tobacco smoke. 2010.

(2) Afshari A, Ardkapan SR, Bergsøe NC, Johnson MS. Technical solutions for reducing indoor residential exposures to ultrafine particles from second-hand cigarette smoke infiltration. 2010.

(3) Shi B, Ekberg LA, Afshari A, Bergsøe NC. The effectiveness of portable air cleaners against tobacco smoke in multizone residential environments. 2010.

(4) Preining O. The physical nature of very, very small particles and its impact on their behaviour. J Aerosol Sci 1998;29(5-6):481-495.

(5) DMI Available at: http://www.dmi.dk/dmi/index/.

(6) Afshari A, Reinhold C. Deposition of Fine and Ultrafine Particles on Indoor Surface Materials. Indoor and Built Environment 2008;17(3):247-251.

(7) Er du udsat for naborøg? - Kræftens Bekæmpelse Available at: http://www.cancer.dk/rensluften/Naborog/er+du+udsat+for+naborog/er+du+udsat+for+naborog.htm?NRMODE=Published&NRNODEGUID={ADA1F4BA-AC27-4D39-A257-82283C166D65}&NRORIGINALURL=/rensluften/Naborog/er+du+udsat+for+naborog/&NRCACHEHINT=Guest.

AIR LEAKAGE OF US HOMES: REGRESSION ANALYSIS AND IMPROVEMENTS FROM RETROFIT

Wanyu R. Chan*, Jeffrey Joh, and Max H. Sherman

Lawrence Berkeley National Laboratory
1 Cyclotron Road, MS 90R3058
Berkeley, CA 94720, USA
* *Corresponding author: wrchan@lbl.gov*

ABSTRACT

LBNL Residential Diagnostics Database (ResDB) contains blower door measurements and other diagnostic test results of homes in United States. Of these, approximately 134,000 single-family detached homes have sufficient information for the analysis of air leakage in relation to a number of housing characteristics. We performed regression analysis to consider the correlation between normalized leakage and a number of explanatory variables: IECC climate zone, floor area, height, year built, foundation type, duct location, and other housing characteristics. The regression model explains 68% of the observed variability in normalized leakage. ResDB also contains the before and after retrofit air leakage measurements of approximately 23,000 homes that participated in weatherization assistant programs (WAPs) or residential energy efficiency programs. The two types of programs achieve rather similar reductions in normalized leakage: 30% for WAPs and 20% for other energy programs.

KEYWORDS

Blower door, fan pressurization measurements, air infiltration, weatherization, retrofit

INTRODUCTION

Residential energy efficiency and weatherization assistance programs (WAPs) have led to many measurements of air leakage being made in recent years. Building envelope airtightness is important because heating and cooling accounts for about 50% of the total energy consumption by US households [1]. Therefore, knowledge on the current state of the US housing stock, and factors that are associated with excessive air leakage, can have substantial energy implications.

In 2011, we collected a large number of air leakage measurements and updated the LBNL Residential Diagnostics Database (ResDB). Our latest efforts not only increased the number of data points, but also improved the spatial representation of the dataset. It is the goal of this regression analysis to identify housing characteristics that explain the observed variability in air leakage of single-family detached homes. In addition, we compared the air leakage measurements of homes before and after retrofit. Insulation upgrades and air sealing are commonly performed in a retrofit. In the US, the expected energy saving in heating and cooling bills from tightening the building envelope and reducing air infiltration is 10% to 20% [2]. But many factors can influence the energy savings and cost-effectiveness of air sealing and other retrofit measures, such as the initial air leakage of the house, and the expected improvement in airtightness from retrofit. This analysis will characterize the airtightness of the current US housing stock, and provide some of the needed data to evaluate the energy saving potential from reducing air infiltration via retrofit.

DATA DESCRIPTION

Data Sources

The newly updated ResDB contains air leakage data from 134,000 single-family detached homes. However, many missing data are present. The handling of these missing data, including year built, foundation type, and duct location, will be explained in greater details below. Overall, forty-three states are represented. The median year built and floor area is 1969 and 140 m^2, respectively.

Income-qualified WAPs are the major sources of data, accounting for about half of the blower door measurements. In prior versions of ResDB [3][4], Ohio was the only WAP present. ResDB now contains WAP data from eleven other states, including Arkansas, California, Iowa, Idaho, Minnesota, Montana, Pennsylvania, Utah, Virginia, Washington, and Wisconsin.

Residential energy efficiency programs are another major sources of data. For example, the Home Performance with ENERGY STAR program[1] is implemented in over 30 US states to improve energy efficiency of homes. New Jersey and Minnesota are the two states with the most pre- and post-retrofit blower door measurements in ResDB. There are also many data from programs in Vermont, Indiana, California, and Georgia.

Other sources that contributed air leakage and other diagnostic measurements include new homes that were tested to obtain an energy efficiency rating, or to demonstrate that they met an airtightness guideline. Homes are identified as energy efficiency rated according to the programs that collected the data, so there are likely some differences in rating criteria between the energy efficient homes. Moreover, there are also data that were collected for research studies or other purposes. Sources voluntarily contributed data to ResDB. Therefore, even though ResDB contains a large volume of data, the self-selected samples are not representative of the homes in US.

Normalized Leakage

Most of the air leakage data in ResDB are blower door measurements at 50 Pa pressure difference. Air leakage measurements are converted to normalized leakage (NL) for this analysis, as follows:

$$NL = 1000\left(\frac{ELA_{4\text{ Pa}}}{Area}\right)\left(\frac{H}{2.5\text{ m}}\right)^{0.3} \quad \text{where} \quad ELA_{4\text{ Pa}} = \sqrt{\frac{\rho}{2\times 4\text{ Pa}}}\left(Q_{50\text{ Pa}}\right)\left(\frac{4\text{ Pa}}{50\text{ Pa}}\right)^{n} \qquad (1)$$

$ELA_{4\text{ Pa}}$ (m^2) is the effective leakage area at 4 Pa, $Area$ (m^2) is the dwelling floor area, H (m) is the dwelling height, $\rho = 1.2$ kg/m^3, and $Q_{50\text{ Pa}}$ (m^3/s) is the airflow rate at 50 Pa measured by the blower door. NL is roughly lognormal distributed, with a geometric mean of 0.61 and a geometric standard deviation of 2.5. ResDB contains 7,000 measurements of pressure exponent, n, which are used to compute NL when available. The distribution of n is roughly normal with a mean of 0.65, and a standard deviation of 0.06. n is assumed to be 0.65 for all other cases [5].

---

[1] http://www.energystar.gov/index.cfm?c=home_improvement.hm_improvement_index

If H is not provided in the data, we assumed 2.5 m for each story, and an additional 0.5 m for ground level and inter-floor framing. In some cases where both the number of story and house height are unknown, we assumed that houses <200 m^2 are single-story, and >200 m^2 are two-story. This simple allocation based on 200 m^2 as the reference point is the same as used in previous analyses of ResDB [3][4]. About 80% of single-story detached houses in US are <200 m^2, but only half of the multi-story detached houses are <200 m^2 [6]. Our method of using the house size to approximate number of story is reasonable, but it is a source of uncertain.

Multiple blower door measurements exist for some homes in ResDB. If additional tests were performed to verify a measurement, then the average value is used. If a house was tested under different configurations, then the one that best described the occupied condition is used, i.e., exclude attic, but include or exclude basement depending on the normal winter condition.

REGRESSION MODEL

The multivariate regression considers the relationship between NL and these housing characteristics:

- Floor area $Area$ (m^2)
- House height H (m)
- Year built $\overleftrightarrow{I_{year}}$: before 1960, 60–69, 70–79, 80–89, 90–99, 2000 and after
- IECC climate zones $\overleftrightarrow{I_{cz}}$: 12 categories
- Homes participated in WAP: $I_{LI} = 1$
- Homes rated for energy efficiency: $I_e = 1$
- Foundation type: I_{slab}, I_{floor1}, or I_{floor2}
- Duct location: I_{cond}, I_{duct1}, or I_{duct2}

$Area$ and H are continuous variables, and all the remaining ones are indicator variables. Twelve of the 16 IECC climate zones[2] are represented: humid (5), dry (3), marine (2), and Alaska (2). The climate zone is determined by the house location, which is typically available by state and county, and the climate zone is identified correspondingly. For WAPs and other data with measurements before and after retrofit, the before values were used in the regression below. Homes are identified as energy efficiency rated by the programs that contributed the data.

Most of the data are missing foundation type and duct location. As a result, we first preformed the regression without these two parameters, as shown in Eq (2).

$$\ln(NL) = \beta_{area}Area + \beta_h H + \overleftrightarrow{\beta_{year}}\,\overleftrightarrow{I_{year}} + \beta_{LI}I_{LI} + \beta_e I_e + \overleftrightarrow{\beta_{cz}}\,\overleftrightarrow{I_{cz}} \qquad (2)$$

Using the coefficient estimates from Eq (2), the model residuals NL' are computed as follows:

$$\ln(NL') = \ln(NL) - \left[\beta_{area}Area + \beta_h H + \overleftrightarrow{\beta_{year}}\,\overleftrightarrow{I_{year}} + \beta_{LI}I_{LI} + \beta_e I_e + \overleftrightarrow{\beta_{cz}}\,\overleftrightarrow{I_{cz}}\right] \qquad (3a)$$

We then considered the effects of foundation type and duct location on the model residuals to estimate their influence on NL. Only the data with known foundation type or duct location is considered in Eq (3b) and (3c) respectively, so the values of NL' used for the regression are different in the two equations.

[2] See http://energycode.pnl.gov/EnergyCodeReqs/ for IECC climate zone classification.

$$\ln(NL') = \beta_{slab} I_{slab} + \beta_{floor1} I_{floor1} + \beta_{floor2} I_{floor2} \quad (3b)$$
$$\ln(NL') = \beta_{cond} I_{cond} + \beta_{duct1} I_{duct1} + \beta_{duct2} I_{duct2} \quad (3c)$$

From our previous work, we expected NL to be strongly correlated with year built [3][4]. To maximize the number of data considered in the regression, we categorized year built by decades from 1960 and onwards. But even when year built is treated as a categorical variable, one-quarter of the data are still missing this information. For these data, we imputed a year built category as follows. We first performed a regression by using three-quarters of the data with no missing data (i.e., year built is known). From this regression, we determined that ln(NL) decreases at an average rate of 0.14 from one year built category to the next newer category. Using this result, we imputed a year built category such that the predicted ln(NL) would best fit the measurements that contain missing data. The results are shown in Figure 1.

The imputation does not change the portion of homes in the different year built categories (Figure 1(a)). Homes that are built before 1960 and after 2000 remain the most common. This imputation method allows more data to be included in the regression model. Otherwise, homes in dry climate zones B-4, 5, and 6, and in marine climate zones C-3 and 4, would not be sufficiently represented in the regression.

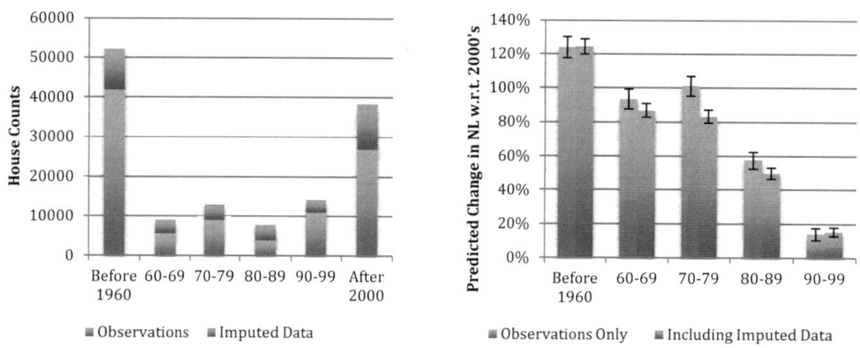

Figure 1 (a) Observed and imputed year built categories of single-family detached homes considered in the regression analysis. (b) Comparison of the predicted change in NL with respect to homes that are built in 2000's with and without imputation. Error bars show the 95% confidence interval.

Table 1 shows the regression results using the imputed data. Homes located in climate zone A-6,7 are selected as the reference, but other choices would give the same relative results. The model explains about 68% of the observed variability. The residuals ln(NL') are normally distributed: mean = 6.2e-17, and variance = 0.20.

One drawback of the imputation method used is that it can lead to underestimation of the differences between the observed and predicted values. In this case, however, the fit of model with (R^2=0.683) and without (R^2=0.682) the imputed data was essentially unchanged. With the imputed data, the predicted differences in NL for homes in different year built also remain roughly the same, as shown in Figure 1(b).

Figure 1(b) shows that homes built from more recent years have lower NL. This indicates that new homes were built with a more airtight building envelope compared to homes dated from earlier years. A recent study of new homes in California that are built between 2002 and 2004 also found that homes are built tighter compared to homes built in the 1980s and 1990s [7]. In addition to improvement in construction practices leading to tighter building envelope, it is also possible that there is a relationship between NL and house age. Older homes have higher NL not only because they were constructed that way, but also because the building envelope became more leaky over time. Both of these factors together likely explain a significant portion of the variability in NL among houses. Further analysis to isolate these two factors will be discussed in future analyses of ResDB.

Explanatory Variable	Coefficient Estimates	Standard Error	Pr(>|t|)	95% Confidence Interval
Area (m^2)	-0.00208	0.0000179	< 2e-16	-0.00211; -0.00204
Height (m)	0.064	0.00125	< 2e-16	0.061; 0.066
Year: Before 1960	-0.250	0.00705	< 2e-16	-0.264; -0.236
1960-69	-0.433	0.00811	< 2e-16	-0.449; -0.417
1970-79	-0.452	0.00762	< 2e-16	-0.467; -0.437
1980-89	-0.654	0.00836	< 2e-16	-0.670; -0.637
1990-99	-0.915	0.00816	< 2e-16	-0.931; -0.899
After 2000	-1.058	0.00748	< 2e-16	-1.073; -1.043
WAP Homes (pre-weatherization)	0.420	0.00428	< 2e-16	0.411; 0.428
Energy-Efficient Homes	-0.384	0.00453	< 2e-16	-0.393; -0.375
Humid A-1,2	0.473	0.01015	< 2e-16	0.453; 0.493
A-3	0.253	0.00653	< 2e-16	0.240; 0.266
A-4	0.326	0.00586	< 2e-16	0.315; 0.338
A-5	0.112	0.00551	< 2e-16	0.101; 0.123
A-6,7	0	--	--	--
Dry B-2,3	-0.038	0.00759	7.57e-07	-0.052; -0.023
B-4,5	-0.009	0.00684	2.00e-01	-0.022; 0.005
B-6	0.019	0.00988	4.91e-03	0.00008; 0.039
Marine C-3	0.048	0.01407	6.02e-04	0.021; 0.076
C-4	0.258	0.01133	< 2e-16	0.236; 0.281
Alaska AK-7	0.026	0.00589	1.42e-05	0.014; 0.037
AK-8	-0.512	0.00938	< 2e-16	-0.530; -0.439

Table 1 Results of regression model (β's in Eq. (2)) without considering foundation type and duct location.

All the coefficient estimates from the above regression are statistically significant at the 95% confidence interval, with the exception of climate zone B-4,5. This means that homes in climate zone B-4,5 tend to be less leaky than in the reference zone A-6,7, but the difference is small, and we cannot exclude the possibility that this apparent difference occurs only by chance in our data sample. We observed no effect on the overall model fit if homes B-4,5 and A-6,7 are grouped together or separately. Since these two climate areas are geographically far apart, for completeness we decided to keep all 12 climate zones in the model.

Foundation Type and Duct Location

For foundation type and duct location, we performed the regression analyses using a subset of the data, and assumed that the coefficient estimates also apply to the larger dataset. There are 12,500 houses with known foundation types: $I_{slab} = 1$ means house is built on slab, $I_{floor1} = 1$ means conditioned basement or unvented crawlspace, and $I_{floor2} = 1$ means unconditioned basement or vented crawlspace. These categories are chosen because after adjusting for the other parameters using Eq. (3a), homes with slab have the lowest NL, followed by homes where $I_{floor1} = 1$, and homes with $I_{floor2} = 1$ have the highest NL (Figure 2(a)).

Figure 2(b) shows a similar comparison but for duct locations using another subset of the data where this information is available. Homes with ducts located inside the conditioned space have the lowest NL, followed by homes with ducts located in the unconditioned attic or basement, and homes with ducts located in the vented crawlspace have the highest NL. However, the comparison by duct location is uncertain because it is based on very few data (526 houses).

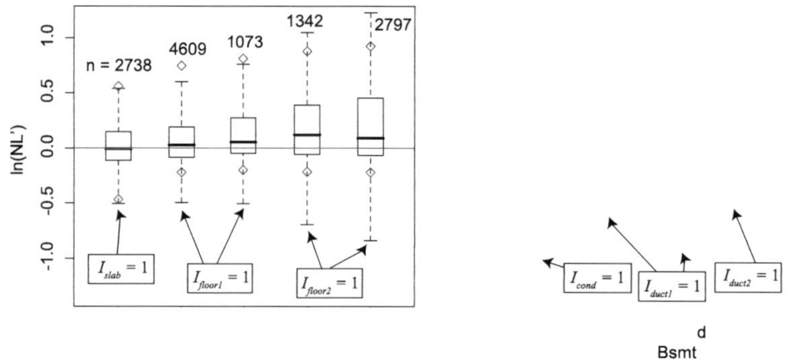

Figure 2 Model residuals of NL, computed using Eq. (3a), for homes with known (a) foundation type and (b) duct location (n = house counts). ln(NL') > 0 means that houses have NL higher than is predicted by Eq. (2).

Results of the regression (Table 2) show that the indicator variables considered are all statistically significant at 95% confidence interval. The coefficient estimates, β's, describe the influences of foundation type and duct location on NL as illustrated in the residual plots (Figure 2). Houses that are built on a slab and have ducts located inside the conditioned space tend to have the lowest NL. On the other hand, houses that have a vented crawlspace tend to have the highest NL, especially if the ducts are located in the crawlspace.

	Coefficient Estimates	**Standard Error**	**Pr(>|t|)**	**95% Confidence Interval**
(a) Foundation Type				
β_{slab}	-0.037	0.00709	1.85e-07	-0.051; -0.023
β_{floor1}	0.109	0.00492	< 2e-16	0.099; 0.118
β_{floor2}	0.180	0.00577	< 2e-16	0.169; 0.192
(b) Duct Location				
β_{cond}	-0.124	0.0255	1.53e-06	-0.174; -0.074
β_{duct1}	0.071	0.0339	3.59e-02	0.0047; 0.138
β_{duct2}	0.181	0.0383	2.98e-06	0.106; 0.256

Table 2 Results of regression model considering the effects of (a) foundation type and (b) duct location.

RETROFIT IMPROVEMENTS

There are 23,000 houses with pre- and post-retrofit blower door measurements. Paired data that showed no improvements (462 homes) or increase in NL (449 homes) were excluded from this analysis. It is likely that those records reflect cases where retrofit did not include air sealing or other work that would reduce air leakage. In homes where NL increased, the percent change from the pre-retrofit measurement is <10% in half of the homes.

There are many differences in how WAPs and residential energy efficiency programs are implemented. WAPs use the minimum ventilation rate limit without mechanical ventilation (based on ASHRAE 62.2) as the target. The resulting savings-to-investment ratio must be greater than one for the work to qualified as allowable expenditures. On the other hand, energy efficiency programs, typically sponsored by utilities, tend to offer rebates and other financial incentives for homeowners to preform an energy audit, and to follow through with its recommendations.

Figure 3 shows a larger reduction in NL pre- and post-retrofit from WAPs overall, compared to the residential energy efficiency programs. When the two programs are considered together, the median ΔNL is -25%. Aside from differences in how the two types of programs are implemented, there are also other state-by-state differences in the kinds of retrofit measures performed, and how the air leakage measurements were collected and documented. As a result, there can be many explanations for the differences between the two programs.

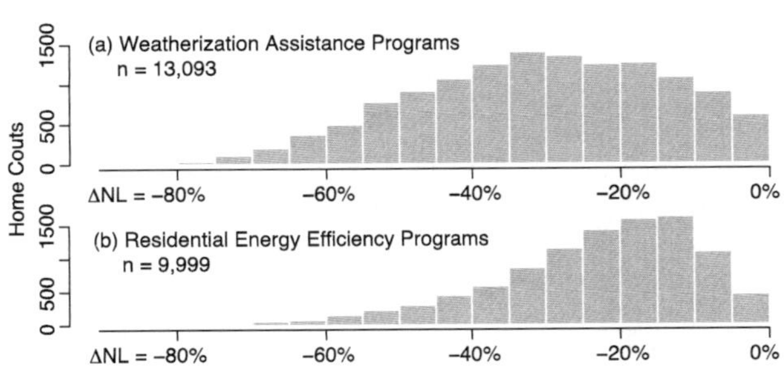

Percent Change in NL:

$$\Delta NL = \left(\frac{NL_{post}}{NL_{pre}} - 1\right) \times 100\%$$

ΔNL	(a) WAPs	(b) E. Prog.
Median	-30%	-21%
25th to 75th Percentiles	-18 to -43%	-13 to -31%
5th to 95th Percentiles	-5 to -61%	-5 to -50%

Figure 3 Reduction in NL as a result of retrofit from (a) WAPs and (b) residential energy efficiency programs.

As shown by the regression model, WAP homes tend to have a higher NL pre-weatherization. This may be one of the reasons why WAPs appear to achieve a higher reduction in NL. It is easier to reduce obvious air leakage pathways that exist in leaky homes, than to make significant improvements in homes that are more airtight to begin with. To test this hypothesis, we considered the relationship between ΔNL and NL_{pre}, and also with other variables, including: climate zone, house dimensions, and year built. Regression analysis suggests that for WAPs, only NL_{pre}, floor area, and height are useful parameters in explaining ΔNL, but not climate zone or year built. However, this relationship does not hold for houses that participated in residential energy efficiency programs, where the regression analysis shows that none of the parameters considered are useful in explaining ΔNL.

RESULTS AND DISCUSSION

Figure 4 compares the potential influence of the various explanatory variables on NL predictions, including:

(a) Other climate zones compared with respect to A-6,7
(b) WAP homes versus non-WAP; homes rated for energy efficiency or not
(c) Floor area increased by 100 m^2; height increased by 2.5 m
(d) Other foundation types: conditioned basement or unvented crawlspace (*floor1*), or unconditioned basement or vented crawlspace (*floor2*), compared with respect to slab (*slab*)
(e) Duct located inside conditioned space (*cond*) or in unvented crawlspace (*duct2)* versus in unconditioned attic or basement (*duct1*)

The percent change in NL is computed using the coefficient estimates of the regression model, as shown in Table 1 and 2. For example, Figure 4(a) shows that houses in climate zone A-1,2 are 60% higher in NL than homes in A-6,7. This is computed by exp(0.473) – 1 = 0.6. The effects of year built are shown in Figure 1(b), and are not repeated here.

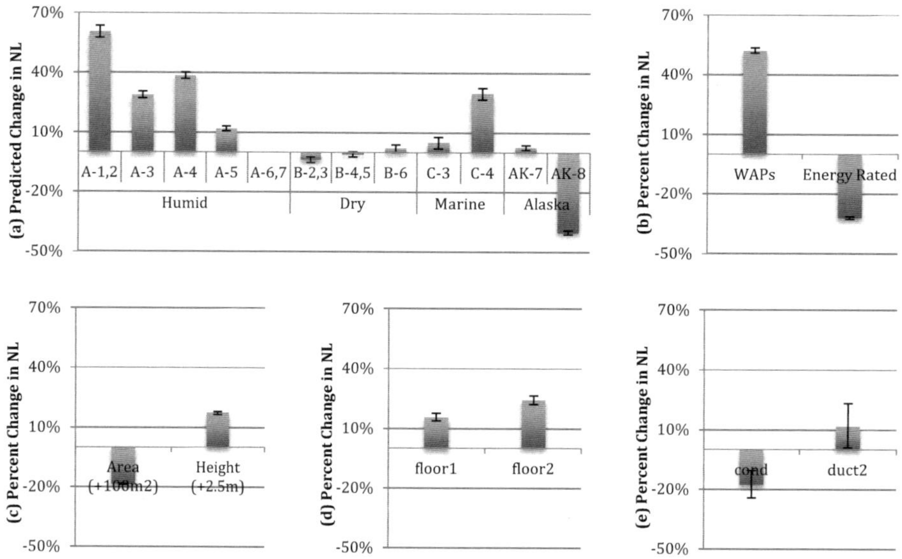

Figure 4 Predicted change in NL for homes in (a) difference climate zones with respect to climate zone A-6,7, and (b–e) other variables considered in the regression. Error bars show the 95% confidence interval.

Much of the variability observed in NL is associated with (a) climate zone, and (b) whether the houses are participants in WAPs or are energy efficiency rated homes. The difference in NL between the two extreme climate zones, A-1,2 and AK-8, is a factor of 2.7. The remaining factors, namely: (c) floor area and house height, (d) foundation type, and (e) duct location, each explain some differences in NL in the 10% to 20% range. In comparison, their importance is secondary for predicting NL. Overall, year built remains an important attribute to consider for predicting NL (see Figure 1(b)). The difference in NL between homes that are built before 1960 and after 2000 is a factor of 2.2.

The regression model presented here gives an estimate of NL based on a number of housing characteristics. For a housing stock, the model can explain 68% of the observed variability. However, the model is much more uncertain when it is applied to one house. This is because of the residual term. For example, the model predicts NL = 0.47 for a 150 m^2, single-story house built in 1990s that is located in climate zone C-3. The 95% confidence interval of this prediction is 0.44 to 0.50. However, the model residual ln(NL') has a variance of 0.20. This means that there is only about 10% probability that a house with the exact characteristics will have NL between 0.44 and 0.50. For this one house, the model predicts there is a 95% probability that its NL is between 0.2 and 1.1. But, for many homes with the same characteristics, the regression model predicts with 95% confidence that the values of NL will likely center in between 0.44 and 0.50.

CONCLUSION

Many blower door measurements have been added to LBNL Residential Diagnostics Database from housing units across the US. Regression analyses were preformed on 134,00 single-family detached homes to describe the relationships between NL and house characteristics. By improving the spatial coverage of ResDB, more meaningful relationships were observed with climate zones. The predictive model explains about 68% of the observed variability, most of which are explained through year built, climate zone, and whether the houses are part of a WAP or energy efficiency rating program. Houses that are older, located in hot and humid areas of the US (climate zone A-1,2), and are occupied by households eligible for WAPs based on income are likely to have higher NL. Other characteristics that are associated with higher air leakage include houses with a vented crawlspace, and especially when ducts are located in the crawlspace as well. This information is useful for estimating the air leakage baseline of US homes, and can be used to target homes that would likely benefit the most from airtightness improvements to lower their energy costs.

Comparison of the before and after retrofit blower door measurements shows a reduction of NL in the 20% to 30% range. WAPs achieved somewhat higher reduction in NL than other residential energy efficiency programs, likely because WAP homes were more leaky pre-weatherization. The current data show comparably reduction in NL across all retrofit programs regardless of house location or year built. This is important because construction methods and practices vary greatly in the US. This analysis suggests that improvement in airtightness is possible across the US housing stock.

ACKNOWLEDGEMENTS

We greatly appreciate organizations and individuals who shared their blower door and other diagnostic data with us. This work was supported by the California Energy Commission Public Interest Energy Research Program award number CEC-500-07-006 and the Assistant Secretary for Energy Efficiency and Renewable Energy, Building Technologies Program, of the U.S. Department of Energy under contract No. DE-AC02-05CH11231.

REFERENCES

[1] RECS. 2005. Table US12. Total Consumption by Energy End Uses, 2005. Residential Energy Consumption Survey, U.S. Energy Information Administration. http://www.eia.gov/consumption/residential/data/2005/c&e/summary/pdf/tableus12.pdf

[2] Beachler M.C., Love, P.M. 2010. Retrofit Techniques and Technologies: Air Sealing, A Guide for Contractors to Share with Homeowners. Building America Best Practices Series, prepared for the U.S. Department of Energy. PNNL-19284. http://apps1.eere.energy.gov/buildings/publications/pdfs/building_america/ba_airsealing_report.pdf

[3] McWilliams, J., Jung, E. 2006. *Development of a Mathematical Air-Leakage Model from Measured Data*, LBNL-59041, Lawrence Berkeley National Lab, Berkeley, CA.

[4] Chan, W.R., Nazaroff, W.W., Price, P.N., Sohn, M.D., Gadgil, A.J., 2005. Analyzing a database of residential air leakage in the United States. *Atmospheric Environment* 39, 3445-3455.

[5] Orme, M., Liddament, M.W., Wilson, A., 1994. *An Analysis and Data Summary of the AIVC's Numerical Database*, AIVC Technical Note. Air Infiltration and Ventilation Centre.

[6] U.S. Census Bureau, Current Housing Reports, Series H150/09, American Housing Survey for the United States : 2009, U.S. Government Printing Office, Washington, D.C.

[7] Offermann, F.J. 2009. *Ventilation and Indoor Air Quality in New Homes*. California Air Resources Board and California Energy Commission, PIER Energy-Related Environmental Research Program. Collaborative Report. CEC-500-2009-085.

LESSONS LEARNED ON VENTILATION SYSTEMS FROM THE IAQ CALCULATIONS ON TIGHT ENERGY PERFORMANT BUILDINGS

Xavier BOULANGER[1] (air.h), Laure MOURADIAN[2] (CETIAT), Charles PELE[3], Pierre Yves PAMART[3](CSTB), Anne-Marie BERNARD[4] (ALLIE'AIR)

1 air.h
C/O CETIAT
Domaine Scientifique de la Doua
25, avenue des Arts
BP 52042
69603 VILLEURBANNE Cedex, France

2 CETIAT
Domaine Scientifique de la Doua
25, avenue des Arts
BP 52042
69603 VILLEURBANNE Cedex, France

3 CSTB
84 avenue Jean Jaurès
Champs-sur-Marne
77447 Marne-la-Vallée Cedex 2, France

4 ALLIE'AIR
4 clos Ballet
01800 Meximieux, France
Annemarie.bernard@allieair.fr

ABSTRACT

During the project QUAD-BBC, several ventilation systems have been studied in residential (individual house and collective dwellings) and non-residential (school, offices) and assessed by the evaluation of an IAQ multi-criteria.

These calculations have shown some typical evolution of pollutants in very tight low consumption buildings and can alert on some possible effects.

For instance, formaldehyde and VOCs criteria are increasing at night when ventilation is shut off which indicates that passive measurement methods are in this case evaluating an average exposure not representative of occupation. It also shows how much airflow should be maintained to reduce the exposure to these pollutants or how much time before occupation the system should be started. Other lessons can be learnt from the pollutions in the kitchen during cooking, the humidity of drying clothes in houses and the impact of occupant behaviour…

Humidity evolution in case of insufficient ventilation in rooms with a high density of occupation (classrooms, meeting rooms…) has also a much stronger impact in a very tight building. The study also shows that the ventilation performance can be improved, especially in main rooms when improving building airtightness. While we could fear the contrary, improved airtightness appears to be beneficial to IAQ in our test cases.

This paper presents the main lessons learnt from the calculation analysis in these buildings.

KEYWORDS

IAQ, Air tightness, ventilation efficiency, simulation

INTRODUCTION

Low-energy buildings are built with well-insulated envelope in order to reduce the energy demand. In France, the conventional primary energy consumption should be inferior to 50 kWh/m^2/year for the residential low-energy buildings. This consumption takes into account the energy demand for heating, which includes the energy demand of ventilation and infiltration, space lighting, air-conditioning, ventilation auxiliaries and hot water production.

The up-coming French thermal regulation RT2012 will apply this specification for the new built buildings. Nevertheless, the regulation on ventilation does not specially deal with low energy buildings. One then can wonder about the energy impact of ventilation in such buildings. This concept of buildings brings out additional questioning on the link between innovative ventilation systems and indoor air quality (IAQ). The main concern is: which ventilation systems are suitable for low-energy buildings? In these conditions, the part of ventilation in the energy consumption would increase in these buildings The adequate ventilation system should meet the energy requirements while providing acceptable indoor air quality.

During the project QUAD-BBC[1] several ventilation systems have been studied in residential (individual house and collective dwellings) and non-residential (school, offices) and assessed by the evaluation of an IAQ multi-criteria.

This criteria takes into account:

- Humidity (number of hours above 80%) linked to occupants and their activities

And 4 specifics indexes related to common activity or impact

- A CO_2 as index of confinement linked to occupation,
- B NO_2, SO_2 (dwellings) and O_3 (offices) linked to occupant activities,
- C CO and 7 VOC linked to materials, activities and occupants behaviour,
- D PM $_{2.5}$ and PM $_{10}$ linked to activities.

The specific indexes are built for each group of pollutants.
For example, the index for occupant activities (Group B) is calculated as followed:

$$I_B = \frac{[NO_2]}{ref\ value\ NO_2} + \frac{[SO_2]}{ref\ value\ SO_2}$$

The lowest index correspond to the best Indoor Air Quality regarding this type of pollutant. These calculations have shown some typical evolution of pollutants in very tight low consumption buildings and can alert on some possible effects.

METHOD

We use, for the simulations, SIMBAD, a Building and HVAC Toolbox developed by CSTB (Husaunndee et al., 1997)[2]. This tool implements multizone and nodal building models in MATLAB/Simulink environment by combining heat and mass transfer phenomena. On the one hand, the thermal model is composed of detailed wall models describing the material layers and their properties, window models, heating and cooling devices, lighting systems, etc. It so deals with conduction, convection and radiation phenomena for calculating surface temperatures, mean radiant and indoor air temperatures.

On the other hand, the airflow model is made of airflow paths. In order to assess the performances of ventilation systems, this model includes the following systems: balanced ventilation with heat recovery and free-cooling, demanded-controlled ventilation based on humidity, CO_2 concentration or presence detection, and natural ventilation. It also deals with the characteristics of fans, ducts, heat exchanger, filters, and airflow paths.

The coupling of both models is done through the "ping-pong" method in which both models run in sequence and each model uses the previous time step results of the other (Hensen, 1999)[3]. The obtained simulation tool is then able to predict energy consumption and indoor air quality according to the pollution schedule and the ventilation system.

MATERIAL EMISSIONS

In offices and school, all ventilation systems simulated show an increase of formaldehyde concentrations at night and during week-end, when ventilation is shut off after occupation. Figure 1 shows for the classroom the fluctuation of formaldehyde concentration on one typical week.

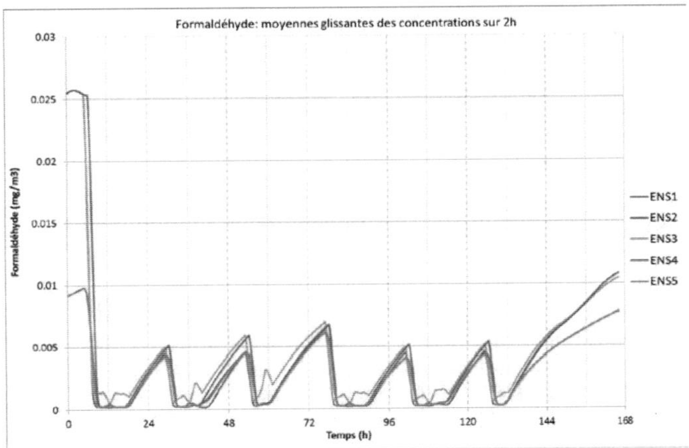

Figure 1 : Evolution of formaldehyde in a classroom on one week for the 5 ventilation systems simulated, all shut off at night.

The systems ENS 2 and ENS 5 are supply and exhaust mechanical ventilation (blue and red curves). Figure 2 shows the formaldehyde fluctuation when 10% of nominal airflow is kept at night to deal with the material emissions.

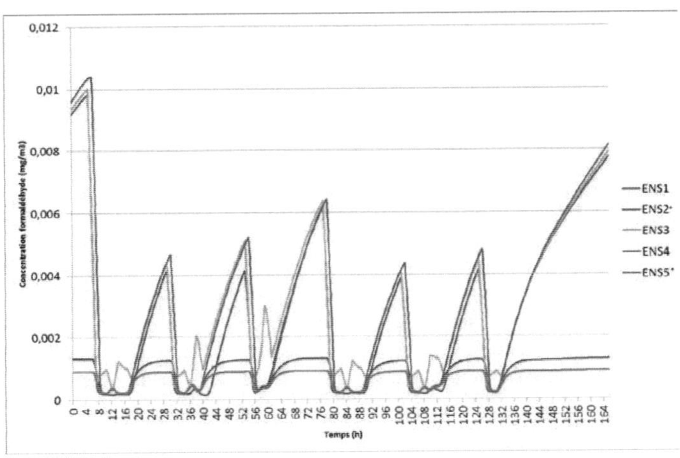

Figure 2 : Evolution of formaldehyde in a classroom on one week for the 5 ventilation systems simulated, ENS 2* and ENS 5* kept at 10% nominal flow, others shut off at night.

When a minimum airflow is maintained at night, the fluctuation is strongly reduced. Yet from the energy point of view, starting ventilation 1 hour before occupants' arrival leads to the same result and will spend less energy.

In France, a new law[4] plans to reduce by 2015 the maximum value of formaldehyde and to measure on site the results. At the moment, passive tubes are considered as the most reliable measurement and tubes are left generally 3 to 5 days on site. Due to the possible losses during this duration on site, the values can be underevualeted in regard of the real values.

It is often asked for DCV (Demand Control Ventilation) if the decrease of airflow when occupants are absent is consistent with maintaining of a good air quality taking into account the emission of materials. In France to get a technical agreement, these systems must have a clock to restart before occupation, stop after occupation and maintain 10% of nominal airflow when occupants are absent of the room but still during the occupation hours of the building. We note that these requirements are enough to correctly deal with the material emissions that have been chosen for our calculations. Recently, the Observatory of IAQ in France has launched a campaign on schools. On site measurements are in average at higher concentrations that in our calculations which would indicate that our scenario of emission may be underestimated. Yet to reach this average amount of formaldehyde in school (the limit from the law is at a concentration of 30 µg/m^3 since 2013), we note that the systems answering the technical agreement are still satisfactory. We note also that satisfactory target values of formaldehyde for health can be achieved at much lower flow than those indicated for low polluting materials in the Perceived Air Quality method of EN 15251[5].

AIRING

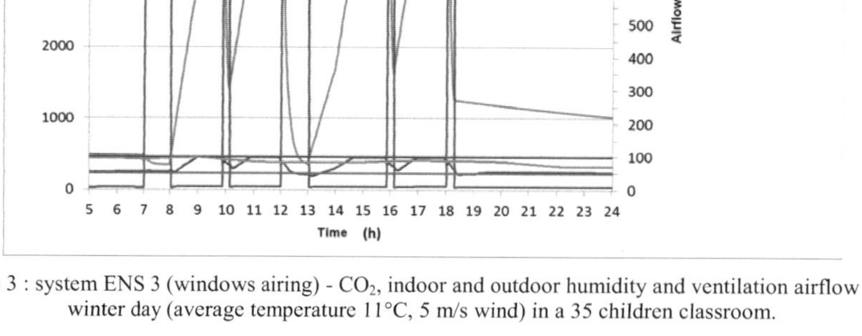

Figure 3 : system ENS 3 (windows airing) - CO_2, indoor and outdoor humidity and ventilation airflow on a winter day (average temperature 11°C, 5 m/s wind) in a 35 children classroom.

Using window airing in school is not sufficient to ensure a correct IAQ. Windows are manually opened from 7h till 8 h, from 12h till 13h and one quarter during the breaks of 10 and 4 pm. Confinement is too high (35 students in a 60m2 room) and the different indexes are incorrect. CO_2 levels reach 6000 ppm 15 minutes after the window is closed; this has already been shown many times. But in this quite tight building (1,7 $m^3/h/m^2$@4 Pa), it is important to note that indoor humidity increases strongly. On the full year, more than 3000 hours are reported over 75% HR, which shows that in highly occupied rooms of tight building, there is a severe risk of condensation which is both unhealthy for occupants and doesn't preserve the building itself.

BUILDING AIR TIGHTNESS

Enhancing building air tightness can improve the ventilation system performance, mainly for single exhaust, by improving air transfer and allowing air to enter where it's planned by design.

For both systems (single exhaut, supply & exhaust), in individual house for instance, improving air tightness slightly improves air quality by decreasing the pollutant concentration.

In this 2 floors' house, air entering the first floor goes down to ground floor to be extracted in the kitchen when air tightness is between 0,3 and 0,6 $m^3/h.m^2$ @4Pa while above, stack effect leads and air going up from the ground floor reduces the entrance of fresh air in the first floor bedrooms.

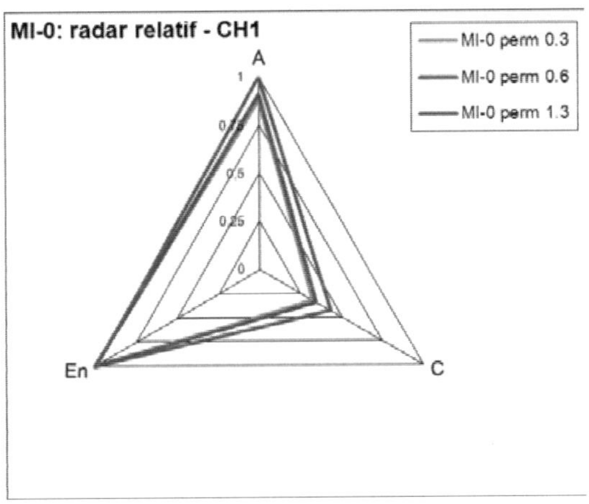

Figure 4 : example of savings on IAQ indexes (A and C) and energy for single exhaust (MI-0) (1st floor of the individual house studied)

It is interesting to note that improving air tightness doesn't reduce IAQ and on the opposite, tends to reinforce the designed air transfer and the efficiency of ventilation in the house. For instance, for single exhaust system in the house, the IAQ linked to material emissions is increased by 20 to 25 % (index C reduced from the same pourcentage as shown in figure 4 MI-0) and index A (concerning CO2 concentration) by around 10%.

In collective dwelling, we had similar conclusion: when leakages are reduced from 1,7 down to 0,3 m3/h/m²@4 Pa, index A decreased by 11% and index C decreased by 17%, which represent better IAQ.

KITCHEN VENTILATION

5 ventilation systems (single exhaust and supply & exhaust) have been studied in the house, 2 of them (LC3 and LC4) including a kitchen hood with a specific air inlet in the kitchen, opened only when hood is switched on. Figure 5 shows that the ventilation system (single or balanced) has no influence in the kitchen, but the presence of hood is efficient on combustion products.

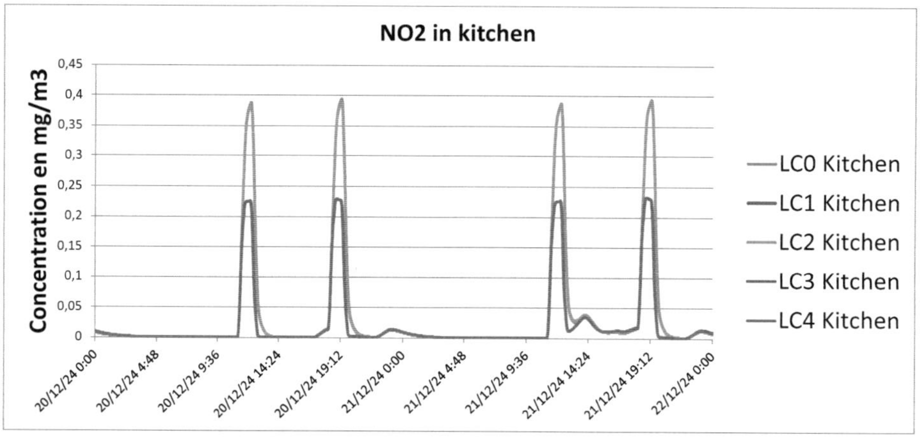

Figure 5 : NO_2 Concentrations in the collective dwelling kitchen for the various ventilation systems studied.

In airtight houses and dwellings, the boost airflow in kitchen and the presence of hood with integrated air inlet are absolutely needed to deal with pollutants load. This conclusion is obvious on combustion products (but also depend on emissions scenarii) but also valid for formaldehyde and material emissions in the kitchen.

OCCUPANTS BEHAVIOR

The use of encens or tobacco has much more impact (more than 10 time bigger) on IAQ indexes than material emissions. Figure 6 shows this effect on formaldéhyde only. A better knowledge of real emissions indoor is needed because today, only a few studies exist and show a lot of discrepancies in their results.

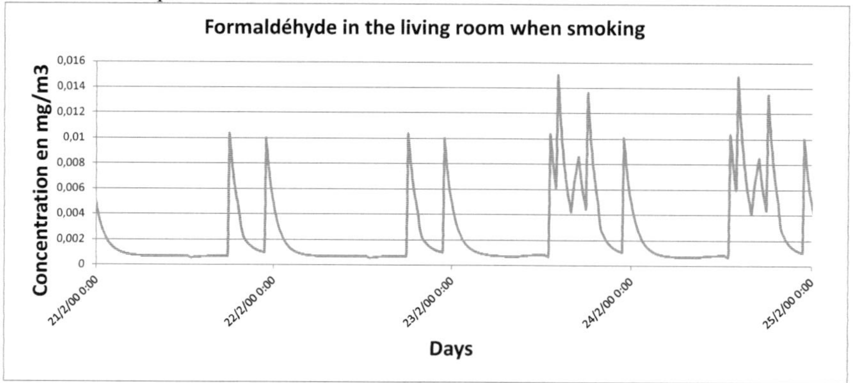

Figure 6 : Formaldéhyde concentration in the collective dwelling living room when 2 cigarets are smocked per day and 6 during the week-end

CONCLUSION

As we can note, some interesting conclusions are possible from this study based only on simulations. The absence of measurements need however to be careful of the impact of hypotheses assumed on emissions scenarii. The evolution of building toward more tightness can be an asset for the performance of ventilation but also need to design and install correctly the ventilation system to rely on it. Humidity may be the first adverse effect visible in case of low ventilation in a tight building before any increase of other pollutants may be noticed.

ACKNOWLEDGEMENTS

QUAD-BBC project partners are: air.h, CETIAT, CSTB, LEPTIAB, INERIS, ALLIE'AIR and we thank ANR, the French Agency for Research for its financial support.

REFERENCES

[1] Koffi, J., Allard, F. 2011. *Control and performance of innovative ventilation systems in low energy buildings.* Proceedings of 32nd AIVC Conference and 1st TightVent conference, October 12-13, 2011, Brussels, Belgium.

[2] Husaunndee, A., Lahrech, R., Vaezi-Nejad H., Visier J.-C., (1997), *SIMBAD a simulation toolbox for the design and test of HVAC control systems*, Proceedings of the 5th IBPSA Conference, Prague, Czech Republic, Vol. 2, pp269–276.

[3] Hensen, J., (1999), A comparison of coupled and de-coupled solutions for temperature and air flow in a building, University of Strathclyde (Scotland) - Energy Systems Research Unit.

[4] Décret no 2011-1727 du 2 décembre 2011 relatif aux valeurs-guides pour l'air intérieur pour le formaldéhyde et le benzène

[5] EN 15251, Indoor environmental input parameters for design and assessment of energy performance of buildings addressing indoor air quality, thermal environment, lighting and acoustics

INFLUENCE OF IMPROVEMENT OF AIR-TIGHTNESS ON ENERGY RETROFIT OF SOCIAL HOUSING, A CASE STUDY IN A MEDITERRANEAN CLIMATE

Jesica Fernández-Agüera[*1], Rafael Suárez[1], Per Heiselberg[2]

1 University of Seville, University Institute of Architecture and Building Science
Av. Reina Mercedes, 2,
41012 Seville, Spain
** jfernandezaguera@us.es*

2 Aalborg University
Division of Architectural Engineering
Department of Civil Engineering
Sohngaardsholmsvej 57
DK-9000 Aalborg, Denmark

ABSTRACT

In Spain, the residential sector is the third principal source of energy consumption; many of these dwellings are obsolete and do not have optimal conditions of comfort. For this reason, their energy retrofitting means an enormous step towards the energy efficiency. Under the general intervention strategies, the study and analysis of the air-tightness of the building envelope (as measured by the degree of infiltration) is a fundamental factor, because of its impact on energy efficiency, thermal comfort of occupants and indoor air quality. For this purpose, it has become a regular research field in other European countries and the USA. However, there is a lack of studies with adequate roominess to allow a proper analysis and interpretation of what happens in our regional climate and construction typology.

The aims of this paper is presenting a case study for the energy retrofit of 68 social multi-dwelling units in Cordoba (Southern Spain) evaluating their global energy demand and analysing the importance of air-tightness.

An in-situ air-tightness measurement campaign was carried out in these multi-dwelling units, before and after retrofitting, using Blower Door equipment. The best method for obtaining these parameters is pressurization/depressurization tests. It has been effectuated some modifications on façades and windows in order to obtain a better air-tightness.

The energy consumption was evaluated for the different levels of air-tightness by some tests which have allowed models to be generated. These models have been analyzed using Design Builder Energy Simulation software program, based on the DOE 2.2 calculation engine, obtaining predictive energy consumption, before and after retrofitting, including only air tightness changes and other retrofitting improvements (insulation, solar protection, U-transmittance in windows and facades) for the dwelling-units during a typical year.

KEYWORDS
Energy efficiency; building retrofitting, social housing buildings; energy consumption; air-tightness.

INTRODUCTION

In Spain, construction and operation of residential buildings accounts for the third highest energy consumption, after traffic and industry. The increased consumption that occurs in homes built in Spain is due to the climate. According to sources at the Institute for Diversification and Saving of Energy [1] air-conditioning accounts for approximately 49% of that consumption.

Social housing represents a significant proportion of the residential building stock of Southern Europe, which, when added to the socioeconomic characteristics of their occupants, necessitates special consideration of the methods to be used for the reduction of their energy consumption, especially that associated with their thermal comfort.

The most effective route for the reduction of energy consumption derived from the control of the energy demand associated with the transfers through the envelope using passive strategies. This intervention can be approached in two ways: the efficient construction of new buildings and the energy retrofit of existing residential buildings, a field of action that presents great potential for energy saving due to the importance of the housing stock built, and in use, in the last 50 years, it is also encouraged by the recent release of the Energy Performance of Buildings Directive, EPBD [2].

The air-tightness of building envelopes is one of the aspects which most affects the hygrothermal performance, indoor air quality and energy consumption of the building. In multi-story dwellings it contributes significantly to the overall demand for heating or cooling. The magnitude of the effects of air-tightness depends on many factors such as weather conditions (temperature, wind speed and direction), the design and geometry of the building and especially the quality of construction (design and execution), which complicates the analysis procedure [3,4,5].

As a result of the impact of the air-tightness of the building envelope (as measured by the degree of infiltration) on energy efficiency, thermal comfort of occupants and the indoor air quality, its study and analysis has become a regular field of research in other European countries and the USA. However, in our regional area, studies lack adequate breadth to allow a proper analysis and interpretation of the impact of air-tightness, in the context of the climate and construction types of the region.

The work of M. Sherman and the Lawrence Berkeley National Laboratory (LBNL) must be highlighted as a main reference in this field. It was associated with construction programmes in the USA which focused, primarily, on the housing field, and carried out extensive characterization campaigns and the development of predictive and calculation models for the phenomena of natural ventilation and uncontrolled infiltration processes. Our goal is to transfer and adapt these techniques and methodologies to our constructional reality and building processes in Spain.

To establish correlations between air flow and the energy performance of residential buildings, reference is again made to Sherman's work, with the Energy Performance of Buildings group of the LBNL [6] and those of Liddament [7] from the Air Infiltration and Ventilation Centre. This work investigates how current levels of ventilation affect energy demand and the estimated energy savings involved in adapting that ventilation to an appropriate indoor air quality, using the ASHRAE Standards 62, 119 and 136 to estimate the ventilation requirements and energy consumption.

The objective of this work was to analyze the importance of the infiltration in the energy demand reduction included in the residencial retrofit sector, analyzing a case study of multi dwelling units in the Mediterranean area. This was carried out on a building for which architect Rafael Suarez, co-author of this paper, designed a recently completed retrofit, boosting saving and energy efficiency.

CASE STUDY

The object of the study and analysis is a building of 68 social housing units, all of them rented, located in the city of Córdoba (Figure 1) in the south of Spain.

This building is a symmetrical U-shaped block five stories high, with housing units and an underground car park. Its construction dates from 1994 and it was retroffited in 2011. The retrofit project was promoted by the Córdoba Town Council and financed by the State Fund for Employment and Local Sustainability [8].

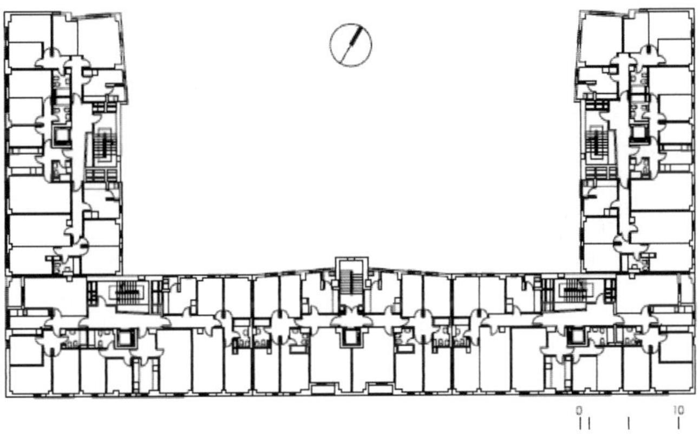

Figure 1. Floor plan

The thermal envelope of the building (table 1) presents low insulation levels, particularly on façades and floors in contact with the exterior, without any type of insulation, and in openings with single glazing.

Building element		U(W/m2K) Original	Retrofitting	Retrofit improvement
Facades	24 cm porous ceramics bricks with exterior rendering and interior plastering	0.94	0.33	Ventilated facade with 6cm Mineral Wood
Openings	Anodised aluminium frames with 5mm single glazing	5.70	3.8	Double glazed 4+6+4
Roof in contact with outdoor	Ceramic tiles, key mortar, brick board bedded on sand, slopes formed with 10 cm cellular concrete and 5 cm extruded polystyrene.	0.47	0.47	
Floor in contact with car parking	Unidirectional framework 25+5 semi-resistant joists finished with terrazo flooring and plaster	2.25	0.54	5cm extruded polystyrene insulation, air chamber and metal false ceiling.

Table 1. Characterization of the thermal envelope

Climate

The climatic profile used comes from the EnergyPlus weather files (EPW) database, part of the energy simulation software created by the U.S. Department of Energy. The file selected for Córdoba, CÓRDOBA SWEC (Spanish Weather for Energy Calculations), was created from the data originating from the Spanish National Institute of Meteorology (table 2)
Location: Córdoba (Spain) (N 37° 53 ') (W 4° 54 ') (GMT +1.0 Hour)
Elevation: 90 m above sea level: Standard atmospheric pressure: 100953 Pa

	Jan	Feb	Mar	Apr	May	Jun	Jul	Aug	Sep	Oct	Nov	Dec
Average monthly temperature (^{0}C)	9.2	10.9	13.5	15.4	19	23.5	27.3	27.2	24	18.5	13.2	10.2
Average monthly wind velocity (km/h)	8	8	9.6	10.6	8.9	9.6	9.5	9.6	8.7	8.5	8	8.3

Table 2. Climate values in Cordoba

The climate is sub-continental Mediterranean with warm summers, very high temperatures (maximum average temperatures of 36 °C) and an average of over 300 hours of sun per month from June to September. The winters are mild and last from November to March, with short springs and autumns.

METHODS

Passive strategies through the envelope are the most effective strategies to reduce the energy demand in the residencial buildings. The original state of the building performance must be known to calculate the potential reduction of the energy demand.

In the analysis of the original state of the building the solar radiation was studied with Ecotect Analysis version 5.50, the thermal bridges and common air leakage paths with an infrared camera, the air tightness with the blower door , the air and superficial temperature with a data logger and energy analysis with Desing Builder. Based on the study of data obtained, it is possible to elaborate a profile of the energy demand and set strategies to improve the energy consumption and thermal conditions. The most efficient strategies was chosen and their demand energy reduction was calculated.

Air tightness measurements using Blower Door

To know the original and retrofitting air tightness of the residential building, pressurization and depressurization tests were carried out using Blower Door equipment, which provides airtightness to the dwelling unit, the Air Leakage rate at 50 Pascals which take place as a result of the infiltrations through the building envelope.

Figure 2. Blower Door Test

In order to carry out these tests a blower door fan was placed at the external door of the housing unit, in order to extract (depressurization) or introduce (pressurization) air into the unit until a negative or positive pressure of 50 Pa was reached and the airflow was measured.

As this was a multifamily residential building, it was not measured as a single space, since staircases, lifts and other elements of the communal areas are not airtight and create air currents that are too large to ensure reliable measurements. Measurements were carried out on individual dwelling units, measuring at least one of them on the top floor, one on an intermediate floor and another on the ground floor.

These tests were executed in the original conditions to locate the main routes of air leakage with infrared thermography. After the retrofitting the tests were carried out in order to prove their improvement.

Energy models

To establish the energy performance of the original and retrofitted building the computer program Design Builder version 2.2.5.004 was used, whose simulation engine, Energy Plus, methodology developed by the United States Department of Energy and recognized by the International Energy Agency, enabled the authors to obtain precise data on annual or monthly demand for its original condition and for the retrofit project.

Each dwelling unit in the model was considered as a single space to be climatized. The official protocol for conditions of use and operation in Spain for the use of alternative energy simulation programs was followed [9].

Activity	Period	Value	Schedule					
			Weekdays		Weekends		Holidays	
Occupation	Winter	0.056 pers/m2	00:00 a 07:00 07:00 a 16:00 16:00 a 23:00	100% 25% 50%	00:00 a 24:00		50%	
	Summer				00:00 a 24:00	100%	00:00 a 24:00	0 %
Equipment & Lighting	Winter	8.88 W/m2 (4.44x2)	00:00 a 08:00				10%	
	Summer		08:00 a 19:00 19:00 a 20:00 20:00 a 23:00 23:00 a 24:00				30% 50% 100% 50%	
Ventilation	Winter	3 ac/h	00:00 a 24:00				0%	
	Summer		00:00 a 08:00 08:00 a 24:00				100% 0%	

Table 3. Protocol for conditions of use and operation in Spain.

RESULTS AND DISCUSSION

Retrofitting

Simulations were produced for the original conditions as well as for each of the intervention solutions proposed in order to obtain increased improvements in the energy demand of the building for its retrofit [10].

The program calibration was carried out in the EFFICACIA Project [11].

After analyzing the original state of the case study, the principal paths and factors where the building was losing energy were found and the main strategies in the retrofit proposal were:

- Encouragement of airflow, mainly through natural ventilation at night during the summer depending on exterior conditions.
- Energy conservation, improving insulation, and the accumulation of energy through thermal inertia. To guarantee complete efficiency in the summer time the thermal mass must be in contact with the night airflow to ensure passive cooling, while in winter the wall must receive solar radiation.
- Solar Radiation and Solar Control, capturing solar radiation in winter and ensuring suitable protection from radiation in summer (solar protection of the openings with the most solar exposure, depending on orientation, using sliding, folding, or fixed slat systems. East and west windows are protected by external movable shading devices which are activated during the cooling period).
- Thermal envelope insulation (Table 1) using a ventilated façade system, with a ceramic or metal finish. This system reduces thermal bridges in beams and pillars and along the joints between bricks and load-bearing structure.
- Thermal transmittance of windows, incorporating double glazing and improving insulation on external framework.

Figure 3. Case study, before and after retrofit.

Air tightness retrofitting

The values of the Blower Door tests can not be used directly for determining the annual infiltration value, because it responds to conditions of depression and pressure differential inside / outside very high, wich fundamental mission is the determination of Air-Tighness at 50 Pa.

Attributed to (and often denied by) Kronvall [12] and Persily [13], there was a rule of thumb that seemed to relate Blower-Door data to seasonal air change data in spite of its simplicity

$$ACH = ACH_{50}/20 \quad (1)$$

That is, the seasonal amount of natural air exchange could be related to air flow necessary to pressurize the building to 50 Pascals, where “ACH” is the natural air changes per hour and “ACH50 ” are the air changes induced by a 50 Pa pressure using a fan. We assume the uncertainty in the calculations using the following correction factors [14]:
-Dwelling units are 1 storie, their height correction factor is 1.
-Dwelling units are situated in the city, surrounding of other buildings, their shielding correction factor is 1.
-N of leakages is about 0,7, their lakiness correction factor is 1

V_{50} and ACH_{50} values obtained from the Blower Door tests and MDU characteristics, before and after retrofitting are represented in table 4.

Before retrofitting, ACH varies between 0.500 and 0.638, its averages is 0,550 and its standard desviation is 0.054. Although the construction system is the same, there are a significant degree of dispersion in air tightness tests, it can be due to constructive problems.

In most European countries the minimum ventilation standard ranges between 0.35 and 0.5 air changes per hour. However, in Spain, from the approval of the Technical Building Code in 2006, the requirements are much higher and the amount increases to 0.9-1 air changes per hour in new residential buildings.

After retrofitting, ACH varies between 0.426 and 0.536, its averages is 0.467 and its standard desviation is 0.041.
It is observed that the dispersion is lower than before retrofitting.
In both cases there was not relation of facade area and window area with air tightness.

These infiltration values aren´t for envelope, they are for whole dwelling units.

	Volume (m^3)	Facades	Facade Area (m^2)	Window Area (m^2)	Indoor Temperature (°C)	Outdoor Temperature (°C)	V_{50}	ACH $_{50}$	ACH
P1-V1	177,72	3(N,W,E)	64,16	12,61	32 (30)	34 (35)	1825 (1529)	10,27 (8,60)	0,513 (0,430)
P3-V2	176,53	1(0)	34,69	7,92	30 (29)	30 (36)	1966 (1695)	11,14 (9,60)	0,557 (0,480)
P3-V3	178,98	2(E,O)	39,94	10,33	32 (31)	32 (37)	2019 (1713)	11,28 (9,57)	0,564 (0,479)
P4-V12	177,72	3(N,W,E)	64,16	12,61	36 (31)	37 (34)	1778 (1513)	10,01 (8,51)	0,500 (0,426)
P4-V13	176,53	1(E)	34,69	7,92	(36)	(39)	(1603)	(9,08)	(0,454)
P5-V1	133,78	3(N,W,E)	58,31	9,28	32 (33)	33 (35)	1707 (1433)	12,76 (10,71)	0,638 (0,536)
Average							1859 (1631)	11,09 (9,35)	0,550 (0,467)
Desvest							130 (109,7)	1,08 (0,81)	0,054 (0,041)

Table 4. Multi dwelling unit characteristics and Blower Door test before and (after retrofitting)

This infiltration reduction is due to the thermal insulation that seal the joint between the facade and the window frame (figure 4) and the sustitution of the simple glass from the window for a double glass 4+6+4 and the improvement of the lock of the frame

Figure 4. Envelope retrofit

Energy demand

The study and analysis of the demand of the existing building revealed major energy losses in winter, due mainly to infiltrations, glazing and lack of insulation on façades,while in summer the main gains resulted from transmissions through openings and infiltrations. Windows play a very important role in thermal operation as they are elements for direct solar capture, natural ventilation and let in daylight.

Of the calculated overall annual demands, most correspond to heating (58% vs. 42% of cooling) resulting from the building orientation, the generation of its own shade, its shape (0.30) and the deficient insulation of the thermal envelope which translate into major energy losses. This demand differs considerably in relation to orientation, so that correction measures take this factor into account.

After retrofitting, the total energy demand reduction varies between 34 % and 46%, while the heating demand reduction is bigger varies between 7% and 41% , the cooling energy demand is lower varies between 1,6% and 16%.

There is a notable improvement of the energy efficiency after the energy retrofit. That is entailed in a consumption energy reduction of 42% respect to the initial stay, while a reduction of 17% respect to original conditions is due to actions that influence the air-tightness level of multi-dwellings. This fact emphasizes the importance of this parameter on consumption reduction.

Uncertainly and errors in calculations are due to the simplify transfer from ACH 50 to ACH, as discussed above, but this affects both states, before and after retrofitting. The airtightness change is due only to the improvement of the envelope.

	Multi Dwelling Units	P1-V1	P3-V1	P3-V2	P3-V3	P4-V12	P4-V13	P5-V1	Average
Total (kWh/m2)	Original conditions	67	73	68	63	82	96	91	77
	Infiltration retrofit	63	62	55	52	65	68	79	63
	Additional retrofit measures	41	40	38	35	48	52	60	45
Cooling (kWh/m2)	Original conditions	21	30	41	29	36	41	31	33
	Infiltration reftofit	20	27	37	26	31	34	28	29
	Additional retrofit measures	14	21	25	19	25	29	24	23
Heating (kWh/m2)	Original conditions	46	43	26	34	47	58	60	45
	Infiltration retrofit	43	35	18	26	34	34	51	34
	Additional retrofit measures	27	19	12	15	23	23	35	22
Increasing about total original conditions (%)	Infiltration retrofit	-5	-15	-19	-18	-20	-29	-12	-17
	Additional retrofit measures	-39	-45	-44	-45	-42	-46	-34	-42
Increasing about cooling original conditions (%)	Infiltration retrofit	-2	-9	-11	-10	-12	-16	-9	-10
	Additional retrofit measures	-34	-29	-38	-32	-31	-30	-20	-31
Increasing about heating original conditions (%)	Infiltration retrofit	-7	-19	-32	-24	-28	-41	-14	-24
	Additional retrofit measures	-42	-56	-53	-55	-51	-60	-41	-51

Table 5. Energy demand based on calculations

Energy demand results was obteined with Desing Builder because their retrofit works have just finished, but a monitoring campaigne is planned for the building.

CONCLUSION

Following the intervention proposal considerable improvement was observed in the thermal behaviour of the building for all energy models simulated and analyzed, more so in winter conditions, with a 51% reduction in demand, contrasting with a reduction of the demand for cooling of 31%, with an estimated reduction of total demand of 42%. Moreover, this reduction in demand translates into an improvement of thermal stability and reduction of temperature oscillations, with considerable repercussions on the increase of internal thermal comfort [10].

A great part of this energy demand reduction is due to an improvement in the air tightness after retroftting, a 40% on the total reduction, it is most important in heating 46% than in cooling 32%. For this reason, testing and review processes are essential to identify the common leakaege paths trought the envelope. This studies have to be presented in such a way that they can be easily put into practice as protocols to control the construction quality and reduce the energy demand in residential buildings.

REFERENCES

[1] Ministerio de Industria, Turismo y Comercio. PROYECTO SECH-SPAHOUSEC Análisis del consumo energético del sector residencial en España IDAE, 2011

[2] Energy Performance of Buildings Directive 2010/31/UE.

[3]Carrié, R., Jobert, R., Fournier, M. & Berthault, S., 2006. *Permeabilite' a` l'air de l'enveloppe des batiments. Generalites et sensibilisation*, CETE de Lyon.

[4]Sherman, M.H. & Dickerhoff, D. J., 2002, *Air Tightness of U.S. Dwellings*, US, LBNL-48671.

[5] Sfakianaki, A., Pavlou, K., Santamouris, M., Livada, I., Assimakopoulos, M. & Mantas, P., et al., 2008. *Air tightness measurements of residential houses in Athens, Greece*, Building and Environment, 43, (4) 398-405.

[6]Sherman, M. & Matson, N., 1997. *Residential ventilation and energy characteristics*, ASHRAE Transactions, Lawrence Berkeley Laboratory, 103, (1) 717-730.

[7] Liddament M., 2006. *Achieving Natural and Hybrid Ventilation in Practice*, International Journal of Ventilation, 115-130.

[8] Oficial State Bulletin: Royal Decree 13/2009 of October 26 2009 (http://boe.es).

[9] Ministerio de Industria, Turismo y Comercio. *Condiciones de Aceptación de Procedimientos Alternativos a LIDER y CALENER*. Madrid : IDAE, 2009.

[10] Suárez, R., Fernández-Agüera, J., 2011. *Retrofitting of Energy Habitability in Social Housing: A Case Study in a Mediterranean Climate*, Buildings, 4-15.

[11] Sendra, J.J.; et al. 2011. *Proyecto Efficacia: Optimización energética en la vivienda colectiva*. 1st ed.; Emvisesa, Sodinur and Universidad de Sevilla: Sevilla, Spain.

[12] Kronvall, J., 1978. *Testing of homes for air leakage using a pressure method, ASHRAE Trans*., 84 (I) 72-79.

[13] Persily, A., 1982. *Understanding Air Infiltration in Homes*, Report PU/CEES =/=129, Princeton University, 79-81.

[14] Sherman, M.H., 1987. *Estimation of infiltration from leakage and climate indicators*, Energy and Buildings, 10, (1) 81-86.

AIRTIGHTNESS OF OFFICE AND EDUCATIONAL BUILDINGS IN SWEDEN – MEASUREMENTS AND ANALYSES

Åke Blomsterberg*[1], Stephen Burke[2]

1 WSP Sweden
Jungsmansgatan 10
Malmö, Sweden
**Corresponding author:*
ake.blomsterberg@wspgroup.se

2 NCC Construction Sweden
Östra Varvsgatan 4
Malmö, Sweden

ABSTRACT

The airtightness of office and educational buildings influences energy use and thermal comfort. A leaky building is likely to have a high use of energy and thermal discomfort. The knowledge of real airtightness levels of entire buildings and their impact on the energy use is very low, except for a study carried out in the USA. Therefore two different methods of airtightness testing were applied to six entire Swedish office and educational buildings built since 2000. The first method involves using the ventilation system of the building and the second one to use a number of blower doors. Information on 30 other airtight tests was collected. During the airtightness testing the air leakage paths were detected using infrared scanning and smoke sticks.

The two methods are useful for testing entire office buildings, apartment buildings, industrial buildings and other premises.

The thirty-six tested buildings show a very good airtightness level, close to the Swedish passivehouse requirements. All previously tested office buildings in the USA, Canada and the UK are much leakier. The tested buildings showed some leakage paths, which could easily have been taken care of during construction, but are rather difficult to stop now.

The paper describes and evaluates the airtightness tests of thirty-six Swedish office and educational buildings and their implication for energy use.

KEYWORDS

Airtightness, blower door, energy, measurement, office building, pressurisation, school.

INTRODUCTION

It is well-known that the building sector plays an important role in the work towards sustainable development. The sector represents extensive economic, social and cultural values, at the same time as it causes extensive environmental impact due to its high use of energy and materials. An important part of the energy use within the building sector is related to office and educational buildings. The total energy use of an average Swedish office building is 220 kWh/(m^{2}year) (heated usable floor area) of which electricity stands for 108 kWh/(m^{2}year). Of this 108 kWh/(m^{2}year), 57 kWh/(m^{2}year) is due to office equipment, of which 23 kWh/(m^{2}year) is lighting. This was shown in a study of 123 office and administrative buildings of different ages [1]. Of the floor area in all office buildings, 69 % is heated by district heating and the average use of district heating energy is 110 kWh/(m^{2}year) [2]. Both new and old office buildings have a substantial potential for energy savings and improvement of indoor climate. While many new office buildings may have a low energy use for heating compared with older office buildings, they may have a higher electricity use. This is due to a high use of electricity for ventilation, cooling, lighting and office equipment. An important parameter affecting the energy use for space heating and cooling, and thus the indoor climate, is the airtightness of the building envelope. In a leaky building the energy use increases due to uncontrolled infiltration/exfiltration. The air leaking in and out through the building envelope increases the energy use as it, for example, does not pass through a heat recovery unit. The uncontrolled air leakage can contribute to discomfort such as draught, which can result in the indoor temperature being raised to improve the comfort, causing an increased energy use from the user's behaviour.

Unfortunately, there is no simple and accurate method of relating the airtightness of a building to the air leakage for an office or educational building in operation. This is due to difficulties in determining the location and characteristics of all leakage paths and determining the wind pressure coefficients [3].

The aim of this project [4] was:

- To use different measuring methods for determining the airtightness of educational and office buildings,
- to determine the airtightness for modern educational and office buildings,
- to determine the influence of airtightness on the energy use for space heating.

METHOD

The hypothesis is that, in many cases, the airtightness can be measured using the ventilation system of the building. Two different methods were used:

- Airtightness testing using a number of blower doors (portable fans), www.energyconservatory.com. European standard 13829, Method B was applied [5].
- Airtightness testing using the ventilation system of the building. Canadian standards were applied [6, 7].

The measurements involve pressurizing or depressurizing the entire building and measuring the corresponding air flow to maintain the different pressure differences between inside and outside. Ventilation openings and lead-throughs are sealed before the measurements. Thus the airtightness of the building envelope is determined. The location of leakage paths are determined using thermography and smoke.

When using the ventilation system of the building the following has to be investigated before:

- Exploring the building automation system to ensure that the ventilation air flows can be controlled and that it is likely to arrive at the necessary air flows. It is usually easier if the building has a demand controlled ventilation system.
- Ensuring that the air flows can be measured and that it can be done with adequate accuracy.

Within this project three schools and three office buildings were tested. An additional 31 tests had been carried out before by other Swedish organizations.

To determine the air infiltration/exfiltration rate from the results of pressurization tests there are different ventilation models. The ventilation models can be divided into: "air change" methods, reduction of pressurization test data, regression techniques, theoretical network methods, simplified theoretical methods [8]. The first three models are empirical techniques, which tend to be loosely based on the physical principles of air flow. The other models are theoretical models, which are based on a much more fundamental approach involving the solution of the equations of flow for air movement through openings in the building envelope. Empirical methods are usually straightforward to use, but tend to be unreliable and have a limited field of application. On the other hand, theoretical models have a potentially unrestricted applicability but are often demanding in terms of data and computer execution time. Theoretical calculation techniques can divided into: single zone network models, multi zone network models and simplified theoretical techniques. These models require a lot of information e.g. wind pressure coefficients, air leakage distribution for the building envelope, local wind speed, geometry of the building. Due to the limited amount of information on the tested buildings the method using reduction of pressurization test data was chosen in order to determine an order of magnitude for the average infiltration/exfiltration rate.

The reduction of pressurization test data method does nevertheless provide valuable information concerning the average infiltration performance of the building. The artificial pressurisation/depressurisation of a building to determine air leakage performance is now fairly common practice. The test only provides data regarding the "leakiness" of the building. The result provides no information on the distribution of openings or on how infiltration will be affected by wind, temperature, terrain, or shielding. However, several experimental results have shown that the approximate air infiltration rate will be of the order of one twentieth of the measured air change rate at 50 Pa [9], i.e.:

$$Q_{inf} = Q_{50}/20 \tag{1}$$

where Q_{inf} = infiltration rate (h^{-1})
Q_{50} = air change rate at 50 Pa.

Calculations have shown that the ratio can vary between 6 and 40 depending upon the house, the climate and the shielding [3].

To determine the energy use caused by air infiltration/exfiltration the infiltration/exfiltration rate was first calculated from the pressurization tests and then the energy use was calculated using degree days for Stockholm. For most of the buildings the only available information was the floor area, the volume, type of ventilation system and type of building technology, the results of a pressurization test.

TESTED BUILDING

The aim was to test educational and office buildings built after the year 2000 with a floor area preferably larger than 1000 m². It should be a mix of buildings with specific airtightness requirements and without.

Outside this project

31 buildings had been tested by different organisations e.g. the Technical Research Institute of Sweden, Akademiska Hus, Skanska, WSP. All the buildings were built between 2007 and 2012. The buildings are mainly schools and offices, but also homes for the elderly, shops, and sports centres (see table 1). The smallest building has an floor area of 800 m² and the biggest 17 000 m². All buildings have balanced mechanical ventilation with heat recovery. The building envelopes vary, ranging from prefabricated concrete to stud walls.

Type of building	Year	Number of storeys above ground	Floor area, m²	Volume, m³	Building envelope
Shop	2011	1-2	8,221	61,090	Prefabricated concrete/stud wall
Sport Centre	2011	1-2			Prefabricated concrete/stud wall
Office	2008		1,950	5,250	Sheet metal/expanded plastics elements
Office	2010		3,905		Prefabricated concrete
Office	2010	6	4,094	15,171	Some kind of facade system, TRP
Office	2010	5	17,000		
Office	2007	5	8,574	25,722	Prefabricated glass façade/stud wall
Office/industry	2009		379/1,269		Plannjaelement/stud wall
Storage/workshop /office	2011	1-2			Stud wall, TRP
Food store	2011		1,540	8,000	Concrete PPM, plaster expanded clay, TRP plastic foil
School	2009				
School	2011	1-2			Stud wall
School	2008	1-2			Stud wall
School	2008	1	1,840		Stud wall
School	2009	1-2	1,134		Stud wall, TRP
School	2008	1-2			
School	2010	1-2	973		Stud wall
School	2010	1-2	973		Stud wall
School	2010	2-3			Concrete/stud wall
School	2010	1-2	761		Stud wall
School	2010	2	3,425	13,500	Infill wall, loft ceiling beams of HDF
School	2011	1-2	959		Stud wall
School	2011	1-2	2,250	8,995	Stud wall, TRP
School	2007	2			
School	2011		880	3,300	PPM, stud wall, plasict foil,
School	2010		800	3,190	Prefabricated concrete/infill wall
School	2011		2,950	11,500	PPM, stud wall, TRP tak, plastic foil
School	2010		3,340	14,000	PPM, stud wall /TRP, plastic foil
Home for the elderly	2012	7	4,762	14,800	Prefabricated concrete
Home for the elderly	2011		4,200	11,000	Infill walls, concrete roof

Table 1. Description of tested buildings. Year refers to year of construction.

Within this project

Five buildings were tested for the purpose of this project, three office and three educational buildings. All the buildings were built between 200x and 2011 (see table 2). The smallest building has an floor area of 800 m² and the biggest 20 000 m². All buildings have balanced mechanical ventilation with heat recovery. The building envelopes vary, ranging from prefabricated concrete to stud walls.

Type of building	Year	Number of storeys above ground	Floor area, m^2	Volume, m^3	Building envelope
Exhibition/office	2011	1-2	20,000	204,000	Prefabricated concrete sandwich elements, HDF-floor structure
Office	2009	6	12,000	48,000	Façade bricks and glass
Office	2009	10			Prefabricated glass facade
School	2007	2	2,628	8,600	Light-weight concrete block wall
School	2011	1	1,030	2,967	Wood-framed wall
School	2009	2	2,098	7,148	Wood-framed wall

Table 2. Description of tested buildings. Year refers to year of construction.

RESULTS

All the buildings tested outside the project are very airtight (see table 3). The average airtightness was 0.3 l/sm² @ 50 Pa which is equivalent to the voluntary Swedish requirement for passive houses [10]. The best building had a value of 0.1. For most of the buildings airtightness requirements were made ranging from 0.2 to 0.8 l/sm² @ 50 Pa, which can be compared with the requirement of the previous Swedish building code (before year 2006), 1.6 l/sm² @ 50 Pa. Only two buildings did not meet their requirement. The current building code does not have any specific requirement. All previously tested office buildings in the USA, Canada and the UK are much leakier [11]. Common leakage paths were exterior doors and connections between façade elements and floors/roofs, most of which would be difficult to tighten afterwards. Most buildings were tested with blowerdoors covering most of the buildings. Some were tested with the ventilation system.

Type of building	Year of construction	Test method	Envelope area, m^2	Airtightness require-ment, l/sm² @ 50 Pa	Measured air tightness, l/sm² @ 50 Pa	Main leakage paths
Shop	2011	Blowerdoors, three fans, the whole building	18,721		0.18	Concrete element joints, exterior doors
Sport Centre	2011	Blowerdoors, two fans, the whole building	6,616	0.4	0.44	Exterior doors etc.
Office	2008	Ventilation system	2,580		0.34	Entrance parts/windows /exterior doors
Office	2010	Ventilation system		0.4	0.27	

Office	2010	Blowerdoors, one fan, the whole building	4,237	0.5	0.43	Connections between floor and wall
Office	2010	Blowerdoors, three storeys, one at a time	14,610	0.6	0.55	
Office	2007	Ventilation system/two blowerdoors		0.8	0.7	Connection between facade elements, facade and roof elements
Office/in dustry	2009	Ventilation system	4,560	0.25	0.26	Connection between ceiling and wall/workshop – exterior doors
Storage/ workshop /office	2011	Blowerdoors, two fans, the whole building	10,034	0.3	0.29	Exterior doors
Food store	2011	Blowerdoors, two fans, the whole building	3,995	0.8	0.62	TRP/expanded clay, windows, Entrance parts
School	2009	Ventilation system, the whole building excl. basement	4,912	0.5	0.36	
School	2011	Blowerdoors, one fan, the whole building	2,607	0.2	0.13	
School	2008	Blowerdoors, one fan, the whole building	3,335	0.45	0.41	
School	2008	Blowerdoors, one fan, the whole building	5,180	0.4	0.21	
School	2009	Blowerdoors, one fan, the whole building	2,832	0.6	0.27	Exterior doors
School	2008	Blowerdoors, one fan, the whole building	2,414	0.3	0.26	
School	2010	Blowerdoors, one fan, the whole building	2,460	0.6	0.23	Exterior doors and windows
School	2010	Blowerdoors, one fan, the whole building	2,460	0.6	0.19	Exterior doors and windows
School	2010	Blowerdoors, one fan, the whole building	2,182	0.6	0.57	
School	2010	Blowerdoors, one fan, the whole building	2,054	0.5	0.38	Exterior doors

School	2010	Blowerdoors, one fan, the whole building	5,513	0.2	0.09	No major leakage paths
School	2011	Blowerdoors, one fan, the whole building	2,520	0.40	0.28	Exterior doors
School	2011	Blowerdoors, one fan, the whole building	4,973	0.25	0.17	Exterior doors
School	2007	Blowerdoors, one fan, the whole building	3,941	0.4	0.45	Exterior doors etc.
School	2011	Blowerdoors, one fan, the whole building	2,261	0.6	0.48	Roof, windows and doors
School	2010	Blowerdoors, one fan, the whole building	2,295	0.3	0.4	Connection wall-ceiling, exterior door
School	2011	Blowerdoors, one fan, the whole building	4,822	0.2	0.16	
School	2010	Blowerdoors, three fans, the whole building	5,641	0.8	0.88	
Home for the elderly	2012	Blowerdoors, one fan, the whole building	4,081	0.3	0.20	Exterior doors
Home for the elderly	2011	Blowerdoors, x fans, the whole building	3,900	0.2	0.14	
Average				0.44	0.30	

Table 3. Measured air leakage and leakage paths.

For twelwe of the buildings information on the volume was available and the airtightness could be recalculated to ach @ 50 Pa (see table 4). A comparison of the buildings is now different due to different ratios between volume and envelope area. Using a simple method of calculating the infiltration (see Method) an average infiltration rate was estimated. The result was an average air infiltration rate during the heating season of 0.03 ach (air changes per hour), varying between 0.01 and 0.06. This is equivalent to an energy use for space heating of 4 kWh/m^{2}year. If the buildings would have only met the requirements of the previous building code the energy use might have been five times higher i.e. 20 kWh/ m^{2}year.

Type of building	Year	Measured airtightness l/sm² @ 50 Pa	Measured airtightness ach @ 50 Pa	Infiltration/exfiltration, ach	Energy use for heating infiltration, kWh/m²year
Shop	2011	0.18	0.20	0.01	2
Office	2008	0.34	0.60	0.03	3
Office	2010	0.43	0.43	0.02	3
Food store	2011	0.62	1.11	0.06	10
School	2010	0.09	0.13	0.01	1
School	2011	0.17	0.34	0.02	2
School	2011	0.48	1.18	0.06	7
School	2010	0.4	1.04	0.05	7
School	2011	0.16	0.24	0.01	2
School	2010	0.88	1.28	0.06	9
Home for the elderly	2012	0.20	0.20	0.01	1
Home for the elderly	2011	0.14	0.18	0.01	1
Average		0.34	0.58	0.03	4

Table 4. Measured air leakage and calculated energy use for heating infiltrating air.

Also the recently tested five buildings were fairly airtight, but not as airtight as the previously tested buildings (see table 5). For the sixth building the result was not available at the time of writing this report. One contributing factor might be that there were only two buildings which had a specified airtightness requirement.

Type of building	Year of construction	Test method	Envelope area, m²	Airtightness require-ment, l/sm² @ 50 Pa	Measured air tightness, l/sm² @ 50 Pa	Main leakage paths
Exhibition/ office	2011	Ventilation system, the whole building	40 400	0,4	0,39	Connection between façade elements and columns, between facade and roof, exterior doors.
Office	2009	Ventilation system, storey 3, back pressure storey 2, 4, atrium and staircase	5 600		0,85	Connection between infill walls and steel columns, windows
School	2007	Blower Door	3 923	-	0,87	Lead-throughs, windows
School	2011	Blower Door	2 775	-	0,45	Doors, windows
School	2009	Blower Door	4 307	-	0,62	Lead-throughs, windows, doors.
Average					0,64	

Table 5. Measured air leakage and leakage paths.

For the recently tested buildings, information on the volume was available and the airtightness could be recalculated to ach @ 50 Pa (see table 6). The comparison of the buildings is now different due to different ratios between volume and envelope area. Using a simple method of calculating the infiltration (see Method) an average infiltration rate was estimated. The result was an average air infiltration rate during the heating season of 0.05 ach (air changes per hour), varying between 0.01 and 0.08. This is equivalent to an energy use for space heating of 6 kWh/m²year. If the buildings would have only met the requirements of the previous building code, the energy use might have been three times higher 20 kWh/ m²year.

Type of building	Year	Measured airtightness l/sm² @ 50 Pa	Measured airtightness ach @ 50 Pa	Infiltration/exfiltration, ach	Energy use for heating infiltration, kWh/m²year
Exhibition/office	2011	0,39	0,28	0,01	5
Office	2009	0,85	0,36	0,02	2
Office	2009				
Education	2007	0,87	1,44	0,07	8
Education	2011	0,45	1,51	0,08	7
Education	2009	0,62	1,34	0,07	8
		0,64	0,99	0,04	6

Table 6. Measured air leakage and calculated energy use for heating infiltrating air.

CONCLUSION

This study clearly shows that it is possible to build very airtight educational and office buildings. Most likely, the energy use for infiltration in these buildings are almost negligible i.e. in the order of magnitude of a couple of kWh/m²year. This number can be compared with the total energy use for space heating for a typical average Swedish office building of 110 kWh/ m²year, where infiltration might account for 10-20 kWh/m²year if only the airtightness requirement of the previous building code is fulfilled, which is likely.

Two different methods of measuring the airtightness of entire buildings have been used, using the building's ventilation systems and using a number of blower doors. Both methods can be used and combined. The choice of method depends on the prerequisites of the test object. For big buildings using the ventilation system can be preferable. Tests during construction, which are recommended to ensure good airtightness, can often only be carried out using blower doors. The two methods can be applied to office buildings, apartment buildings, industrial buildings and other premises. For apartment buildings the blower door technique is often the only method as the ventilation system often has insufficient capacity, unless the building is very airtight. Complete testing includes determination of the location of leakage paths.

ACKNOWLEDGEMENTS

The project was funded by SBUF (Development Fund of the Swedish Construction Industry), NCC and WSP.

REFERENCES

[1] Persson, A. (2007). Improved energy statistics for premises – Report for year 1 – Inventory of offices and administrative buildings (in Swedish). The Swedish Energy Agency, rapport ER 2007:34.

[2] SCB (2006). *Energy statistics for premises 2005.* Swedish Statistics, EN16SM0603, www.scb.se.
[3] Blomsterberg, Å., (1990). Ventilation and airtightness in low-rise residential buildings - Analyses and full-scale measurements. Swedish *Council for Building Research, D10:1990, ph.d. thesis.
[4] Blomsterberg, Å., Burke, S. (2012). Real airtightness of large buildings – Measurements and calculations (in Swedish). SBUF (Development Fund of the Swedish Construction Industry), NCC and WSP.
[5] CEN (2000). Thermal performance of buildings - Determination of air permeability of buildings – Fan pressurization. European Committee for Standardization, EN 13829.
[6] CGSB (1986). Determination of the Airtightness of Buildings envelopes by the Fan Depressurization Method Using the Buildings's Air Handling Systems, Canadian General Standards Board, Standard 149.10-M86, National Standard of Canada.
[7] CGSB (1996). Determination of the Overall Envelope Airtightness of Buildings by the Fan Pressurization Method Using the Buildings's Air Handling Systems, Canadian General Standards Board, National Standard of Canada.
[8] Liddament, M. (1986). Air Infiltration Calculation Techniques – An Application Guide. AIVC.
[9] ASHRAE (2009). ASHRAE Handbook Fundamentals. ASHRAE, Atlanta, USA.
[10] Nollhus, 2012, Kriterier för passivhus m.m. (Criteria for passive houses etc.), www.nollhus.se .
[11] Blomsterberg, Å. (2009). Airtightness and energy use of an Swedish office building – measurements and calculations. Presented at the 30th AIVC- and the 4th BUILDAIR-Conference in Berlin.

ABSTRACT SUBMISSION FORM FOR THE 2012 AIVC-TIGHVENT CONFERENCE

Send your abstract using this form at info@inive.org

PAPER TITLE
An Investigation of the Relationship between Dwelling Permeability, Operational Adventitious Air Leakage, and Winter Space Heating Demand

MAIN AUTHOR
Benjamin M. Jones

OTHER AUTHORS
Robert J. Lowe, Payel Das, Clive Shrubsole Michael Davies

ABSTRACT

The importance of operational adventitious air leakage and its reduction in order to save energy is highlighted by the appropriate standards of many countries. It occurs when the internal pressure of a building differs from the external pressure and is caused by the action of the wind around a building and the difference between internal and external air densities. This operational air leakage is often predicted by the measurement of air permeability, a physical property of a building that indicates the resistance of its fabric to air leakage. Knowledge of the permeability of a building provides a measure of the flow rate through its envelope at a constant pressure differential across the envelope of 50 Pascals. However, operational pressure differences are dynamic and typically an order of magnitude lower than 50 Pascals. Thus there is much uncertainty when using a value of permeability in an attempt to predict operational air leakage.

Powerful simulation systems can model the ventilation rates found in a building in great detail, yet these complex systems contrast with the much simpler tools that are used frequently to estimate annual energy consumption for space heating in dwellings. For example, the UK Standard Assessment Procedure, a government tool used to compare the energy and environmental performance of dwellings, assumes a simple fixed relationship between air permeability measured at high pressure and mean background infiltration during the heating season; the so-called *rule of 20*.

This paper takes a fresh look at this rule-of-thumb. Firstly, a theoretical model of adventitious air leakage for a two-storey terraced house is presented. Secondly, the predictions of the model are compared against those of CONTAM, a multizone airflow analysis tool, for an identical building and environmental conditions. Thirdly, the model is used to predict the mean infiltration rate and the corresponding heating energy required to replace heat lost via infiltration during the heating season for a terraced house located in 14 different UK cities. Finally, the predictions of the model are used to develop a relationship between the adventitious air leakage at pressure, operational infiltration, and energy consumption during the heating season. The relationship is used to discuss the accuracy and applicability of the *rule of 20* and its use by simple modelling approaches such as the Standard Assessment Procedure.

STRATEGIES FOR CONTROLLING THERMAL COMFORT IN A DANISH LOW ENERGY BUILDING: SYSTEM CONFIGURATION AND RESULTS FROM 2 YEARS OF MEASUREMENTS

Peter Foldbjerg [*1], Amdi Worm [2], Thorbjørn Asmussen [1], and Lone Feifer [1]

1 VELUX A/S, Daylight Energy and Indoor Climate,
Ådalsvej 99, 2970 Hørsholm, Denmark
**Corresponding author: peter.foldbjerg@velux.com*

2 Esbensen Consulting Engineers
Silkeborgvej 47, 8000 Aarhus, Denmark

ABSTRACT

The thermal comfort of the residential building Home for Life is investigated with a particular focus on the strategies used to achieve good thermal comfort, and the role of solar shading and natural ventilation. Home for Life was completed in 2009 as one of six buildings in the Model Home 2020 project. It has very generous daylight conditions, and is designed to be energy neutral with a good indoor environment.

The kitchen/dining room has a large south-facing window area and is selected for the detailed analyses. The thermal environment is evaluated according to the Active House specification (based on the adaptive method of EN 15251), and it is found that the house reaches category 1 for the summer situation. Some undercooling occurs during winter, causing the room to achieve category 2 if the entire year is considered. The undercooling is due to the occupants' preferred balance between indoor temperatures and heating consumption.

It is found that ventilative cooling through window openings play a particularly important role in maintaining thermal comfort.

KEYWORDS
Thermal comfort, ventilative cooling, residential buildings, natural ventilation, solar shading.

INTRODUCTION

Background
Five single-family houses in five European countries were built between 2009 to 2011 as a result of the Model Home 2020 project. The first house (Home for Life, Denmark), was completed in spring 2009 and has been occupied by two different families, of which the last family has bought the house. Measurements were performed for a full year during the occupancy of the two families. The results from year one have been reported already and compared to simulations [1, 2].

The six houses follow the Active House [3] principles which mean that a balanced priority of energy use, indoor environment and connection to the external environment must be made. In practice this means that the houses should have both an excellent indoor environment and a very low use of energy. There is a particular focus on good daylight conditions and fresh air from natural ventilation. The calculated daylight factors are seen at Figure 1.

Figure 1. Calculated daylight factors in "Home for Life". Ground floor on the left, and upper floor on the right. The dotted rectangle indicates the kitchen and dining room which will be investigated in details.

Measurements of IEQ include light, thermal conditions, indoor air quality, occupant presence and all ocupant interactions with the building installations, including all operations of windows and solar shading. Further, an anthropological study of the family's experiences in the house was performed. Measurements of energy performance include space heating, domestic hot water and electricity for appliances, lighting and technical installations. The occupants also reported their own observations in a diary, and an anthropologist has followed the project and made structured interviews with the family.

The present paper describes the design and setup of the systems that have an influence on the summer, winter and intermediate situations, particularly the natural ventilation system and the solar shading.

The presented results focus on thermal conditions, effectiveness and experience with the applied strategies. Recent examples of demonstration houses in Scandinavia have experienced problems with overheating, often due to insufficient solar shading, [4], [5].

Technical systems

Home for Life is an experiment and the hypothesis is that a synergy between a low CO_2 emission and a good IEQ can be achieved through optimal window layout (40% of façade area) distributed towards all orientations and the roof for both good daylight conditions (daylight from all orientations) and good natural ventilation performance, through hybrid ventilation and through automatic control of solar shading, heating and ventilation. It is a 1½-storey house with a total floor area of 190 m^2.

The ventilation system is hybrid, i.e. natural ventilation is used during the summertime and mechanical ventilation with heat recovery during the wintertime, while hybrid ventilation is used spring and fall. The switch between mechanical and natural ventilation is controlled based on the outdoor temperature. The setpoint is 12,5 ^{o}C with a 0,5 ^{o}C hysteresis. Below the setpoint the ventilation is in mechanical mode, above the setpoint the ventilation is in natural mode. In both natural and mechanical mode, the ventilation rate is demand-controlled. CO_2 is used as indicator for the Indoor Air Quality, and a setpoint of 850 ppm CO_2 is used. Besides that, relative humidity is also used as indicator. When RH is 60% or higher, ventilation is increased step-wise to maximum ventilation, which is used when RH is 80% or higher.

There is external automatic solar shading on all windows towards South, and overhangs are used where appropriate. The solar shading was initially controlled based on the indoor temperature. But based on the responses from the occupants, this control stategy seemed to react too slowly to prevent overheating, and the control was changed so that external solar radiation was used instead. On the con side, using solar radiation can cause unnecessary use of solar shading in the winter period, leading to an increased heating demand.

Each room is an individual zone in the control system, and each room is controlled individually. There are sensors for humidity, temperature, CO_2 and presence in each room. The solar shading is controlled by external solar radiation for each facade, and not at room level.

The building occupants can override the automatic controls, including ventilation and solar shading at any time. Override buttons are installed in each room, and no restrictions have been given to the occupants. As house owners they have reported a motivation to minimise energy use on an overall level, and to maximise IEQ on a day-to-day basis.

Figure 2. "Home for Life". South and east facades (left). Concept for daylight, ventilation and energy (right)

Data recording

The data from the sensors that are used for the controls of the house is recorded. The IEQ data is recorded for each individual zone as an event log, where a new event is recorded when the value of a parameter has changed beyond a specified increment from the previously recorded value. The event log files are automatically converted to data files with fixed 15-minute time steps, which are used for the data analysis.

Data analyses

The recorded temperature data is evalueted according to the Active House specification[3], which is based on the adaptive approach of EN 15251 [6].

RESULTS

The results presented here are based on the measurements and analyses for the second year of occupancy. The main part of commissioning and adjustment of all systems took place during the first year occupancy, and year two is for that reason considered the most representative. Figure 3 shows the outdoor conditions during the reported period.

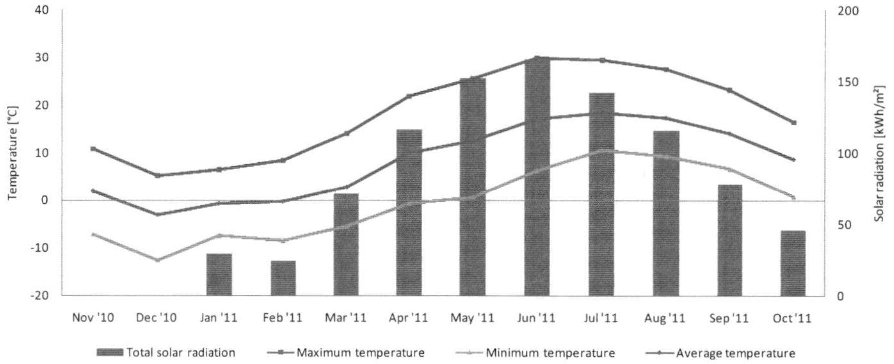

Figure 3. Outdoor conditions during the measurement period. The pyranometer was installed in January 2011, and therefore data is not available for the previous months.

Figure 4 shows that five rooms achieve catogory 2, while six rooms achieve category 4. It is clear from the figure that the majority of the hours in category 2, 3 and 4 are caused by low temperatures, i.e. undercooling rather than overheating. When undercooling is disregarded, all rooms except bedroom and scullery achieve category 1.

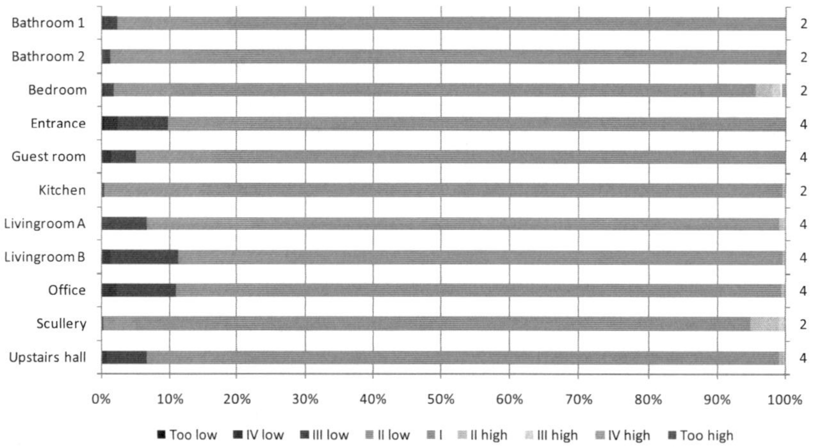

Figure 4. Thermal comfort for each of the rooms evaluated according to Active House specification (based on adaptive method of EN 15251). Criteria are differentiated between high and low temperatures. The number at the right side of the diagram indicates the score for each room (1 to 4).

The focus of the present paper is on the performance related to ventilative cooling and potential overheating. The further analyses will focus on the performance of the kitchen, which is a combined kitchen and dining room with a large south-facing window section, Figure 2.

The distribution of categoeries between months are seen at Figure 5. As expected from Figure 4, the undercooling is an issue in 4 winter months from November to February. From April to October, category 1 is achieved.

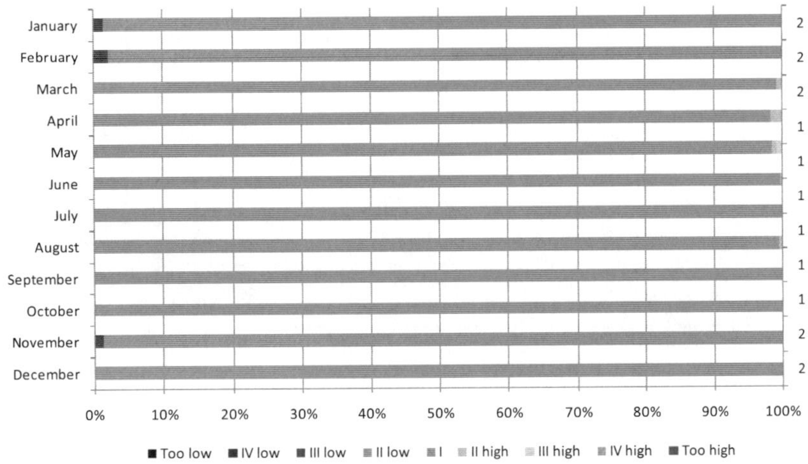

Figure 5. Thermal comfort categories for each month of the year for the kitchen and dining room. The number at the right side of the diagram indicates the score for each month (1 to 4).

Figure 6 shows the indoor temperature at each hour of the year plotted against the running mean outdoor temperature as defined in EN 15251. It is seen that temperatures below the category 1 limit (21 °C) occur both in winter and in the transition periods, but only with a few hours below the category 2 limit (20 °C). It is suspected that the episodes with temperatures below 21 °C are caused by either manual or automatic airings, or more simply by user preference. The occupants have not reported discomfort due to undercooling in their diaries.

Some episodes with temperatures above 26 °C are also seen during winter and in the transition periods, auggesting large variations in temperature during short periods of time. This is suspected to be due to solar gains. The automatic control of window openings and solar shading is setup to prevent overheating, but especially during winter the system will accept high solar gains to reduce the heating demand.

During summer the system prioritizes to maintain thermal comfort, and Figure 6 shows very limited overheating during summer, with only a few episodes with temperatures above category 1. Relatively low temperatures are observed during summer, with episodes with temperature drops below 21 °C. This is suspected to be caused by night cooling, where the temperature decreases during the night to reduce overheating the following day, which in some situations lead to temperatures in the morning between 20 °C and 21 °C.

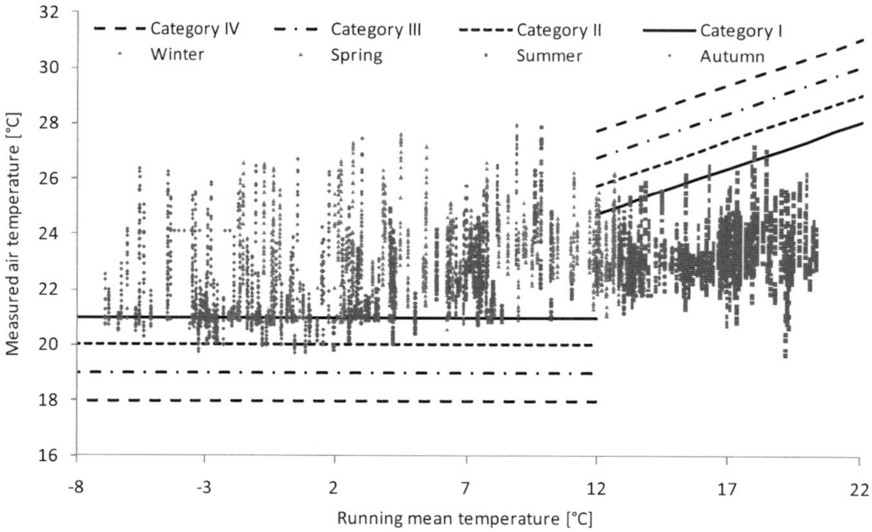

Figure 6. Indoor temperatures in the kitchen plotted against running mean temperature for each hour of the year including the Active House category limits. The dots are coloured to represent a season.

The variation over time-of-day and time-of-year is further investigated in Figure 7. It is seen that the episodes during winter with temperatures below category 1 can last for several days during the winter, but that in many of the episodes, the temperature reaches category 1 between 12:00 and 20:00, possibly due to solar gains.

During summer, only few episodes with temperatures beyond category 1 are observed.

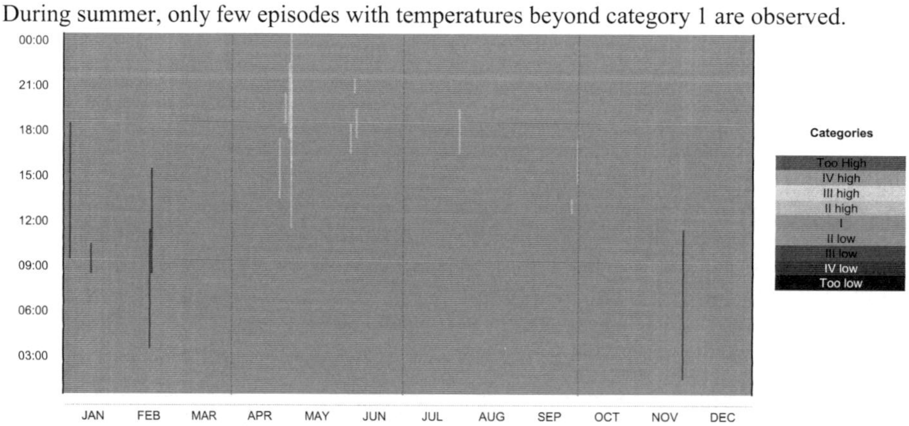

Figure 7. The comfort category of each hour of the year is plotted as a temporal map (kitchen and dining room)

To investigate the role of window openings in maintaining comfort, Figure 8 is used. A rather strict comfort definition is imposed for the sake of the analysis (category 2 was the design target), where only category 1 is considered comfort. The figure also shows if windows where active during each hour.

Figure 8 shows that windows were not open during the winter episodes with temperatures below category 1 (orange), indicating that these episodes were not caused by airings. The heating system during winter is controlled in such a way that the supply temperature for the floor heating system is set at the heat pump control. The lower the supply temperature, the better the system efficiency. The occupants have reported that they set the supply temperature so that the room temperature would reach 20-21 °C to reduce heating consumption. The episodes with winter temperatures below category 1 can thus be attributed to user preferences.

A few episodes with red colour are seen during summer in the late afternoon, indicating that overheating occurred and that windows were opened, but that this was not sufficient to maintain category 1.

Figure 8 further shows that during the summer, windows are almost permanently open between 9:00 and 22:00 and that category 1 is maintained during these hours (green). The figure shows many episodes with open windows between 22:00 and 9:00 (green), which can be assumed to be caused by automatic window opening for night cooling. Also in the transition periods (March to May and September to October) windows are used to a large extent, with openings between 12:00 and 18:00 as a typical episode (green).

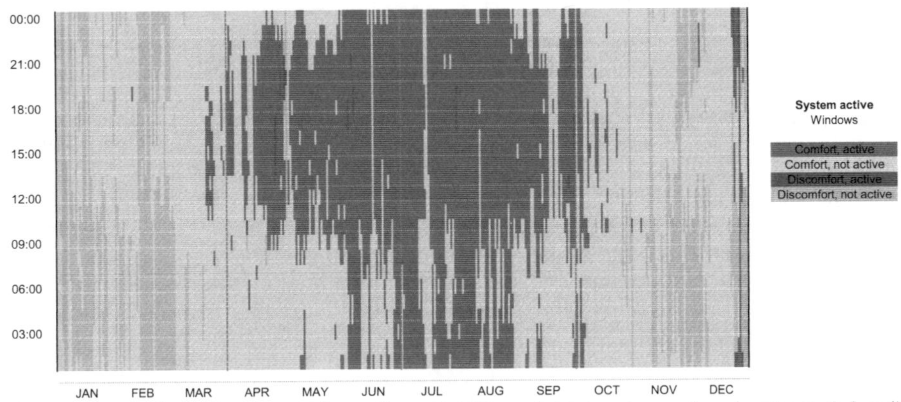

Figure 8. Temporal map showing comfort or discomfort (discomfort is here temperatures in category 2, 3 or 4) and if windows were open or closed (active or not active). Figure shows kitchen and dining room.

The occurrence of windows in relation to outdoor temperature is further investigated at Figure 9. It is seen that windows are generally closed (red dots) when the running mean temperature is below 10 °C. When the running mean temperature is above 12 °C, windows are generally opened when the indoor temperature exceeds 22-23 °C, which is in accordance with the control strategy.

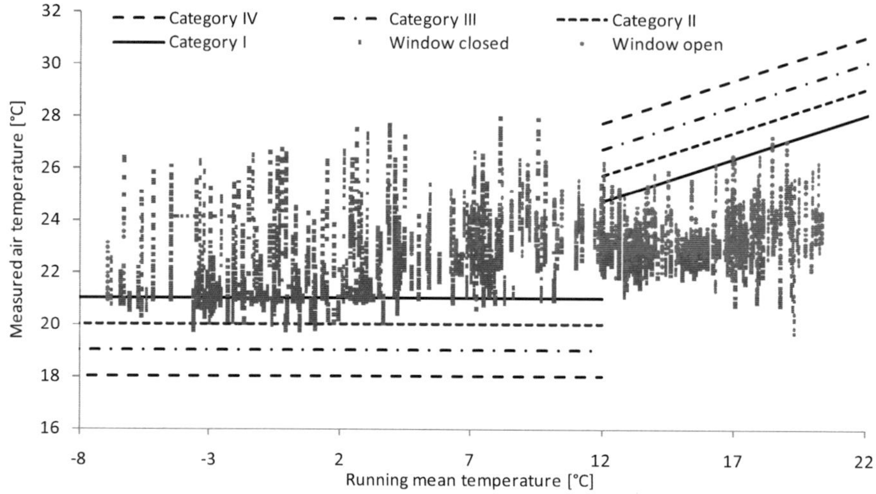

Figure 9. Indoor temperature vs. running mean outdoor temperature. The colour indicates if windows were open (green) or closed (red) for each hour (kitchen and dining room)

The role of the external solar shading is investigated at Figure 10. The figure shows no clear correlation between use of shading and indoor temperature nor running mean temperature. The external shading is activated when the external solar radiation exceeds a threshold and so is not controlled by temperatures, but some relation between temperature and use of shading could have been expected, as was the case for window openings. A further explanation is that solar shading can also be activated if glare is experienced, and also for privacy.

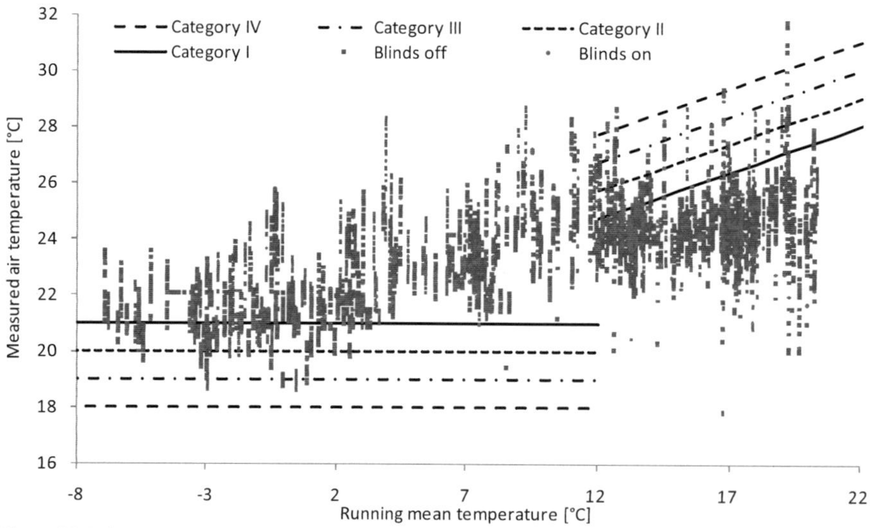

Figure 10. Indoor temperature vs. running mean outdoor temperature. The colour indicates if external solar shading was active (green) or inactive (red) for each hour (kitchen and dining room).

To further investigate the temporal use of shading, the use of shading is plotted at Figure 11, where colour codes indicates the use of shading and the comfort categories. The use of shading is increased during summer (green dots). The use appears to be distributed evenly over the day, due to the control being based on external radiance. The consistent use of shading during an hour in the morning and the late afternoon/evening has been investigated, but no clear explanation is found. It is not suspected to be caused by user controls, but rather as a result of the automatic control algorithms.

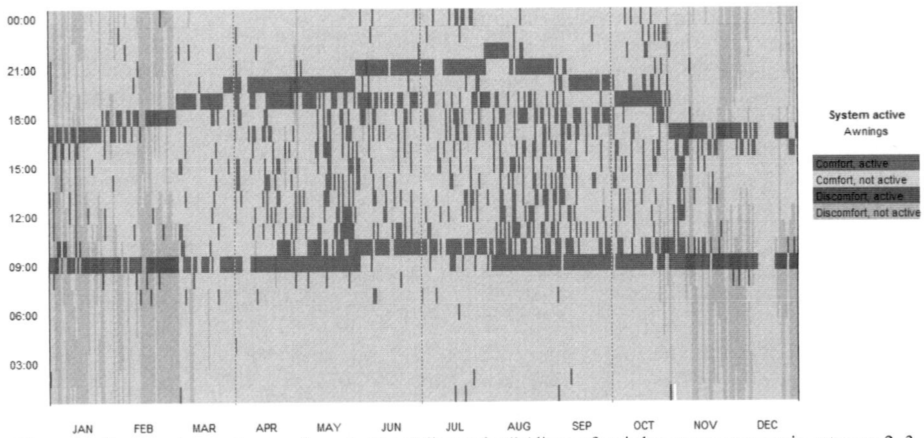

Figure 11. Temporal map showing "comfort" or "discomfort" (discomfort is here temperatures in category 2, 3 or 4) and if shading was active or not active (kitchen and dining room).

DISCUSSION

For the rooms in Home for Life, half fall in category 2 and the other half in category 4 with regards to thermal conditions, when evaluated according to the Active House specification, which uses the same methodology and criteria as EN 15251 with regards to thermal comfort. The hours not in category 1 are mainly hours with undercooling, while overheating is rare. If undercooling is disregarded, the primary rooms of the house achieve category 1. For low energy houses, overheating should be prevented by the building design, as overheating may require substantial measures if handled after completion. Home for Life thus meets the category 1 with regards to overheating, which is very satisfactory, given the generous daylight conditions.

The episodes with undercooling could be caused by insufficient heating capacity, window airings, poor building airtightness or occupant preferences. It was found that there was no correlation between window openings and undercooling. The airtightness has been verified by a blowerdoor test. The heating system is known to have a sufficient capacity, but the supply temperature was actively reduced by the occupants to reduce the heating consumption. Undercooling in Home for Life is therefore explained by occupant preferences.

In the kitchen/dining room, a correlation between window openings and the combination of high indoor and outdoor temperatures was found. Further, a clear correlation between window openings and acceptable thermal comfort was found. This indicates that window openings have contributed to achieving and maintaining good thermal conditions.

No clear correlation between use of external solar shading and temperature. Users may often have used the override function to deactivate the automatic control of solar shading, which could explain the missing correlation between use of shading and the combination of high indoor and outdoor temperatures.

In conclusion, Home for Life achieves a good thermal performance in real use, which should be seen in connection to the high daylight levels of the building. The good performance is achieved with automatic control of window openings and solar shading, where especially the ventilative cooling from open windows was important.

ACKNOWLEDGEMENTS
The project Minimum Configuration and Home Automation (MCHA) has contributed to the analyses presented here. The MCHA project is funded by the Danish Enterprise and Construction Authority.

REFERENCES

[1] Foldbjerg, P., Hansen, E.K., Feifer, L., and Duer, K. 2011. *Measurements of indoor environmental quality and energy performance of 6 European zero carbon houses – a case study from the first house.* Proceedings of Indoor Air 2011, Austin, Texas
[2] Foldbjerg, P., Hammershøj, G.G., Feifer, L., and Hansen, E.K. 2011. *Indoor Environmental Quality of the first European ModelHome 2020 : Home for Life.* Proceedings af CISBAT 2011, Lusanne, Switzerland.
[3] Active House Alliance. 2011. *Active House – Specification* (last accessed September 2012). http://activehouse.info/about-active-house/specification. Active House Alliance.
[4] Larsen T.S., and Jensen R.L. 2012. *The Comfort Houses - Measurements and analysis of the indoor environment and energy consumption in 8 passive houses 2008-2011.* Aalborg University. (last accessed September 2012) http://vbn.aau.dk/files/65267145/ The_Comfort_Houses.pdf
[5] Isaksson, C and Karlsson, F, 2006. *Indoor Climate in low-energy houses – an interdisciplinary investigation.* Building and Environment, Volume 41, Issue 12, December 2006
[6] CEN. 2007. *CEN Standard EN 15251:2007, Indoor environmental input parameters for design and assessment of energy performance of buildings addressing indoor air quality, thermal environment, lighting and acoustics.* European Committee for Standardisation.

VENTILATED COURTYARD AS A PASSIVE COOLING STRATEGY IN THE HOT DESERT CLIMATE

Mohamed Hssan Hassan

Department of Architecture - Faculty of Engineering – Aswan University
AboElrish, 81542 Aswan, EGYPT
mamado_moh@yahoo.com

ABSTRACT

Traditional architecture gives ideas to enrich modern architecture. In traditional architecture, local materials and renewable energy resources have been used. The courtyard was one of the traditional architecture solutions as a climate modifier. The inclusion of an internal courtyard in buildings design is attributed to the optimization of natural ventilation in order to minimize indoor overheating conditions.

The paper investigates the potential of a ventilated courtyard for passive cooling in a small building in a hot desert climate. The analyzed model is one of the low-income housing models in New Aswan City - Egypt. Which was built and provided with main services by the government, these housing models were characterized by their improper design in many cases, especially, concerning with climatic design.

To evaluate the performance of a ventilated courtyard, building simulation software TRNSYS 16 (The coupling between TRNSYS and COMIS) was used. The courtyard parameters considered were the courtyard orientation and the courtyard geometry. To evaluate the performance of a ventilated courtyard, the average monthly indoor air temperature for the purposed building determined in the overheating summer season depended on the weather data for the building site. The results of the investigations of the courtyard parameters indicate that there are some important parameters and other are of less significance which affect the thermal performance of the courtyard building model.

KEYWORDS: Courtyard, hot desert climate, natural ventilation, low-income housing.

INTRODUCTION

Natural ventilation is one of the natural passive cooling strategies recommended for hot desert regions. In desert areas, this strategy is utilised to conserve energy while maintaining appropriate thermal comfort for inhabitants inside the building.

The aim of this study is to further the understanding and optimisation of natural ventilation cooling in buildings that have a ventilated courtyard. The performance of the courtyard was evaluated by using of TRNSYS 16 simulation tool, in a multi-story house located in New Aswan City, which is about 850 km South of Cairo, Egypt.

BRIEF LITERATURE REVIEW

Many researches such as Dunham, Fathy, Givony, Hinrichs, Konya, Lippsmeier, Olgyay and Saini reached the conclusion that a building with a courtyard is a desirable concept in the hot dry climate [1].

In a recent study, Al-Hemiddi describes an experiment to investigate the effect of a ventilated interior courtyard on the thermal performance of a single-family house in a hot–arid region.

Statistical analysis of data recorded during the summer of 1997 was carried out. The results indicate that the courtyard gives high efficiency in providing cool indoor air through cross-ventilation [2].

In another study, a comparison between different geometries of courtyards in terms of wind flow characteristics and indoor air speed is performed based on the validation of Computational Fluid Dynamics (CFD) simulations with 2D published wind-tunnel experiments. In addition, assessment of thermal comfort is made inside a number of selected dwelling rooms facing different courtyard geometries. It is confirmed that rooms with cross ventilation have higher indoor air speed values and therefore a better thermal comfort than with single-side ventilation. The courtyard dimensions, the position of the room and the orientation are important aspects influencing the indoor air speed and thermal comfort [3].

In their research, Rajapakshaa, Nagaib and Okumiya investigate the potential of a courtyard for passive cooling in a single storey high mass building in a warm humid climate. From the results of thermal measurements, a significant correlation between wall surface temperatures and indoor air temperatures is evident. A reduction of indoor air temperature below the levels of ambient is seen as a function of heat exchange between the indoor air and high thermal mass of the building fabric. However, this behaviour is affected by indoor airflow patterns, which are controlled through the composition between envelope openings and the courtyard of the building. From a computational analysis, several airflow patterns are identified. A relatively better indoor thermal modification is seen when the courtyard acts as an air funnel discharging indoor air into the sky, than the courtyard acts as a suction zone inducing air from its sky opening. The earlier pattern is promoted when the courtyard is ventilated through openings found in the building envelope [4].

The passive cooling effects of a courtyard of a small building were determined numerically by safarzadeh and Bahadori in their paper, employing energy-analysis software developed for that purpose. The passive cooling features considered were the shading effects of courtyard walls and two large trees (of various shapes) planted immediately next to the south wall of the building, the presence of a pool, a lawn and flowers in the yard, and the wind shading effects of the walls and trees. It was found that these features alone cannot maintain thermal comfort during the hot summer hours in Tehran, but reduce the cooling energy requirements of the building to some extent. They have an adverse effect of increasing the heating energy requirements of the building slightly. The same savings in cooling energy needs of the building can be obtained through many features such as wall and roof insulation, double-glazed windows, and special sealing tapes to reduce infiltration. They all save on heating energy requirements as well [5].

It is clear from the foregoing review that the courtyard building form, although being described as a suitable solution in hot dry climate, has not investigated with regard to the multi-storey residential building and evaluation concerning the interaction between the courtyard geometry and the orientation. Consequently, the importance and significance of the present study are apparent.

ANALYSIS OF THE CASE STUDY

Climatic features of New Aswan City

New Aswan City is located on the west bank of the Nile, 10 Km northern of the present Aswan City and 850 km South of Cairo, so it is located in the south of Upper Egypt region which characteristic of a hot, dry climate, with a very wide difference between day and night temperatures. During the summer season, the day-by-day mean of maximum outdoor air

*** Formatting Issue - Best Available Paper/Graphic ***

temperature reaches 35° C in the north of Upper Egypt, and 41° C in the south. In winter season, the day-by-day mean of minimum outdoor air temperature reaches 6° C in the north of Upper Egypt, and 10° C in the south. In general, the Upper Egypt has a typical desert climate with large variations between seasons and between day and night temperatures [1].

The level of humidity in Upper Egypt is relatively low, especially during the summer months. The yearly mean of humidity in the Upper Egypt differs from 55% to 20%, north to south respectively. In New Aswan City relative humidity records the lowest value in May and June (12%), whereas reach to the highest value in December and January (36%, 34% respectively). On the other hand, The annual wind rose for New Aswan City indicates that most winds blow from north, northwest, and northeast (49.2%, 21%, and 12.9% respectively) [1].

Building description

The use of courtyards in residential buildings in Egypt, and other countries in the Middle East, is many centuries old. The courtyards provided security and privacy for the residents, and daylight for the rooms which were built around them. By building a pool, fountain and planting trees in the yard, the architects created a very pleasant space for the residents to spend a portion of their time during summer months in the yard.

In Egypt at the present time, the courtyard usually employed in buildings to collect sanitary pipes of the service's rooms (w.c, kitchen), which were built around it. Also, provide security, privacy and daylight for the residents in these rooms.

The original residential building was chosen for the study consists of three floors, and has not a courtyard as shown in Fig. 1. It was constructed in 2005 within one of national housing projects was built and provided with main services by successive governments. Such a policy was adopted and applied since the fifties and is up until now with the main care devoted to produce as many residential units as possible, with less care of units' quality. Consequently, these projects were characterized by their improper design in many cases, especially, concerning with climatic design.

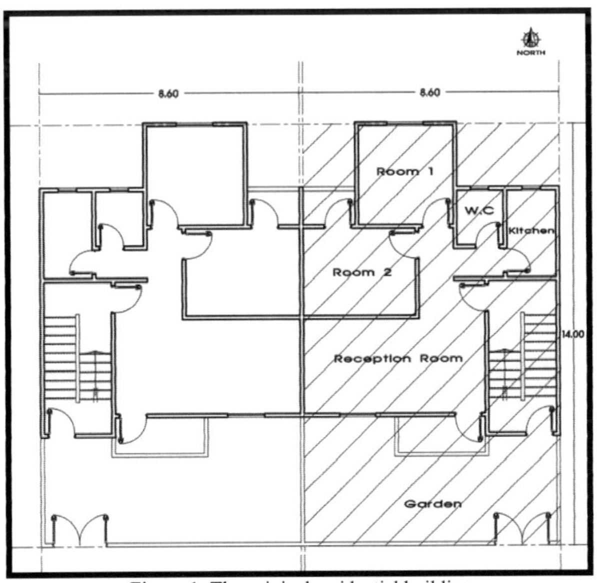

Figure 1. The original residential building

*** Formatting Issue - Best Available Paper/Graphic ***

The building does not have insulation on the walls and roof. It is constructed using standard local building materials. The external walls of the rooms are constructed with single bricks, 16 cm in thickness. The roof is flat and made of a concrete slab, 12 cm in thickness. External walls are painted light brown.

In a recent study TRNSYS 16 simulation tool was used to predict the hourly indoor air temperature which shows that the results indicates that the maximum temperature obtained is 43.6° C in the ground floor, while, in the first floor the maximum temperature obtained is 43.9° C, and in the last floor the maximum temperature obtained is 44.9° C.

In addition, the predicted average monthly indoor air temperature in the summer season (as shown in Table 1) indicates the overheating conditions which exceed the comfort limits to some extent [6].

Month		**Jun.**	**Jul.**	**Aug.**
Temperature (deg. C)	Ground floor	36.6	37.1	37.5
	First floor	36.5	37.0	37.4
	Last floor	37.1	37.7	38.0

Table 1. The predicted average monthly indoor air temperature in the summer season

The previous results declare the need for cooling system to exceed climatic difficulties, so, the results guide to the importance of employing solutions such as courtyard in buildings to act as a climate modifier.

The modified building has a courtyard in the central area of the house. The same area and dimensions were preserved; the courtyard opened to sky, and has ventilation open, and connected with the rooms around it with windows.

In order to determine the efficiency of the courtyard, the simulation process will be carried out for the last floor which considers the worst case. On the other hand, the courtyard parameters investigated will be in different four scenarios as follow.

Case 1: rectangular courtyard with ventilation open facing the north orientation (Fig. 2).
Case 2: rectangular courtyard with ventilation open facing the south orientation (Fig. 3).
Case 3: square courtyard with ventilation open facing the north orientation (Fig. 4).
Case 4: square courtyard with ventilation open facing the south orientation (Fig. 5).

Figure 2. Rectangular courtyard with north orientation

Figure 3. Rectangular courtyard with south orientation

Figure 4. Square courtyard with north orientation

Figure 5. Square courtyard with south orientation

Fig. 6 shows the cross section of the purposed building.

Figure 6. Cross section through the building

TRNSYS program

The present research uses the computer simulation program TRNSYS 16 (The coupling between TRNSYS and COMIS).

TRNSYS (the TRaNsient SYstem Simulation program) is a flexible tool designed to simulate the transient performance of thermal energy systems. TRNSYS can be connected to COMIS (Conjunction Of Multizone Infiltration Specialists) through use of an add-on link component called Type157. This type recasts COMIS as a TRNSYS component. In this case, the COMIS input file is generated not using a separate graphical interface but using the TRNSYS simulation studio itself [7].

RESULTS

Effect of the orientation

The comparing between the north and the south orientations clear that:

In the case of the courtyard which facing the north, the indoor air temperature recorded either in room 2 or in reception room values less than the indoor air temperature which recorded in the case of the courtyard facing the south (Fig 7). Also, the same results were recorded in the case of the square courtyard (Fig 8).

The square courtyard is more efficient than the rectangular courtyard.

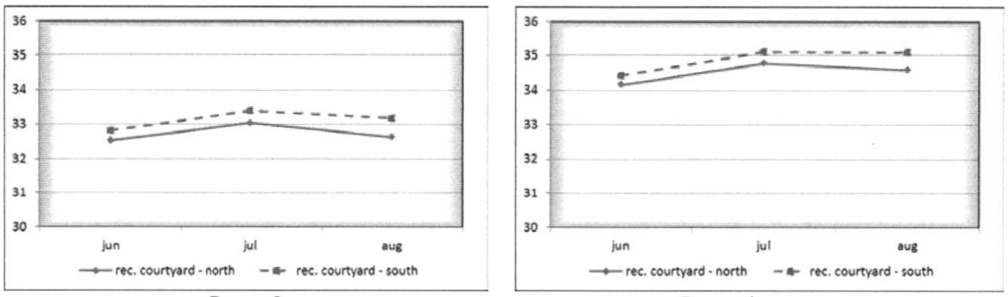

Room 2 Reception room

Figure 7. Results recorded in the case of rectangular courtyard

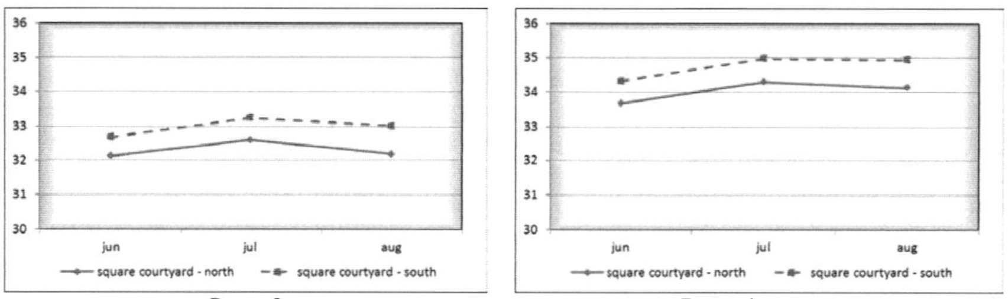

Room 2 Reception room

Figure 8. Results recorded in the case of square courtyard

Effect of the geometry

On the other hand, the comparing between the rectangular and the square courtyards indicates that the square courtyard is the best in all cases (Fig 9 – Fig 10).

In the same time, the north orientation is more efficient than the south orientation.

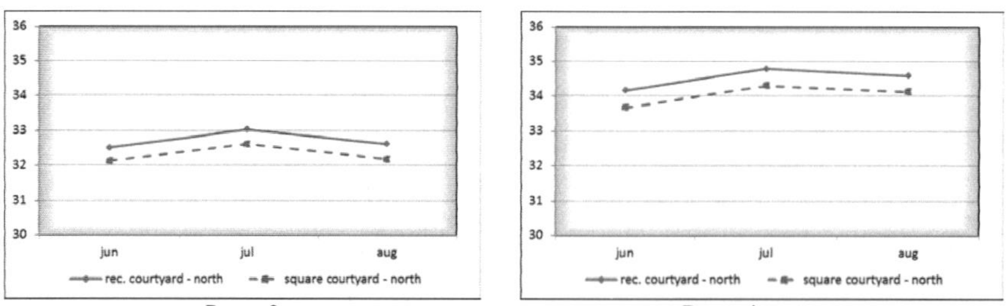

Room 2 Reception room

Figure 9. Results recorded in the case of courtyard facing the north

*** Formatting Issue - Best Available Paper/Graphic ***

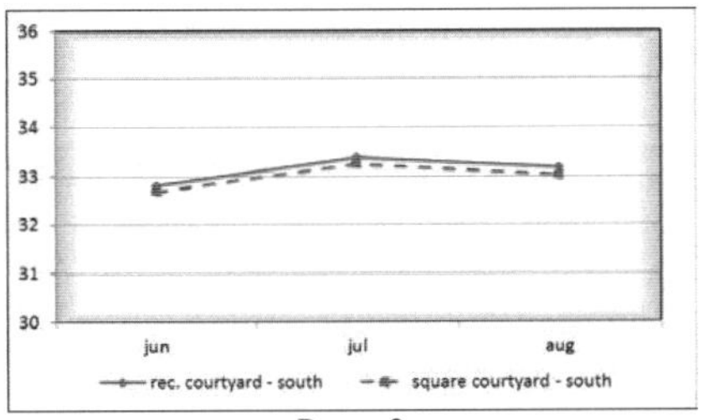

Room 2

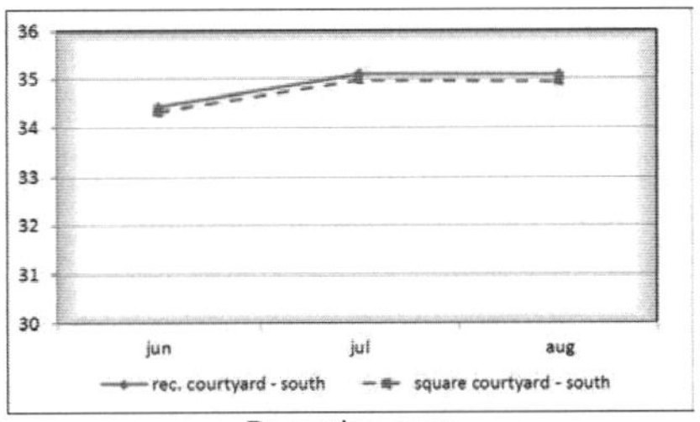

Reception room

Figure 10. Results recorded in the case of courtyard facing the south

In general, it is clear to note that employing the courtyard in the building reduce the indoor air temperature in all cases by 3°C : 5°C (as shown in Table 2).

	Indoor air temperature in room 2			Indoor air temperature in reception room		
	jun	jul	aug	jun	jul	aug
Building without courtyard	37.1	37.7	38.0	37.1	37.7	38.0
Rectangular courtyard facing the north	32.5	33.0	32.6	34.2	34.8	34.6
Rectangular courtyard facing the south	32.8	33.4	33.2	34.4	35.1	35.1
square courtyard facing the north	32.1	32.6	32.2	33.7	34.3	34.1
square courtyard facing the south	32.7	33.3	33.0	34.3	35.0	34.9

Table 2. The comparison between the four scenarios

The reception room records more values of indoor air temperature than the room 2.

The square courtyard which facing the north orientation is the best case correspond to the rooms around the courtyard.

In all cases, it is noted that the reduction of the indoor air temperature is not enough to achieve the thermal comfort during the summer season.

CONCLUSION

The results showed that it is possible to get considerable reduction in the indoor air temperature with natural ventilation by using the ventilated courtyard in hot desert climate. However to ensure a good performance designers and other professionals should pay attention to some parameters like the orientation and the geometry of the courtyard.

The courtyard is an applicable design strategy for a building from the perspective of climatic and cost–benefit analyses. These strategies can be applied to single storey or multi-storey building, also, the use of the courtyard with a pond, fountain and trees could be an applicable strategy during the hottest periods.

Finally, the development of the above formulae helps designers and architects to predict the thermal performance of courtyard buildings cooled by natural ventilation.

However, this approach only helps to predict the indoor air temperature of the purposed building, under given climatic conditions. Therefore this approach is not universally applicable, and values and judgements based upon it could change when we explore the effects of other designs and/or climatic conditions on courtyard performance.

ACKNOWLEDGEMENTS

The author would like to thank Aswan University for the funding of this research

REFERENCES

[1] Aly, A.M. 1994. *The effect of courtyard on the human thermal comfort inside residential building spaces in Upper Egypt*, Ph.D. thesis, Assiut University.
[2] Al-Hemiddi, N.A., Al-Saud, K.A. 2001. *The effect of a ventilated interior courtyard on the thermal performance of a house in a hot–arid region*, Renewable Energy 24, 581–595.
[3] Tablada, A., Blocken, B., Carmeliet, J., De Troyer, F., Verschure, H. 2005. *The influence of courtyard geometry on air flow and thermal comfort: CFD and thermal comfort simulations*, PLEA2005 - The 22nd Conference on Passive and Low Energy Architecture. Beirut, Lebanon.
[4] Rajapaksha, I., Nagai, H., Okumiya, M. 2003. *A ventilated courtyard as a passive cooling strategy in the warm humid tropics*, Renewable Energy 28, 1755–1778.
[5] Safarzadeh, H., Bahadori, M.N. 2005. *Passive cooling effects of courtyards*, Building and Environment 40, 89–104.
[6] Hassan, M.H., *Development of Building Simulation Model for Passive Cooling in Hot Desert Climate*, 2010. Ph.D. thesis, South Vally University (Egypt) on behalf of Dortmund University (Germany).
[7] http://www.trnsys.com/about.htm

TOWARDS THE AERAULIC CHARACTERIZATION OF ROOF WINDOWS ?

Bruno Peuportier[1*], Berenger Favre[1], Eric Vorger[1], Olfa Mejri[1], Michael Cohen[1], Karsten Duer[2], and Nicolas Dupin[2]

1 ARMINES, CEP
5 rue Léon Blum
91120 Palaiseau, France
**Corresponding author: bruno.peuportier@mines-paristech.fr*

2 VELUX A/S
Ådalsvej 99
DK-2970 Hørsholm, Denmark

ABSTRACT

Low energy buildings, being highly insulated, are subject to important overheating risks. Thermal simulation as well as experimental studies have shown the large potential of ventilative cooling. One barrier against this approach is the difficulty of evaluating air flows. Appropriate calculation methods and characterization of openings are needed, so that these systems can be dealt with in design, regulation and certification tools.

The present study is based upon the monitoring of a 135 m2 zero energy house situated near Paris. Temperature profiles have been measured when varying the ventilation pattern, i.e. opening or closing vertical windows, roof windows and internal doors. A dynamic thermal simulation tool is used to evaluate temperature profiles in the house under the climatic conditions corresponding to on site measurements (external temperature and solar radiation). The model accounts for the conductive, radiative and convective heat transfer, as well as energy storage in the building envelope related to solar and internal gains. In this highly insulated new construction, the most uncertain parameter is the natural ventilation flow rate. This parameter, and the related aeraulic characteristics of the openings, can be calibrated by minimizing the discrepancy between calculated and measured temperature profiles. Given the small window height, and the large height between ground floor and roof windows, a one way flow model is considered.

The roof window characteristics will also be evaluated in a laboratory benchmark. A cell (3m x 3m x 2m) is divided into two compartments by a slanting wall including the window. A ventilator blows air into one space and the pressure difference is measured between both sides of the window. Varying the air flow rate allows a relationship between the flow rate and the pressure difference to be identified. This relationship may depend on the pressure difference between both sides of the openings, therefore calibration using on site measurement is helpful. The air exchange rate estimated by this method will be compared to measurements using tracer gas, performed in the house as well as in the laboratory benchmark. The possibility to use anemometers will also be tested.

The method proposed here, combining a benchmark in a laboratory with numerical simulation and on site monitoring may bring a supplementary input, complementing the existing knowledge in the field of passive cooling of buildings. The feasibility of using this method in order to prepare appropriate input data for numerical models implemented in regulation, design and certification tools will be studied.

KEYWORDS
Ventilative cooling, energy performance, roof windows, characterization

INTRODUCTION

Low energy buildings, being highly insulated, are subject to important overheating risks. Thermal simulation as well as experimental studies have shown the large potential of ventilative cooling [1]. One barrier against this approach is the difficulty of evaluating air flows. Appropriate calculation methods and characterization of openings are needed, so that these systems can be dealt with in design, regulation and certification tools.
The present study is based upon the monitoring of a zero energy house situated near Paris. Temperature profiles have been measured when varying the ventilation pattern, i.e. opening or closing vertical windows, roof windows and internal doors. Measurements will be compared with dynamic thermoaeraulic simulation results in order to identify an air change rate and calibrate aeraulic characteristics of the openings. These characteristics will also be evaluated in a laboratory benchmark, but may depend on the pressure difference between both sides of the openings, therefore calibration using on site measurement may be helpful. The air exchange rate estimated by this method will be compared to measurements using tracer gas, performed in the house as well as in the laboratory benchmark. The possibility to use anemometers will also be tested.

1 DESCRIPTION OF THE ZERO ENERGY HOUSE

1.1 Architectural concept

The 130 m² floor area extends over one and a half storeys, with the spaces under the roof put to full use. Maison Air et Lumière, using a design principle that integrates architectural quality and energy efficiency, manages to place the emphasis on interior comfort whilst respecting the energy and environmental objectives for new detached houses for 2020. A model of the building is shown in Figure 6

1.2 Daylight

Particular attention has been paid to daylight to ensure the physical and psychological health and well-being of the residents, and to enlarge the visual perception of the indoor spaces whilst saving energy by reducing the need for artificial lighting. The amount of daylight and the quality of its distribution have been carefully studied using VELUX Daylight Visualizer 2.

1.3 Ventilation

According to the season and weather conditions, ventilation is provided by a hybrid system that combines the advantages of mechanical ventilation with heat recovery in winter and, in summer, natural ventilation by window opening (supplemented by mechanical extraction in bathroom and kitchen).

1.4 Energy design

The energy concept of Maison Air et Lumière is based on the maximum use of renewable resources (solar energy, natural light, passive cooling) in order to minimise the need for air conditioning in summer, to reduce heating in winter and to reduce artificial lighting and energy use for ventilation. The combination means a neutral environmental impact and maximum comfort for the residents. The house, which is built on a concrete slab on an earth platform insulated on the underside, is constructed with a well-insulated wooden frame and with a window-floor ratio of nearly 1:3. All windows are equipped with dynamic solar protection and the operation of all systems in the building (heating, ventilation, shading window-opening, lighting etc) is fully automated.

With its interplay of roof structures, the building is compact and very well insulated and, in order to create a stable and comfortable room temperature, the interior walls are lined with terracotta tiles, appreciably improving the thermal mass of the building. Heating and hot water are provided by a heat pump connected to thermal solar panels and a low-temperature underfloor heating system. The artificial lighting, domestic appliances and multimedia equipment were selected on the basis of their low consumption. Moreover, to reduce electricity consumption further, the washing machine and dishwasher can be directly connected to a cold and hot water inlet. All electric power consumption will be offset by the contribution from 35 m² of photovoltaic panels integrated in the roof. In normal use of the building, the overall annual energy balance is positive.

2 EXPERIMENTAL PROTOCOLE AND DESCRIPTION OF THE MONITORING

2.1 Measurements in a laboratory facility

A benchmark is installed at the CEP laboratory in order to identify the characteristics of a roof window. A ventilator is used to create a pressure difference in a test cell divided in two compartments. The roof window is installed on a 45° sloped wall between these compartments. The pressure difference (P_2-P_1) is measured, as well as the air flow rate Q: a diaphragm and Pitot tubes are used in the inlet air pipe in order to get a reference value of the flow rate (see fig. 1).

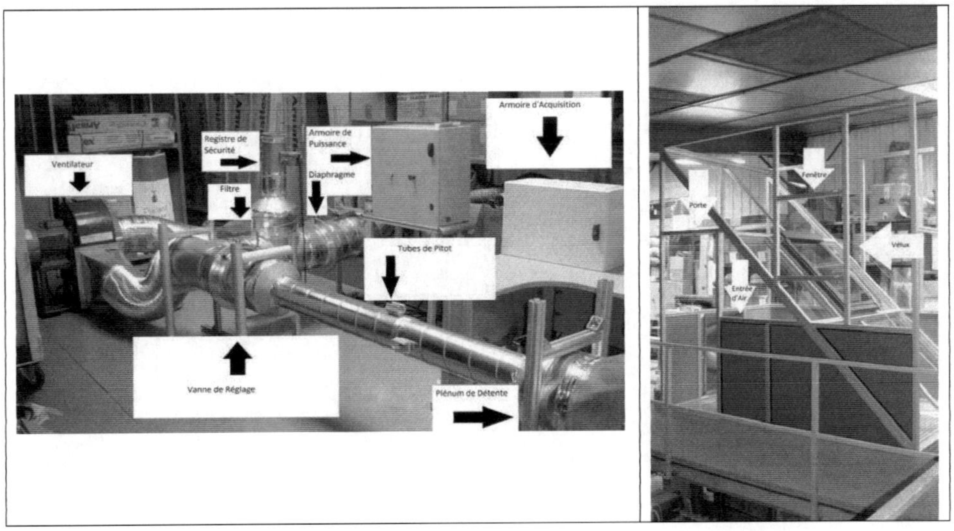

Figure 1. Air inlet of the laboratory benchmark, including flow rate measurement

This reference value will be compared to the air flow rate obtained from analyzing CO_2 concentration profile (tracer gas method). First a profile will be measured with the window closed in order to identify the air infiltration flow rate. Then the window will be open and the additional flow rate will be derived by difference.

A one way flow is assumed in the conditions of this experiment. The ventilation flow rate and therefore the pressure difference will be varied in order to draw a curve relating the flow rate Q to the section S and pressure difference, and to derive characteristics of the roof window Cd and n [2]:

$$Q = S.Cd\,(P_2 - P_1)^n \qquad (1)$$

The section S considered is the geometrical opening section. The difference $P_2 - P_1$ will be in the range from 0.05 to 1 Pa. In the real house, it may be lower, depending on wind conditions, but it is hoped that the values of Cd and n will not vary too much. The laboratory test will therefore provide two parameters that can be used in the analysis of on site measurements, and that can be refined using a calibration step.

Anemometers will be used in order to study the possibility to measure the air speed at different locations of the opening, and to derive the flow rate. A CFD model of the benchmark has been developed [3] in order to know if some position of the anemometers leads to a more precise evaluation of the flow rate, see Fig. 2. The velocity is assumed uniform on the inlet section (0.1 m/s), which corresponds in the experiment to the use of a honeycomb structure.

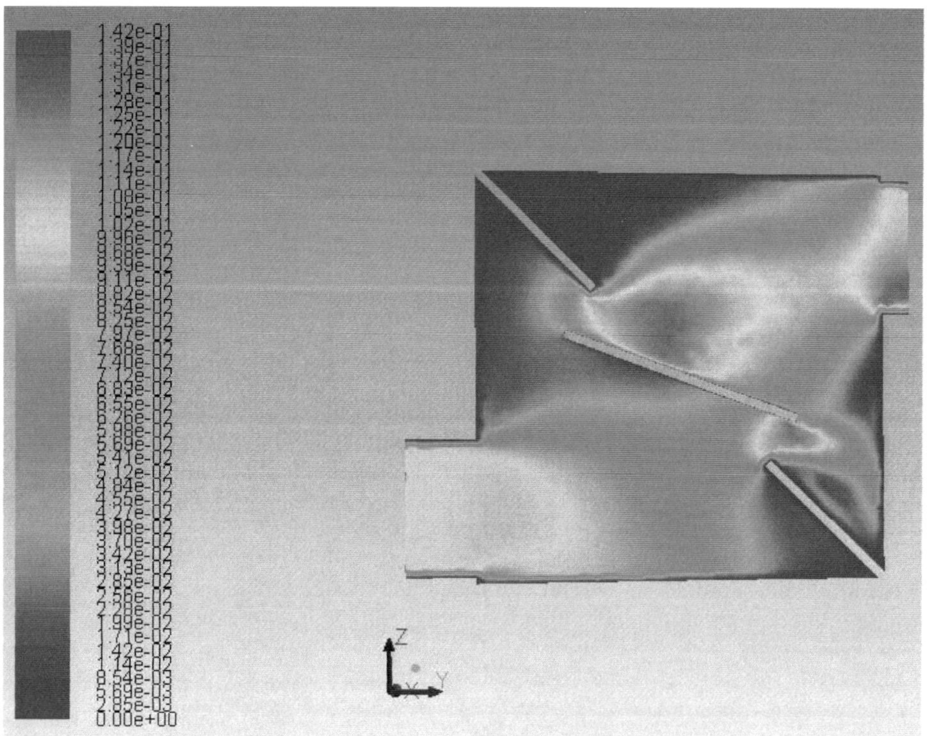

Figure 2. Example CFD model of the bench mark and air velocity results

CFD can also be helpful to choose the position of pressure gauges, see Fig. 3.

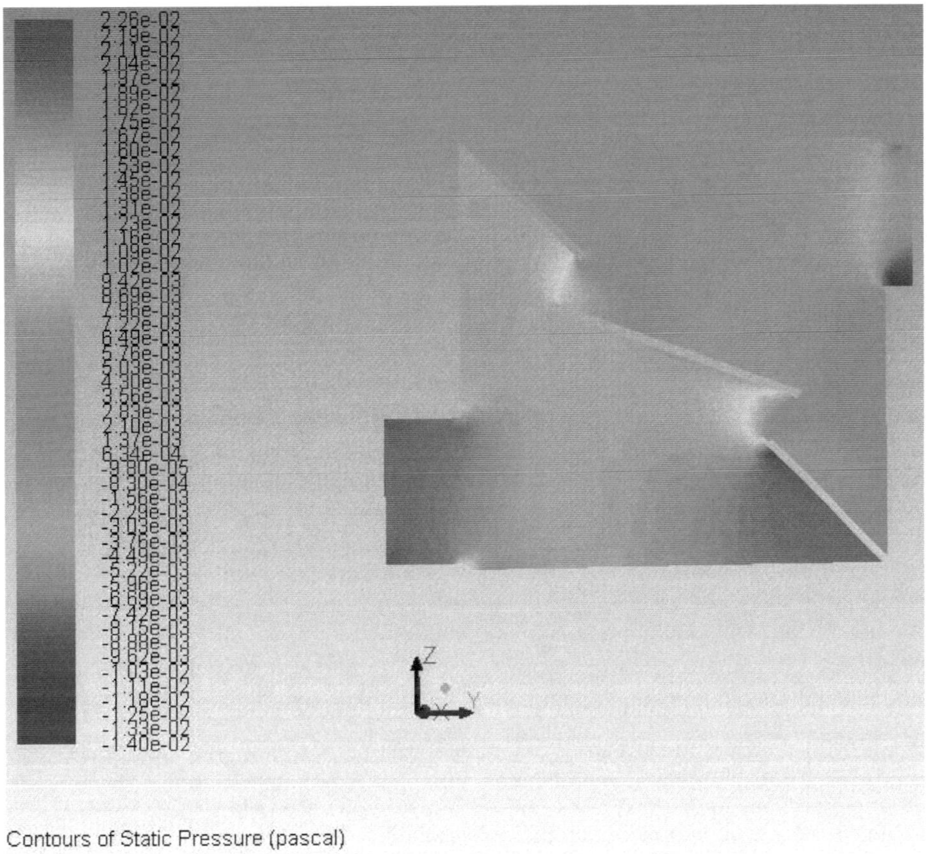

Figure 3. Example CFD results, Static pressure in the benchmark

2.2 Monitoring of the house

The measurements have been performed from 20 July to 20 August 2012, according to four scenarios successively:

- Without natural ventilation (all windows are closed), in order to obtain a reference,
- With natural ventilation (all roof windows and top vertical windows are open), without movable shading and with internal doors open (to get the maximal effect of natural ventilation),
- With natural ventilation, with movable shading and internal doors closed (to get the minimal effect of natural ventilation),
- With natural ventilation, with controlled movable shading and internal doors closed (to get the more realistic effect of natural ventilation).

Figure 4 shows temperature profiles in different rooms without ventilation (first period), with uncontrolled ventilation (second period, doors open), and with controlled ventilation (fourth period).

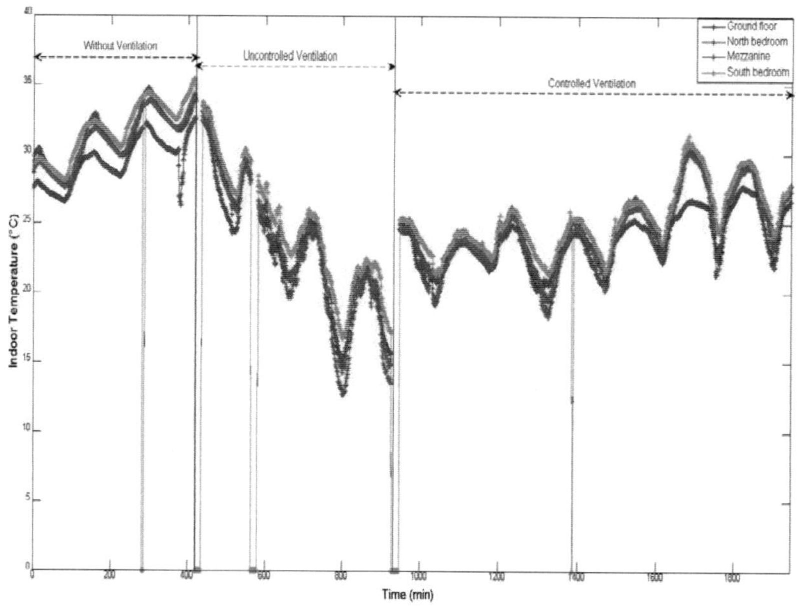

Figure 4. Example measured temperature profiles in Maison Air et Lumière

Pressure differences between outside and inside will be measured at certain times (not continuously) through windows situated on the different facades. Wind velocity and direction are also measured, so that pressure coefficients can be derived, allowing pressure on the different facades to be estimated over the whole period.

Given the small window height, and the large height between ground floor and roof windows, a one way flow model is considered. Discharge coefficients Cdi and exponents n, identified for roof windows using the laboratory benchmark presented above, or collected in the literature for other windows, allow the air flow rates through the different windows to be evaluated in terms of the internal pressure. The sum of inlet and outlet flow rates being zero, this internal pressure can be derived. An air exchange rate can then be evaluated.

The global air exchange rate can also be evaluated using a tracer gas method, which provides a second estimation of this parameter, but the precision is also questionnable as it will be addressed in the discussion §. In a homogeneous zone (volume V), the internal concentration Cint depends on the emission from the source So, the fresh air flow rate Q, internal and external air densities ρint and ρext, and external concentration Cext [4]:

$$\rho int\ V\ dCint/dt = So - \rho ext\ Q\ (Cint - Cext) \qquad (2)$$

In a steady state, the concentration is constant (dCint/dt = 0) so that:

$$So = \rho ext\ Q\ (Cint - Cext) \qquad (3)$$

Due to the high flow rate in the house when opening all roof windows, using the steady state option would require a large quantity of gas. It seems therefore preferable to inject a certain quantity of gas and to measure the decrease of gas concentration. Equation (2) allows the air flow rate to be identified.

Example measurement results are shown in Fig. 5 for two configurations: internal doors being closed or open. The logarithm of the concentration has been derived, so that the air flow rate Q can be identified using a least square method. In the example below, the result is around 4.5 air change per hour (ach) when the doors are closed and 5.5 ach doors open.

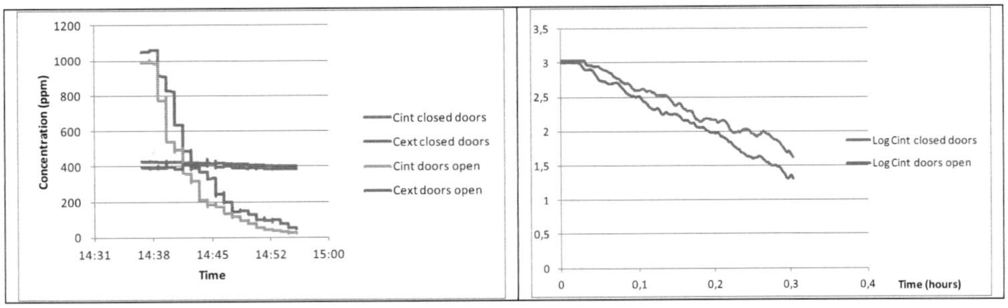

Figure 5. Example tracer gas concentration measurements in Maison Air et Lumière

3 DYNAMIC THERMOAERAULIC SIMULATION

Complementing monitoring results, numerical simulation constitutes another way to better understand the behaviour of a building. A dynamic thermal simulation tool is used to evaluate temperature profiles in the house [5], which has been modelled using 7 thermal zones : the living space (on two levels), three bedrooms (with different orientations), a garage, and other rooms (ground floor and first floor), see Fig. 6.

3D model	Thermal zones

Figure 6. Thermal model of Maison Air et Lumière, graphic modeler ALCYONE

The model accounts for the conductive, radiative and convective heat transfer, as well as energy storage in the building envelope related to solar and internal gains. In this highly insulated new construction, the most uncertain parameter is the natural ventilation flow rate. Figure 7 shows simulation results for a typical summer week in Greater Paris Area, considering three levels of ventilation flow rate (no ventilation, 5 ach and 20 ach).

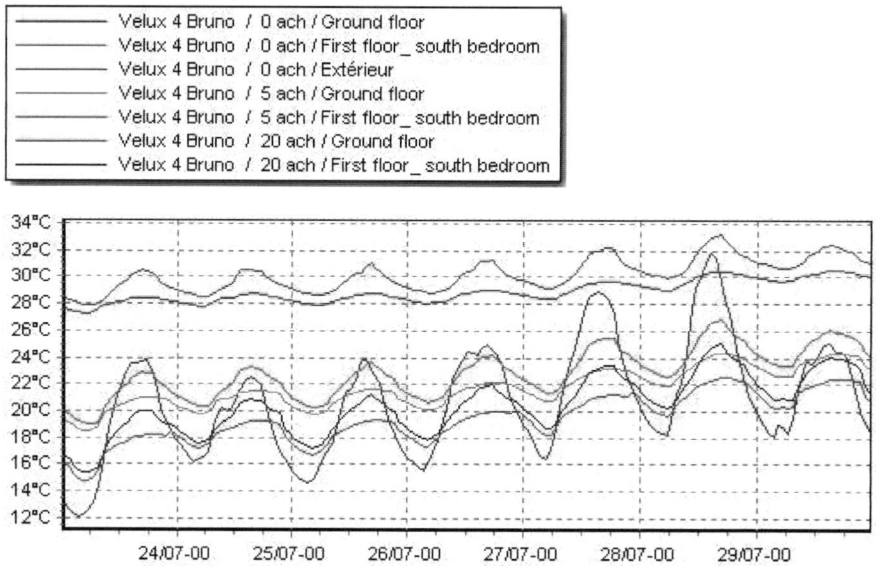

Figure 7. Example simulation results of Maison Air et Lumière, dynamic thermal simulation tool COMFIE

According to these results, this parameter has a large influence on the temperature profiles. It can therefore be calibrated by minimizing the discrepancy between measured temperature profiles and simulation results [6] using the climatic data corresponding to on site measurements (external temperature and solar radiation). This constitutes a third way to evaluate the global air exchange rate, and may be helpful to refine the characteristics evaluated in the laboratory benchmark, by taking into account the actual conditions in the real house.

Another added value of numerical simulation is, once the model has been calibrated, to compare, under the same climatic conditions, temperature profiles with and without ventilative cooling in order to evaluate the benefit of this approach. Such evaluation would otherwise require the construction of two identical houses, which is technically difficult and of course expensive.

4 DISCUSSION, STUDY OF A CHARACTERIZATION METHOD

Evaluating the interest of ventilative cooling requires simulation methods with at least hourly time steps because monthly or annual methods are not able to evaluate temperature profiles. Such methods need input data regarding the aeraulic properties of openings. The most common models for one way flows consider two characteristics: a discharge coefficient and an exponent, relating the air flow rate to the pressure difference on both sides of the opening (see equation 1).

As seen in § 2, using on site monitoring in a real house to determine these parameters is very difficult due to very low pressure differences on both sides of windows, therefore measurable with a high uncertainty. Air flow rates are also difficult to measure: velocities can be measured, but a flow rate is an integration of these velocities over the opening section. Anemometers can be used to measure velocities, but they have to be placed in specific points in order to obtain an average value through the whole section. The precision of such measurement is then very low. If both the pressure difference and the flow rate are not precisely known, it is very difficult to derive the two parameters Cd and n of equation (1).

A benchmark in a laboratory has two advantages:

- the flow rate can be higher than in the real building, so that the pressure difference is higher and therefore easier to measure;
- the flow rate can be measured using e.g. a Pitot tube, constituting a reference value.

The exponent n of equation (1) may be somewhat different in real conditions, because it depends on the pressure difference. On site measurement is therefore needed to calibrate this parameter. At the moment the tracer gas method is used in practice, but there is a large uncertainty due to possible uncomplete mixing following the density difference between the air and the most common gases (CO_2, SF_6, N_2O). These gases being heavier, they accumulate in the lower part of the building so that their concentration is not varying as modelled in equation (2). The measured air change rate may therefore be underestimated. Other gases like some VOCs might be used in the future, provided that they are not emitted by building elements (e.g. painting, glue etc.).

Calibrated window characteristics allow the ventilative cooling potential in a building to be evaluated, using thermoaeraulic simulation. This procedure may be validated thanks to a comparison between calculated and measured temperature profiles. Calibration of the simulation model using the measured temperature profiles may also constitute an alternative to using tracer gases: measuring temperatures is simpler and cheaper than measuring gas concentration.

CONCLUSIONS

The method proposed here, combining a benchmark in a laboratory with numerical simulation and on site monitoring may bring a supplementary input, complementing the existing knowledge in the field of passive cooling of buildings. The feasibility of using this method in order to prepare appropriate input data for numerical models implemented in regulation, design and certification tools will be studied.

ACKNOWLEDGEMENTS

This study has been performed with the support of VELUX.

REFERENCES

[1] Ghiaus C., Allard F., 2005, Natural Ventilation In The Urban Environment: Assessment And Design, *Earthscan, London, 241 p*

[2] Axley J. W. Emmerich S. J. and Walton G. N., 2002, Modeling the Performance of a Naturally Ventilated Commercial Building with a Multizone Coupled Thermal/Airflow Simulation Tool, *ASHRAE Transactions 2002, V. 108, Annual Meeting. Proceedings. Honolulu, 1-16 pp*

[3] http://www.ansys.com , accessed September 2012

[4] Sherman M.H., 1990, Tracer gas techniques for measuring ventilation in a single zone, *Building and Environment 25-4 365-374*
[5] Peuportier B. and Blanc Sommereux I., 1990, Simulation tool with its expert interface for the thermal design of multizone buildings, *International Journal of Solar Energy 8 109-120*
[6] Mejri O., Palomo Del Barrio E., Ghrab-Morcos N., 2011, Energy performance assessment of occupied buildings using model identification techniques, *Energy and Buildings 43 285-299*

AIR TURBULENCE INTENSITY INFLUENCE ON THE THERMAL COMFORT EVALUATION FOR DIFFERENT VENTILATION STRATEGIES

Cristiana Verona Croitoru[1], Ilinca Nastase[2*], Florin Bode[3]

1 CAMBI Research Center
Technical University of Civil Engineering in Bucharest, Building Services Faculty, Avenue Pache Protopescu 66, Bucharest, Romania

2 CAMBI Research Center
Technical University of Civil Engineering in Bucharest, Building Services Faculty, Avenue Pache Protopescu 66, Bucharest, Romania
**Corresponding author: ilinca.nastase@cambi.ro*

3 Technical University of Cluj-Napoca, Mechanical Engineering Department,B-dul Muncii nr.103-105, D03, Cluj-Napoca, Romania
CAMBI Research Center, Technical University of Civil Engineering in Bucharest, Building Services Faculty, Avenue Pache Protopescu 66, Bucharest, Romania

ABSTRACT

Thermal comfort is a subjective term, closely related to the sensation of warm or cold for the occupants, defining the state of mind of humans that expresses satisfaction with the surrounding environment. Since we spend more than 90% of our time inside buildings or vehicles, achieving a good thermal quality of this enclosed environment is vital. An optimal thermal comfort prediction can lead to „bien-être", efficiency in our work, unaltered health and even energy economy.
Among the newest and highly used approaches for thermal comfort studies are CFD methods, giving the possibility of extended parametric studies and in depth analysis of every thermo-fluidic parameter involved.
In this paper, several ventilation strategies were tested, in order to evaluate the thermal comfort of the occupants. We are comparing two mixing ventilation, and two displacement ventilation cases. This study is a part of larger experimental and numerical campaign intended to evaluate the influence of several flow parameters, such as the turbulence intensity at the inlet of the terminal air diffusion devices, on the local draft sensation and thermal discomfort of ventilation users. A numerical thermal manikin was placed in the centre of a test cell where different strategies of ventilation can be used. The results of CFD simulations were validated in a previous experimental campaign where a thermal manikin was used. While the PMV, and the PPD indexes seem to be less sensible to the value of the local turbulence intensity, the heat fluxes exchanged between the different body parts and their environment are highly varying between the studied cases. Local correlations between the turbulence intensity and convective heat flux of the body were found.

KEYWORDS

Thermal comfort; CFD; turbulence intensity; ventilation strategies;

INTRODUCTION

Maintaining thermal comfort for occupants in buildings in extreme climatic conditioning requirements and irrespective of the environmental outside conditions has been the main focus for the Heating Ventilation and Air Conditioning (HVAC) design engineers and systems developers. We observe, however, that contemporary techniques of air flow diffusions are not optimized simultaneously for these two inseparable goals: thermal comfort and energy savings. This observation can be applied to both buildings and vehicles fields. This paradox is due on one hand to bad diffusion of cold air, and on the other hand, to weakness of the conception of these systems. Behind this, the use of air flow models that are not fully adapted to conditions in the buildings and all other interior spaces can be found. On one hand, in our opinion, this issue finds a theoretical response to the adaptation of existing theoretical models for different indoor (building or other enclosures) conditions, in terms of human thermal comfort.

Nowadays, we have the possibility of using advanced methods and devices both in terms of computing capabilities and experimental techniques. The existing thermal comfort models are all built with simplified assumptions, often limited because of available resources when they were conceived – over 30 years ago for the most used. We have today the opportunity to validate these models by taking into account the variation of several parameters, we also have the opportunity to correct them and to propose new models. On the other hand, a technical answer may come from the conception of the air diffusion devices which have to be optimized for improving mixing between supplied flows and their ambient in order to improve thermal. Nevertheless, this technical direction of research has to be preceded by the theoretical advances in improving the existing comfort models which seem to be inappropriate in many situations [5-8].

In an article from 2001 [3], the Professor Fanger, founder of the first "school" of thermal comfort research and "father" of this scientific field, indicated that the thermal comfort standards are outdated and following their prescriptions cannot lead to acceptable conditions for most users: "We need to reconsider the concept related to our comfort to achieve excellence in environmental quality. Our goal should be essential to provide fresh air, accompanied by a pleasant feeling, refreshing, without any adverse health effect and a comfortable thermal environment for all users." said the Professor in [3].

At the same time if we consider two bibliographic articles at a distance of 20 years - [4] and [5] – we can see that nothing has changed in definition and use of these models and evaluation indexes of interior ambiental comfort.

In this context this study is a part of a larger experimental and numerical campaign which is intended among other directions to study the influence of the turbulence intensity at the exit plane of the terminal air diffusion devices on the local draft sensation and thermal discomfort of mixing and personalized ventilation users.

We put this question to what extent the turbulence intensity of the flows generated by various air diffusion devices can affect the comfort and also what are the consequences of an "incomplete" assessment based on existing models? How is affected the design of ventilation and air conditioning due to the use of these models for pre-evaluation of interior parameters?

EMPLOYED METHOD

Numerical case study

Numerical simulations using a CFD approach using a RANS (Reynolds Averaged Navier Stokes) model were performed to study the airflow and heat transfer around a human body for several air diffusion strategies and for different values of the turbulence intensity at the jet inlet (see Table 1). The virtual thermal manikin was placed in a test cell (Fig. 1a) which is a reproduction of a real laboratory facility [10]. Figure 1a, gives the dimensions of the test cell and position of the thermal manikin. In Figure 1b are presented different inlet and outlet conditions studied through numerical simulation corresponding to the four cases, marked below and where the inlets are coloured with blue and the outlets with red. We define two virtual median plans of the manikin: - the median plane, passing through the heart (in green in Fig. 1 b) – **coronal plane**, and the transverse median plan, symmetric – **sagittal plane**.

Case 1 and Case 2 (**M1** and **M2)** are mixing ventilation distributions where the inlet device is placed at the upper part of the room and the outlet at the lower part (M1) and, vice-versa, the inlet at the lower part and the outlet at the upper part of the room respectively (M2). Case 3 **(D1)**, represents the simulation of a displacement ventilation strategy, the inlet device being considered as the entire duct surface, indicated with yellow in Fig. 1 a and in blue in Fig. 1c. For Case 4 **(D2)**, we considered the inlet on the whole front wall. Mixing ventilation implies supplying an air jet with relatively high velocity and high air change rates. The velocities in the occupancy zone and the noise level must be under certain limits. In the displacement case, the inlet is positioned directly in the occupied zone, with low air velocity and temperatures. Hot and polluted air is evacuated in the upper part of the room. The efficiency of the ventilation system depends on the interior configuration. Piston-type ventilation strategy is a particular case of displacement ventilation, where the whole surface of a wall is used as an inlet device. The principal demand is to have low velocities and turbulence intensities. For our case, the air exchange rate is over 200 h^{-1}. This strategy is applied for white chambers, operating rooms etc. [13] . All studied inlet velocity and turbulence conditions are given in Table 1

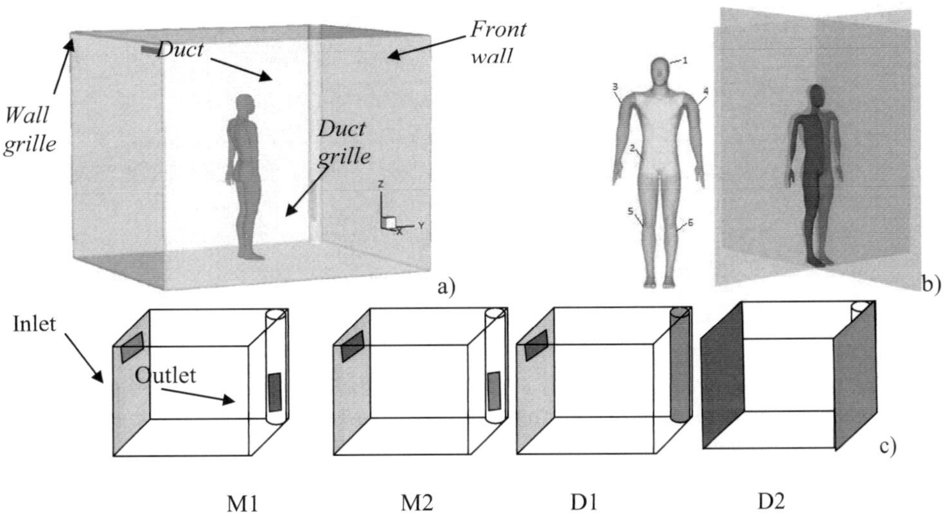

Fig.1: a) Possible inlets in the test celle, b) Human body parts: 1- Head, 2- Torso, 3-Right arm, 4-Left arm, 5-Left Leg, 6-Right Leg, sagittal plan (purple) and coronal (green)c) Inlet and Outlet definition for the four studied cases

The accuracy of a CFD simulation depends, in a high percentage, on the way of reproducing the geometry that defines the calculation domain and the heat sources. The approach used here is based on freely available geometry from the MakeHuman project [11] that was meshed in an automatic unstructured grid generator HEXPRESS from Numeca [12]. The virtual manikin has a height of 1.88 m and a body surface of 1.9 m^2. This way, the virtual manikin has 18 distinctive segments which can be "controlled" separately. However in this study we grouped the segments by three in order to obtain larger regions which are similar to the ones of the experimental thermal manikin. Three segments from one region have the same number in Fig.1b (i.e. head, body, arms and legs). The meshing process consisted of 5 steps: initial mesh generation, mesh adaptation, mesh snapping, mesh optimization and viscous layer insertion. During the final step, a viscous layer was inserted over the surface of the human, using 20 layers with a first layer thickness of 0.2 mm and a growth factor of 1.13 (Figure 1 c) [13]. The result was a mesh with a total of 2.2 million cells which was imported in Fluent 12. A mesh dependency study was conducted and showed that this grid was fine enough in order to obtain stable results.

The boundary conditions used during the numerical study considered the walls being at a constant temperature of 20°C. The surfaces of the different segments of the virtual manikin were considered as having temperatures which were previously determined on a thermal manikin using an infrared camera (see Table 2).

Ventilation strategy	Inlet position	Imposed inlet velocity	Imposed turbulence velocity	Inlet type	Outlet type
M1	• Upper part-wall grille	2 m/s	0%, 3%, 10%, 30%, 50%	Velocity inlet	Pressure outlet
M2	• Lower part-duct grille	1 m/s	3%, 10%, 30%		
		2 m/s	3%, 10%, 30%		
D1	• Duct	0.3 m/s	3%, 10%, 30%		
D2	• Front wall	0.2 m/s	0%, 3%, 10%, 30%, 50%		
		0.3 m/s	0%, 5%, 3%, 10%, 15%, 20%, 25%, 30%, 50%		

Table 1: Studied cases: velocity and turbulence intensity values imposed at the inlet and other boundary conditions

Body surface				*Walls*
Head (1)	Body (2)	Arms (3, 4)	Legs (5, 6)	
34.2°C	31.9°C	30°C	26.8°C	t=20°C

Table2: Temperature boundary conditions for the CFD study

A "second order discretization" method was used for the calculation of convective terms and the SIMPLE algorithm for pressure-velocity coupling. The chosen turbulence model was k-ω standard [14] with low Reynolds number correction. The radiation flux was calculated by using the "surface-to-surface" model which allows setting the radiation surfaces. Convergence solution is assumed to be achieved when the dimensionless residuals of the flow equations are less than 10^{-3}.

Experimental validation

In this study, Particle Image Velocimetry measurements were used for validation. They targeted experimental validation of the velocity distribution obtained by numerical study of convective flow above the manikin's head and of the airflow released by an inlet device from a classical mixing ventilation case. The validation test consisted in compare a mean field of the velocity magnitude from PIV measurements (the employed PIV system is described in [10]) in the sagital plane above the manikins head, and the numerical velocity field, in the case M1. In Fig.3 we compared the two components of velocity fields W and U and the plan distribution of velocity magnitude obtained from PIV measurements and numerical simulations using the k-omega SST model. In this figure, we can see the existence of a smaller convective flow in the direction of the forehead, which meets the main convective flow. This secondary flow seems to meet with another flow coming from behind the manikin. General shape of the three fields is similar and we emphasize that k-omega SST model satisfactorily reproduces the dynamics of convection plume.
Even if the stagnation region is more extended in the experimental case, the two contour plots display some common characteristics: the presence of a smaller flow coming along the forehead which is reuniting with a stronger plume above the head. This latter flow seems to be connected with another plume coming from the back of the manikin. The values of the velocity magnitude from both numerical and experimental fields are very similar to the ones presented by Sorensen and Voigt [2].

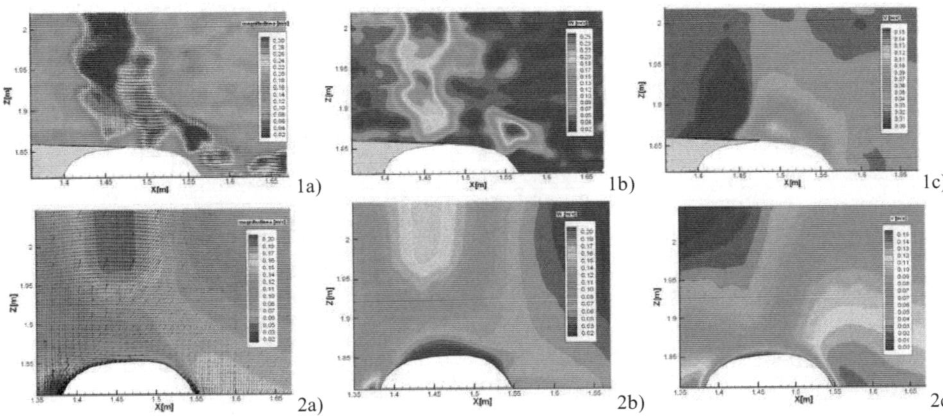

Fig 3. Velocity field comparison –sagittal plan- with jet (Tu=10%): 1 – experimental, 2 –numeric
a) Velocity magnitude [m/s] b) W component [m/s] c) U component [m/s]

RESULTS AND DISCUSSION

In the design of indoor environment, it is still not acknowledged that convection flows caused by heat sources like the human body plume may significantly affect the flow distribution in rooms [1]. Generally, attention is given only on the flow generated by the air diffusion terminal devices. As shown by Kosonen et al [1] the point of occurrence of the maximum air velocity in the occupied zone depends on the heat source strength and its distribution in the room. Thus, the air flows interaction in ventilated rooms is of great importance when estimating occupants' comfort. In this context, the second goal of this campaign is to take into account the presence of the human body thermal plume in order to obtain realistic conditions in both experimental and numerical investigations of air distribution in ventilated rooms.
As it has been shown by Fanger [14], the velocities and the turbulent characteristics of the flows may generate a thermal discomfort translated by the sensation of "draught" as "an undesired cooling of the human body caused by air movement" [14, 15]. This way, we wanted to check first, the influence of the variation of the jet initial turbulence intensity on the behaviour of the global temperature and velocity fields inside the test cell. Therefore, in Fig. 4 and 5 we are presenting the temperature and velocity fields in both coronal and sagital planes for all studied cases, for two values of the turbulence intensity.
In the mixing ventilation cases, M1 and M2, the air surrounding the manikin seems to be well mixed, allowing nevertheless the observation of the thermal plume. Slight differences between temperature or velocity fields, for the two initial turbulence intensities might be observed, especially in the region of the convective plume flow. The turbulence intensity value seems to influence more the velocity and temperature distributions for the displacement ventilation case D1. As for the piston case D2, there is no obvious influence of the turbulence intensity on the fields in Fig. 4 and 5.
The CFD simulation results allow us to easily evaluate the PMV index as defined by Fanger. In Fig. 6 are given the corresponding distributions of the global PMV and PPD of the room for each studied case and each turbulence intensity level. While these indexes should be sensitive to velocity and temperature fluctuations, we couldn't find any variation with of their values with the turbulence intensity imposed at the several studied air diffusers.

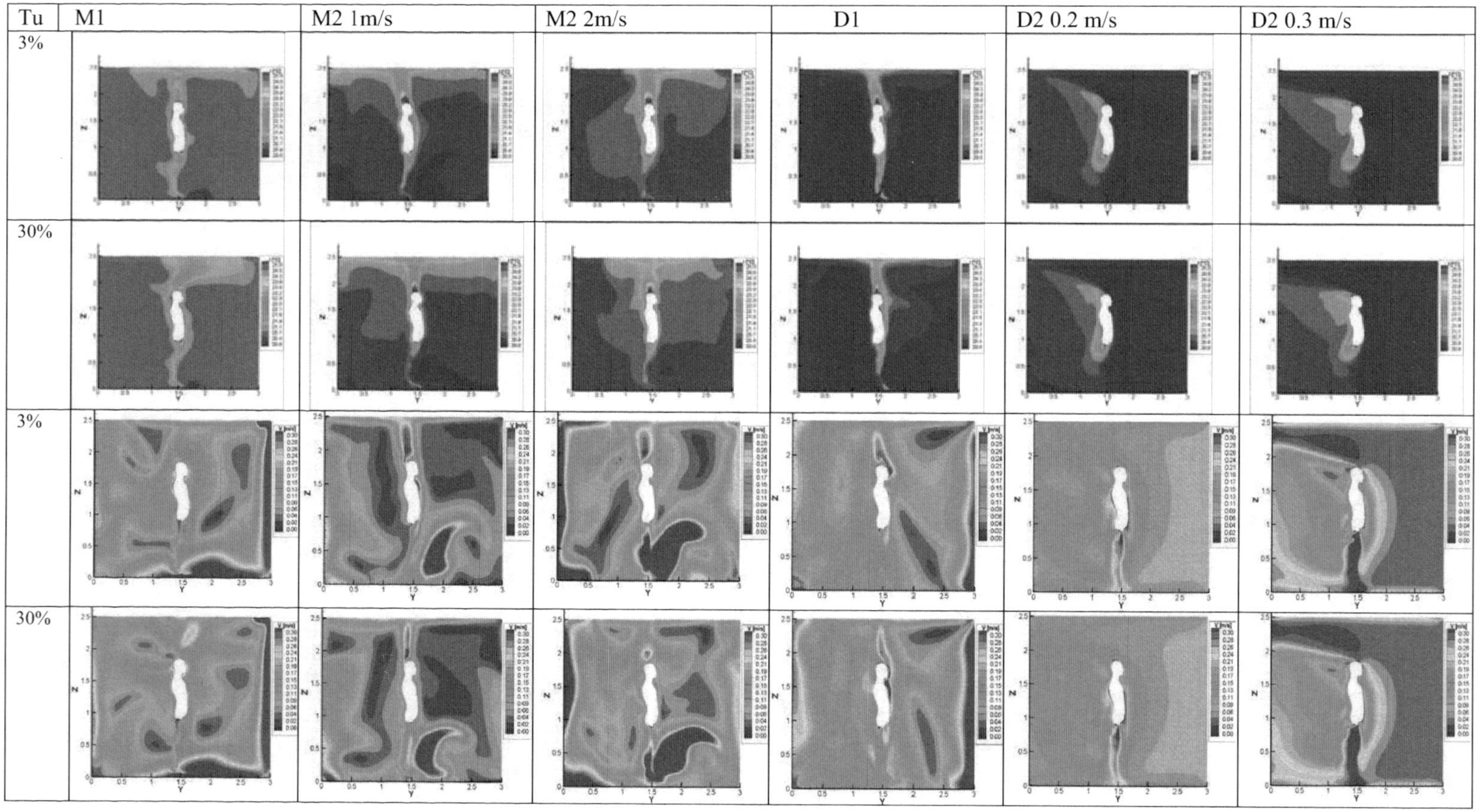

Fig.4: Temperature and velocity fields in the sagital plane for Tu=3% and Tu=30% for all studied cases

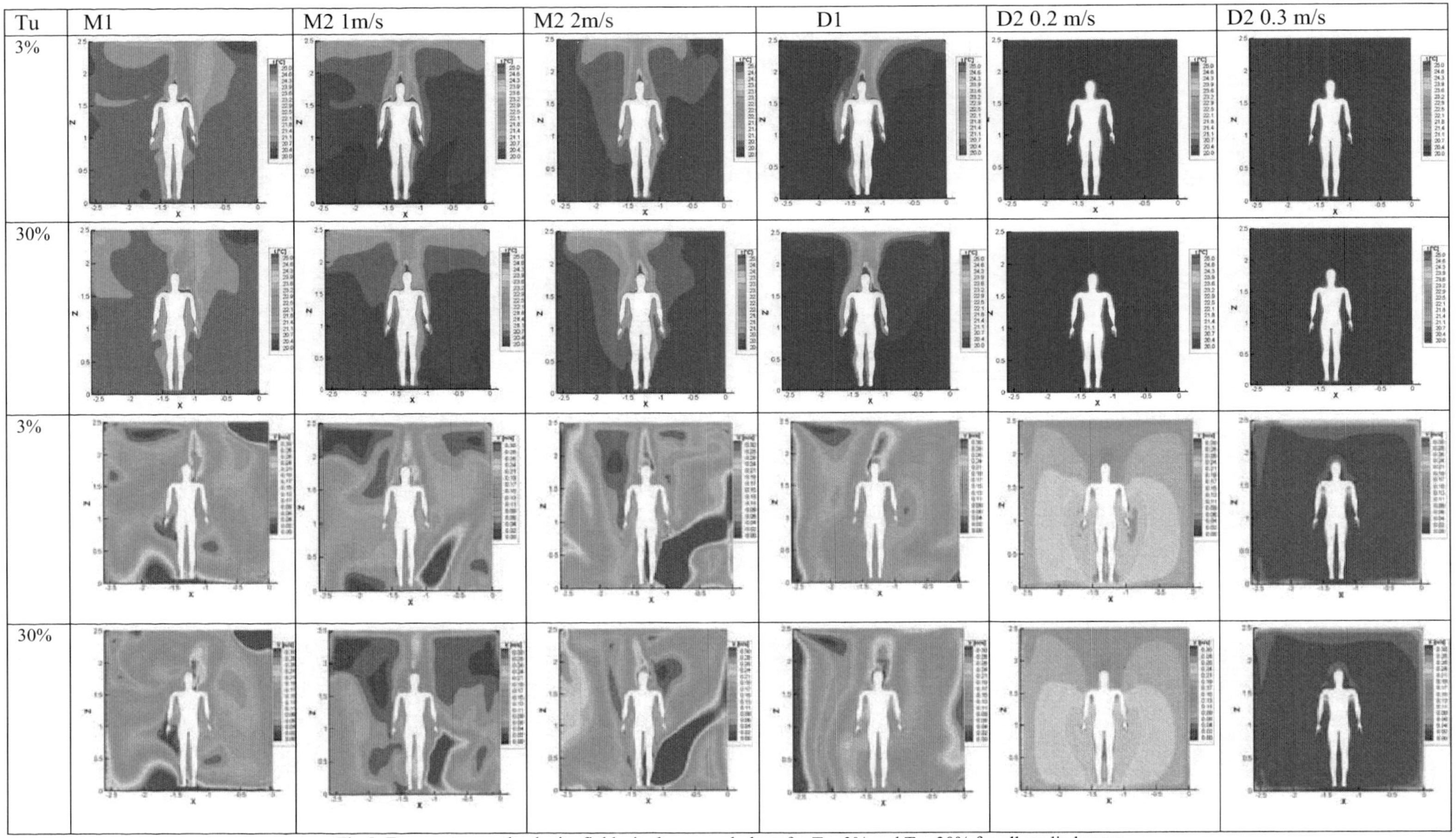

Fig.5: Temperature and velocity fields in the coronal plane for Tu=3% and Tu=30% for all studied cases

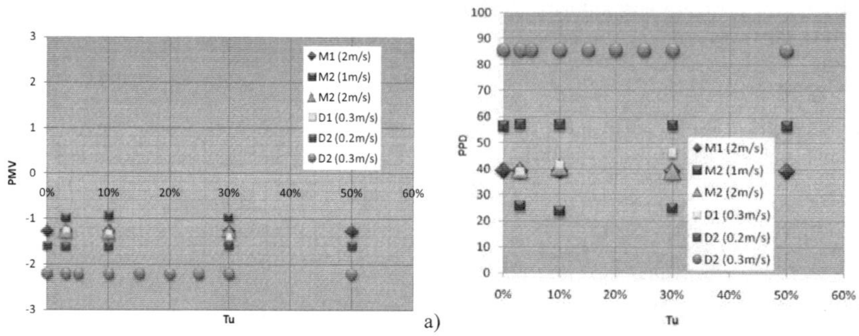

Fig. 6: Global PMV(a) and PPD (b) of the room for each studied case

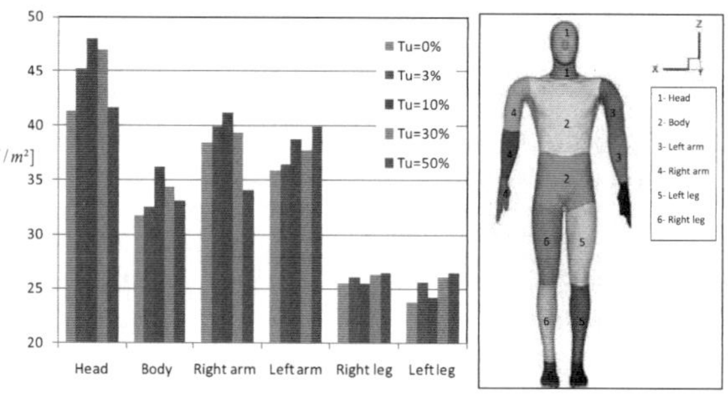

Fig. 7: Average convective fluxes of each body part for M1 case

Another way of studying the thermal comfort implications in changing the initial turbulence level at the inlet of the jet flow is looking to the heat transfer between the manikin's body and its environment. This way we represented in Fig.7 average convective fluxes for different body segments, for the M1 case. From both figures it can be observed that a more intense heat transfer occurs for the head and **body** of the manikin in the case having the initial turbulence intensity *Tu=10%*. As the fluctuations of the convective heat flux for different parts of the body (Fig.7) are quite important with the variation of the initial turbulence intensity at the jet inlet, they have to interfere with the local thermal comfort of the body. In the same time we saw that the PMV and PPD values are not sensitive to these fluctuations.

As it is obvious that the virtual body suffers a non-uniformly distributed convective heat transfer, with high variations between the different studied cases, it is necessary to employ another quantity being able to quantify these variations. A thermal comfort index employed in nonuniform thermal environments is the equivalent temperature t_{eq} which takes into account the combined effect of the local air temperature, local thermal radiation, local air velocity based on the local heat transfer rate at the skin surface [24]. The charts of the body parts sensations in Fig. 8, are showing strong differences between the considered turbulence intensities values in the case M1.

Given this dynamics of the convective fluxes we wanted to get insight into the local phenomena governing the thermal transfer occurring at local scales. Indeed, if we question ourselves which is the level of the local turbulence intensity influence on the thermal comfort and thermal transfer, the most appropriate approach would be to search possible correlations between the local turbulence intensities and local convective fluxes.

Figure 9 is an example of such type of correlation that can be done between the turbulence intensity values and the local convective heat flux. As we can see in Fig. 9, in the region behind the legs, the convective heat flux is in the range of 25 W/m to 48 W/m, for all the mean turbulence intensities around the legs. For each case we observe an increase of approximately 14% on convective heat flux, with a maximum of 18% for the case mixing ventilation with inlet at the lower part of the room (Case 2B). The fact that the turbulence intensity has an important impact on heat transfer between the body and the environment is well known, but this hypothesis has never been quantified before in the literature. *We can observe that only by varying the inlet imposed turbulence intensity of the air we obtain important differences by quantifying the thermal effect.* This conclusion opens new research directions in which this parameter is seriously taken into consideration for evaluating thermal comfort.

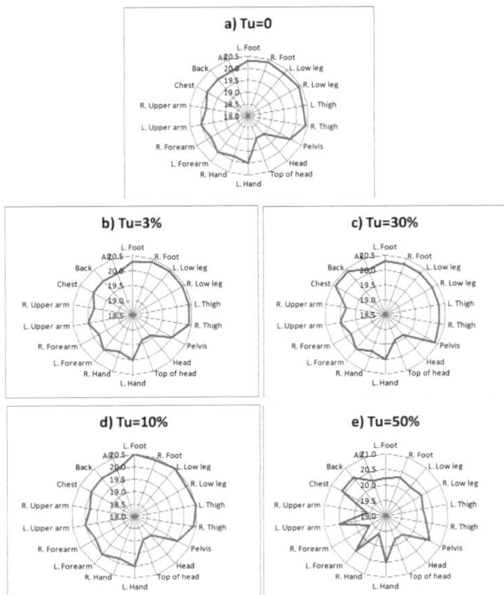

Fig.8: Distributions of the equivalent temperature t_{eq} for M1 case: a) Tu=0% b) Tu=3%, c Tu=30%, d) Tu=10%, e) Tu=50%

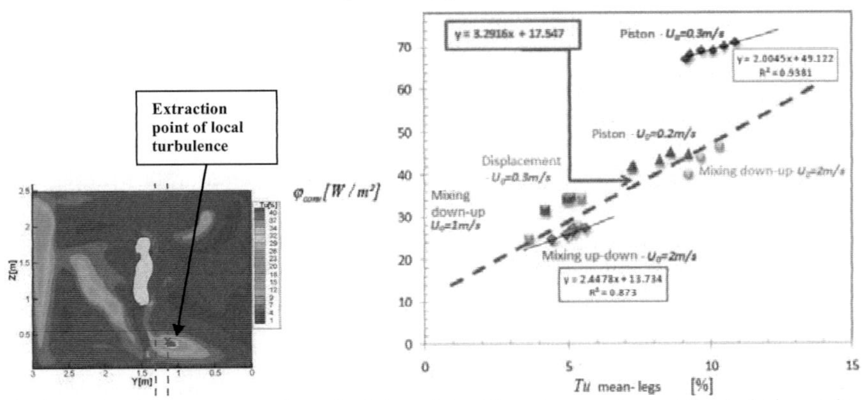

Fig. 9: Correlations between mean values of turbulence intensities and convective heat flux in the legs region for all the studied cases

CONCLUSION

The four strategies of ventilation (up-down, down-up, displacement and piston) are studied using classical thermal comfort indexes and original correlations between the degree of local turbulence and the convective heat flux exchanged between the human body and its environment. For each of these cases, we observe an intensification of convective heat transfer expressed by an average increase of 14% of convective heat flux with a maximum increase of almost 18% for the strategy of mixing ventilation with the inlet at the lower part of the room. Varying only one parameter (air turbulence intensity imposed at the inlet device) we obtain high differences on convective heat loss, fact that implies a change in thermal comfort state. This approach can determine an energy efficiency and economy in exploitation of HVAC systems. In conclusion, we can say that the analysis of convective heat flux distributions on the surface of the thermal manikin heat reveals the importance of a detailed study of the dynamics of convective flows around the body for evaluating thermal comfort. This finding is reinforced by the strong correlation between the local convective flux and local turbulence intensity.

ACKNOWLEDGEMENTS

„This work was supported by the grants of the Romanian National Authority for Scientific Research, CNCS – UEFISCDI, project numbers: PN-II-PD-PCE-2011- 3-0099, PN-II-ID-PCE-2011-3-0835 and PN-II-PT-PCCA-2011-3.2-1212".

REFERENCES

1. Abadie, M., *Contribution à l'étude de la pollution particulaire : rôle des parois, rôle de la ventilation*, in *LEPTAB*. 2000, Université de La Rochelle.
2. Meslem, A., *Contribution à l'étude du couplage thermique entre un jet et un local climatisé.* 1997, INSA Lyon.
3. Fanger, P.O., *Human requirements in future air conditioned environments* International Journal of Refrigeration, 2001. **24**: p. 148-153.
4. Fountain, M.E., *Laboratory studies of the effect of air movement on thermal comfort: a comparison and discussion of methods.* ASHRAE Transactions, 1991. **97**(1): p. 863-873.
5. Djongyang, N., R. Tchinda, and D. Njomo, *Thermal comfort : A review paper.* Renewable and Sustainable Energy REviews, 2010. **14**: p. 2626-2640.
6. Gao, N. and J. Niu, *CFD study on micro-environment around human body and personalized ventilation.* Building and Environment, 2004. **39** p. 795 - 805.
7. Cropper, P.C., et al. *Exchange of simulation data between CFD programmes and a multi-segmented human thermal comfort model.* in *Air Conditioning and the Low Carbon Cooling Challenge*. 2008. Windsor, UK.
8. Inard, C., H. Bouia, and P. Dalicieux, *Prediction of air temperature distribution in buildings with a zonal model.* Energy and Buildings, 1996. **24**(2): p. 125-132.
9. Inard, C., P. Depecker, and J.-J. Roux, *Un modèle simplifié pour la prédiction du champ de température dans les bâtiments - A simplified model for the prediction of air temperature distribution in buildings.* Révue Générale de Thermique, 1997. **36**(2): p. 113-123.
10. Inard, C., et al., *Structure moyenne et anlyse intégrale du panache thermique des convecteurs électriques.* Révue Générale de Thermique, 1997. **36**: p. 495-509.
11. Sorensen, D.N. and L.K. Voigt, *Modeling airflow and heat transfer around a seated human body by computational dynamics.* Building and Environment, 2003. **38**(6): p. 753-762.
12. Meslem, A., et al., *Numerical study a turbulent jet flow issued from a lobed orifice.* Scientific Journal of Technical University of Civil Engineering in Bucharest - Series: Mathematical Modelling in Civil Engineering, 2010. **6**(3): p. 42-51.
13. Awbi, H.B., *Ventilation of Buildings*. 2003, London, U.K.: E&FN SPON. 313.
14. Fanger, P.O., et al., *Air turbulence and sensation of draught.* Energy and Buildings, 1988. **12**(1): p. 21-39.
15. Fanger, P.O., *The new comfort equation for indoor air quality.* Ashrae Journal, 1989. **31**(10): p. 33-38.
16. Catalina, T., J. Virgone, and F. Kuznik, *Evaluation of thermal comfort using combined CFD and experimentation study in a test room equipped with a cooling ceiling.* Building and Environment, 2009. **44**(8): p. 1740-1750.
17. Zhang, Y. and R. Zhao, *Overall thermal sensation, acceptability and comfort.* Building and Environment, 2008. **43**(1): p. 44-50.
18. Zhang, Y. and R. Zhao, *Effect of local exposure on human responses.* Building and Environment, 2007. **42**(7): p. 2737-2745.

REDUCING ENERGY CONSUMPTION IN AN EXISTING SHOPPING CENTRE USING NATURAL VENTILATION

Gitte T. Tranholm[1], Jannick Karsten Roth[*1] and Lennart Østergaard[1]

1 WindowMaster A/S
Skelstedet 13
Vedbæk, Denmark

**Corresponding email: jkr.dk@windowmaster.com*

ABSTRACT

The energy consumption needed for establishing a good indoor climate in shopping centres is often very high due to high internal heat loads from lighting and equipment and from a high people density at certain time intervals. This heat surplus result in a need for cooling during most of the year, typically also during the winter, and often the needed cooling is provided by a mechanical ventilation system with integrated mechanical cooling.

However, for certain areas, especially the hallways connecting the individual shops, natural ventilation might be an energy efficient alternative or supplement to the traditional mechanical system.

This paper presents a case study on an existing shopping centre in Copenhagen, Denmark (Fields shopping centre). The building owner's key reason for considering natural ventilation was a desire to improve the thermal indoor climate in the hallways, and, at the same time, reduce the energy consumption for ventilation.

On this background, WindowMaster conducted a number of simulations in the dynamic simulation program BSim2002. These calculations suggested a significant energy saving potential (60% reduction) and a significant improved thermal indoor climate (70% reduction of annual hours above 28 °C) by adding natural ventilation to the ventilation strategy.

Thus, in the beginning of 2011 the building owner decided to install automatically controlled natural ventilation in the hallways in the shopping centre in addition to the existing mechanical ventilation system. The basic control idea was to use the natural ventilation system in the summer and transient seasons and the mechanical ventilation system in the winter (hybrid ventilation).

Measurements of the thermal indoor climate in the first year (September 2011 - august 2012) show that the indoor climate has improved significantly. In this year, the actual results outperform the expected results from the simulations, and the building owner has expressed his satisfaction with the improvements in the thermal indoor climate.

KEYWORDS
Thermal building simulation, energy savings, natural ventilation, hybrid ventilation, natural cooling, shopping centre, case study, log-data analysis.

INTRODUCTION

As of today, only few shopping centres have adopted natural ventilation in the hallways. This might be caused by a limited number of reference cases thus creating uncertainty for the building owner about the potential benefits in terms of reduced CO_2 emissions, running costs and improved indoor climate.

It has however been demonstrated in a number of practical cases e.g. Ernst-August-Galerie in Germany [1] and Green Light House in Denmark [2] that the use of natural ventilation in combination with mechanical ventilation might give significant improvements in the thermal indoor climate and the energy consumption in comparison with pure mechanical ventilation. Both cases have a DGNB certificate showing high performance on ecological quality and indoor climate.

In general, natural ventilation is preferable to mechanical ventilation in multi-storey rooms as the heat and stale air generated in the air volume is easily expelled through automatically opened roof windows. Such rooms exist not only in the hallways of shopping centres, but also in many other buildings, like atriums in office buildings, hallways in airports, exhibition halls, production halls and in sport facilities. Thus, the results and principles given in this paper are useful in a number of building types.

The Fields shopping centre, see Figure 1 was completed in 2004 and it is as of today the biggest shopping centre in Denmark and one of the largest in Scandinavia. It is located in Ørestad, Copenhagen and the total size of the centre is 115,000 m^2 with a total shopping area of 65,000 m^2. The centre contains more than 140 retailers and it is the workplace for around 2,500 employees. After the completion, the building owners have been committed to implementing new technologies for reducing the building energy consumption and increasing the shopping experience of the customers.

Originally the centre was equipped with a full mechanical ventilation system. However, as explained in the Abstract, in the beginning of 2011 it was decided to install automatically controlled natural ventilation in the hallways in the shopping centre to support the mechanical ventilation (hybrid ventilation). The purpose was twofold – in order to reduce energy consumption and to improve the thermal indoor climate for the benefit of customers and employees.

Automatically controlled natural ventilation is controlled depending on variations in the outdoor conditions measured from a weather station and the indoor climate measured from for example temperature/CO_2/humidity sensors and information about the usage of the building. Automatically controlled natural ventilation is highly advanced ventilation and should not be confused with manually controlled natural ventilation.

This paper presents the dynamic building simulations which were used as basis for the decision to implement natural ventilation in the building, together an analysis of the log-data of the indoor climate obtained during the first year of service. Results for the thermal indoor climate and energy consumption for ventilation are presented.

Figure 1. Picture of the shopping centre

BUILDING AND VENTILATION PRINCIPLES

The roof above the hallways is made of glazed sections as illustrated in Figure 2 and Figure 3. In the glazed areas automatically controlled window openings are established. Through these openings air enters and leaves the hallways. Based on WindowMaster's experience, the hallways have been partitioned into eight zones in the simulations – divisions which are later followed in the design of the system implemented in the building.

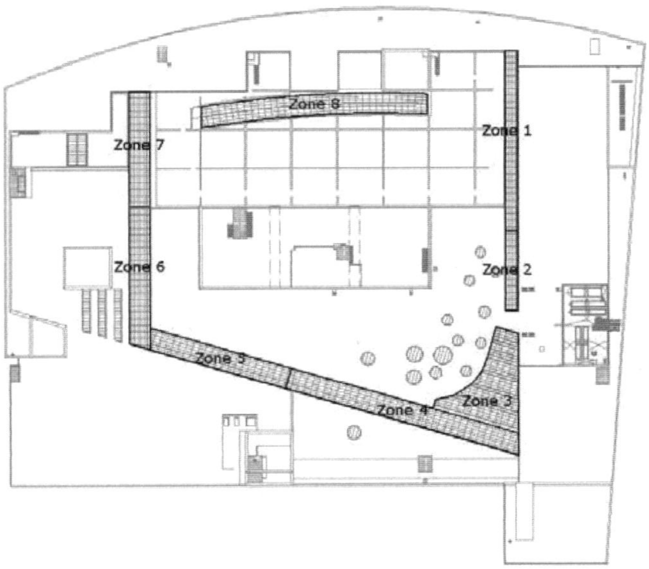

Figure 2.Drawing of the shopping centre roof including marking of the eight zones.

Figure 3. Picture of hallways from the inside and from the roof.

Natural ventilation uses the natural driving forces (wind and thermal buoyancy) for moving air and is therefore able to ventilate without any energy consumption for air movement (SFP=0). Therefore, in situations where the room or building in consideration needs cooling, natural ventilation is able to supply fresh air without energy consumption. This is an advantage for the shopping centre since warm air can be removed, while at the same time the energy consumption for air movement is reduced.

In Figure 4 the natural ventilation principle for the hallways is illustrated.

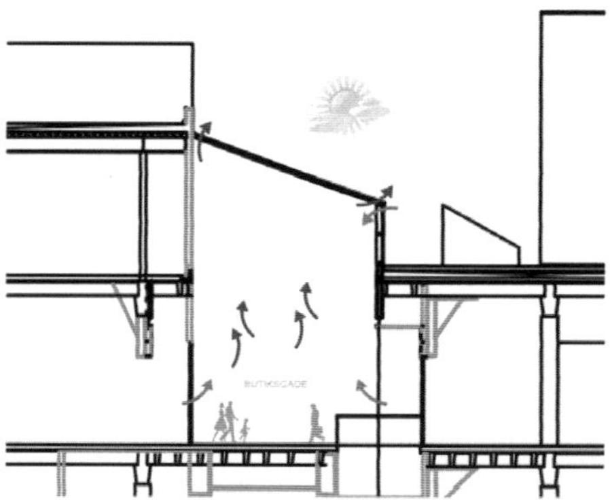

Figure 4. Ventilation principle for the natural ventilation system in the shopping centre

The primarily ventilation principle is stack-ventilation. This type of ventilation is primarily driven by warm air rising to the top whereby it creates a pressure difference which drives the ventilation.

Most of the opening area for natural ventilation is established in the glass roof above the hallways - only about 2% of the opening area is established in the facade. This distribution is not optimal, but caused by the fact that the natural ventilation system is retrofitted to an existing building. For a new shopping centre, where natural ventilation is included from the early design phase, it is recommended that the opening area is more evenly distributed between facades and roofs in order to achieve an optimal placement of the neutral pressure level. Nevertheless, the natural ventilation system capacity in the Fields shopping centre is still high as openings are placed in different levels in the roof and since it was possible to place some openings in the facade.

SIMULATION TECHNIQUES AND ASSUMPTIONS

The method used is thermal building simulation of the hallways made in the dynamic simulation program BSim2002 [3]. The software simulates the thermal indoor environment and energy performance during a year based on the specific constructions, usage and the outdoor weather conditions. Weather data from Denmark is used (DRY [4]). Based on steady state calculations, made prior to the dynamic simulation, the capacity of the natural ventilation system is determined. The methods are described in [5] and [6].

The natural ventilation system modelled is demand controlled, as the heat loads from especially people and sun changes a lot during the opening hours.

Figure 5 shows an illustration of the 3D section of the hallways which was modelled in the simulation software. This section is representative for all the hallways since these were sufficiently similar with regard to their design, internal heat loads and solar gain.

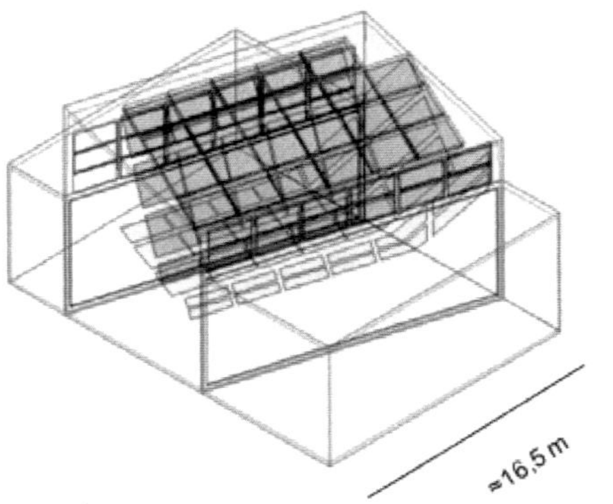

Figure 5. Calculation model for the BSim2002 simulations

Three different ventilation strategies were modelled to investigate the thermal indoor environment and the energy saving potential:

Model 1: Mechanical Ventilation (MV)

- In this model, the already existing system with mechanical ventilation was modelled in detail
- No natural ventilation system was added
- In the calculations it is assumed that the mechanical ventilation system is a CAV system with a ventilation rate of 4 h^{-1} and a heat recovery coefficient of 0.7
- The inlet air is at least 17°C
- The mechanical ventilation system is turned off during the night

Model 2: MV in winter, NV in summer (hybrid ventilation)

- In the second model, the mechanical natural ventilation was used in the winter only (week 1-15 and 46-52, in total 22 weeks of the year)
- A new natural ventilation system was implemented in the transient season and during the summer time (week 16-45, in total 30 weeks of the year)
- The natural ventilation system is set to have a maximum capacity of 4 h^{-1} and it is assumed that the natural ventilation is automatically controlled and utilized for both day ventilation and night cooling.

Model 3: MV all year, NV in summer (hybrid ventilation)

- The third model allowed the natural ventilation and mechanical ventilation to run at the same time in the summer (same weeks as given above)
- The purpose is to further reduce the number of warm hours in the hallways
- During summer mode the mechanical ventilation system is activated when the operative indoor temperature reach 26°C

For all three calculations the opening hours is Monday to Friday 10AM - 8PM and Saturday - Sunday from 10AM to 5PM. The people load is 5 m^2/person and the usage level is 100 % Monday - Friday from 2PM - 6PM and 50% in the rest of the time on weekdays. Saturday and Sunday the usage level is set to 90 % from 10AM - 5PM. The lightning level is set to 30W/m^2 in the opening hours. The g-value of the glass in the roof is 0.3.

SIMULATION RESULTS

Figure 6 shows the results of the thermal building simulations for the three models. The figure shows the total number of hours in the hallways for which the operative roomtemperature is above 28°C, 30°C and 32°C respectively. Note that only hours within the opening hours of the shopping centre have been counted. It is clearly demonstrated that usage of a natural ventilation system in the summer and transient seasons will reduce the number of "warm hours" significantly – from 677 hours to 182 hours above 28°C. It is noted that the results after installing natural ventilation would have been even better if the system had been implemented in the building from the early design phase.

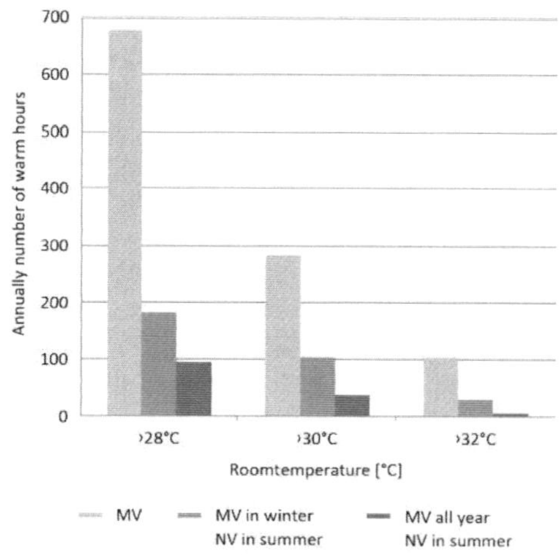

Figure 6. Annual number of hours with roomtemperature above 28, 30 and 32 °C.

Based on the simulations, calculation of the electrical energy consumption for air transport has also been made, see Figure 7. Energy consumption for heating is not included as the natural ventilation system is only used when there is a need for cooling (week 16-45). It is seen that the introduction of natural ventilation has a potential for reducing the energy consumption for ventilation by almost 60% (~ 40 kWh/m^2 year). To achieve these savings in practice, this requires that the mechanical ventilation system can be switched off – this is therefore an important design criterion for hybrid ventilated buildings.

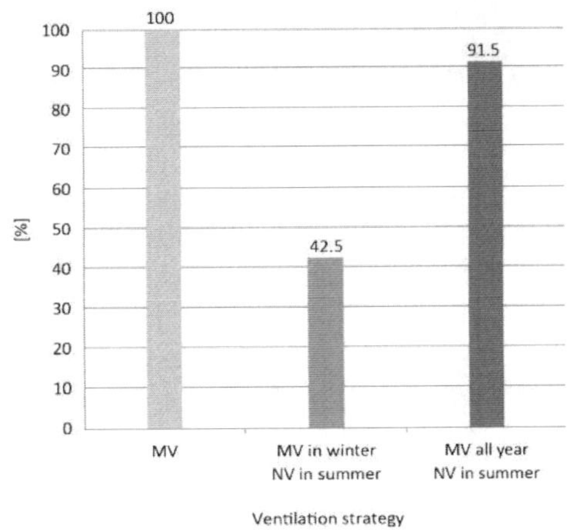

Figure 7. Energy consumption for air transport (indexed).

The results also show that in Model 3, where both natural and mechanical ventilation is used during the summer an additional reduction of 87 hours above 28°C could be achieved. This strategy does of course increase the energy consumption for the mechanical ventilation system significantly, but an 8.5% reduction compared with the pure MV system is still achieved. This is due to the fact that it is possible to turn off the mechanical system in selected periods.

The results suggest that there is a significant potential for reducing energy consumption in hallways of shopping centres by the use of natural ventilation. It is noted that the savings would have been even higher if the natural ventilation system had been designed into the building from the early design phase.

MEASUREMENTS

In reality the hallways are split into eight indoor climate zones. In each zone four or five combined temperature and CO_2 sensors are placed. These sensors form the basis for calculation of an average value for the temperature and the CO_2 level in the actual zone. The values are used for regulating the natural ventilation system and deciding whether the mechanical system should be turned on or off.

Figure 8 shows an example of a sensor placement.

Figure 8. Example of placement of temperature/CO2 sensor and detail showing the sensor.

Figure 9 shows actual measured temperatures in four representative hallways for one of the warmest summerweeks in Denmark, 2012 (week number 33). In this week, the hallways were ventilated using both the natural and the mechanical ventilation system (hybrid ventilation).

Figure 9 clearly shows that the natural ventilation system is capable of conditioning the thermal indoor climate in the hallways during hot periods in the year. The local outside temperature (measured on the roof) in the end of the week is close to 35 °C while the indoor temperature, despite the high heat loads, is kept below 28 °C in general.

Note also that the room temperature experienced by the occupants will feel lower than the calculated temperature due to the air speed which it is possible to establish when natural ventilation is used. The air movement may reduce the experienced temperature with up to 2-3°C. This additional cooling effect is beneficial during hot periods and will be experienced as a cool breeze by the occupants.

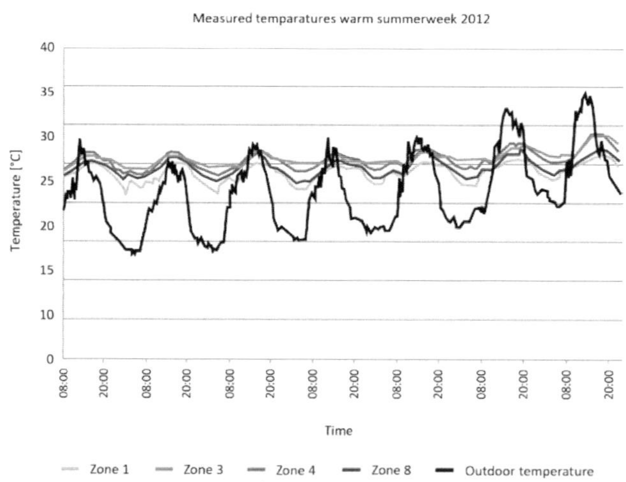

Figure 9. Measured temparatures in four representative hallways in a warm summerweek in august 2012.

Figure 10 shows the number of warm hours for one of the hallways - zone 3. The temperature has been logged for the period 01.09.2011 - 31.08.2012 – i.e. a whole year. It is noted that zone 3 is one of the warmest areas - which is also indicated in Figure 9. During this period, the system is ventilated using both the natural and mechanical ventilation system. Note, that Figure 10 compared with Figure 6 use a different temperature interval for the bars. This has been chosen since there are no temperatures above 30°C in the building – in fact the temperatures are almost entirely below 28°C.

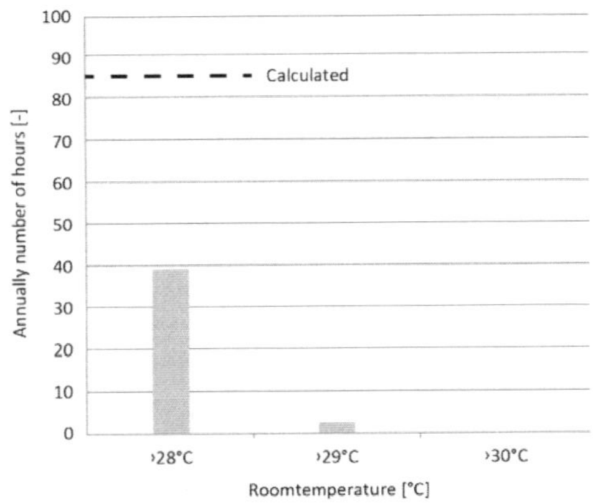

Figure 10. Annually number of hours above 28, 29 and 30 °C for the period 01.09.2011 - 31.08.2012 in hallway 2.

The number of hours with room temperatures above 28°C is approximately 40 for the year 2011 - 2012. In the dynamic simulation the annual number of hours was approximately 85 for simulation model 3 where mechanical ventilation was active all year and natural ventilation during summer and transient seasons. It is noted, that no recordings of the temperature levels in the hallways before the installation of the natural ventilation systems were made. Thus, the measured data from the building may only be compared with the simulation results.

However, based on the results after the installation of the natural ventilation system, it is clearly indicated that the theoretical calculations are accompanied by a real improvement in the indoor thermal climate. It is also noted that employees working in the centre has expressed that the indoor thermal climate has improved.

It is unfortunately not possible to present data on the electricity consumption as this was not recorded before the implementation of the natural ventilation system, and due to the fact that after the implementation a number of other energy optimization projects have been carried out. However the measured temperatures indicates a dramatic improvement in the thermal indoor climate and since natural ventilation is able to supply fresh air without energy consumption in situations where the room or building in consideration needs cooling the indoor climate has improved without increasing the energy consumption.

DISCUSSION

This particular case study is a renovation project where natural ventilation is established in hallways with existing mechanical ventilation. Natural ventilation was therefore not included from the early design phase. For a new shopping centre, where natural ventilation is included from the early design phase, the effect of the natural ventilation system can be even higher. For instance more openings can be established in the facades to increase the capacity of the ventilation and the relation between inlet air and the building constructions can be optimized for optimal night cooling.

For a new shopping centre it might as well be possible to include the shops in the natural ventilation strategy maybe for night cooling. E.g. small openings in the facades can let the outdoor air into the shops, where it cools down the construction and inventory during the night so the energy demand for cooling is reduced the following day.

For many low-energy buildings we see an almost year-round cooling demand. For these buildings we expect natural ventilation to have an even greater potential since the energy demand for ventilation almost can be eliminated as the electrical energy consumption for natural ventilation is zero.

Looking at a building in a life cycle and life cost perspective (LCA and LCC) natural and hybrid ventilation systems have a highly performance as well partly because the material consumption for a natural and hybrid ventilation system often is much less than for a mechanical ventilation system. This has been demonstrated in recent analysis of office buildings [7] [8].

CONCLUSION

This case study indicates a large potential for natural ventilation and hybrid ventilation in hallways in shopping centres as well as in similar rooms and buildings.

The dynamic building simulation shows that adding natural ventilation to the existing mechanical ventilation system can reduce the number of hours with operative room temperatures above 28°C with more than 80%. Measurements of the thermal indoor climate in the first year (September 2011 - august 2012) support the building simulation. In this year, the actual results outperform the expected results from the simulations, and the building owner has expressed his satisfaction with the improvements in the thermal indoor climate.

The dynamic simulation further shows that improving the thermal indoor climate can result in energy savings since the electrical energy consumption for air transport can be reduced. The size of the energy saving potential depends on the quality of the thermal indoor climate desired by the building owner – and they will be by far greatest if the mechanical system is switched off during the summer months. It has been demonstrated that doing so might result in an annual reduction of hours with operative room tempareatures above 28°C of approximately 70% and energy savings of almost 60%.

REFERENCES

[1] Shopping centre in the heart of Hanover. 2008. See ECE - sustainable - success. 2009/2010. Page 12.

[2] Office building in Copenhagen. 2009.

[3] BSim2002. Building simulation program developed by SBi, Denmark. Package of easy-to-use and flexible programs for evaluating the indoor climate and energy conditions as well as the designing of the heating, cooling and ventilation plants.

[4] DRY: Design reference year.

[5] SBi. By og Byg Anvisning 202. Naturlig ventilation i erhvervsbygninger 1. udgave, 2002.

[6] CIBSE. AM10. Natural ventilation in non-domestic buildings.

[7] Drivsholm C. and Tranholm, G.T. 2008. Naturlig ventilation kan reducere CO_2 udledningen med 40%. HVAC 7 2008, 28. (English titel: Natural ventilation can reduce CO_2 emmision with 40%).

[8] Larsen, A. F., Tranholm, G. T. and Hagelskjær, S. 2009. Hybridventilation er fremtiden i kontorbygninger. HVAC 6 2009. 52 - 55. (English titel: Hybrid ventiation is the future in office buildings).

FAÇADE-INTEGRATED VENTILATION SYSTEMS IN NORDIC CLIMATE

Michael Gruner[1] and Matthias Haase[*2]

1 Link Arkitektur
Street address
Oslo, Norway

2 NTNU
Alfred Getz vei 3
7465 Trondheim, Norway
**Corresponding author: Matthias.haase@sintef.no*

ABSTRACT

The work evaluates the applicability of façade-integrated ventilation systems in a Nordic climate. For this purpose the state of the art of façade-integrated ventilation (FIV) and demands for ventilation system in Norway and criteria for an comprehensive evaluation are identified. In this framework agreements between national requirements and system-specific performance are assessed. The evaluation investigates indoor environment and comfort with focus on aspects of indoor air quality. Energy efficiency and emission efficiency are evaluated by comparison with a centralised ventilation system.
The used tools in this work include "ESP-r", "Simien" for dynamic simulation of building performance and energy performance
The results of the evaluation show that current systems do not comply with all requirements of the Norwegian building code and related regulations. Some aspects need adaptation to local requirements. However, good performance and many possibilities can be expected in other fields e.g. indoor environmental comfort and user satisfaction since advanced principles are exploited.
The technology has an enormous potential. It might be an alternative if there is demand for high expectations on indoor environment and conventional ventilation systems are not applicable.
The technological limits of façade-integrated ventilation are not reached yet. Possibilities of further development of the concept itself and related technologies are outlined in the work

KEYWORDS: FAÇADE, INTEGRATION, VENTILATION, SIMULATION

INTRODUCTION

As buildings become more and more thermally super-insulated and airtight, ventilation accounts for an increasing portion of heat loss and energy consumption. Strategies have been developed to decrease the energy demand connected with ventilation. In case of mechanical ventilation systems the focus lies on the optimisation of transported volume, treatment and distribution of air within the system. The discussion about minimised airflow rates and demand controlled operation is in the centre of attention [1].
Other concepts like hybrid ventilation integrate ´free´ natural ventilation to provide a robust and sustainable strategy. Aim is to benefit from the best of both, mechanical and natural ventilation. Highlighted is the positive response from users, the possibility of individual control and the transparency of the ventilation's response. Less dependency on mechanical systems and increased flexibility while reducing costs are strong motivations in this respect [2].

Primary goal of most of these concepts is to ensure a comfortable indoor environment with acceptable indoor air quality. Especially in non-domestic buildings high requirements on the indoor air quality meet a high energy demand [1].
Decentralised ventilation systems are proposed to be a state-of-the-art technology which has shown favourable performance in the field. Flexibility and individual control are combined with high energy efficiency and reduced need for space. In the last decade the technology has proven to be mature and successful in operation. This good performance was experienced so far only in the context of Central Europe. If it can be also an option in another climate like Norway will be investigated in this work.

METHODOLOGY

The building is a low-energy office building with ambitions for lowest possible energy demand. Oslo as standard reference location was selected as geographical position. For the "Simien" simulation could be drawn on the standard climate condition for Oslo/Blindern [9]. This climate data is not available for "ESP-r" where the climate file for Oslo/Fornebu is used which has been retrieved from the US DoE website. Here seasons and typical weeks have been determined in agreement with the procedure described in [3].

Ventilation

Centralised systems with demand-controlled ventilation suffer from constraints mainly related to the throttle valves in the ductwork which are necessary to control the air distribution. The dampers in ventilation systems can neither be completely opened nor closed. As a consequence the airflow rate through the damper is limited to a range of 30 to 80% of the maximum airflow rate. Furthermore, the control of dampers within the distribution network of a centralised DCV system is very complex as discussed in length by Grini and Wigenstad (2011) [4].
The common practice to deactivate the ventilation outside occupancy is not in compliance with TEK 10 § 13-3 [5]. A minimum airflow rate of 0.7 $m^3/(h \cdot m^2)$ is required to remove material and interior related emissions. Grini and Wigenstad (2011) argue that the background for this regulation is not clear. The airflow rates were not related to EN 15251 regarding the procedure described in Annex B.1.2 [6]. However, a requirement for an airflow rate of 0.1 to 0.2 $l/(s \cdot m^2)$ can be found in EN 13779 as secondary option to ventilation with two air volumes prior to the occupancy [7]. In the related standard NS 3031 this is taken into account as the operation time for ventilation starts 6:00 to supply fresh air two hours before working hours (Dokka, Berg and Lillelien, 2011) [8;9]. Additionally the minimum airflow rate outside occupancy is required which poses the question why both measures, morning boost and basic ventilation rate are used in NS 3031 [8].

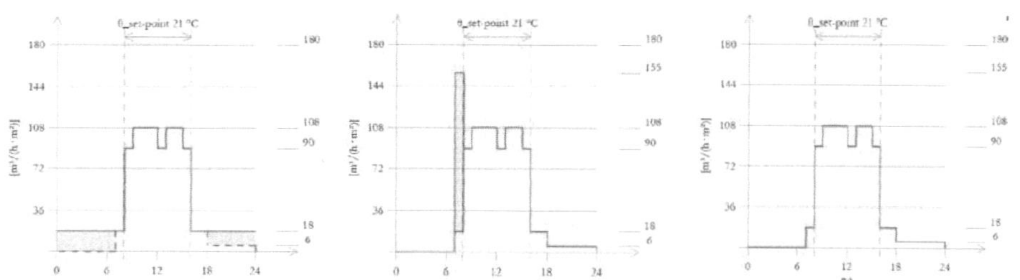

Figure 1. Illustration of different ventilation strategies.

Preventing growth of micro-organisms in the air handling unit is considered a reason for continuous operation. Reviewed research on the alternative operation strategies (continuous vs. night shut-off) shows no clear advantage for one strategy [4].
However, the necessity of regular cleaning and maintenance was highlighted in this context. These issues do not apply to FIV. Units can run for short periods only outside occupancy which is advised by EN 13779 as third alternative. Running one of the two units for 18 minutes an hour at an airflow rate of 60 m^3/h can provide the required 0.7 $m^3/(h \cdot m^2)$ for the exemplar room, for instance. Furthermore, dampers and ductwork are not present and hence problems linked to them can be neglected.
The influence of the different strategies on the thermal comfort and energy performance was investigated. The continuous operation is represented in the base case of the "ESP-r" model and will be compared with the alternatives of a morning boost and total deactivation outside occupancy (figure 1).

Modelling and simulation

For whole year energy performance simulations the software "Simien" version 5.010 is used. "Simien" is based on the the method for dynamic simulation as described in NS 3031 and has been validated against EN 15625 [10].
The detailed simulation and parametric studies are performed with "ESP-r" utilising the airflow network model. "ESP-r" was developed at the Energy Systems Research Unit (ESRU) at the University of Strathclyde [11]. In this work the most recent distribution 11.11 for GNU Linux was used. Optical properties for the glazing and shading ("complex fenestration constructions") are obtained from "GSLedit" using specifications for glazing and shading from "Pilkingtion" and "Warema" [12].
One single exemplar room is simulated instead of an entire building. This approach reduces complexity, improves quality control and allows flexibility for multi-criteria assessments [3]. (Hand, 2010). The exemplar room represents a generic office room for three persons with two FIV units in the uppermost floor of a fictitious office building in Norway. The generic room is not project specific allowing the simulation of conditions in new buildings as well as in energy-focussed refurbishment of existing buildings. The setting can be considered as worst-case scenario. The room is exposed to outdoor conditions both, on the façade and via the ceiling (roof) while having a high occupancy (8.64 m^2 per person).

Construction

Buildings elements facing outside are super-insulated. Walls and the roof feature U-values of 0.08 $W/(m^2K)$ and the U-value of the window glazing and framing is 0.8 $W/(m^2K)$. As typical in Norway, the wall construction is a light-weight timber construction with low heat storage capacity. The size of windows is guided by TEK 10 § 14-3 limiting the window area to 20 % of the usable area BRA [5]. The triple-pane glazing has a g-value of 0.45 and external shading with 80 mm Venetian blinds and automatic control triggered 200 W/m^2 insolation limits unwanted solar gains. The concrete slab at the ceiling is fully exposed. A raised floor with parquet flooring is assumed for the energy calculations and results in a light-weight floor construction (in the detailed simulation the floor is a separate zone). Internal walls are light-weight partitions with gypsum plasterboard finish. Thermal bridges are considered with 0.03 $W/(m^2K)$. All building materials are considered normal-emitting.
The value for airtightness n50-value is 0.4 h-1 which is considered easily achievable with proper quality of the construction. For "ESP-r" the n50-value must be converted to infiltration in normal pressure conditions. Therefore 0.4 h-1 is multiplied with the screening factor e = 0.07 in agreement with NS 3031. The resulting value for n4 is approximately 0.03 h-1 and applied to the zone of the raised floor only.

Model

The façade of room is 4.80 m wide representing four common Norwegian façade raster of 1.20 m [13]. To investigate different usage scenarios the exemplar room spans over two single office room modules with 2.40 m width. The partition walls and the connecting central wall depend on the usage. The height of the room is 3.00 m.

The determination of the room depth and the eventual total space volume is a result of considerations related to the FIV system to find an optimum starting point for a parametric study. The room could be designed according to heating, cooling or ventilation capacity of the FIV units. However, heating and cooling are not crucial parameters because firstly, the units are powerful enough to cover the loads and secondly, compensation with auxiliary systems is possible. In contrast, designing a system according to indoor air quality dictates high benchmarks when using TEK 10 and EN 15251.

Most commercial systems have three-stage pre-sets. Common settings are 60, 90, 120 m^3/h outdoor air supply. A first literature review in DeAL (2008) shows that airflow rates higher than 90 m^3/h may cause acoustic issues [14]. Therefore this value is considered as design airflow rate during occupancy. Up to 120 m^3/h airflow rate may be used outside operation hours e.g. for night cooling or higher occupancies.

The dimensioning airflow rates follow TEK 10 § 13-3 – during occupancy 26 m^3/h per person and 3.6 m^3/h per square meter floor area for materials and outside occupancy 0.7 m^3/h per square meter floor area. The size of the room was chosen to find reasonable matches between the required airflow rates and the air volumes of the FIV units. In the base case which will be used in the energy calculations two units with each 90 m^3/h and a total of 180 m^3/h supply air to an office with an occupancy of three persons representing a demanding setting with 8.64 m^2/person.

In summary, the dimensions of the exemplar room are 4.80 x 5.40 x 3.00 metres, the floor area is 25.92 m^2, and the heated air volume 77.76 m^3. Windows are treated differently in "Simien" and "ESP-r". In "Simien" each window measures 2.20 x 1.20 m. In "ESP-r" a window measures 2.40 x 1.20 m whereof 10 cm frame all around result in a glazing area of 2.20 x 1.00 m. The exterior surface/volume ratio A/V equals 0.52 m-1, which makes the room an ambitious environment for studies.

Schedules and internal loads

Various schedules and loads are used for different simulations due to their significant impact on the performance. Initial input values for the energy performance simulations with "Simien" have been reviewed since the test room does not include secondary spaces like corridors, storage, etc. [15].

"ESP-r" allows very dynamic schedules and loads. This will be taken into account for the detailed simulation focussing on transient conditions. Design values for human loads and equipment from EN 13779 will be used. The possibility to express the delay between presence and actuation (See also further discussion of the time delay TD-OFF in Halvarsson, 2012 was discarded to reduce the complexity but also because instant response from FIV system can be expected [16].

Schedules

The occupied period in NS 3031 is scheduled as 12 hours a day five days a week in 52 weeks per year. Default settings in "Simien" are set from 6:00 until 18:00 and are used for the simulations there. Since normal working hours are from 8:00 until 16:00 this results in two hours overhang before and after the working hours. It is assumed that the ventilation system starts two hours before the occupancy begins at 8:00 and remains active one hour after the occupancy has ended at 17:00 [9]. For an advanced representation of DCV a scenario was developed with an occupied period from 8:00 until 17:00. This schedule is assigned to the

case studies with index "b". ′Unforeseen′ occupancy is assigned then in the hour between 16:00 and 17:00. This is in good agreement with with measured data for operation hours in offices by Halvarsson (2012) [16].
The presence during the occupied period is estimated in NS 3031 by reducing the standard values by 20 %. PR 42 assumes 60 % actual presence to establish input values for lighting loads and ventilation rates. The results of Halvarsson (2011) reinforce this assumption. Based on his profile for high occupancy on page 164 a slightly modified and simplified occupancy pattern was developed for the "ESP-r" simulations. The assumption of variable occupancy equivalent to 2 minutes per hour in the evening was kept.

Equipment
Dokka et al. (2009) assumes the use of equipment (a total of 100 W / person, same as used in EN 13779 for design load) for 80 % of the occupied period in the primary area while in the secondary area the load is assumed as 2 W/m² during the occupancy. Therefore the gains in the scenarios PR42 and FIV of the energy simulations are also decreased to 80 W per person. In the detailed simulations the loads follow the percentages of the schedule assuming 100 W/person as 100%.

Lighting
Dokka et al (2004) refers to "Lyskultur" and uses 6.4 W/m² as initial value for the primary area without control and 100 % occupancy. In the case study PR42_0 this values is used adjusted for 60 % presence. For case studies with higher precision (PR42_a, PR42_b, FIV_a, FIV_b) loads for the single months are respecified using an "Ecotect"/"Radiance" daylight simulation of the test room in combination with EN 15193. In a second step the values have been altered depending on presence (60 %). The last column is the reference value from PR 42 with static values for each month. (Table 1)
For "ESP-r" 6 W/m² are assumed in case of 100 % presence throughout the year. The hourly values depend on the dynamic schedule.

month	1	2	3	4	5	6	7	8	9	10	11	12	mean	default
based on 100% presence	6.1	5.6	5.0	4.5	4.5	4.5	4.5	4.5	5.0	5.5	6.0	6.2	5.2	6.4
Adjusted for 60% presence	3.7	3.4	3.0	2.8	2.7	2.7	2.7	2.7	3.0	3.3	3.6	3.7	3.1	3.8

Table 1. Monthly internal gains from lighting according to EN15193.

RESULTS

Energy consumption

A FIV system with two different schedules (FIV_a and FIV_b) based on detailed design input parameters is compared with an ideal TEK 10 compliant centralised system. Two reference settings are based on standard values in the corresponding sources NS 3031 and PR 42 (TEK10_DCV, PR42_0). They are adjusted for DCV and primary area. Others reference cases (PR 42_a and PR 42_b) use design values equivalent to the FIV case study for a direct comparison of performance. Cases FIV_a and PR 42_a will be in the main focus as both allow a direct comparison.

Heat loss
Figure 2 shows the heat loss budgets for the case studies FIV_a and PR 42_a.

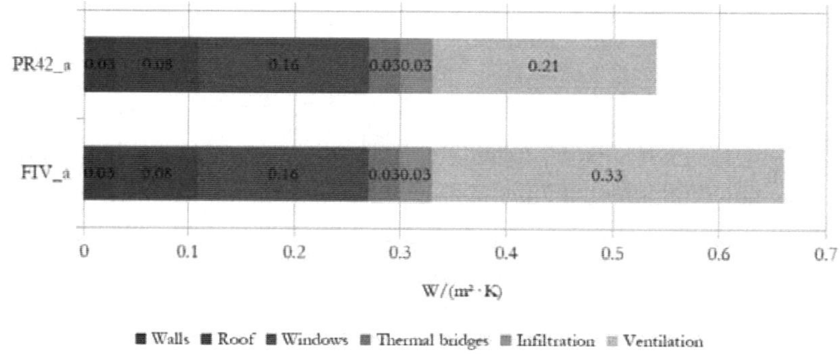

Figure 2. Heat loss of the two cases (PR42_a and FIV-a).

A heat loss factor of 0.5 W/(m^2·K) as specified by PR 42 for office buildings cannot be achieved with either system due to the location of the exemplar room and the resulting heat loss to the roof. Case study FIV_a reaches a heat loss factor of 0.65 W/(m^2· K) while the reference scenario PR42_a reaches 0.53 W/(m^2· K). The ventilation heat loss of the FIV system accounts for half of all losses.

Heating and cooling

An overview of the simulation results is given in Table 2.

	Specific energy demand			Specific installed capacity		
	Space heating [kWh/(m2a)]	Ventilation heating [kWh/(m2a)]	Ventilation cooling [kWh/(m2a)]	Frost protection [W/m2]	Ventilation heating [W/m2]	Ventilation cooling [W/(m2]
TEK10_DCV	7.8	11.4	3.8	-	19.5	23.0
PR42_0	9.1	9.1	4.0	-	21.8	24.2
PR42_a	6.2	10.5	3.4	-	17.5	19.5
PR42_b0	9.4	11.7	3.1	-	17.0	19.5
FIV_a	5.3	24.3	2.4	12.6	36.2	18.7
FIV_b	9.4	20.3	2.7	9.8	36.2	18.7

Table 2. Specific energy demand and required heating/cooling capacities.

The space heating demand is approximately equal for both systems. However, the heating demand of FIV units is significantly higher than centralised systems. The ventilation heating demand is up to 2.5 times higher due the low efficiency of the heat recovery. This also affects the power requirements. The FIV case studies require 46 to 49 W/m^2 installed ventilation heating capacity where 36 W/m^2 account for the heat exchanger compared to the less than 18 W/m^2 of the comparable centralised system. However, even assumed 50 W/m^2 installed heating capacity result in approximately 650 W per unit which does not affect the applicability of FIV as the nominal heating capacities for FIV units begin at ca. 800 W (Appendix B). Furthermore, it has to be considered that these values do not include frost protection.

In general, the cooling demand is very low compared to the heating demand due to the night cooling strategy. The necessary capacities for corresponding centralised and decentralised systems are close to 20 W/m^2. The dependency of the cooling demand on the transported air volume is apparent comparing the results for FIV_a and FIV_b. If the schedules are optimised to the occupancy as in FIV_b then both, the energy demand for cooling and the installed power of centralised systems is lower than for the comparable case studies PR42_a and PR42_b.

Electricity demand for fans
The difference of the specific fan power between centralised and decentralised ventilation system is evident in the results for the electricity consumption of the fans (figure 3).

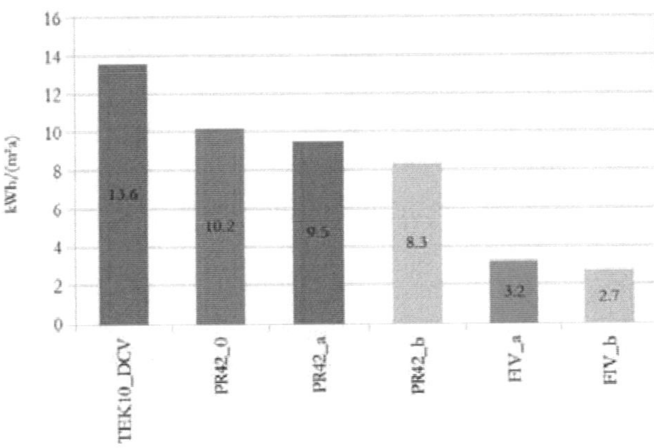

Figure 3. Specific energy demand for fans.

Decentralised systems consume on average only one third of the respective centralised system. Advantageous for the FIV systems are the low SFP of 0.6 kW·s/m³ due to the use of energy-saving EC fan motors.
Also noticeable are the quite different results for the case studies of the centralised ventilation system depending on the used input parameters. Using design values shows in general lower demands than the default values. Separate calculations for clearly defined primary and secondary areas instead of assumptions for the whole building (60 % primary area in Dokka et al. 2009 or 65% in Dokka, et al. 2011) might show more refined results.

Net energy demand
Figure 4 shows the total energy budgets for the investigated case studies.

Within the framework of this study the examined FIV case studies show higher net-energy demands than the centralised ventilation systems. The energy demand of FIV_a is 8% higher than the equivalent case study PR42_a. However, considering the quick response which is typical for FIV the comparison between PR42_a and FIV_b is possibly more realistic. The reduced occupancy from 12 to 9 hours leads to reductions of 11 to 13% energy demand. The detailed energy budgets for PR42_a and FIV_a are presented in figure 5.

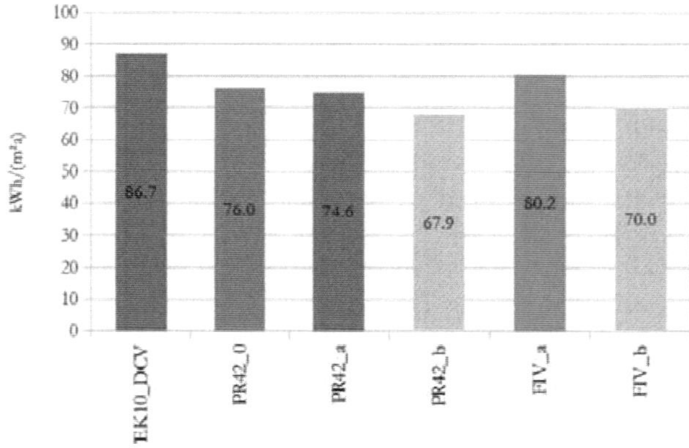

Figure 4. Specific net energy demand.

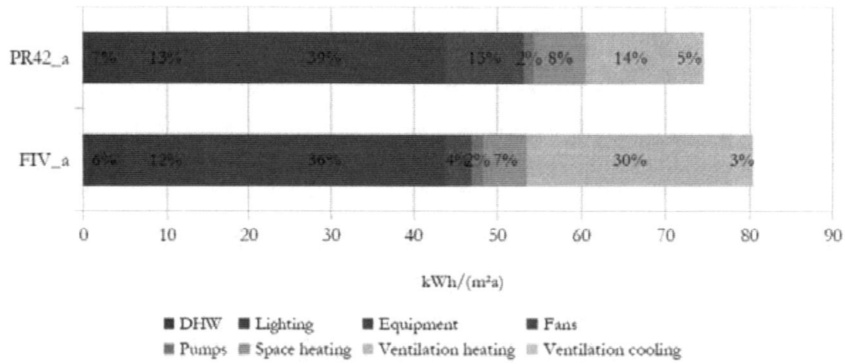

Figure 5. Specific net energy demand.

In both cases almost 60% of the energy budget is related to domestic hot water, lighting, and equipment where lighting is already adjusted to every single month. In case of the FIV scenario the positive impact of the little energy demand vanishes considering the high demand for heating which accounts for 37% of the total energy budget. In contrast, the centralised system has both, a low energy demand for fans and heating due to demand controlled ventilation. Apparent is the little demand for cooling. Reasons may be found in the impact of adjusted input parameters for schedules and internal loads due to demand control.

Heat recovery

Key issue to an improved energy performance of FIV systems is heat recovery. Most FIV units have heat recovery with low efficiencies due to the use of plate heat exchanger. Higher efficiencies have not been requested in previous applications. If a rotary heat recovery would be used then efficiencies can be further increased and frost protection is not required. The integration of rotary heat recovery in FIV is technologically possible as one system uses already rotary heat recovery with up to 62 % efficiency in a 160 mm thin FIV unit (LTG FVM, 2009).

This has been investigated further with case study FIV_a using PR42_a as reference. Figure 6 shows the net-energy demand for FIV_a equipped with either plate heat exchangers or rotary heat recovery with various heat recovery efficiencies. The maximum limit of the plate heat exchanger is assumed as 70 % efficiency.

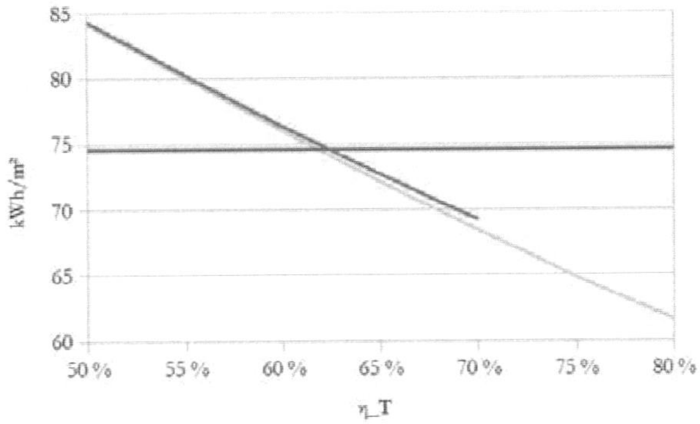

Figure 6. Specific net energy demand.

If the heat recovery of the FIV case study would reach an efficiency of approximately 62% then the net-energy demand equals the reference case with 80% efficiency. A rotary heat recovery unit with 70% efficiency results in an annual energy demand of 92% of the corresponding centralised system. With the same efficiency heat recovery as the reference the net-energy demand of a FIV system would be only 83% of the reference.

CONCLUSION

Aim of the work was to evaluate the applicability of façade-integrated ventilation in a Norwegian context. It can be seen that currently the reviewed FIV systems designed for Central European conditions do not comply with all requirements of the Norwegian building code.

Adaptations to Nordic conditions are necessary leading to custom-made systems which are possible if a demand arises. For some aspects a good performance can be expected also in Norway while other topics require upgrades.

Energy efficiency with respect to the net-energy demand is critical in Nordic conditions. Competitiveness with centralised systems cannot be achieved in this respect at the moment. If also the energy supply and related emission factors are considered then both systems can reach similar performances.

High indoor comfort and indoor air quality can be expected as state-of-the-art concepts and mature high-performance components are utilised. Acoustic and humidity conditions are critical where the first may be solved by proper planning. The possibilities of individual control and usability will results in high user satisfaction. On the other hand, cost and maintenance can be an issue especially in Norwegian conditions. If must be decided in the individual case if feasibility is given when the expenditures are traded off against user satisfaction.

However, the concept of façade-integrated ventilation has an enormous potential as the technology has not reached its limits yet. More research and development regarding design of components and operational strategies of the technology is necessary. Also the potentials of application have to be explored further. FIV can be a competitive alternative when conventional systems fail or are not applicable, in retrofitting for example.

ACKNOWLEDGEMENTS

The paper is based on the work done during a one semester master thesis in Sustainable Architecture at NTNU. Support from the Centre for zero emission buildings (www.zeb.no) is highly appreciated.

REFERENCES

[1] Hegger, M., Fuchs, M., Stark, T., Zeumer, M., 2008. Energie Atlas; Nachhaltige Architektur. Basel: Birkhäuser.

[2] Heiselberg, P. ed., 2002. Principles of Hybrid Ventilation . Aalborg: Aalborg University [pdf] Available at: <http://www.hybvent.civil.aau.dk/puplications/report/Principles%20of%20H%20V.pdf> [accessed 18 January 2012].

[3] Hand, J., 2010. The ESP-r cookbook. Glasgow: University of Strathclyde [pdf] Available at: <http://www.esru.strath.ac.uk/Documents/ESP-r_cookbook_dec_2010.pdf> [accessed 29 August 2011].

[4] Grini, C., Wigenstad, T., 2011. LECO; Behovstilpasset ventilasjon; Hvordan får man alle brikkene på plass?. Oslo: SINTEF Byggforsk [pdf] Available at <www.sintef.no/Publikasjoner-SINTEF/Publikasjon/?pubid=SINTEF+A20783>[accessed 26 January 2011].

[5] Direktoratet for Byggkvalitet, 2012. Veiledning om tekniske krav til byg gverk . [pdf] Available at: <http://byggeregler.dibk.no/dxp/content/tekniskekrav.pdf> [accessed 3 May 2012].

[6] Standard Norge, 2007. NS-EN 15251:2007; Indoor environmental input parameters for design and assessment of ener g y performance of buildings addressing indoor air quality, thermal envir onment, lighting and acoustics . Lysaker: Standard Norge.

[7] Standard Norge, 2007. NS-EN 13779:2007; Ventilation for non-residential buildings; Performance requirements for ventilation and room-conditioning systems . 2nd ed. Lysaker:

[8] Standard Norge, 2011. NS 3031:2007+A1:2011. Beregning av bygningers ener giytelse: Metode og data. Lysaker, Standard Norge.

[9] Dokka, T. H., Berg, M. O., Lillelien, E., 2011. Oppdragsrapport; Underlagsmaterialet for prNS 3701. Trondheim: SINTEF Byggforsk.

[10] Programbyggerne, 2008. SIMIEN. [online] Available at: <www.programbyggerne.no/SIMIEN/> [accessed 1 May 2012].

[11] ESRU, 2012. ESP-r. [online] (www.esru.strath.ac.uk/Programs/ESP-r.htm) [accessed 01.05.2012].

[12] Lomanowski, B. A., 2008. Implementation of Window Shading Models into Dynamic Whole-Building Simulation . Waterloo: University of Waterloo [pdf] Available at: <http://www.uwspace.uwaterloo.ca/bitstream/10012/4164/1/Lomanowski_Bartosz.pdf>[accessed 1 April 2012].

[13] Byggforsk, 2004. Byg gforskserien 344.110; Tilpasningsdyktige kontorbygninger . 2nd ed. Oslo: Norges byggforskningsinstitutt.

[14] Mahler, B., Himmler, R., Silberberger, C., Matt, C., 2008. DeAL; Evaluierung dezentraler außenwandintegrierter Lüftungssysteme; Abschlußbericht . Stuttgart: Steinbeis-Transferzentrum und Steinbeis Forschungszentrum Energie-, Gebäude- und Solartechnik [pdf] Available at: <http://www.stz-egs.de/wp-content/uploads/2008/09/bericht-deal-080820-_final.pdf> [accessed 12 January 2012].

[15] Dokka, T. H., Klinski, M., Haase, M., Mysen, M., 2009. Prosjektrapport nr. 42; Kriterier for passivhus og lavener gibyg g . Oslo: SINTEF Byggforsk [pdf] Available at <http://www.sintef.no/upload/Byggforsk/Publikasjoner/SB%20prrapp%2042.pdf>[accessed 13 May 2011].

[16] Halvarsson, J., 2012. Occupanc y Patter ns in Office Buildings; Consequences for HVAC system design and operation . Doctoral thesis. Norwegian University of Science andTechnology.

SWEDISH EXPERIENCE WITH AIR TIGHT TESTING: OVERALL SCHEME, TEST PROTOCOL, AND PRACTICAL EXAMPLES

Johnny Andersson
Technical Director
Ramboll
P.O.Box 17009
SE-104 62 Stockholm, Sweden

ABSTRACT

Starting already 1950 – i.e. for more than 60 years back in time – we have been using a probably quite unique quality assurance system in Sweden covering all aspects of building and installation technologies. Practically all buildings and their installations are performed according to the quality requirements in the AMA specification guidelines (General Material and Workmanship Specifications). The AMA requirements are made valid when they are referred to in the contract between the owner and the contractor.

The HVAC-part of AMA included requirements for tight ventilation ductwork systems already in the early sixties. Sweden has thus a long and unbroken tradition of demanding tightness of ventilation ductwork. During this long period, since 1966, the AMA tightness requirements have been raised in tact with technology improvements and increased energy costs.

But requirements and demands can be worthless unless they are controlled. The AMA requirements thus also include demands for tightness testing of the ductwork. The result of the tightness test has to be reported on standard protocol forms signed by the testing contractor.

And this has been shown to be very effective in raising the quality of ductwork. As e.g. shown in two EU-projects this long time focus on ductwork quality in Sweden has resulted in very low air leakage in normal Swedish duct installations.

And there are several reasons that justify the requirements for tight duct installations:

- Many studies have identified defective ventilation systems and insufficient airflows as a main reason for occurrence of sick buildings - the supply air needed to assure a good air quality should thus reach the areas where it is needed and not disappear along its transport through the building.
- The supply air flow has to cover the sum of total nominal air flow and the leak flow. With leaky ductwork this will lead to a considerable and costly increase of the needed fan power.
- Duct leaks can result in disturbing noise.
- When leaky supply and extract air ducts are installed above a false ceiling part of the air will take the simplest way, from the supply duct direct to the extract duct without bothering to pass through the connected rooms.

Swedish industries, building owners and authorities work together with the object to increase the quality of ventilation systems. Parallel with the voluntary AMA demands (i.e. voluntary until the contract has been signed) Swedish authorities 1991 thus started a compulsory system for ventilation control (OVK) in Sweden with aim to control and improve the function of ventilation installations. According to the ordinance (1991:1273) a control of the ventilation in most types of buildings has to be made before the installations are taken into operation and then regularly at recurrent inspections.

TESTING OF DUCTWORK AIR TIGHTNESS: OVERALL SCHEME

AMA – a sixty year old system, easy and accepted tool for specifying quality

The future building proprietor and his consultant use the AMA system (General Material and Workmanship Specifications) as a tool to specify the requirements for a new project. AMA covers all aspects of building and installation works of various kinds – e.g. buildings, installations, roads, and tunnels – and is split up in several parallel main parts covering these aspects, from building foundations to HVAC and electrical installations. The AMA requirements are made valid when they are referred to in the project contract between the owner and the contractor.

It all started way back in 1950 when AMA was born with "House AMA" and "Pipe AMA". During the following decades the group was extended with other AMA books for Ventilation, Ground works, Heating installations, Electrical installations and Refrigeration. Today these areas have been collected in four books, one of these is the HVAC AMA covering among many other aspects ventilation ductwork and e.g. protocol forms for reporting the result of tightness tests of ductwork.

Requirements in AMA are accompanied by advices in RA

Each of these AMA books (specifying the requirements) are accompanied by a parallel book (e.g. "RA – Advices and Instructions") comprising advices to the consultant on how to specify and quantify systems and components. In many cases they also give advice on how to choose a quality level. These RA-books serve also as check list for how to write a complete specification where the demands on the tenderer/future contractor are clearly shown in a way enabling him to calculate the cost for his contract commitments.

A common AMA-rule states that these requirements shall be expressed in measurable terms combined with control methods with known (and possible low) measurement errors. Another AMA-rule is that the cost for fulfilling the demands shall be calculable for the tenderers.
The AMA books are shown in Fig. 1.

Figure 1. The AMA family (VVS = HVAC), 1998 edition.

Requirements are raised in tact with technical progress and when economically motivated

The level of the AMA quality requirements are based on a kind of "80/20"-type rule. They should be suitable for most of the applications ("80 %") while for the rest they are either too high (the project, e.g. a building, has a very short planned life span and thus does not need the normal AMA quality) or too low (for projects where a higher quality is needed, e.g. laboratories and hospitals).

The AMA quality requirements are lifted when possible by technology progress and when found profitable for the owner on a Life Cycle Cost basis. Proposed increased requirements are established after they have been referred for consideration to a large number of owners, manufacturers, contractors, consultants and other interested parties. Wherever possible, AMA refers to relevant national Swedish standards and European norms. Twice a year the AMA requirements can be updated through the AMA-nytt (AMA News) Journal and added to computer-based specification tool used by the consultants. AMA is published by The Swedish Building Centre, a non-profit organization).

The AMA system follows the project through all phases of the building project – from design (supplying advices to the designer), to tender documents with specifications (these include references to relevant AMA clauses and advices on how to quantify), to installation (stating quality requirements for material/components and workmanship e.g. for duct connections, insulation of ducts or soldering of copper pipes), testing (e.g. measurement methods, protocols, e.g. for tightness test of ductwork), and maintenance (e.g. labelling and marking of components, cleaning of ductwork).

AMA vs. regulations

AMA is a voluntary complementary to Swedish statutory rules, regulations and specified building standards laid down by the authorities. Even if AMA and the regulations have common interest areas in securing sustainability and low energy use there is a difference between the two - the statutory rules are normally mostly focussed on reducing the risk of injuries to workers and users while AMA (not having to deal with those aspects) is focussed on reducing property damages and LCC-costs.

The AMA demands on ductwork tightness

Specifying requirements on ductwork tightness has a long story in Sweden; it has been specified as part of building specifications since the AMA edition 1966.

As described the AMA quality requirements are raised when possible by technology progress and when found profitable for the owner on a Life Cycle Cost basis. This is also true for ductwork tightness requirements:

In AMA version 1966 two "tightness norms" A and B, were defined. They were to be spot checked by the contractor; minimum tested duct surface area was 10 m^2.

In AMA 1972 the requirements were transformed into two "tightness classes" A and B (same as the EUROVENT classes today). Class A was the basic requirement for the complete duct system in the air handling installation (i.e. including dampers, filters, humidifiers and heat exchangers). It was advised to raise the requirement to meet Class B when the system operates for more than 8 hours/day or the air is treated (cooling, humidification, high class filters etc.).

In AMA 1983 a new tightness Class C was added to be used round ductwork larger than 50 m^2. Class B was to be used for round duct systems having a surface area smaller than 50 m^2 and also for rectangular ductwork.
Class A was accepted for visible supply and exhaust ducts within the ventilated room (i.e. not hidden above false ceiling).

In AMA 1998 a new tightness Class D was added being 3 times tighter than Class C. The use is not specified. It is an optional requirement for larger circular duct systems and where leakage can lead to hazards.

In AMA 2007 also rectangular ductwork has to meet tightness class C.

In AMA 2011 the ductwork tightness requirements are the same as in 2007.

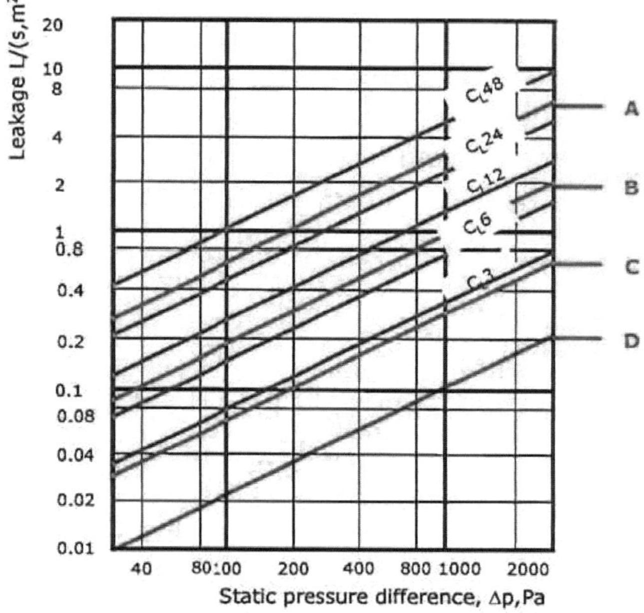

Figure 2. Eurovent Tightness Classes A – D and ASHRAE Classes C_L 48 - 3.

Often the duct manufacturers initially objected to these increased demands but as soon as one of them quickly announced that e.g.: "We can meet the new AMA requirements", the rest of the gang was forced to follow.

Specify what you can control – and do it!

Figure 3. Express your demands in measurable units and measure it! The building owner's consultant points out what part of the ductwork that shouyld be tested by the contractor

A ductwork system should not be specified to be tight – instead the permissible leakage rate at a specified test pressure is stated – that is possible to measure! Unless otherwise specified the tightness classes are to be in accordance with AMA demands (as stated above). AMA also states the requirements for the testing of ductwork tightness.

The general AMA rules stated above are thus relevant for ductwork tightness: "Express your requirements in measurable terms and control that you have got it!" and the other: "The costs and risks for the contractor to fulfil the requirements in the contract should be possible to calculate".

To ensure the quality of the duct system the leakage has to be verified; this is normally done either by the contractor himself or by a specialist engaged by the contractor. This is included in the contract and the cost is thus covered by the contractor. This test is undertaken as a spot check where the parts to be checked are chosen by the owner's consultant. For round duct systems 10 % and for rectangular ducts 20 % of the total duct surface normally has to be verified as specified in AMA.

Should the result of this test however show that the leakage is higher than allowed for the tightness class specified the contractor has first to tighten the leak points until the tightness requirement is fulfilled as verified by a new test of the same part of the ductwork – the contractor has consequently to redo his job until found OK! But in addition to this, another part of the same duct installation (e.g. another 10 % of a round duct installation) has to be checked. If this is also shown to be leakier than allowed the whole duct installation (i.e.100 %) has to checked and tightened until accepted.

And this increased testing, tightening and retesting can be costly for the contractor who is responsible for delivering an installation fulfilling the specification requirements. Quite naturally this has led to high quality tight ductworks – instead of risking this costly and time-consuming additional work the contractor aims to do a good job right away. Even though the tightness requirements have been raised during the past years, the new types of rubber gasket provided ducts and duct components have made the duct installation job easier, cheaper and more reliable than before.

The contractors do their best to avoid costly setbacks from inferior duct quality. The duct manufacturers are competing in inventing and marketing tight duct systems that are easy to install. Both circular and rectangular duct connections are provided with rubber gaskets that are very tight compared to older (and foreign) systems. New types of duct joints have reduced earlier laborious installation works.

In summary: the costs for the tests – the first 10 %, then another 10 % if not accepted and then at the end the whole system - is part of the contract and thus to be covered by the contractor. The mechanical contractor can either make the tightness test with his own personnel, provided he has equipment and skilled personnel to do that, or he can have it done by another specialized contractor. In both cases he has to cover the costs which can be quite considerable if the tests have to be repeated due to bad test results. The result of the leakage test shall be reported on AMA standard protocols and handed over to the owner.

Is it worthwhile to require and control the ductwork tightness? Yes!

There are several reasons that justify the requirements for tight duct installations:

Many studies of SBS, the Sick Building Syndrome, have identified defective ventilation systems and insufficient airflows as a main reason for the occurrence of sick building problems. The required supply air flow needed to assure a good indoor air quality should of course be delivered to the areas where it is needed and not be allowed to disappear along its transport through the building. This requires tight ducts!

In order to guarantee that the correct air flow is delivered to the room the supply air flow from the fan has to cover both the sum of the total nominal air flow and the disappearing leak flow. With leaky ductwork this will lead to a considerable and costly increase of the needed fan power (that has to be raised with up to third power of air flow increase).

Ductwork leak points can result in disturbing high frequency noise.

If leaky supply and extract air ducts are installed above a false ceiling part of the supply air will take the simplest way, from the supply duct with overpressure direct to the extract duct with underpressure without bothering to pass through the connected rooms.

And the AMA system has been shown to be very effective in raising the quality of ductwork. When compared to the result of tightness test of ductwork in Belgium and France as shown in two EU-projects this long time focus on ductwork quality in Sweden has resulted in very low air leakage in normal Swedish duct installations.

Comparison of the results from an EU project – Ductwork in Sweden was 25-50 times tighter!

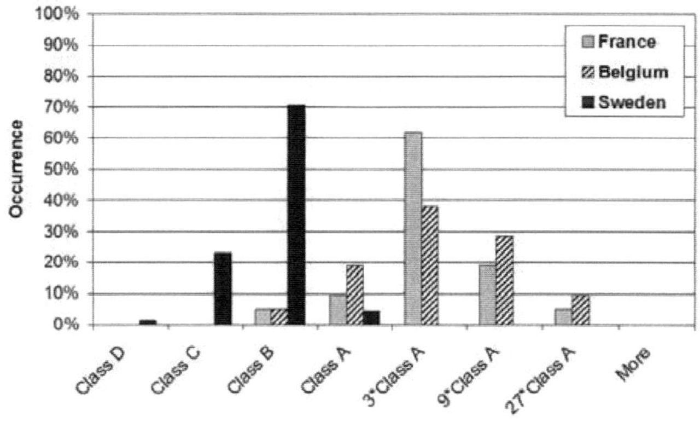

Figure 4. Comparison of the results from an EU project – Ductwork in Sweden was 25-50 times tighter!

As shown above duct leakage is detrimental to energy efficiency, comfort effectiveness, indoor air quality, and sometimes even to health. However, in most countries designers, installers, building managers and building owners, often ignore the benefits of airtight duct systems. Furthermore, as there are no incentives in most countries, over the years, this has (probably) lead to poor ductwork installations in a large fraction of the building stock.

In these countries ductwork installation is often undertaken using conventional in situ sealing techniques (e.g. tape or mastic), and therefore the ductwork airtightness is very much dependent upon the workers' skills..

The measurements and literature review performed within the EU-project SAVE-DUCT found that duct systems in Belgium and in France are typically 3 times leakier than EUROVENT Class A, see Figure 4. Typical duct systems in Sweden fulfilled the requirements for EUROVENT Class B and C and were thus between 25 – 50 times tighter than those in Belgium and France.

The answer to the question "Why this large difference between the countries?" is most probably that Sweden has required tight ducts, i.e. specifying how much they are allowed to leak at a certain test pressure, since the early sixties whereas in the two other countries tightness of ductwork is normally neither required nor tested.

CONCLUSION

Duct leakage is detrimental to energy efficiency, comfort effectiveness, indoor air quality, and sometimes even to health.

The Swedish long-time experience of quality approach to ductwork airtightness has shown that tight ductwork systems are cost effective and sustainable.

REFERENCES

AIVC-TightVent workshop in Brussels, 28-29 March 2012

ASHRAE. 2012. ASHRAE handbook—Fundamentals, Chapter 34 Duct Design. Atlanta: American Society of Heating, Refrigerating and Air-Conditioning Engineers, Inc.

Carrié FR, Andersson J, Wouters P. 1999. Improving ductwork. A time for tighter air distribution systems. AIVC, Air Infiltration and Ventilation Centre.

Malmstrom T, Andersson J, Carrié FR, Wouters P, Delmotte Ch. 2002. AIRWAYS Source book on efficient air duct systems in Europe. Belgian Building Research Institute.

VVS AMA 98. Allmän material- och arbetsbeskrivning för VVS-tekniska arbeten. AB Svensk Byggtjänst. Stockholm 1998. Copyright 1998.

VVS&Kyl AMA 07. Allmän material- och arbetsbeskrivning för VVS-tekniska arbeten. AB Svensk Byggtjänst. Stockholm 2007. Copyright 2007.

KEYWORDS

AMA, Ductwork, Tightness,

NATURAL VENTILATION AND PASSIVE COOLING SIMULATION IS NOT ANY MORE A PRIVILEDGE OF EXPERTS

Flourentzos Flourentzou[1], Bernard Paule[*1] and Samuel Pantet[1].

1 Estia SA,
Parc Scientifique EPFL
1015, Lausanne
**Corresponding author: flou@estia.ch*

ABSTRACT

Natural ventilation and dynamic temperature simulation of buildings was until now a priviledge of highly skilled building physicists. Combined simulation of both is even rarer.

A new software approach guiding the user to construct rapidely, easyly and intuitively step by step a valid thermal model, makes passive cooling design affortable to any building designer, even to a non building physics experts. Dial+ software offers the possibility to sumulate for the same room natural ventilation air flow, dynamic thermal behaviour and natural lighting performance. Behind its intuitive interface, Radiance for natural lighting, Cocroft algorithm for natural ventilation and EN ISO 13791 indoor air dynamic simulation give answers to the designers questions in order to optimise design and meet passive building specifications.

The article presents how the software was employed to optimise two similar office buildings situated one in Geneva and one in Nicosia. Both buildings consume less than 40 kWh/m2 of primary energy. The results show how essential is night cooling in both mediteranian and central Europe climates and quantify the contribution of passive techniques (insulation, glazing performance, solar shading, thermal mass etc) to the high energy performance. Both buildings were constructed and the performances verified in practice.

KEYWORDS

Potential for ventilative cooling strategies - Design approaches for ventilative cooling and case studies - Ventilative cooling in energy performance regulations - Summer comfort and ventilation.

INTRODUCTION

Estia SA in collaboration with the Laboratory of Urban Architecture and Energy Reflexion (LAURE/EPFL) has developed a suite of software tools aiming to simultaneously assess:

- Daylighting: **Dial+Lighting,**
- Natural window ventilation: **DIAL+*Ventilation*,**
- Thermal dynamic behaviour for winter or summer conditions: **DIAL+*Cooling***.

The possibility to combine natural and artificial lighting analysis with natural ventilation and dynamic indoor temperature analysis existed in the past, but it was a privilege for researchers, or building physics experts. The corresponding tools did require a high level of expertise, many weeks of training and many working hours to build credible models able to calculate correctly a given reality. Thanks to its intuitive interface, DIAL+ innovates and gives to non-expert users the opportunity to quickly model a complex room and to calculate the following results:

- Dynamic indoor temperature (ISO 13791)
- Number of overheating hours (SIA 382/1, EN 15251)
- Solar gains (taking into account fixe or movable shading and horizon obstacles)
- Stack effect natural ventilation airflow rate (dynamically hour by hour)
- Annual energy demand for heating and cooling (EN 15255, EN 15265)
- Daylight factor values (BREEAM, CERTIVEA)
- Daylighting autonomy (MINERGIE ECO, SMEO, LEED)
- Annual energy consumption due to artificial lighting (SIA 380/4, MINERGIE)

The three software modules are designed with a particular focus on user-friendliness. User data input is simple and straightforward and thus allows planners to correctly model the room and quickly analyse lighting and thermal indoor comfort, even without profound knowledge of building physics.

The software suite DIAL+ therefore represents a simple and efficient professional tool, ideal not only for proving the conformity to various norms and building labels, but also for optimising building design, especially in early planning phases.

Software interface description, methods and more references may be found in [1]. The software possibilities are presented through optimisation in the design process of two similar office buildings, one situated in Geneva and one in Cyprus.

PRESENTATION OF THE TWO BUILDINGS AND SOME INTERFACE WINDOWS OF THE SOFTWARE DIAL+

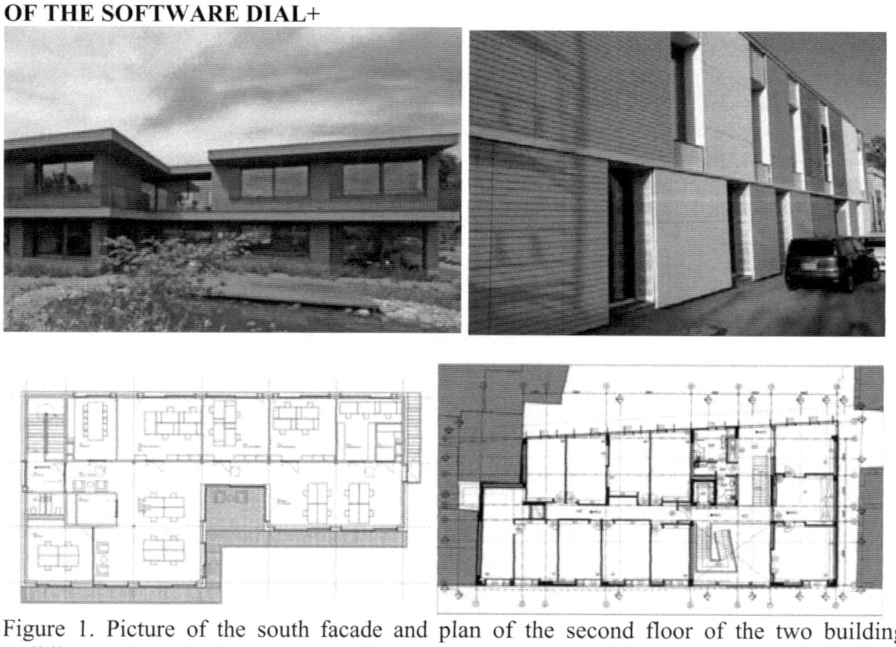

Figure 1. Picture of the south facade and plan of the second floor of the two buildings. Building Ge is on the left and the building Ni on the right.

Building Ge is situated in Geneva suburbs - Switzerland in an industrial area, while building Ni in Nicosia city centre – Cyprus. Both of them are low energy office buildings, privileging passive strategies and technical sobriety rather than high standard complex technical installations. Both buildings are naturally ventilated and heated / cooled with a standard solution common in the local market. Both buildings rely on a high energy-performance envelope, passive solar gains and passive cooling. Inauguration of building A took place in 2009, and of building B in 2012.

	Building Ge	Building Ni
Thermal insulation position	external	external
Wall thermal insulation thickness	16 cm	10 cm
Roof thermal insulation thickness	20 cm	10 cm
South glazing g value	0.6	0.4
North glazing g value	0.4	0.4
Window U value	1.3 W/m^2K	1.3 W/m^2K
Glazing light transmittance	O,7	0.7
Ventilation opening dimensions	55X170	40/60X300
South glazing dimensions	340 X 170	140X300
North glazing dimensions	340 X 200	2 X 140X300
Static solar shading south	100 cm top	60 cm top, 60 cm sides
Static solar shading north	-	Vertical fin 60 cm
South movable solar shading	G shading 0.2	-

Table 1. technical characteristics of the two buildings.

To compare the behaviour of the buildings we isolate a section of the building of 3.6 m large

Some windows from the software interface to enter the space characteristics

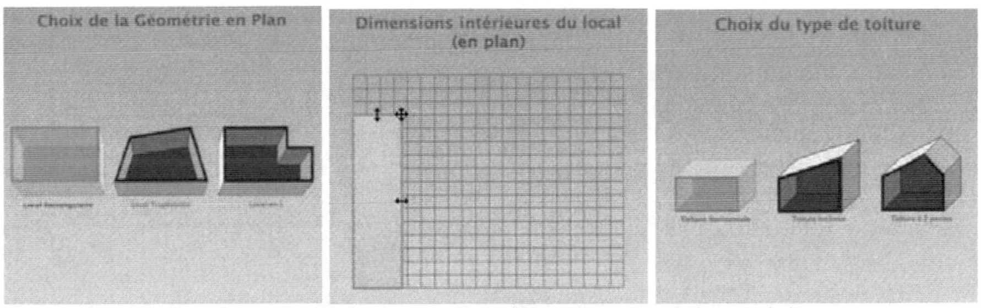

Figure 2. Intuitive interface makes it easy to enter the building dimensional characteristics.

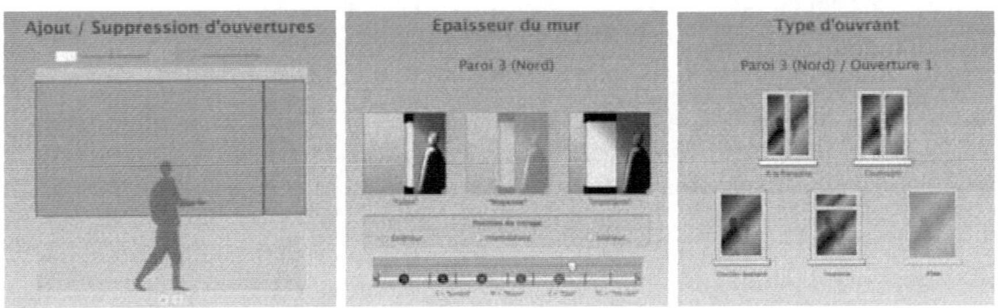

Figure 3. Some window characteristics.

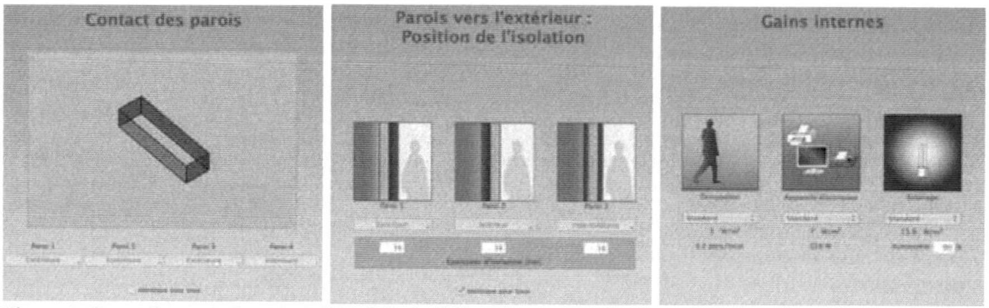

Figure 4. Some thermal characteristics of the building elements and some schedules of conditions of use.

These are some of the interface windows showing the intuitive spirit of the software. For every quantitative value, a qualitative description or a picture help the users to introduce a correct model corresponding to reality, even when if they are not specialists.

THERMAL BEHAVIOUR OF THE BUILDINGS SIMULATED BY DIAL+

Both buildings are simulated in the Swiss climate and in the Cyprus climate. Simulations consider no special free cooling strategy except of opening the windows when it is too hot.

Overheating hours

Overheating hours are calculated whithout and mechanical cooling.

In Swiss climate

In Cyprus climate

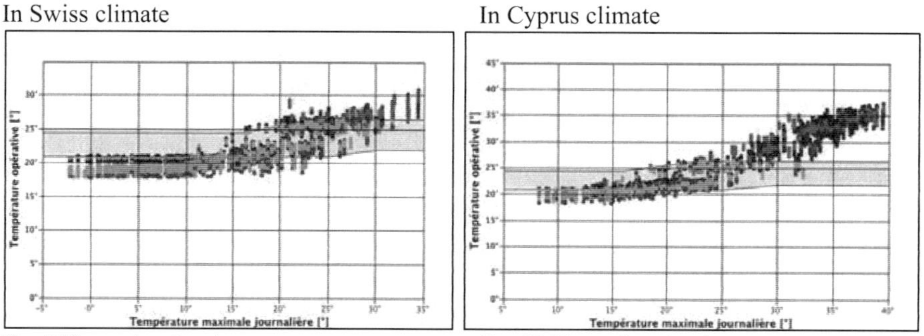

Ge: 240 overheating hours
Ni: 191 overheating hours

Ge: 1251 overheating hours
Ni: 1262 overheating hours

Figure 5. Temperature distribution of the buildings in Geneva and in Nicosia.

The behaviour of the two buildings is similar for both climates. The Ni building presents less overheating hours than the Ge building because the glazed part is less in the south façade. Although the Ge bulding has external solar protection, the Ni building is better protected by the sun.

Heating demand

Heating demand depends on thermal insulation and on solar gains.

In Swiss climate | In Cyprus climate

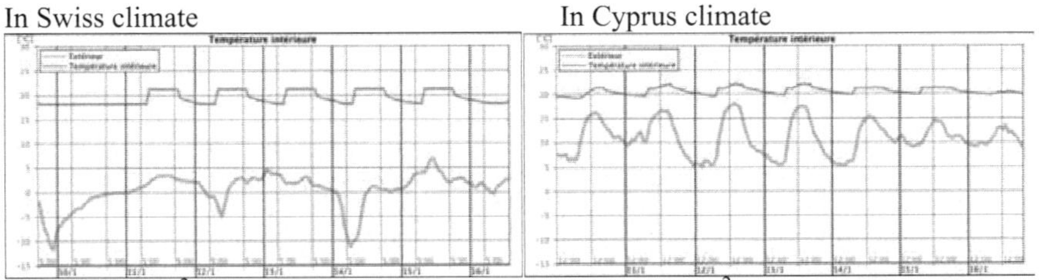

Ge: 56 kWh/m^{2}y of heating demand | Ge: 7 kWh/m^{2}y of heating demand
Ni: 65 kWh/m^{2}y of heating demand | Ni: 8 kWh/m^{2}y of heating demand

Figure 6. Heating demand of the buildings in Geneva and in Nicosia.

The Swiss building with 16 and 20 cm of thermal insulation saves only 1 kWh/m2 in the Cyprus climatic conditions. This means that over the passive standards of 10 cm of insulation more insulation has very little effectiveness. The Ni building with 10 cm of insulation would consume 16% more than the Ge building. This means that additional insulation in the Swiss climate makes sens.

Cooling demand

Cooling demand is calculated with a conventional air conditioning functioning during working hours to bring temperature down to 26.5°C.

In Swiss climate | In Cyprus climate

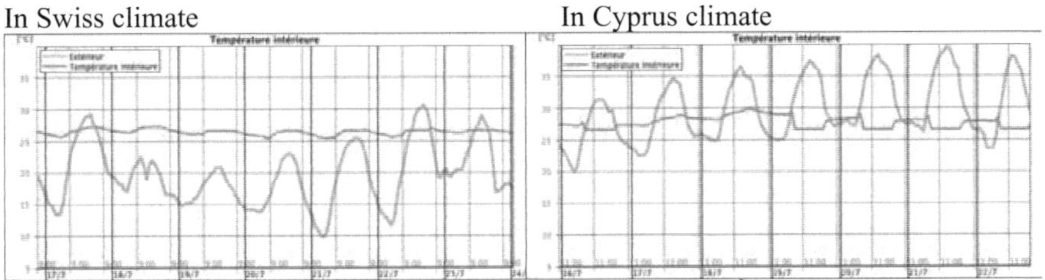

Ge: 3 kWh/m^{2}y for cooling demand | Ge: 65 kWh/m^{2}y for cooling demand
Ni: 3 kWh/m^{2}y for cooling demand | Ni: 35 kWh/m^{2}y for cooling demand

Figure 7. Cooling demand of the buildings in Geneva and in Nicosia.

These results show as for the calculation of the overheating hours that the Ni building is better adapted to the hot climate because of its smaller glazing at the south façade.

OPTIMISATION STRATEGIES

Night cooling in Switzerland

Opening the window during working hours when it is too hot for Ge building is not a sufficient strategy. Although opening the windows reduces the overheating hours from 491 to 240 compared with a scenario with the windows closed, poor comfort is still present. The solution is a mechanical cooling, consuming 3 kWh/m^{2}y of a free natural cooling by night ventilation. Night ventilation reduces to 5 the overheating hours, offering satisfactory comfort conditions.

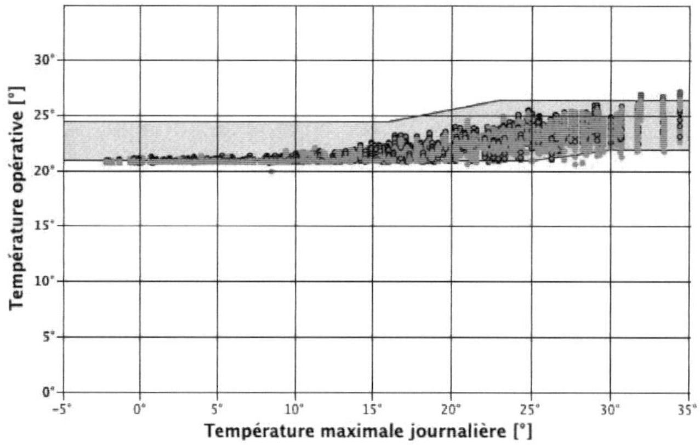

Figure 8. Overheating hours for Ge building with night ventilation strategy for cooling. Overheating hours are reduced to 5.

Night cooling in Cyprus

Night ventilation without air conditioning reduces overheating hours from 1262 to 707. It is a considerable reduce but still uncomfortable. However, with the use of air conditioning, night ventilation reduces the cooling demand to 16.5 kWh/m^2y instead of 36.5. Cooling power of night ventilation goes up to 1500W or 32 W/m^2.

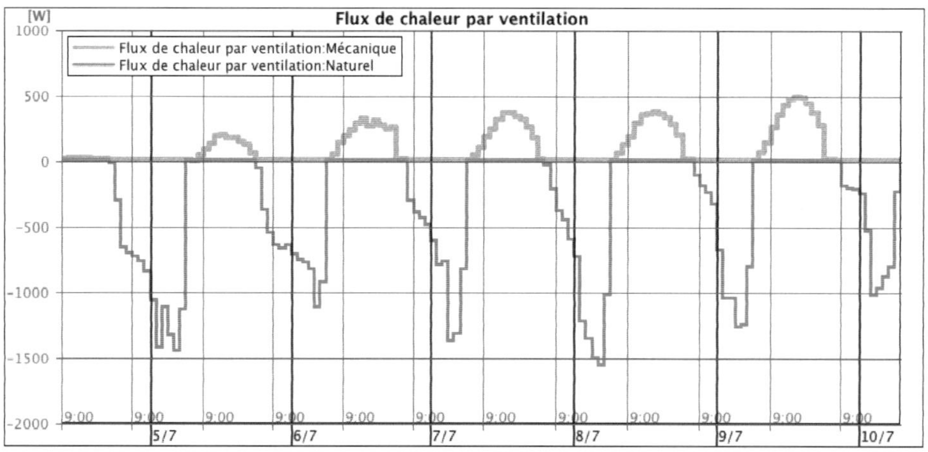

Figure 9. Free cooling power of night ventilation.

The presence of a ceiling fan, rising the tolerated highest temperature to 28.5 °C, before switching on air conditioning, reduces cooling demand even more at 6 kWh/m^2y instead of 16.5 kWh/m^2y with set temperature 26.5°C.

Heat recovery in Switzerland and in Cyprus

In Cyprus a heat recovery system of 80% efficiency reduces heating demand to 4.5 and cooling demand to 33 kWh/m^{2}y instead of 8 and 35 kWh/m^{2}y respectively. The total reduction is 5.5 kWh/m^{2}y or 13%. However, this increases energy consumption to run the fans. The total balance of primary energy is not positive.

In Switzerland, the same system reduces heating demand to 35 kWh/m^{2}y instead of 56. Cooling demand change is not significant. The reduction of 21 kWh/m^{2}y or 37.5% is significant and may pay the primary energy necessary to run the fans, especially when the recovered energy if of high primary energy content.

CONCLUSION

In order to illustrate the use of DIAL+ software we compared the dynamic thermal behaviour of tow similar office buildings in Cyprus and in Switzerland. This comparison shows the potential of the software, but it also shows that passive technics pay differently in a hot and in a cold climates. For hot climates, instead of airtightness and heat recovery, common in north and central Europe countries, natural ventilation and night cooling are much more efficient. Without night cooling strategy it is difficult to meat passive house standards.

REFERENCES

[1] Bernard Paul et al, 2012. *DIAL+Suite : a new suite of tools to optimize the global energy performance of room design*, Status Seminar, Zurich.

More references about DIAL+ methods are found in reference 1.

PROPOSAL FOR UPDATING FRENCH REGULATION CONCERNING AIRTIGHTNESS MEASURING EQUIPMENTS' CALIBRATION

Florent Boithias*[1], Sarah Juricic[1], Sylvain Berthault[1]

1 CETE de Lyon
Département Laboratoire d'Autun
Boulevard Bernard Giberstein
71404 AUTUN CEDEX

**Corresponding author: florent.boithias@developpement-durable.gouv.fr*

ABSTRACT

French standard for airtightness measurements is NF EN 13829. It is completed by French application guide GA P50-784, to set calibration rules more precisely, among other issues. This guide was published in 2010. To answer measurers' remaining questions, a Frequently Asked Questions web site was created by CETE de Lyon. Today, some weaknesses of French GA P50-784 have been clearly identified. It was therefore planed to update it, taking into account the experience gained in the last few years in dealing with airtightness in France, measurers frequently asked questions and ISO 9972 standard requirements in revision.
This article presents the conclusions of a working group created in 2011, lead by CETE de Lyon, which was in charge of updating French philosophy about calibration rules by optimizing the compromise between calibration's precision and costs. It is also planned to improve precision requirements for low air flow rates measurements. The content of this article must not be considered as the future French regulation; it is only the final proposal of the working group.

KEYWORDS

Airtightness, building, calibration

INTRODUCTION

Airtightness measurements in France are based on three complementary references:

- European standard NF EN 13829, published in 2001 [1] ;
- French application guide GA P50-784 of European standard, published in 2010 [2] ;
- On-line Frequently Asked Questions (FAQ) on CETE de Lyon's website.

First, NF EN 13829 details the protocol to be followed to measure airtightness. It also explains calculations which must be done to get the air-leakage flow rate at the considered pressure level, as well as the conversion of air-leakage flow rates into one single airtightness indicator : n_{50} [vol/h].
Then, GA P50-784 explains how to determine French legal airtightness indicator $Q4Pa_{surf}$ (also written Q4Pa_surf) and associated uncertainty. It gives sampling guidelines for grouped

and collective housing. Calibration rules are also specified in this application guide. It is published under French Association for Normalization (AFNOR) copyright.
Finally, answers to measurers' Frequently Asked Questions can be found on CETE de Lyon's website. These answers are not based on any reference. They complete and precise NF EN 13829 standard and GA P50-784. Authorized measurers agreed by French Ministry in charge of Construction must follow the FAQ's specifications, in case of doubt on some issues dealt with in the standards.
Today, some weaknesses of French GA P50-784 have been clearly identified. It was therefore planed to update it, taking into account the experience gained in the last few years in dealing with airtightness in France, measurers Frequently Asked Questions and future ISO 9972 [3] standard requirements, which should replace NF EN 13829.
This article presents the first conclusions of a working group created in 2011, lead by CETE de Lyon, which is in charge of updating French philosophy about calibration rules. The content of this article just gives some elements of the final proposal of the working group and should not be considered as part of the future French regulation. It begins by reminding today's calibration rules; then, weaknesses of these rules are detailed; finally, tracks of progress are proposed.

CALIBRATION RULES APPLICABLE TODAY

Calibration rules given in GA P50-784 are shown in Table 1.

Device	Measuring range	Required precision	Calibration or verification frequency	Authorized organisms
Barometer	900 – 1100 hPa	± 2 hPa	3 years	- Manufacturer - COFRAC accredited organism
Pressure-gauge	0 – 100 Pa	± 2 Pa	1 year	- Manufacturer - COFRAC accredited organism
Flowmeter	Unspecified	± 7 %	1 year	- Manufacturer - COFRAC accredited organism
Thermometer	-30 °C / +50 °C	± 1 °C	3 years	- In-house - Manufacturer - COFRAC accredited organism
Wind gauge	0 – 25 m/s	± 0,5 m/s	3 years	- Manufacturer - COFRAC accredited organism
Distance measuring equipment	0 – 20 m	± 1 cm	3 years or replacement for electronic devices	- In-house - Manufacturer - COFRAC accredited organism
Adjustable diaphragm	Unspecified	Unspecified	1 year	- Manufacturer - COFRAC accredited organism
Fixed diaphragm	Unspecified	Unspecified	5 years or less according to manufacturer specifications or in case of damage	- Manufacturer
Fan	Unspecified	Unspecified	5 years or less according to manufacturer specifications or in case of damage	- Manufacturer
Pipes' shape	Unspecified	Unspecified	Unspecified	- Manufacturer

Table 1. Calibration rules given by French application guide GA P50-784

The certificate delivered after calibration must describe the calibration protocol and state whether the device is in conformity or not. For non-measuring devices, like diaphragms or fans, it is only asked to report a list of tested components and associated corrections.

MAIN WEAKNESSES

First problematic issue is precision and reliability of calibration: French application guide makes it possible to choose between sending measurement devices either to the manufacturer or to an external COFRAC accredited organism. COFRAC is the French Accreditation Comity, member of EA (European Accreditation) and ILAC (International Laboratory Accreditation). The manufacturer is not able to be as reliable as a COFRAC accredited organism.
Secondly, calibration is not mandatory for fans and associated diaphragms. It is only required to check shapes, rotation speeds and stability, without any precision criteria nor operation range.
Thirdly, full airtightness measurement systems cannot be calibrated at once. Components can only be calibrated separately. However, manufacturers observed in some cases that deviation measured on a full system could be greater than theoretical deviation calculated from known components' deviations.
Fourthly, systems for air-flow rate mesurement have no specified measuring range. They can therfore be calibrated on a given range and then used on a wider range. This point creates difficulties to reach precision especially at low air-flow rates ranges, for which most of blowing-door technologies are usually not calibrated.
Finally, manadatory information given by the calibration certificate is not complete for the measurer to calculate uncertainty. For example, standard reference and calibration uncertainty, as well as devices' resolution, are not specified in the certificate.

TRACKS OF PROGRESS

The aims of the working group created in 2011 are :

- To ensure greatest precision of airtightness measurements;
- To consider any existing or future technology for airtightness mesurement;
- To minimize practical and financial constraints for measurers.

Precision is particularly crucial for future controls of airtightness for dwellings built after january 2013. According to French Thermal Regulation (RT 2012), those buildings will be measured at the end of their construction and their airtightness will have to be below:

- 0.6 $m^3/h/m^2_{cold\ surface}$ at 4 Pa differential pression between indoor and outdoor for individual dwellings;
- 1 $m^3/h/m^2_{cold\ surface}$ at 4 Pa differential pression between indoor and outdoor for collective dwellings.

COFRAC was identified by the working group as the only way to get fully transparent and reliable calibration, with all information needed to calculate uncertainty given on the certificate. It was therefore decided that COFRAC calibration would become mandatory for all components of measurement systems, to ensure excellent precision, except for barometers (for atmospheric pressure measurements), thermometers and distance measuring equipments.

Those equipments were considered as having lower impact on the final result than flowmeters, for example.
To enable calibration of all existing and future measuring systems, not necessarily using pressure-gauges as flowmeters, and also to ensure the consistency of the full acquisition chain, the group recommended that the systems should be calibrated without separating components. However, to limit practical constraints for measurers, who sometimes need to use flowmeters in association with many different fans, it was decided that the possibility of calibrating components separately would be preserved.
A two-possibility solution was therefore proposed:

- First, the full system has to be calibrated when sold to the customer, with a maximal tolerated error of 2 $m^3/h \pm 7$ %;
- Then, two choices are proposed :
  - Either the system is kept full every time it is used and can therefore be calibrated at once every year, with a maximal tolerated error of 2 $m^3/h \pm 7$ %;
  - Or system's components are not always kept together and must be calibrated separately, with frequencies and maximal tolerated errors reported in Table 2.

Device	Measuring range	Required precision	Calibration or verification frequency	Authorized organisms
Barometer	700 – 1100 hPa	± 2 hPa	4 years	- Category 1[1]
Pressure-gauge	{-100, -50, -10, 0, 10, 50, 100} Pa	1 Pa ± 1%[2]	1 year	- COFRAC accredited organism
Thermometer	-20 °C / +40 °C (3 steps on full range)	± 1 °C	4 years	- Category 1
Distance measuring equipment	0 – 20 m (0 – 100 m for telemeters)	± 1 cm		- Category 2[3]
Complete measuring system (full acquistion chain)	6 steps on desired range (3 steps per configuration[4] if many)	2 $m^3/h \pm 7\%$	1 year	- COFRAC accredited organism
Pressure-gauge (flowmeter)	Desired measuring range	1 Pa ± 1%	1 year	- COFRAC accredited organism
Wind gauge	5 steps on desired range	0,5m/s ± 3%	3 years	- Manufacturer - COFRAC accredited organism
Fan and associated aperture or cone	6 steps on desired range (3 steps per configuration if many)	2 $m^3/h \pm 4\%$	1 year or 2 years (still to be discussed)	- COFRAC accredited organism

Table 2. Calibration rules proposed by the working group

[1] Category 1: calibration or verification must be done in conformity to FD X 07-012 or FD X 07-011 (in-house, manufacturer or external organism).
[2] Percentage of measured value
[3] Category 2 : self-control with specific protocol and verification file.
[4] Configurations can be rings with different diameters, for example.

Each component and the entire system must not be used out of the measuring range used for calibration. This ensures precision at low air-flow rates, for example, because systems which are not calibrated on low ranges can no more be used. For very airtight buildings measurements, specific systems should therefore be bought by measurers and calibrated on low air-flow rates ranges.
It was observed that new rules proposed by the working group were quite complex, compared to previous ones. However, they were considered as the only way to meet the three objectives the group had defined. It is still unknown whether one of the two possible solutions for calibrating systems (full system or component-by-component) will be more expensive than the other. This could make measurers prefer one solution than the other on the long term.

POINTS STILL TO BE DISCUSSED

First, frenquency of calibration for fans must be precised: chosen period will be either 1 year or 2 years, depending on observed deviations on Building Services Research and Information Association (BSRIA, UK) database. It was decided that the period would not exceed 2 years in order to keep maximum competitiveness between the two possible calibration solutions.
As airtight door is never verified together with full measuring system, the group also imagined a service control procedure, which could become mandatory to check full system's consistency between two calibrations. A decision must still be made on this point.
Then, after having found the air-flow rate range usually used for dwelling measurement, specific calibration rules should be defined for higher and lower air-flow rates, in order to guide calibration organisms' investments.
Afterwards, fans with uncommon diameter should not be forgotten in calibration rules. Specific rules should therefore also be defined for them.
Finally, as there is no French organism able to do COFRAC calibration for fans at the moment, it was planed to set a deadline for the regulation changes to be applicable, so that organisms have time to adapt. Suitable deadline should also let enough time to measurers to calibrate their equipments and get aligned with new rules. This deadline is still to be defined.

ACKNOWLEDGEMENTS

The authors would like to warmly thank BSRIA for their useful cooperation. They would also like to thank all members of the working group for their interest and cooperation and Valérie Leprince for the work she did to prepare the first meetings. They also thank Florian Desfougères, who did an internship at CETE de Lyon and provided statistics about airtightness measurements in France.

REFERENCES

[1] AFNOR NF EN 13829:2001, Performance thermique des bâtiments – Détermination de la perméabilité à l'air des bâtiments – Méthode de pressurisation par ventilateur
[2] AFNOR GA P50-784:2010, Performance thermique des bâtiments – Guide d'application de la norme NF EN 13829:2001
[3] ISO 9972:2006, Thermal performance of buildings – Determination of air permeability of buildings – Fan pressurization method

OVERFLOW ELEMENTS: IMPACTS ON ENERGY EFFICIENCY, INDOOR AIR QUALITY AND SOUND ATTENUATION

Gabriel Rojas Kopeinig[*1], Mattias Rothbacher[1], Rainer Pfluger[1]

1 AB Energy efficient housing, University of Innsbruck
Technikerstrasse 13,6020 Innsbruck, Austria
**Corresponding author: gabriel.rojas-kopeinig@uibk.ac.at*

ABSTRACT

When planning ventilation systems for energy efficient housing, an appropriate design of the overflow elements between rooms is important as it influences ventilation losses, indoor air quality and sound attenuation between rooms. Based on calculation results of the natural in- or exfiltration rates through the building envelope as a function of the overflow element's flow resistance, this work proposes a maximal pressure drop of 2-3Pa for overflow elements in energy efficient buildings. Measurement data on sound attenuation and pressure drop for simple door or doorframe integrated overflow solutions is presented. It shows that a volume flow of 40m³/h can be achieved with a pressure drop ≤2Pa with simple and cost effective modifications to a standard door construction, while still complying with a normal sound attenuation requirement of 30dB.

KEYWORDS

Overflow elements, sound attenuation, pressure drop, flow resistance, infiltration, exfiltration, air distribution

INTRODUCTION

How does the design of an overflow element (OFE) influence the in- and exfiltration rates, and therefore the energy efficiency, the indoor air quality and the sound transmission? This work establishes pressure drop requirements for overflow elements of energy efficient houses. It also provides so far unavailable experimental data of sound attenuation and pressure drop for various simple and cost effective overflow solutions.

METHOD

Numerical calculations to determine the requirements regarding flow resistance

A numerical multi-zone model is used to calculate the natural and forced in- or exfiltration rates and the supply air distribution depending on the pressure loss of the overflow elements connecting the zones. Results shown herein are based on a 3-zone model in steady state as depicted in Figure 1. The following set of coupled power law equations [1] are solved using Newton's method:

$$\dot{V}_{IN} - \dot{V}_{OFE1} + \sum_j \dot{V}_{AP1j} = 0 \tag{1}$$

$$\dot{V}_{OFE1} - \dot{V}_{OFE2} + \sum_j \dot{V}_{AP2j} = 0 \tag{2}$$

$$\dot{V}_{OFE2} - \dot{V}_{OUT} + \sum_j \dot{V}_{AP3j} = 0 \tag{3}$$

With $\dot{V}_{IN}$ and $\dot{V}_{OUT}$ being the supply and the exhaust volume flow, $\dot{V}_{APij}$ and $\dot{V}_{OFEi}$ being the volume flow through the respective air path or overflow element. They can be calculated as follows:

$$\dot{V}_{OFEi} = C_{OFEi} \cdot \left|p_{Ri} - p_{R(i+1)}\right|^{n_{OFEi}} \cdot Sign\left(p_{Ri} - p_{R(i+1)}\right) \tag{4}$$

$$\dot{V}_{APij} = C_{APij} \cdot \left| p_0 + p_{Wij} - p_{Ri}(h_{ij}) \right|^{n_{APij}} \cdot Sign\left(p_0 + p_W - p_{Ri}(h_{ij})\right) \quad (5)$$

With p_0 being the ambient pressure, p_{Wij} being the wind pressure at the respective air path according to [2] or [3] and $p_{Ri}(h)$ being the pressure in room/zone i at a given height h:

$$p_{Ri}(h) = p_{Ri}(0) - \rho(T_i) \cdot g \cdot h \quad (6)$$

T_i is the temperature in zone i and g the gravitational acceleration. Note that for simplicity, the set of equations is based on a volume flow balance rather than on a physically correct mass balance. The validity of the model and the assumed parameters was checked with transient simulations with CONTAM, a multi-zone airflow analysis software [3]. Figure 1 shows the comparison with the results obtained with CONTAM for a two bedroom flat with 80m² with a transient weather file generated with Meteonorm 6.1 [4] for Vienna. The used parameters for 3-zone model are as presented below.

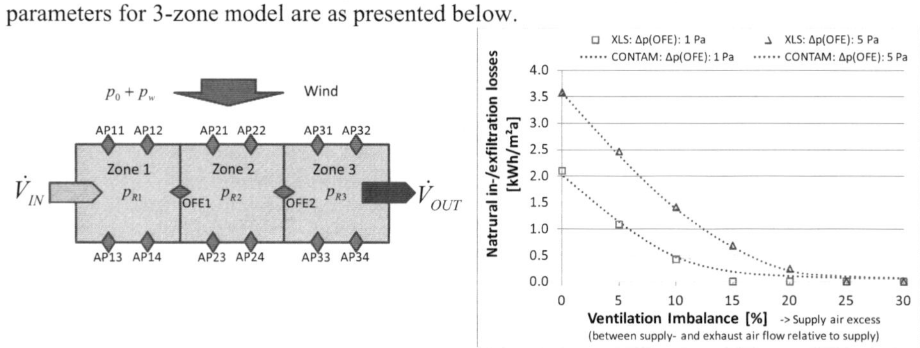

Figure 1: Left: Schematic sketch of 3-zone-model used for airflow analysis. Right: Comparison of natural in-/exfiltration losses calculated with the presented 3-zone-model (steady state) and with CONTAM (transient).

Experimental setup to measure flow resistance and sound attenuation

The tested overflow elements were installed in a double chamber test lab at the University of Innsbruck. The test lab, suited for sound attenuation measurements according to ISO 140-3, was also supplemented for pressure drop measurements. The test samples were mounted in an opening (205 x 85 cm) of the double shell concrete wall. If necessary, the remainig area was closed with a high sound-insulating dry-wall construction. A diffuse sound field was generated in one chamber. The averaged sound pressure levels were measured in both chambers. By determining the absorption area of the receiption chamber, the sound reduction index and the standardized sound level difference can be determined.

To determine the airflow resistance, one side of the opening of the double shell concrete wall was closed with an airtight board, producing a 230 x 140 x 25cm pressurizable air volume as illustrated in Figure 3. A low volume „blower door tester", type J225, from SI-special instruments GMBH, was used to apply a pressure difference across the test objects and to measure the volume flow. If required, an additional ventilator was used to „boost" the volume flow to up to 100 m³/h. The additional flow was determined with a pressure loss measurement over an initially calibrated tube.

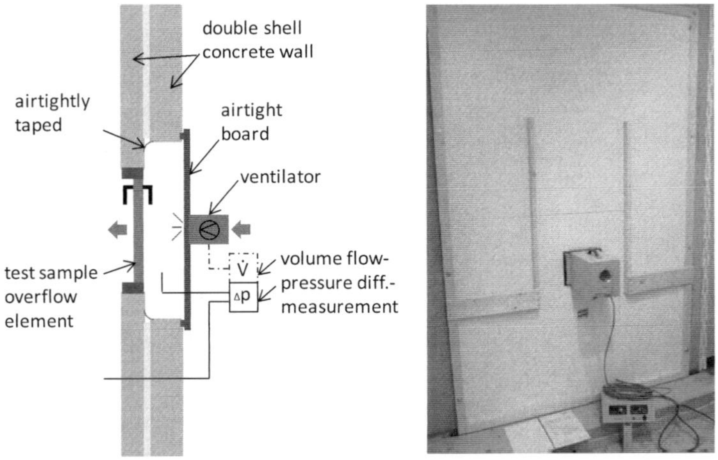

Figure 2: Experimental setup for measurement of the flow resistance

REQUIREMENTS ON OVERFLOW ELEMENTS

Energy efficiency

For the design of the ventilation system one needs to be aware that overflow elements influence the ventilation losses. The ventilation losses for a building with mechanical ventilation are determined by losses due to the mechanical ventilation system (i.e. ineffective heat recovery and forced in-/exfiltration due to imbalanced supply and extract flows), and due to the natural in-/exfiltration through the building envelope. They are shown in Figure 4 in dependence of the ventilation imbalance. The natural in/-exfiltration is driven by the difference of inside and outside pressure. The pressure difference in a natural ventilated building is imposed by vertical density gradients (stack effect) and wind pressure. For mechanically ventilated buildings the pressure difference is also affected by the flow resistance of the overflow elements (see Figure 1). Note, that the in- or exfiltration due to this flow resistance of the overflow elements could, in a semantic sense, be added to the forced in-/exfiltration. But for calculation and further analysis it is advantageous to include it in the natural in-/exfiltration, as it is done for the calculations of this paper.

The natural in-/exfiltration losses as a function of the ventilation imbalance are calculated for different flow resistances at the overflow elements, see Figure 4. The flow resistance is given as pressure drop at nominal volume flow per overflow element.
To determine the pressure drop requirements, the results are plotted versus the nominal pressure drop of the overflow elements in Figure 5. The natural in-/exfiltration rates for an airtightness level of n_{50}=0.5 h^{-1}, increase from 0.02 h^{-1} for a situation with no flow resistance, to about 0.03 h^{-1} for OFEs with a pressure drop of Δp=3 Pa. That is equivalent to an inrease of 10% of the total ventilation losses of an energy efficient system (heat recovery rate of 80%) and therefore tolerable. If the flow resistance across the overflow element is increased further, the in-/exfiltration losses augment even stronger (higher gradient), which is even more pronounced for buildings with little airtightness standards. E.g. if an overflow element with a pressure drop of 5 Pa instead of 2 Pa is used for a building with a n_{50}-value of 0.5 h^{-1}, the total ventilation losses would be equivalent to a building with an airtightness level of 0.8 h^{-1}.

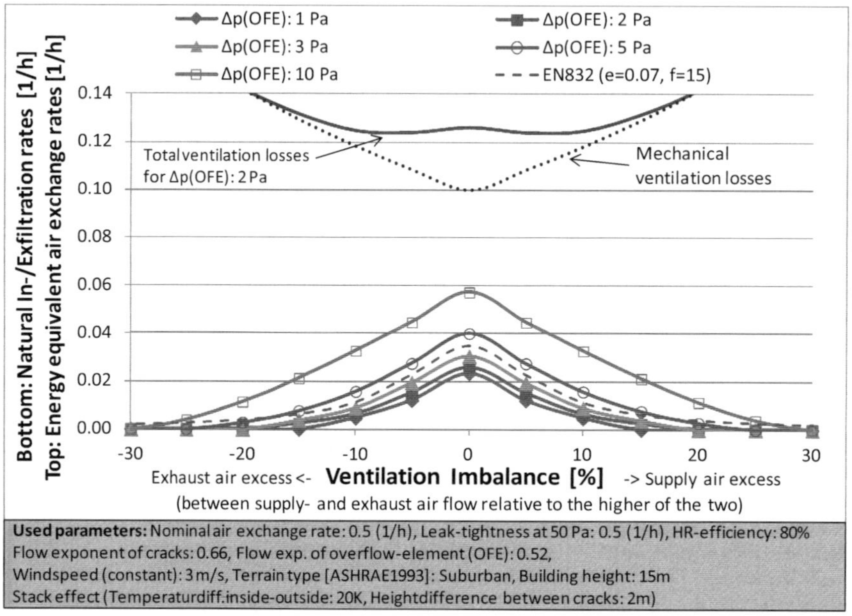

Figure 3: Natural in-/exfiltration rates for different pressure drops across the overflow elements (OFE). Also shown are the mechanical ventilation losses and the total ventilation losses for OFEs with a Δp of 2Pa.

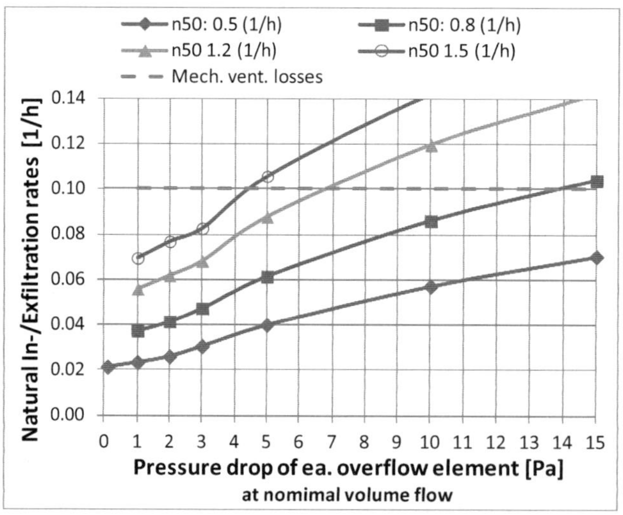

Figure 4: Natural in-/exfiltration as a function of the pressure drop for different airtightness levels.

Indoor air quality

Another aspect to consider when defining the specification for overflow elements is their effect on indoor air quality (IAQ). The supply air distribution is altered if the door to one of the supply air rooms is opened, see Figure 6. This is also true for the exhaust air distribution and exhaust air rooms.

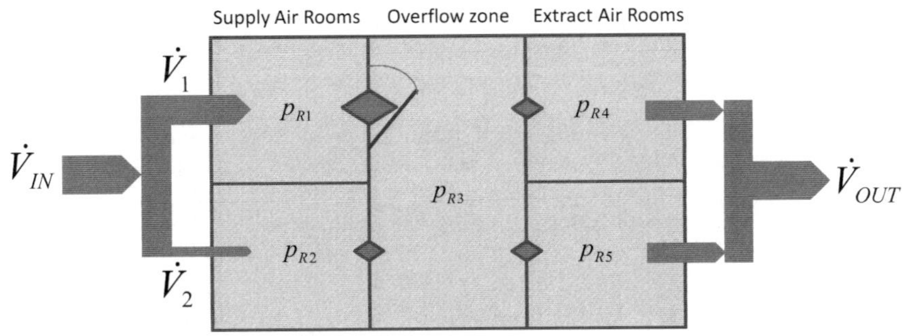

Figure 5: Schematic illustration of change in supply air distribution when a door is opened.

Figure 7 shows the misadjustment of the supply air volume flow in the room with closed door relative to its nominal volume flow. As the misadjustment is also dependant on the pressure drop over the supply air nozzle, the results are shown for different supply air nozzle pressure drops. To minimize electric consumption of the ventilators, the air distribution system for energy efficient housing should have little pressure losses. Recommendations for the pressure drop of supply air nozzles for passive houses are given with 10-15 Pa. If the misadjustment of the supply air is to remain below 10% (for a supply air nozzle with a Δp of 10 Pa), the pressure drop of the overflow element should be kept below 2-4 Pa, depending on the number of supply air rooms.
As a generalizable statemente we recommend a maximal pressure drop of 2-3 Pa.

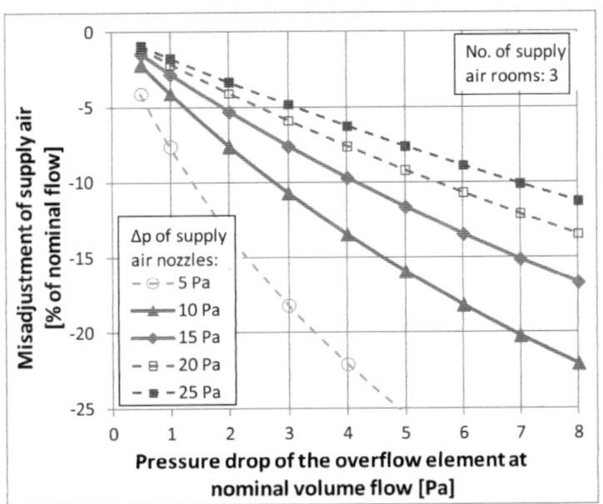

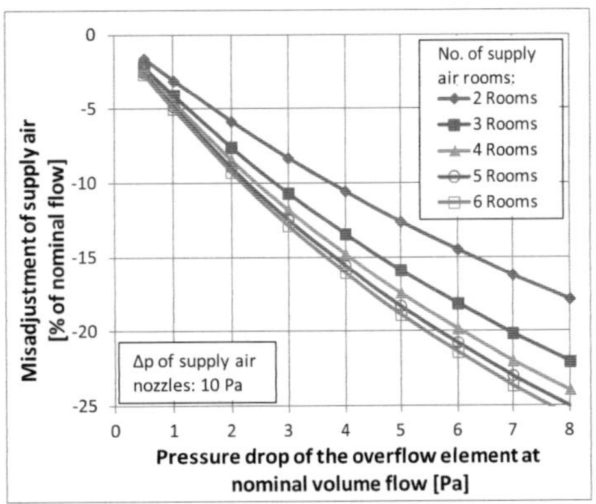

Figure 6: Misadjustment of the supply air flow for various supply air nozzle pressure drops (left), and for different numbers of supply air rooms (right). Shown is always the worst case scenario, i.e. the misadjustment in the room with closed door while the doors of the other rooms are open.

Sound attenuation requirements

In most countries there are no compulsory reglementation for acoustic attenuation requirements within a housing unit. Nevertheless, according to a recommendation in DIN4109 (1989) the acoustic attenuation R'_w between rooms of different usage within in one housing unit (calculated for a 8.26 m² wall and a 1.74 m² door) should be ≥30 dB for normal requirements and ≥35 dB for higher demands.
By nature, sound attenuation requirements are always opposed to low flow resistance requirements. Nevertheless, limited possibilities for increasing the attenuation, while keeping the pressure drop at an acceptable level, exist, even for simple overflow solutions. To improve the acoustical performance, the geometry of the inlet and the outlet of overflow-elements can

be designed to minimize the entry of sound waves. The sound propagation in the flow channel can be reduced by increasing the absorption area and by implementing directional changes.

Other requirements

Further requirements on overflow elements not quantified in this paper could arise from aspects like design, passing light, cleanability, possibilities of short-circuited air flow and draft risk. By limiting the pressure drop and therefore the air speed, a draft risk issue might only arise if the overflow element is placed close to the occupant's regular position, e.g. close to the standing area in the bathroom.

MEASUREMENT OF FLOW RESISTANCE AND SOUND ATTENUATION

To give ventilation planners and architects the necessary data for proper layout of the overflow elements, sound attenuation and pressure drop characteristics of typical and novel door or doorframe integrated solutions were determined experimentally. The tested OFE were selected out of an initial literature and market study, focusing on solutions that can be realized by simple modifications to a standard door construction and for which no measurement data was available. A description of the measured overflow solutions is given below. A summary of the parameterized results are listed in Table 1.

Tested overflow elements

- **Door gap (1):**

  This very simple solution is widely used, but often arbitrarily dimensioned, resulting in high pressure drops or an excessive deterioration of the sound attenuation.

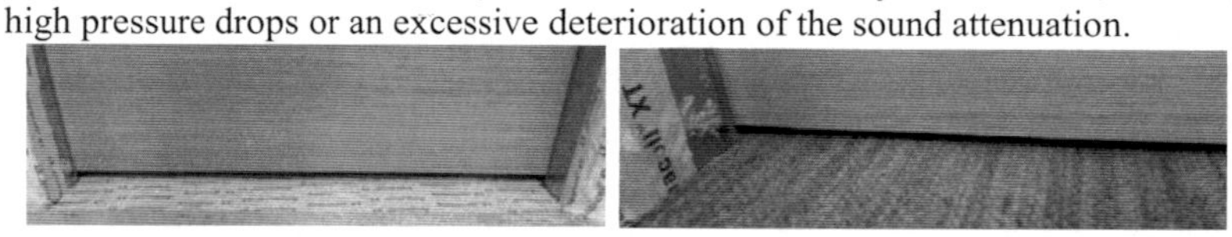

Figure 7: Measured door gap without (left) and with carpet (right).

A series of measurements with gap heights from 5 to 20mm was performed to give the following formulas as a layout aid to calculate the volume flow in *[m³/h]* as a function of the desired pressure difference Δp *[Pa]* as a planning aid:

$$\dot{V} = C * (\Delta p)^n \quad (1)$$

The parameters C $[m^3/h/Pa^n]$ and n can be calculated in dependence of the gap height h *[mm]* and the gap-length l *[m]* (=door width).

$$C = (k * h + d) * l \quad (2)$$

$$n = \frac{A}{h - B} + 0.5 \quad (3)$$

Note that the roughness of the gap surface does have an notable influence. The dependence of the gap depth (=thickness of door) can be neglected.

Modified wooden door frame (2):

This also quite popular solution was first published in [5] and consists of leaving a gap between lintel and wooden frame and to cut air in-/outlets into the top portion of the door frame. (Alternatively or additionally, it could also be realized at the sides of the door frame.)

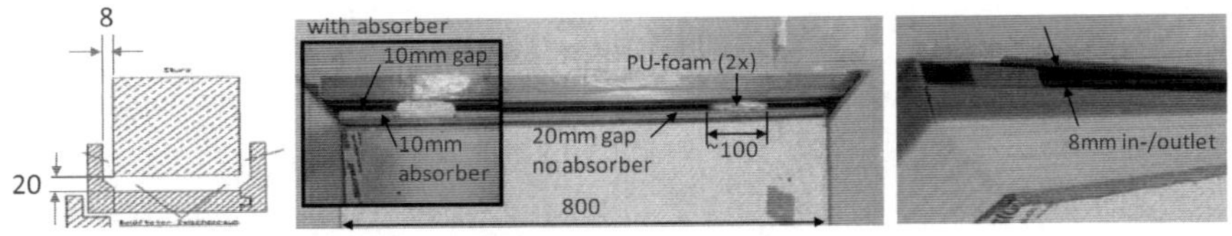

Figure 8: Modified wooden door frame. Left: Schematic drawing.
Right: Pictures of the test situation.

The volume flow *[m³/h]* for the tested design (2) can be estimated for a given pressure difference by

$$\dot{V} = C' * l * (\Delta p)^n \qquad (4)$$

with *l [m]* being the gap length, i.e. the door width. Note that the measured parameters might vary for other frame designs.

- **Modified metal door frame with shadow gap (3):**
This design was developed within this work at the University of Innsbruck and aims to combine an unobtrusive and simple design with a decent sound attenuation.

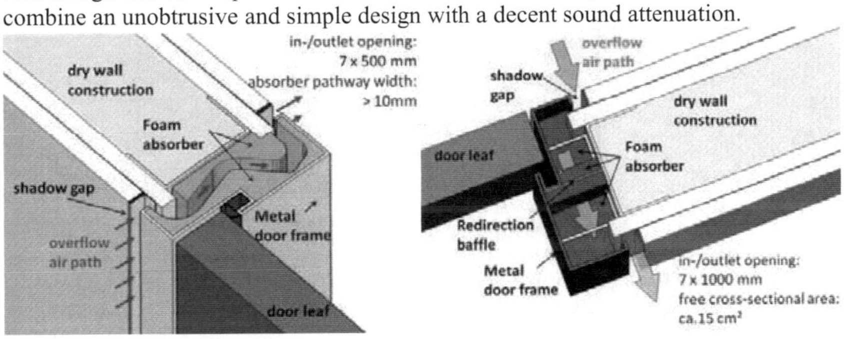

Figure 9: Two different versions of the modified metal door frame, as developed at the University of Innsbruck. The air in- and outlets are hidden within the shadow gap.

The volume flow *[m³/h]* for this design is given by Eqn.(4) with *l [m]* being the gap length. Design details are will be published in the final report of [6].

- **Special ventilation door frame (4):**
A solution developed and manufactured by Keller Zargen AG (Switzerland). So far no measurement data was available. The included supply air outlet was closed during the measurements. For a given pressure difference the volume flow *[m³/h]* can be calculated with Eqn.(1).

- **Rabbet gap (5):**
This solution has a reduced rabbet width at the top, leaving a gap between door-rabbet and door seal. In an "extended" version the rabbet was cut thinner all around, leaving only some short segments for a door to seal contact.

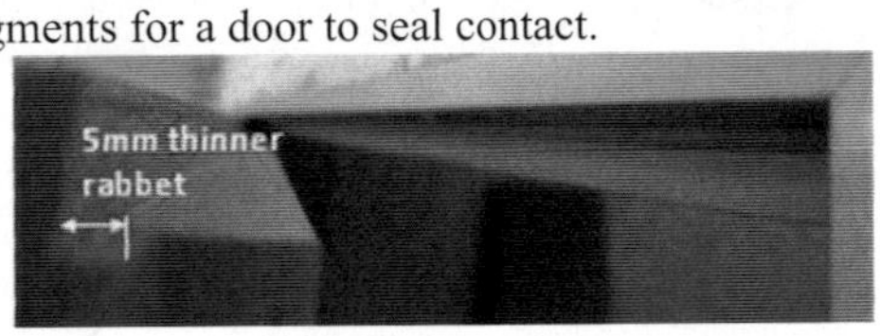

Figure 10: Picture showing the reduced rabbet at the top of the test door, leaving a gap between door and frame when closed.

For a given pressure difference the volume flow *[m³/h]* can be calculated via Eqn.(1).

- **Perforated door (6):**

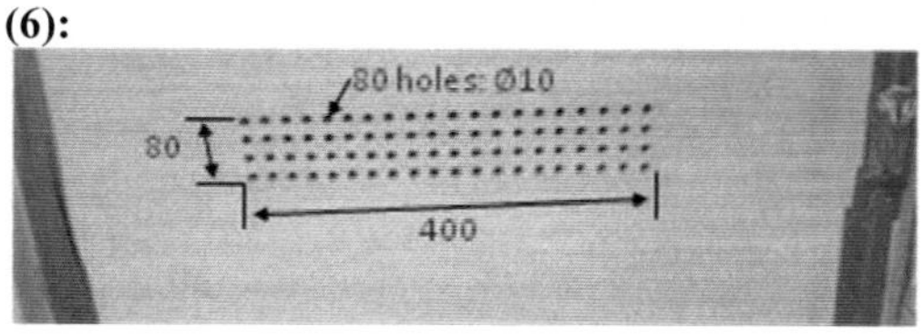

Figure 11: Picture of a door with a number of through holes at the lower region of the door as measured.

This simple solution consists of four rows of each 20 through-holes with Ø10mm. The resulting volume flow *[m³/h]* can be estimated via Eqn.(4) with x being the number of hole-rows (with 20 holes each).

- **Z-Blind (7):**

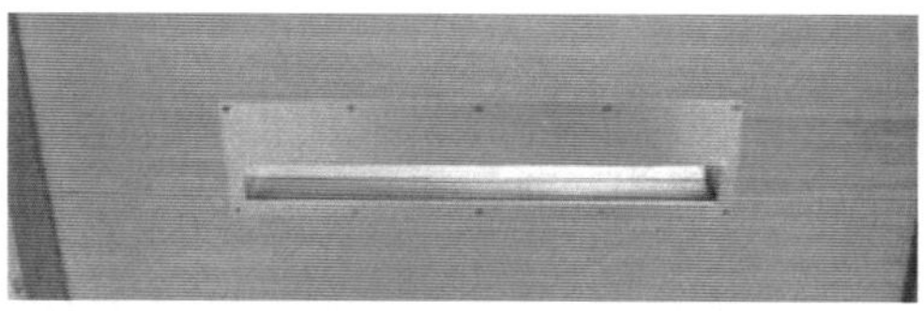

Figure 12: Picture of the measured Z-blind.

This overflow element is also an alternative to regular door grilles, and consists of a door blind (4) where the air flow is directed in z-shaped path supplemented with an 1 cm thick foam-absorber lining. The resulting volume flow *[m³/h]* can be calculated with Eqn.(1).

Measurement Results

The parameters obtained from the pressure drop and sound attenuation measurements are summarized in Table 1.

	Overflow element / Overflow solution	A	B	k	d	C/C'	n	$D_{n,e,w}$ [dB]
1a	Door with bottom gap	0.12	1.17	3.21	-0.08	-	-	-
1b	Door with carpet in bottom gap	0.12	1.17	2.51	-2.17	-	-	-
2a	Wooden door frame	-	-	-	-	20.2	0.58	36
2b	Wooden door frame with foam absorber	-	-	-	-	18.2	0.57	40
3a	Metal frame V1	-	-	-	-	7.0	0.59	37
3b	Metal frame V2	-	-	-	-	8.1	0.49	44
4	Special ventilation door frame	-	-	-	-	11.6	0.50	45
5a	Top rabbet gap with (door B)	-	-	-	-	4.5	0.57	34
5b	All-around rabbet gap (door B)	-	-	-	-	21.8	0.52	30
6	Perforated door (door A)	-	-	-	-	18.7	0.54	32
7	Z-blind (door A)	-	-	-	-	19.8	0.52	30

Table 1: Parameters of pressure drop (A, B, k, d, C, C'and n) and of sound attenuation ($D_{n,e,w}$) measurements.

For better comparison the measured data is used to calculate the resulting acoustic attenuation (R'_w) for a typical construction situation. The resulting R'_w and the achievable volume flow at a pressure difference of 2 Pa are listed in Table 2, Table 3 and Table 4. A typical construction situation was defined here as a 10m² dry-wall construction with a R_w of 41 dB and with a 80 cm wide door, resulting in a 8.26 m² wall and 1.74 m² door surface.

	Overflow in bottom door gap	R'$_w$ of door with gap [dB]	Resulting* R'$_w$ [dB]	Vol. flow [m³/h] at Δp=2Pa
1a	Door with 5mm bottom gap	23	30	18.4
1b	Door with 10mm bottom gap	20	27	36.4
1c	Door with 15mm bottom gap	18	25	54.4
1d	Door with 20mm bottom gap	15	23	72.4
1e	Door with carpet in bottom gap (remaining gap 10mm)	(23)	(30)	26.1

Table 2: Resulting acoustic attenuation and volume flow at given pressure drop for a door with bottom gap. R'$_w$ of door itself is 30dB. *Resulting acoustic attenuation as defined in the text.

	Door or door frame integrated overflow solutions including a 5mm bottom door gap	R'$_w$ of door with gap [dB]	Resulting* R'$_w$ [dB]	Vol. flow [m³/h] at Δp=2Pa
2a	Wooden door frame	23	29	41.5
2b	Wooden door frame with foam absorber	23	30	40.9
3a	Metal frame V1	23	29	39.7
3b	Metal frame V2	23	30	41.3
4	Special ventilation door frame	23	30	34.9
5a	Top rabbet gap with (door B)	23	(28)	19.0
5b	All-around rabbet gap (door B)	23	27	41.7
6	Perforated door (door A)	23	28	45.7
7	Z-blind (door A)	23	27	47.0

Table 3: Resulting acoustic attenuation and volume flow at given pressure drop for various door or door frame integrated overflow solutions in combination with a 5mm wide bottom door gap.
*Resulting acoustic attenuation as defined in the text.

	Door or door frame integrated overflow solutions without bottom door gap	R'$_w$ of door [dB]	Resulting R'$_w$ [dB]	Vol. flow [m³/h] at Δp=2Pa
2a	Wooden door frame	30	33	23.1
2b	Wooden door frame with foam absorber	30	35	22.4
3a	Metal frame V1	30	34	21.2
3b	Metal frame V2	30	36	22.8
4	Special ventilation door frame	30	36	16.4
5a	Top rabbet gap (door B)	31	32	6.7
5b	All-around rabbet gap (door B)	31	29	31.2
6	Perforated door (door A)	30	30	27.2
7	Z-blind (door A)	30	29	28.5

Table 4: Resulting acoustic attenuation and volume flow at given pressure drop for various door or door frame integrated overflow solutions without bottom door gap. This realization would require a floor rabbet seal or a drop down seal. *Resulting acoustic attenuation as defined in the text

CONCLUSION

A proper overflow element layout is important to avoid:

- excessive in- and exfiltration through the building envelope and therefore a decrease in energy efficiency,
- maladjustment of the supply air distribution when interior doors are opened and therefore an impairment of indoor air quality,
- needless deterioration of the sound attenuation between rooms and therefore a lowering of the acoustic comfort.

For energy efficient buildings with mechanical ventilation, overflow elements should be designed and dimensioned for pressure drops <2-3 Pa at nominal ventilation rates.

For a typical construction (10m² dry-wall), the following can be stated:

- Volume flows of 30-40 m³/h can be achieved with simple and cost effective door and door frame integrated solutions while still complying with normal sound attenuation recommendation of 30 dB according to DIN4109 (1989).

- A door gap of 5mm in combination with a modified door frame (No. 2b or 3b) will fulfill these requirements.
- A door gap of 10 mm by itself allows a volume flow of 35 m^3/h, but it reduces the sound attenuation between rooms to about 27 dB.
- Higher acoustic attenuation requirements (≥35 dB) require a door with a floor rabbet seal or a drop down seal in combination with some form of an overflow element. Depending on the desired volume flow and on the desired sound protection a wall or ceiling integrated overflow element might be necessary.

ACKNOWLEDGEMENTS
This work was done within the project "Doppelnutzen" funded by the Austrian Ministry for Transport, Innovation and Technology (bmvit) within the framework of "Haus der Zukunft plus". For further information see: http://www.hausderzukunft.at/results.html/id6370
We also thank the following companies for providing sample materials: Huter & Soehne, JOSKO Fenster und Türen GmbH, Keller Zargen AG, Franz Schneider Brakel GmbH + Co KG.

REFERENCES

[1] D. Etheridge and M. Sandberg, "The power law for narrow openings," in *Building Ventilation - Theory and Measurement*, 1st ed., John Wiley & Sons, 1996, pp. 88–89.

[2] "Airflow around buildings," in *ASHRAE Fundamentals*, Atlanta, GA: ASHRAE, 2009, p. 24.1 ff.

[3] G. N. Walton and W. S. Dols, "CONTAM User Guide and Program Documentation CONTAM User Guide and Program Documentation." NIST, 2010.

[4] "Meteonorm." [Online]. Available: http://meteonorm.com/home/. [Accessed: 31-Aug-2012].

[5] J. Werner and M. Laidig, "Grundlagen der Wohnungslüftung im Passivhaus," in *AKKP 17: Dimensionierung von Lüftungsanlagen in Passivhäusern*, 1st ed., W. Feist, Ed. Passivhaus Institut, 1999, pp. 25–54.

[6] R. Pfluger, "Doppelnutzen: Komfort- und kostenoptimierte Luftführungskonzepte für energieeffiziente Wohnbauten," *Haus der Zukunft, Bundesministerium für Verkehr, Innovation und Technologie, Österreich*, 2011. [Online]. Available: http://www.hausderzukunft.at/results.html/id6370. [Accessed: 30-Aug-2012].

NUMERICAL PREDICTION OF THE AIR EXCHANGE IN THE MUSEUM PREMISES EQUIPPED WITH NATURAL VENTILATION SYSTEMS

Andrzej Baranowski*, Joanna Ferdyn-Grygierek

Department of Heating, Ventilation and Dust removal Technology, Silesian University of Technology
Konarskiego 20
44-100 Gliwice, Poland
**Corresponding author: andrzej baranowski@polsl.pl*

ABSTRACT

Ensuring a proper indoor environment in the museum exhibition rooms requires, among others, the achievement and maintenance of the proper air change rate. It is important because of the minimum rate necessary to remove the excessive heat gains and moisture and energy demand for the ventilation purposes.

Two existing museum buildings were selected for the research purposes. First is located in a historical small castle built in 14th century and the second one is a modern building constructed as the a museum in the 20th century. Both buildings are equipped with the natural ventilation system.

The numerical models of both buildings were created and the simulations of the air changes were performed using the CONTAM software. The simulations were carried out for the winter period on the basis of the measured and recorded at this time weather data. As the result of simulations the air change rate in the selected museum rooms are presented against a background of the external temperature.

KEYWORDS
Natural ventilation; museum building; simulation methods; CO_2 concentration.

INTRODUCTION

Museum buildings belong to the special group of buildings where the indoor air quality (IAQ) is important. Particularly, the certain parameters, such as indoor air temperature, relative humidity and solar radiation decide on the IAQ. Ensuring the proper conditions in connection to the safety of the exhibits sometimes stands in contradiction to the primary function of museum – making the exhibits available for visitors. The large number of visitors can make the IAQ unsuitable because of the sudden growth of the internal heat gains and the air humidity that can be even dangerous for the exhibition [1]. Fluctuations of temperature and humidity in time affect the exhibits even more than their - although unfavourable - but constant level. Recommended parameters of the internal climate in museums are a relative humidity of 50 % and the temperature of 15 °C ÷ 25 °C. Deviations of ± 10 % for humidity are admitted, as well as deviations of temperature by ± 2 K (these ranges concern short fluctuations of the parameters due to instantaneous redundant gains and the non-homogenous environment – gradients of these parameters over the space of the museum) [2].

Additional difficulties in ensuring the proper IAQ in the museum buildings results from their historical value. Very often those buildings are lack of proper heating and ventilating systems, and because of their heritage importance there is no allowance for rebuilding and reconstruction of the HVAC system.

The paper presents the preliminary results of the investigations carried out within the frame of the research project which targets the identification and assessment of the indoor air quality in the museum buildings and points out the possible activities leading to the improvement of it.

BUILDINGS DESCRIPTION

The investigation concerns two museum buildings located in different cities in southern part of Poland in the Upper Silesia region. The buildings differs significantly in relation to the construction, materials used as well as the size and volume (Fig. 1).

The first museum is located in a historical small castle built in 14th century. The exhibition rooms are located on three levels of the building. Besides the exhibition rooms there are also storage rooms, offices and laboratories located on the floors. Internal structure of the building is rather complicated because the castle was rebuilt many times giving the existing state. The building is equipped with the ventilating duct system connecting some rooms with the outlet chimneys on the roof. The fresh air infiltrates through the window cracks. Originally probably all castle rooms were ventilated but now because of some alterations of the building a few exhibition rooms are lacking the ventilation. The total exhibition area amounts to 68 m^2, 104 m^2 and 114 m^2 on the 1st, 2nd and the 3rd floor respectively.

The second museum was erected in 1929 - 1930 and was specially designed to serve exhibition purposes. It is a five-storey, double-winged building with the exhibition rooms on the 2nd, 3rd and 4th storey. The total exhibition area is much greater than in the previous building: 400 m^2, 900 m^2 and 650 m^2 on particular floors. The very unfavourable feature of this building is total lack of any ventilating system. The whole building ventilation is maintained only by infiltration mode.

Figure 1. Selected museum buildings.

METHOD

The aim of the work – assessment of the ventilation in the investigated museums – was realized by the numerical simulation.

The calculations of the ventilation air flows within the considered building was made by CONTAM program. This program is designed for multizone analysis of the ventilation and indoor air quality in buildings [3]. CONTAM can be applied to the global assessment of the ventilation effectiveness in the whole building, search for the time variation of the ventilation air flows in the particular zones or for checking the influence of building air-tightness on the air infiltration. The research program comprises continuous measurement of the main indoor environmental parameters: air temperature and humidity and CO_2 concentration.

The measurement campaign was carried out for whole heating season. In both museum buildings all main exhibition rooms were equipped with the measurement sensors located in selected points of every room. The local weather station provided the set of the meteorological data necessary for simulations.

MODELS

Two numerical models of both museums were built reproducing internal structure of the whole buildings. Figure 2 presents, as an example, the CONTAM models of the third floors of both buildings. All identified flow paths were modelled, mainly through the windows cracks. The stairwells in both building were also included in the model as the important flow path in case when the ventilation is realized by stack effect. For the model of the museum located in the old castle the system of the ventilating ducts was identified and modelled.

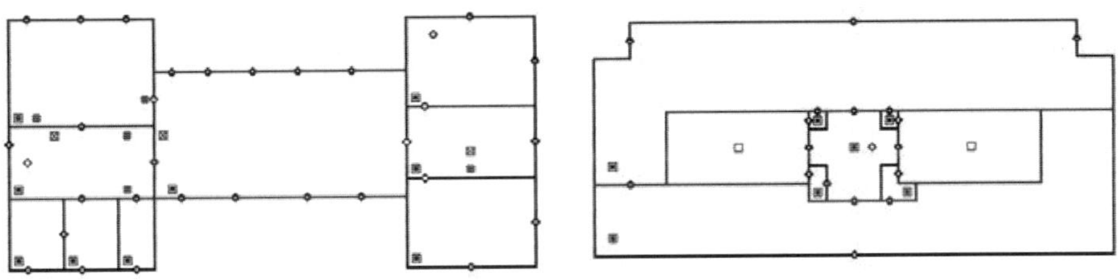

Figure 2. CONTAM models of the museum buildings

One of the biggest uncertainties when the model is created is the value of the air infiltration coefficient which describe the air tightness of the windows. Because of the lack of the windows characteristics the assessment of these parameters were based on the authors experience. In the contemporary building three types of windows were identified: metal, wooden and PVC – all are weather-stripped and double glazed. For the preliminary simulations the air infiltration coefficient was assumed to be equal to 0.2 $m^3/m \cdot h \cdot Pa^{0.67}$ which means that the windows were tight. In the old castle this issue was more complicated: almost all window openings were different and for this reason 16 types of windows were declared. After inspection in the building the air infiltration coefficient was taken on the level of 1 $m^3/m \cdot h \cdot Pa^{0.67}$ because the windows were old and not airtight.

The model was adjusted and tuned using the results of the CO_2 concentration measurements. It was assumed that the CO_2 concentration changes because of the presence of the people in the exhibition rooms. The levels of concentration differs for both buildings and for different storey. Thanks to the recording of the CO_2 concentration variation it was possible to apply the concentration decay method [4] to determine the air change rate in particular exhibition rooms in the buildings. After that the results were compared with the simulation results giving the possibility to correct the air infiltration coefficients for windows.

RESULTS

Measurements

The measurements were performed during the last year's heating season, between October and April. In the paper some results for February and March are presented. From the effectiveness of ventilation point of view the CO_2 concentration was interesting. Figure 3 presents the variation of the CO_2 concentration in the chosen exhibition rooms in both buildings. The repeatable changeability of the CO_2 level can be observed: increase of the concentration during the day results from the presence of the personnel and visitors in the museum. In the museum located in the small castle the actual concentration reached sometimes even more than 2400 ppm. In the largest exhibition rooms of the newer building these concentrations are much lower, the maximum was about 1100 ppm. It can be explain by the greater cubature of these rooms when the source of contamination (number of people) is usually comparable in both museums.

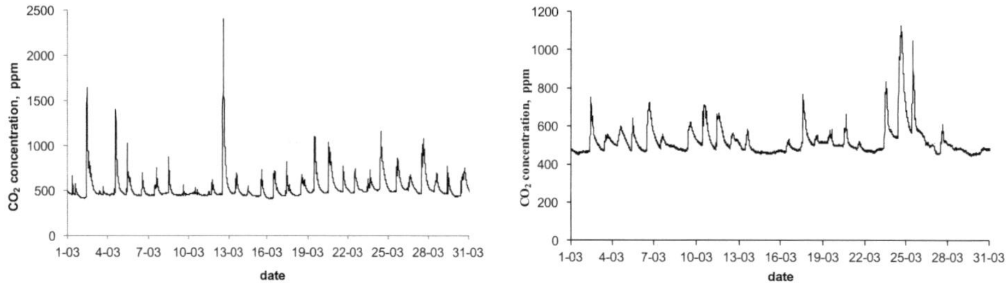

Figure 3. Variation of the CO_2 concentration on the selected storey's of the museum buildings (left – castle, right – modern building)

For several cases the CO_2 concentration decay was calculated giving the air change rate in the particular exhibition rooms. The results could serve for models adjusting and validation.

Simulations

The series of ventilating air flows simulations were performed using the meteorological data recorded during the heating season by the local weather station. The results of the air change rate simulation were compared with the data resulting from the CO_2 concentration decay measurement. After that the tuning of the models was made – that was done mainly changing of the air infiltration coefficients of particular type of windows. Thanks to the number of simulations performed the new value of the air tightness of windows were assumed. In case of the greater building the air infiltration coefficient of the new PVC windows was equal 0.2 $m^3/m{\cdot}h{\cdot}Pa^{0.67}$ and 0.5 $m^3/m{\cdot}h{\cdot}Pa^{0.67}$ for other type of windows (wooden and metal). Regarding the windows in the small castle where the windows were old and not so airtight the air infiltration coefficient was established 0.5 and 1.0 $m^3/m{\cdot}h{\cdot}Pa^{0.67}$.

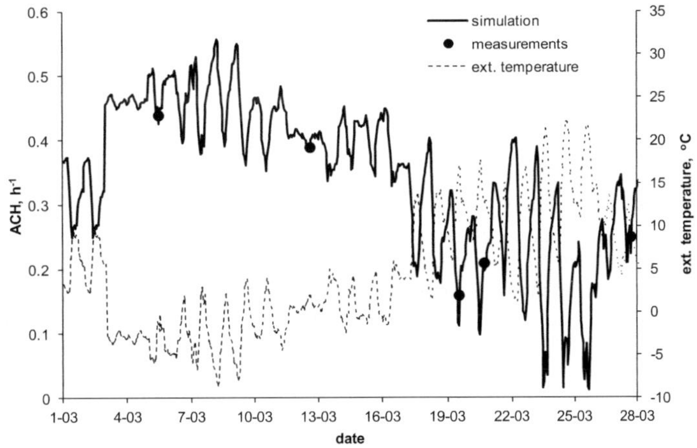

Figure 4. Air change rate in the selected exhibition room in the castle

Figures 4 and 5 show the run of air change rate variation. The measured values of the air change rate were marked with points. In case of the museum located in the old castle quite good compliance was achieved. The results for the second museum are not so good. In fact in the most cases simulation results differ from the measurements. The reason for that is not obviously the fault of the numerical model. The analysis of the measurement data shows not a very big difference between the high level of the CO_2 concentration and the basic level. Hence the calculation of the CO_2 concentration decay can be imprecisely. The second reason of the incompatibility is that in the large cubature of the exhibition room the CO_2 concentration was measured only by one detector. On the other hand the proper use of the concentration decay method consist on the excellent mixing of the contaminant in the whole volume. The ventilation of the exhibition rooms in the considered museum was very poor so the perfect mixing is also uncertain.

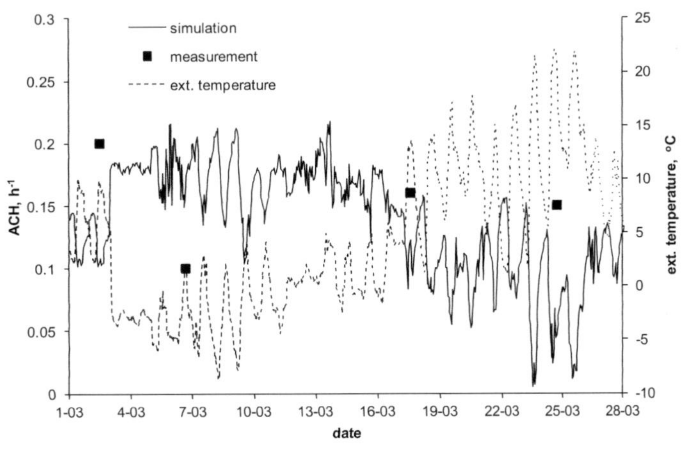

Figure 5. Air change rate in the selected exhibition room in newer building

CONCLUSIONS

Ventilation of museums in most cases is limited to natural ventilation. Sometimes the air is exchanged only by opening windows or doors what may cause rapid change in temperature and humidity indoors and may be dangerous for artefacts. Insufficient ventilation and lack of natural ventilation systems may also create problems by excessive increase in humidity and temperature in the exhibition rooms.

The research performed shoud be treated as the introduction to the complex analysis of the ventilation effectiveness in the selected museum buildings.

The work performed gives only preliminary results concern a short period of time. Based on these results some potential problems with ventilation was indicated.

The numerical models presented here were adjusted and partially validated by measurement of CO_2 concentration. The next analysis for the longer period of time (e.g. for the whole heating season) will give more precise assessment of the reliability of simulation models.

ACKNOWLEDGEMENTS

The work was supported by Polish Ministry of Science and Higher Education within research grant N N523 448136 and by the Polish Ministry of Science and Higher Education – statutory research BK-364/RIE-1/2011.

REFERENCES

[1] Camuffo, D. 1998. *Microclimate for Cultural Heritage*, Elsevier Science.

[2] ASHRAE Handbook—HVAC Applications 2007. Ch.21: *Museums, Galleries, Archives, and Libraries.*, American Society of Heating, Refrigerating and Air-Conditioning Engineers, Inc. Atlanta.

[3] Walton, W.G.N., Dols, S. 2005. *CONTAM 2.4 User Guide and Program Documentation,* National Institute of Standards and Technology, Gaithersburg.

[4] Roulet, C-A. 2008. *Ventilation and Airflow in Buildings,* Earthscan, London.

CONSTRUCTION AND SET-UP OF A FULL-SCALE EXPERIMENTAL HOUSE FOR VENTILATION STUDIES

M. Plagmann*, S. McNeil, M. Bassett

BRANZ Ltd
1222 Moonshine Road
Porirua City, New Zealand
**Corresponding author:*
manfred.plagmann@branz.co.nz

ABSTRACT

This paper reports on the construction, experimental set up and infiltration characteristics of a purpose built full-scale experimental house. The building has been designed as an experimental platform for measuring the moisture removal effectiveness of active and passive ventilation systems with indoor and outdoor climate conditions seen in New Zealand.
The two bedroom building was purchased as a pre-fabricated shell and moved onto the testing site. The inner wall lining was then airtightened following the Canadian "airtight drywall" approach to achieve less than 1 N_{50} (air changes per hour under a pressure difference of 50Pa). We then installed ventilation ports in walls, floor and ceiling so that the airtightness can be adjusted between 1 and 10 N_{50} to cover the current range of new housing in New Zealand. The building is equipped with temperature, relative humidity probes and multi-tracer gas equipment to track inter zonal air and moisture flows.
Early work has measured infiltration rates at four levels of airtightness, some of which are compared to infiltration rates calculated using a zonal model of the building.

KEYWORDS

Ventilation, infiltration, full-scale, moisture removal

INTRODUCTION

The New Zealand Building Code offers an acceptable solution to home ventilation [1, 5, 6] that allows homes to be naturally ventilated through openable window and door openings greater than or equal to 5% of the floor area. Such a simple passive approach has been satisfactory given the temperate climate in New Zealand but this is potentially now not the case for the following reasons:

- The airtightness of new houses in New Zealand has increased over time, even though there is no airtightness requirement in the building code. The average airtightness of houses built pre-war is around N_{50} = 19 ACH and this reduced to N_{50} = 9 ACH for houses built 1960-1990 when large area sheet materials replaced strip lining and flooring. For houses built between 1990 and 2010 the average N_{50} = 4.5 ACH [4, 5, 6]. These changes are a natural consequence of building design and material selection but they have closed down natural ventilation paths that may have added useful ventilation.

- A recent survey of ventilation rates in homes built since 1994 [4] showed that the infiltration minimum is often supplemented by opening windows but that a proportion are exhibiting moisture problems because windows are kept closed for security and privacy. Clearly window opening cannot always be relied on for ventilation.
- Mechanical supply-only ventilation systems have become a popular retrofit solution to indoor moisture in New Zealand homes but with relatively simple controllers, they are not optimised for energy efficiency.

This paper reports on early steps towards trialling a range of ventilation options in a new full scale ventilation research building at BRANZ. Its purpose is to study the effectiveness of ventilation solutions in removing contaminants (particularly moisture), along with their ability to adapt to an occupant that opens windows. The work forms part of a wider WAVE (Weathertightness, Air Quality and Ventilation Engineering) programme at BRANZ. One of the aims of WAVE is to provide guidance on suitable ventilation options that are optimised for moisture control, energy efficiency and the airtightness of the house.
Our intention is to trial ventilation systems similar to those investigated by Yoshino et al. [7] as well as Liu and Yoshino [8] who have studied the performance of different ventilation systems in a full-scale two storey house at a fixed airtightness level without monitoring moisture removal or other contaminants. The effect of moisture buffering and ventilation as studied by Lengsfield et al. [9] and Hasegawa et al. [10] will be of particular interest to our research. Moisture production levels of various domestic activities we intend to simulate are going to be used according to Aizawa et al. [11] or Pallin et al. [12].

THE EXPERIMENTAL BUILDING

The building shown in Figure 1 was constructed as a pre-built shell in 2007 and recently transported on to the research site. The single storey house has a floor area of 91 m^2 and a volume of 206 m^3. The volume of the roof cavity is approximately 45 m^3. The house is a traditional timber frame construction that is clad with painted fibre-cement weatherboard directly fixed over a flexible wall underlay. The gable roof has corrugated iron cladding on timber trusses. The floor is made of particle board which is sealed with polyurethane. The walls and the ceiling are insulated to the requirements of the New Zealand building code [13] with fibre glass. All inner wall surfaces and the ceiling are lined with gypsum based plasterboard which received 3 coats of an acrylic paint.

In order to study ventilation effectiveness at airtightness levels that are present in a large part of the New Zealand housing stock we fitted the house with sealable ports that penetrate the envelope. The ports are located in the floor, the walls and the ceiling connecting the living area to the subfloor, the cavity of the outer walls and the attic, respectively. Our intention was to reach an airtightness level as low as 1 N_{50} (all ports sealed) and an upper level of about 9 N_{50} (all ports open). The pre-fabrication and the fact that it was going to be transported to its location on the research site made it necessary to achieve the airtightness through detailing of the indoor wall linings which were installed after the building reached its destination. We decided to implement the Canadian "airtight drywall" approach. To avoid air leakage from the outer walls through the inner walls into the room we isolated the inner wall by means of applying a 3 mm thick closed foam tape to the corners where the inner walls join onto the outer walls of the building. For the electrical outlets we used flush boxes that have seals at the

Figure 1: The experimental building on site.

cable inlet and where the plasterboard butts on the box rim. To avoid air leakage through gaps between the floor and the ceiling boards the plasterboards were sealed using silicone caulking. Every penetration of the plasterboards for cables, lighting, access hatch to the attic and the like was sealed as best as possible.

Characteristics of the Ports

In order to derive an model of the infiltration, we measured the pressure/flow characteristics of the ports by pressurizing the building. The ports were constructed from PVC tubing with an inner diameter of 38mm and 64mm. The ports were installed in the walls, the floor and the ceiling which has given rise to four different pressure/flow characteristics. These characteristics were determined by fitting a exponential function $Q = C(dP)^n$ to the measured data points. Figure 2 shows the measured data and the graph of the fitted model while the fit parameters are provided in Table 1.

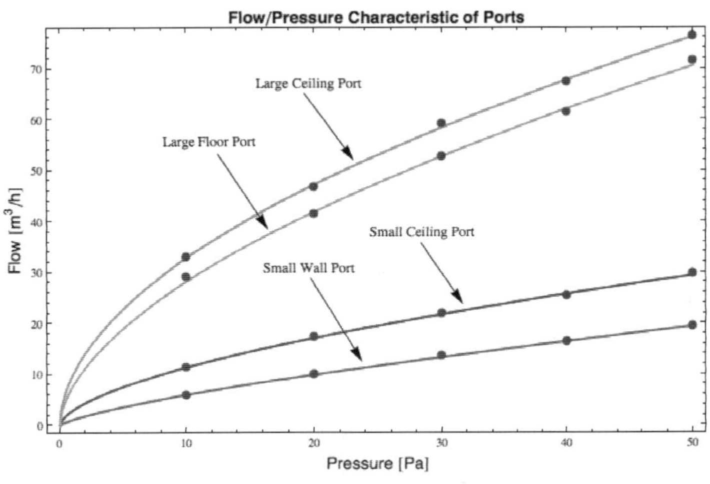

Figure 2: Measured data and fitted function for the flow characteristic of the various ports

The diameter of the port is the dominant factor determining the flow/pressure characteristics. The flow characteristic of the small ceiling ports and the wall ports show comparatively little differences taking into account that those ports lead on one hand into the large attic, while the wall ports lead into the confined space of the wall cavity. This indicates that the outer shell of the walls is not very airtight and that the airtightness level of 1 N_{50} achieved in the house is largely determined by the inner wall lining.

Port location/Size	Coefficient C	Exponent n
Wall Ports	1.1 ± 0.1	0.73 ± 0.02
Floor Ports	7.6 ± 0.5	0.56 ± 0.02
Small Ceiling Ports	2.9 ± 0.1	0.58 ± 0.01
Large Ceiling Ports	9.8 ± 0.4	0.52 ± 0.01

Table 1 Fit parameters of the power law pressure/flow model $Q = C(dP)^n$.

INSTRUMENTATION

The building is equipped with instruments that allow it to run infiltration and contaminant removal measurements in a semi automatic way. All operations are controlled by a computer which controls the indoor climate, the sampling of the tracer gas, the temperature and humidity sensors and writes the retrieved data into a database (see Figure 3). A database table is used to describe indoor climate parameters such as temperature and humidity. The house can be heated and the humidity can be increased but no cooling or dehumidifying is available at this point in time apart from what the installed ventilation system is providing. The airflow through the ventilation system into each zone is measured by means of pressure averaging tubes installed in the ducting.

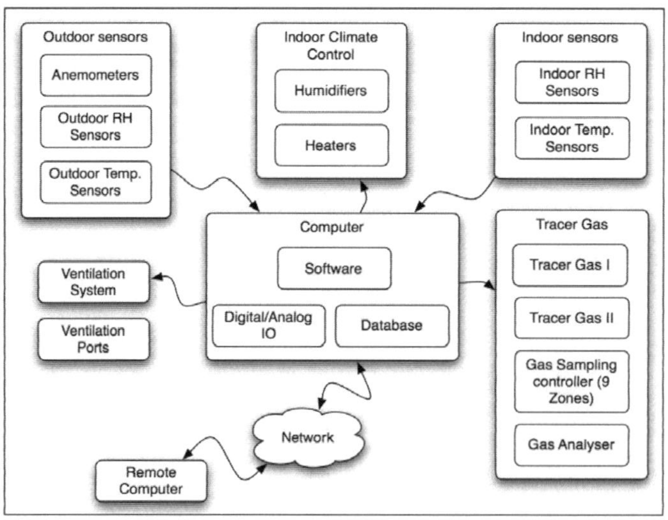

Figure 3: Instrumentation scheme used in the house.

The tracer gas injection rate is manually controlled at this point in time. Flow controllers used in gas chromatography are used to adjust the injection rate of the tracer gases from a few millilitres to hundreds of millilitres a minute. The flow controllers are kept at a few degrees above ambient temperature to minimise drift of the tracer gas injection rate. The flow rate is measured using simple bubble flow meters. An Innova 1412 photo acoustic gas monitor is equipped with filters to detect CO_2, Freon, sulphur hexafluorid (SF_6) and water with detection limits of 3.4 ppm, 0.02 ppm, and 0.006 ppm, respectively. The dynamic range of the gas monitor is typically 4 to 5 orders of magnitude. The target working concentrations of the tracer gas in the zones is usually at least 10 times the detection limit or, in case of CO_2 10 times the background concentration. The tracer gases are sampled from the zones by means of a computer controlled manifold that can switch each of the possible 9 sampling locations onto the gas monitor. Each room, including the attic, is equipped with a number of sampling and dosing tubes. In the living area these tubes are located at approximately 1.5m off the ground. Before the gas monitor analyses the air sample it purges the tubes and the sampling chamber to avoid cross contamination. Measuring each location in turn takes about 10 minutes to process, thus allowing 6 samples to be taken from each location per hour. Wind velocities are obtained from the weather station located next to the house.

INFILTRATION MEASUREMENTS

Before we can determine the performance of various ventilation systems, the infiltration characteristics of the house at different airtightness levels in the absence of a ventilation system had to be established. Our intention is to measure the ventilation effectiveness at the four airtightness levels of 1, 3, 5 and about 9 N_{50}. Various ports in the walls, the ceiling and the floor are opened to achieve these levels of airtightness. A ventilation port plan is used to make sure that only those ports are opened at a given airtightness level that allow for an even distribution of air leakage paths throughout the building. The injection rate of the tracer gas is adjusted in accordance with the set airtightness level to reach a tracer gas concentration of at least 10 times the detection limit, thus allowing for enough dynamic range and lower signal to noise ratio.

Figure 4 and 5 show the hourly averaged infiltration measurements of a single zone i.e., one tracer gas, over 2 - 4 days at different airtightness levels. The measurements were completed during a calm period with average wind speeds of only 2 m/s measured at 10 metres height.

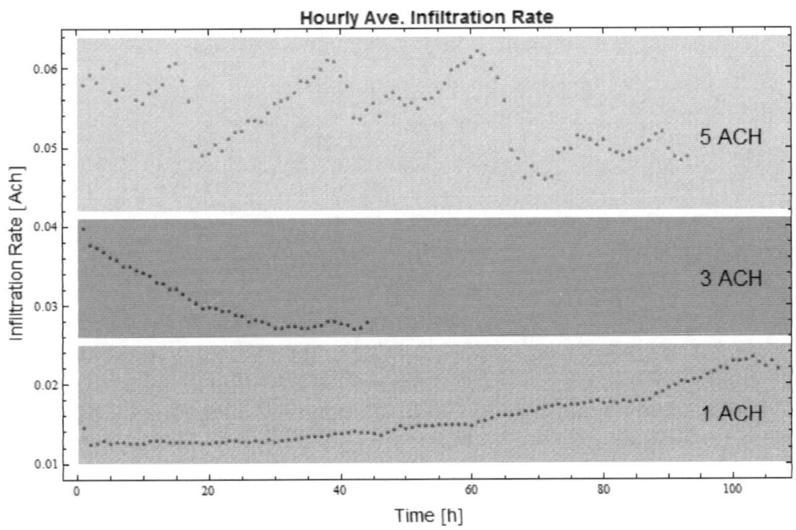

Figure 4: Single zone infiltration rate of the living area at different N_{50} airtightness levels. The graph for the airtightness level of 9 ACH has been moved to Figure 5 due to scaling.

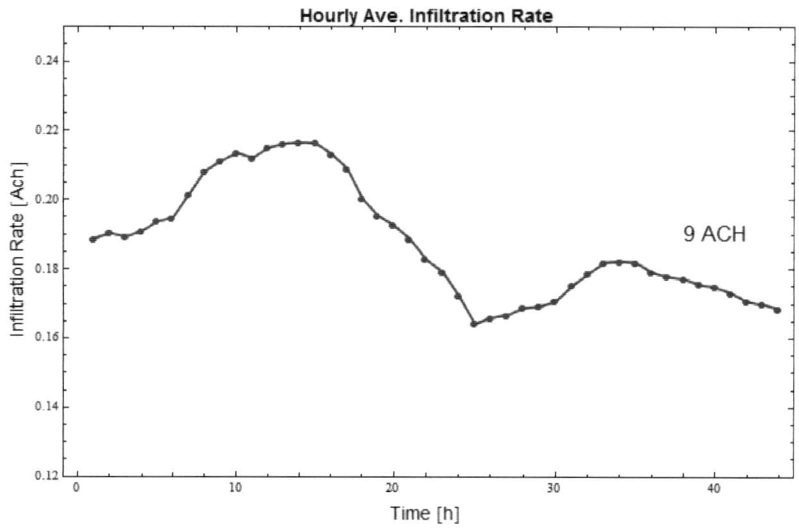

Figure 5: The graph shows the single zone infiltration rate of the living area at about 9 N_{50}. All ports are open at this level of airtightness but windows and doors are closed.

The hourly averaged infiltration rate of a short 2 zone infiltration experiment is shown in Figure 6. One of the bedrooms (Zone 2) of the house was filled with CO_2 while the remaining living area (Zone 1) of the house was filled with SF_6. Both zones were at an airtightness level of about 2 N_{50}. Only wall ports were open during this experiment, therefore, there was no cross infiltration between the two zones via the roof apart from through adventitious openings. Most of the inter zonal infiltration would have taken place through openings under the closed door. The average wind speed during this period was about 1.5 m/s at a height of 10 metres.

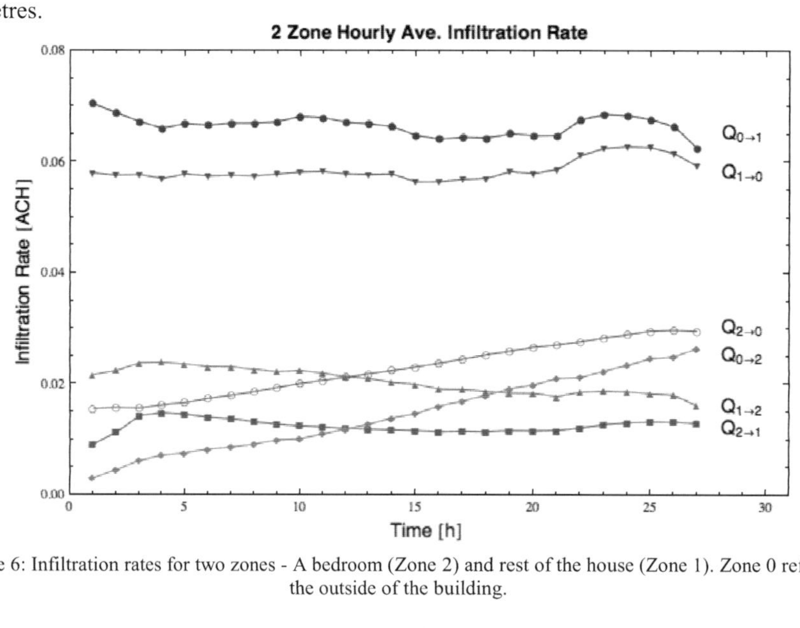

Figure 6: Infiltration rates for two zones - A bedroom (Zone 2) and rest of the house (Zone 1). Zone 0 refers to the outside of the building.

CONTAM MODEL

We have created a model of the infiltration characteristics using CONTAM. This model will be used later in the study to compare the measured performance with a ventilation system in place with what the performance would be without the ventilation system

At this point in time we have developed a single zone CONTAM [14] model using only the wall ports to simulate the ventilation in the living area of the test house. The pressure/flow characteristics of the wall ports (see Figure 3 and Table 1) have been used as parameters describing the flow paths in the model. Wind, outdoor and indoor temperature data were obtained from a weather station and the sensors in the test house. Pressure coefficient values for the building were derived from a wind tunnel measurements published by Tokyo Polytechnic University [15].

The tracer measurement was started on the 28th July and ran continuously till midnight of the 31st July. The hourly averaged simulated (dashed line) and the measured (continuous line) infiltration rate of the living area of the test house is shown in Figure 7. While the infiltration data is reasonable noisy it shows a good agreement between the simulation and the measurement. This indicates that the assumptions made in the model about the buildings pressure coefficients and the calculated pressure/flow characteristics of the wall ports are reasonable. Over time we will compare the simulation output of the model with other infiltration measurements to make the model more robust and show its validity under different wind and temperature conditions.

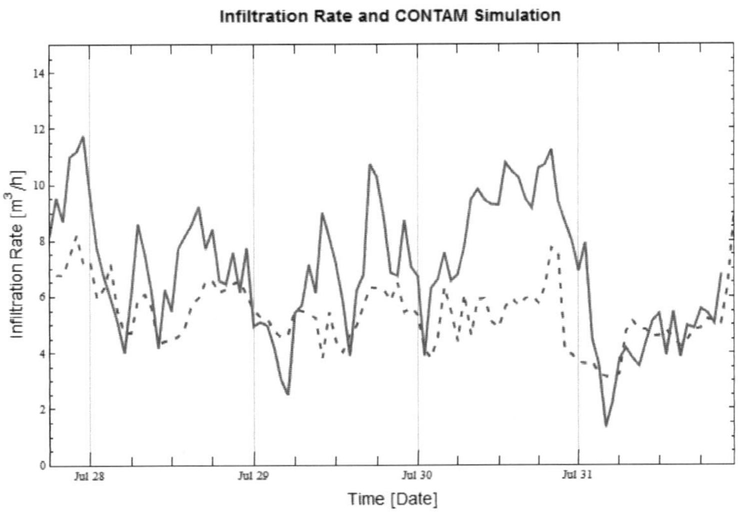

Figure 7: Good agreement is shown between the measured infiltration rate (continuous) and the rate simulated by use of a CONTAM model (dashed).

CONCLUSION

In this paper we have described the set up of a full scale experimental building to study the moisture removal effectiveness of ventilation systems and have presented initial measurements of infiltration data to characterise the building. The experimental setup can measure at up to 9 sample locations at a rate of 6 samples per hour. A single zone CONTAM model was developed and shows good agreement with a data set derived from an initial infiltration rate measurement.

ACKNOWLEDGEMENTS

This work was funded by the New Zealand Building Research Levy and the New Zealand Ministry of Business, Innovation and Employment. The assistance of Mark O'Hare, a postgraduate from Queens University, Belfast, working at BRANZ under the IAESTE scheme, is gratefully acknowledged.

REFERENCES

[1] Department of Building and Housing. *Approved Document G4 Ventilation*. 2008 Wellington.

[2] McNeil, S., Quaglia, L., Bassett, M., Cunningham, M. Overton, G., Plagmann, M., Boulic, M., Leardini, P., Rosemeier, K. and Smith, I. *Trends in airtightness of New Zealand houses*. Proceedings of the 2011 Biometrology Conference. Auckland.

[3] Bassett, M. R. The infiltration component of ventilation in New Zealand houses, Proc. 6th AIC Conference. *Ventilation Strategies and Measurement Techniques*. Southern Netherlands, September, 1985.

[4] McNeil, S., Quaglia, L., Bassett M., Overton G., Plagmann M., Cunningham M., Upton, S., Boulic, M., Smith, I., Bishop, T., Leardini, P., Rosemeier, K., 2012. *A survey of airtightness and ventilation rates in post 1994 NZ homes*, AIVC 33rd Conference: Optimising Ventilative Cooling and Airtightness for [Nearly] Zero-Energy Buildings, IAQ and Comfort, 10-11 October 2012, Copenhagen, Denmark

[5] Standards New Zealand. NZS 4303:1990. *Ventilation for acceptable air quality*. Wellington.

[6] Standards Australia. 1991. AS1668.2 1991. *The use of mechanical ventilation in buildings*, Part 2 Mechanical ventilation for acceptable air quality. Sydney

[7] Yoshino H., Lee J., Wada J. *Performance evaluation on ventilation systems with full-scale test house,* Proceedings of Room Vent 2002 (8th International Conference on Air Distribution in Rooms) - 8-11 September 2002 - Moltke's palace, Copenhagen - Denmark - pp 513-516

[8] Liu J., Yoshino H., *Performance evaluation of hybrid ventilation system in a full-scale test house*, AIVC 21st Conference, proceedings of "Innovations in Ventilation Technology", The Hague, Netherlands, 26-29 September 2000

[9] Lengsfeld, K.; Holm, A.; Krus, M., *Moisture-Buffering Effect – Experimental Investigations and Validation*, 6th International Conference on Indoor Air Quality, Ventilation & Energy Conservation in Buildings IAQVEC 2007

[10] Hasegawa, K.; Yoshino, H.; Mitamura, T., *Experiment for Moisture Buffering and Effect of Ventilation Rate and Volume Rate of Hydrothermal Materials in a Single Room under Steady State Exterior Conditions*, 6th International Conference on Indoor Air Quality, Ventilation & Energy Conservation in Buildings IAQVEC 2007

[11] Aizawa, Y.; Tanabe, S.; Yokoo, K.; Watanabe, K.; Amai, H., *Amount Of Moisture Production And Field Measurement In Dressing Room,* 6th International Conference on Indoor Air Quality, Ventilation & Energy Conservation in Buildings IAQVEC 2007

[12] Pallin, S. ; Johannson, P. ; Hagentoft, C. , *Stochastic Modeling of Moisture Supply in Dwellings Based on Moisture Production and Moisture Buffering Capacity*, Proceedings of Building Simulation 2012 : 12th Conference of International Building Performance Simulation Association, Sydney, 2011

[13] New Zealand Department of Building and Housing, *Compliance Document for New Zealand Building Code Clause H1: Energy Efficiency* , 3rd Edition, 2011

[14] National Institute of Standards and Technology, Indoor Air Quality and Ventilation Group, http://www.bfrl.nist.gov/IAQanalysis/index.htm

[15] Tokyo Polytechnic University, *Aerodynamic database for low-rise buildings*, http://www.wind.arch.t-kougei.ac.jp/info_center/windpressure/

UPDATE OF THE SPANISH REGULATION REGARDING VENTILATION AND INFILTRATION: ANALYSIS, COMPARISONS AND REPERCUSSIONS

Salmerón Lissén José Manuel[1*], Sánchez de la Flor Francisco José[2], Álvarez Domínguez Servando[1], Molina Félix Jose Luis[1], and Macias Olga[3]

1 University of Seville, Grupo de Termotecnia.C/ Avda de los Descubrimientos S/N.41092 Sevilla, Spain
2 University of Cadiz, Dpto. De Máquinas y Motores Térmicos.C/ Chile, 1. 11002 Cádiz, Spain
3 Tecnalia,Parque tecnológico de Bizkaia. E-48160 Vizcaya, Spain
**Corresponding author: jms@us.es*

ABSTRACT

The Spanish Technical Building Code is one of the three royal decrees that were approved in Spain as a consequence of the transposition of the European Directive on the energy performance of buildings (2002/91/EU) [1]. One basic document of the Technical Building Code deals with the limitations in the energy demand of buildings. Nowadays, due to the recast of the European Directive on the energy performance of buildings (2010/31/EU) [2], a revision process of the current regulations has begun, starting with the Technical Building Code, with its first revision envisaged for 2012. In this article we, as collaborators in this updating process, describe and analyse the main changes regarding ventilation and infiltrations that the updated Technical Building Code (TBC) is going to introduce. These main changes were classified in the next aspects, the calculation methodology, the ventilation technologies that can be considered, and the default values and the data that the user can supply to assess its particular case. The article compares the ventilation required in Spain with the mandatory values in other 16 European countries. Finally, the mean air infiltration values are compared to the values of the EN 15242:2007 [3] and to measured air infiltration values in actual Spanish dwellings.

KEYWORDS

Ventilation and infiltration regulations. Air flow rates. Calculation methods. Residential buildings.

INTRODUCTION

This article is going to classify the main changes that appear in the new Spanish regulation regarding ventilation and infiltration in the next topics:

1. Calculation methodology.
2. Ventilation technologies that can be considered.
3. Default values and data requested to the users.

Thus, these topics will be the three main sections of the present paper.

The objective of this paper is then to show how the regulation regarding ventilation and infiltration has been modified in one State Member of the European Union, Spain. Along the paper we will describe the criteria that have been followed to select the new specifications. Finally, we will comment the consequences of these modifications.

CALCULATION METHODOLOGY

Basically, the calculation methodologies regarding air flows implement a multi-zone loop method. Loop methods have been used extensively in the duct networks analysis. They provide an 'exact' analytical approach to size components of natural and hybrid ventilation systems and offer a number of advantages when compared to the node continuity methods [4]. On this basis we are going to describe the particularities of each regulation:

Former Regulation (2006)

The former regulation is based in the EN 13465:2004 [5], thus it implements a loop method with only one indoor pressure and one outdoor pressure by assuming that all the indoor spaces are a single-zone for single family houses and also for blocks of flats.
The Technical Building Code [6] establishes the ventilation rates for indoor air quality in the basic document HS3 on "indoor air quality", and also limits the infiltrations through windows in the basic document HE1 on "energy needs limitation".
Following the HS3 the inlet flows –bedrooms and living-room- and exhaust flows –bath-rooms and kitchen- should be calculated separately and consider the ventilation flow as the higher of both. For calculations the next table should be used:

Type of zone	Per occupant	Per square meter	Other
Bed-rooms	5	-	-
Living-rooms and dinning-rooms	3	-	-
Bath-rooms and toilets	-	-	15 per room
Kitchens	-	2	50 per room
Storage rooms	-	0.7	-
Parkings and garages	-	-	120 per place
Waste storage	-	10	-

Table 1. Minimum ventilation flows required (l/s).

These values can be compared with the ones in the table 1 of the article [7] by Dimitroulopoulou.
The maximum ventilation flow from inlet and exhaust flows is the one considered in a constant schedule of 24 hours a day. Next figure shows this schedule for a building where the maximum air flow is equal to 45 l/s.

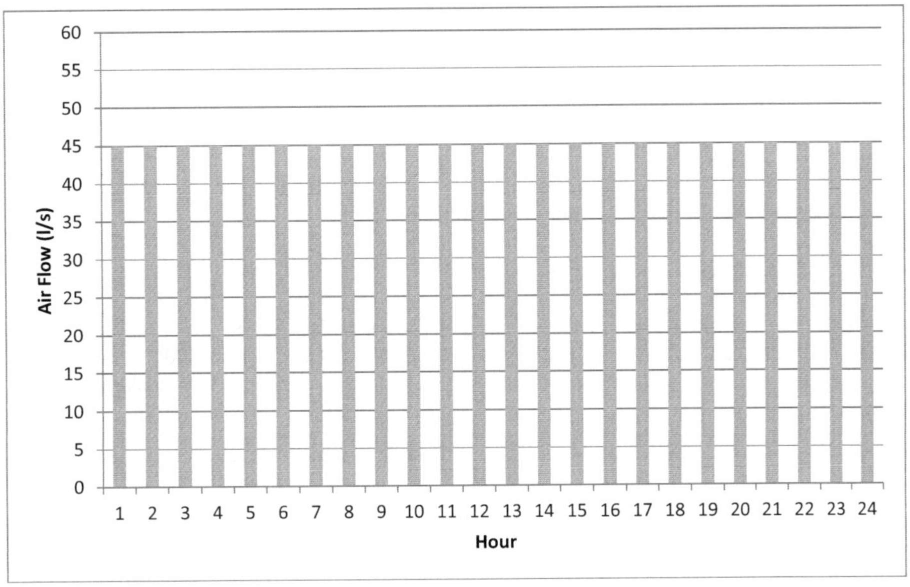

Figure 1. Standard supply air flow schedule.

The air flow of the extraction hood and the eventual air flow requirement of the domestic hot water boiler –or any combustion system- are not envisaged in this regulation.

In order to compare ventilation requirements in different countries, we calculated the same test cases than Brelih and Seppänen in [8]. These tests were developed for two different dwellings, a kitchen, a toilet, a bathroom, a school classroom, a kindergarten playroom, and an office. These cases are very interesting as they represent real-world situations. The authors include the results in 16 European countries, this, allowed us to situate the Spanish ventilation rates in this context. In conclusion, it can be said that in the dwellings the values for Spain are very compared to the average value in European countries. We analyzed the cases and we detected that the origin of this excess is the consideration of the maximum inlet and exhaust flows in the rooms.

The equations for the pressure losses in the cracks, windows and vents are the proposed in the European standard EN 13465:2004 [4].

The default values for the infiltration through opaque elements (cracks) are the next:

- 0.30 air changes per hour at 1 Pa in single-family dwellings. 0.76 at 4 Pa.
- 0.24 air changes per hour at 1 Pa in blocks of flats. 0.61 at 4 Pa.
- 0.10 air changes per hour at 1 Pa in other buildings (tertiary). 0.25 at 4 Pa.

These values are fixed and can not be changed by the building designer.
If we compare these values with the ones that appear in other regulations we can see that these one do not depend on the outdoor or exposed surface of the building like in the French Thermal Regulation [9], on the contrary these are constant values that only depend on the drop pressure between indoor and outdoor, they were obtained from the annex A (not mandatory), tables A.2 and A.3, of the EN 13465:2004 [5].

The permeability of windows is a data that the building designer can modify if the window has a certificate of permeability. The maximum values are 50 m^3/hm^2 at 100 Pa for the south of Spain, and 27 m^3/hm^2 at 100 Pa for the middle and north of Spain. The doors have an invariant permeability coefficient of 60 m^3/hm^2 at 100 Pa due to the absence of a certificate for these elements.

The vents are not defined by the user, they are sized by the official software of the Technical Building Code in order to ensure the air ventilation flow required.

New Regulation (2012)

The proposed new regulation uses the calculation methodology of the EN 15242:2007 [3]. This standard is an actualization of the former and derogated EN 13465:2004 [5]; it is also based on a loop method and keeps the calculation of air infiltrations trough opaque elements, windows and air vents. The novelty is the indoor zoning that allows calculating (using mass conservation laws) the air flow between zones in a similar way that software SIREN [10] does.

The minimum ventilation air flows are not going to be modified by the moment, but futures modifications are not discarded if the practice shows them to be inappropriate to ensure acceptable indoor air quality levels. One of the most important implications of the modification in the required air flow rates is that the energy labelling of a determined building would change as the energy performance scale of new and existing buildings defined in [11, 12] was done with the minimum air flows of the 2006 regulation. This modification in the energy labelling is difficult to justify because it would be a consequence of a change in the calculation methodology –required air flow rates- and not of an improvement of the methodology for the ventilation of the building or its systems or devices.

As referred in [8] lower air flows, in general, are not recommendable. However, the present values used in Spain and in several European countries are very high and a coincidence factor should be applied. As a consequence, a European harmonized regulation theoretically based is necessary.

Regarding indoor air quality, probably a performance-based criterion similar to France, Finland, Norway and Portugal regarding the CO_2 concentration will be used. In Spain, the regulations [13, 14] define the maximum difference between indoor and outdoor CO_2 concentration and the maximum CO_2 concentration in 600ppm and 2500 ppm respectively.

The novelty in the air flows is that the extraction hood in the kitchen is going to run two hours a day in order to take into account the real use of the system, the air flow requirements for combustion (for instance domestic hot water boiler) is going to be considered null because all the new boilers should be airtight.

The default values for the infiltration through opaque elements (cracks) have been changed to fit with the EN 15242:2007 that appeared after the publication of the 2006 regulation. In this standard the Table B.1 of the Annex B (normative) give three values for the permeability of the envelope depending on the air leakage level, in the next point we will justify why we have choose the highest level and thus the default values for the infiltration through opaque elements (cracks) have been fixed to 1.8 m^3/h per square meter of exposed wall surface for single-family dwellings and 1.95 m^3/h per square meter of exposed wall surface for blocks of flats.

As in the former regulation the permeability of windows will be a data that the building designer could modify if the window has a certificate of permeability. The maximum values probably will be more severe than the previous ones.

The software of application of the regulation will size the vents although it will allow to the designer to modify the type –different vent types are listed in the point titled "other energy saving measures"- and probably the nominal air flow in order to give more freedom for the different alternatives.

All these innovations allow to skip the negative effects of the assumptions done on the regulation of 2006: in first term because the calculation of the air flow between zones allows to assign the ventilation energy load to the exact zone that is receiving it, and keep it equal to zero in the zones that only receive the ventilation air from conditioned spaces. A second consequence of calculating the air flow between zones is that it is possible to define the ventilation strategy –demand controlled ventilation, heat recovery ventilation with double flux-. A third consequence is the elimination of a market barrier for the ventilation systems, because now if a designer has implemented a ventilation strategy it can be calculated and it is taken into account in the energy performance of the building, this is very important as the maturity of the ventilation sector is high and the solutions adopted were very basics due to the regulation. Finally, this new regulation open a door for improvements on the tightness of the new buildings as the permeability of the opaque elements is not a default value anymore and now this data that can be given by the designer after a permeability essay, for example a blower door test.

CRITERIA FOR THE NEW SPECIFICATIONS

The criteria for the new specifications have been to keep, when it is possible, the same mean values, updating the reference standards and allowing to the designer a highest degree of freedom in his decisions.

In this sense the main change in the default values have been the change in the opaque permeability value, because is has been moved from a fixed value in air changes per hour – Annex A of EN 13465:2004- to a variable (per square meter of outdoor surface) value from Annex B of EN 15242:2007.
In this section we will explain how we have converted the former values in the new ones keeping the same mean values.

If we compare the values of the EN 13465:2004 with the ones in the EN 15242:2007 there are two main differences: both values are not in homogeneous units -air changes per hour vs. m^3/h per square meter of exposed wall surface-; on the other hand the values of the first standard are referred to the air flow through opaque elements of the building envelope (cracks), while the values of the second one are referred to air infiltration through the whole envelope including opaque elements and window voids.

What we have done is to convert the former regulation values based on the EN13465:2004 into a value comparable to the values of the EN 15242:2007. In order to do that first of all we will multiply the original values with the ratio volume/area of exposed wall surface, thus we will obtain the infiltration through cracks in m^3/h per square meter of envelope surface. After this we will add to this air flow the infiltration through windows using the next equation:

$$Q[m^3/hm^2] = q_{4Pa}^{opaque} \cdot \frac{V}{A^{env}} + perm_{100Pa}^{window} \cdot \left(\frac{4}{100}\right)^{0.67} \cdot \frac{A_{window}}{A^{env}} \quad (1)$$

Where:

$q_{4Pa}^{opaque}[1/h]$: is the infiltration flow through opaque elements at 4 Pa. The normative value following the former regulation is 0.76 for single-family dwellings.
$perm_{100Pa}^{window}[m^3/hm_{window}^2]$: is the window permeability at 100 Pa.
$\frac{V}{A^{env}}[m^3/m^2]$: is the ratio volume/area of envelope surface.
$\frac{A_{window}}{A^{env}}[m_{window}^2/m^2]$: is the ratio window area/area of envelope surface.

Last two parameters depends on the geometry of the building, next tables gives representative values of this for typical constructions in Spain.

Type of building	$\frac{V}{A_{cv}^{exp}}[m^3/m^2]$	$\frac{A_{vent}}{A_{cv}^{exp}}[m_{vent}^2/m^2]$
Single-family dwellings	2.13	0.093
Blocks of flats	3.56	0.138

Table 5. Mean values for the ratios in single-family dwellings and blocks of flats.

The resulting air flow Q is comparable to the EN 15242:2007 because now it includes the same infiltration flows and due to the congruency of unit system. Next table gives the mean values obtained:

Type of building	$Q[m^3/hm^2]$
Single-family dwellings	2.01
Blocks of flats	2.72

Table 6. Mean values for air infiltration flow in single-family dwellings and blocks of flats.

The conclusion of the comparison is that these values are around the maximum value given by the European standard –for the highest level of leakage- that is 2 m^3/h per square meter of envelope surface. Thus we have assumed this figure as the new air infiltration flow in the new regulation. The permeability of opaque elements at 4 Pa that corresponds to this value is 1.6 m^3/h per square meter of envelope surface, this give 1.8 m^3/h per square meter of exposed wall surface –without window voids- for single-family dwellings and 1.95 in the same units for blocks of flats.

<u>Experimental data</u>

In Spain, there are few data about field measurements of the tightness of building envelope, so it becomes difficult to adjust normative with experimental values.

Most of the existing field data corresponds to residential buildings from a region of the north of Spain (Basque Country). Derived from European Directive 93/76/CEE (SAVE), CADEM entity, which belongs to Basque Energy Agency, implanted a voluntary energy certification system for buildings where field tests such as Blower Door Test are mandatory in order to obtain the real energy performance of the certified buildings.

Data provided by CADEM about Definitive Energy Certificates given until June 2010 were:

No. promotions	No. dwellings	Energy savings (Toe)
≈ 200[1]	≈ 13.000	≈ 7.000

Table 7. Data about energy certificates in the Basque Country region.

Since 1994 until 2010 there were carried out about 600 airtightness tests. These field measurement of aightightness were fulfilled following European Standard EN 13829 [15] (previosly International Standard ISO 9972)

There are not accurate stadistical data, but a general relation between air leakage rates and definitive energy rate is observed.

In general, the average air leakage rates obtained from blower door test at 4 Pa for buildings that were certified with highest rate, A, are closed to 0,35-0,40 h^{-1}, whereas for buildings certified with C rate, the average air leakage values are closed to 0,60-0,65 $^{h-1}$.

Energy rating	Air leakage rate at 4 Pa	Air leakage rate at 50 Pa
A	0,35 – 0,40	1,9 – 2,2
C	0,60 – 0,65	3,3 – 3,6

Table 8. Typical air leakage values for A and C energy labelled dwellings.

Comparing Spanish air leakage rates with Dutch data provided by TNO [16] it is possible to conclude that Spanish buildings certified with C energy rate present similar aightightness values than Dutch constructions in 1990, whereas most efficient buildings certified with A get air leakage rates similar to Dutch buildings in 2005.

VENTILATION TECHNOLOGIES

The calculation methodology implemented for the new regulations (2012) allows calculating the outdoor air that gets into the building through the walls (infiltrations), through the windows (infiltrations) and through the specific vents designed for the ventilation purposes. Also the methodology allows following the path of the air assessing the air flow from the inlet zones to the corridor and from there to the exhaust zones (bath-rooms).
Thus this methodology is useful for calculating the next systems:

- Demand controlled ventilation: there are two levels of this technology:
    - Basic: In this case the maximum air flow would be used when the bath-rooms are in use (3 hours a day), and the rest of the time the required air flow would be equal to the minimum.

[1] In each promotion (one promotion corresponds to one energy certificate) an average of three airtightness tests were carried out.

- Advanced: In this case the air flow rate would be the necessary for the occupation hour by hour.

- Double-flux (balanced) system, this system includes a duct network for the inlet and exhaust air flows, so it is possible to pre-heat the inlet air with the exhaust flow. This system includes the installation of a heat recovery unit as exchange unit.

- Combinations of demand controlled ventilation and heat recovery systems.

The technology required for the basic level of demand controlled ventilation consist on presence sensors in the bath-rooms, a double speed exhaust fan and a control device that connect the sensors with the fan. Figure 2 shows the supply air flow hour by hour.

The technology required for the advanced level consist on presence sensors in all the rooms, a multi-speed exhaust fan –it is possible to have more than 6 air flows depending on the number of rooms of the dwelling and the occupancy-, controlled air vents in all the inlet zones – bedrooms and living-room- and an advanced control device that connect all the elements of the system. Figure 4 shows the supply air flow hour by hour.

The previous technologies are based on the presence; alternatively, strategies based on the relative humidity, CO_2 concentration can be used. They are described for residential buildings in [17, 18, 19].

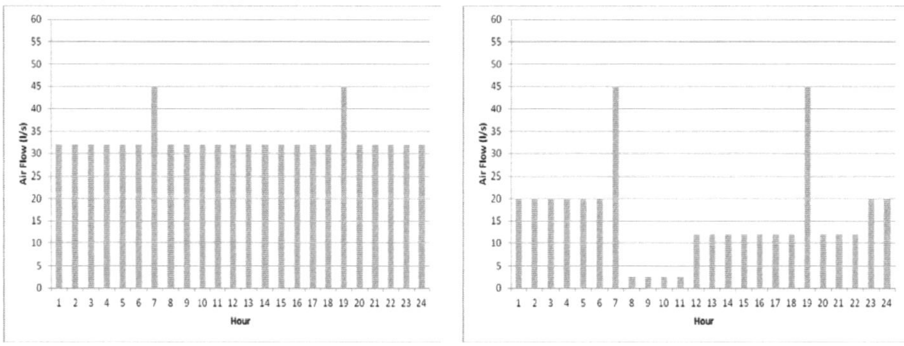

Figure 4. Left: Supply air flow schedule for the basic level of demand controlled ventilation. Right: Supply air flow schedule for the advanced level of demand controlled ventilation.

OTHER ENERGY SAVING MEASURES

Also the new regulation allows calculating the influence of the next energy saving measures in the energy demand:

- Use of self-regulating vents.
- Use of non-return self-regulating vents.
- Use of hygro-regulating vents.
- Improve of the air tightness of opaque elements.
- Improve of the air tightness of windows.

The market of the vents for ventilation is mature and has certain products that could help to improve the energy demand. The former regulation does not give any improved value if the designer chooses one of these products. The new regulation will change the situation giving simulation software that allows calculating the energy consumption and the indoor air quality when using different kinds of vents.

The air tightness of opaque elements is a default value that the designer could not change, but also if a special attention has been put in the construction using vapour barrier and membrane of air tightness for the walls, and using sealants products, silicones, foams or polyethylene tapes for the junction of the walls with the windows or doors openings, the designer could make a permeability test and use the result of this as a input for the software.

The air tightness of windows is a variable value that the designer can move to fulfil with the requirements of the Technical Building Code and also to improve the energy demand or IAQ results.

CONCLUSIONS

The basic documents of the Technical Building Code have to be updated periodically. These actualizations can be motivated for obsolete standards or for the need of contemplate and assess the effect of using certain technologies that previous versions do not deals with. This is the case of the basic document of energy limitation and it relation to the basic document regarding indoor air quality.

Present document shows how we have updated the references to derogated standards using new ones, and simultaneously keeping the congruence of the default values. Also we have used new simulation methodologies that allows to assess the effect of implement a higher number of ventilation technologies.

Basically we have produced a new regulation that assumes one indoor pressure and one outdoor pressure (calculation methodology EN 15242:2007). But keeping and indoor zoning that allows calculating (using mass conservation) the air flow between zones. The calculation methodology permits the possibility of using a control system allowing different flows each hour for a demand controlled ventilation scheme and pre-heating of supply air flow. And finally the new regulation is flexible to allow sizing the vents, using self-regulating vents or changing the permeability of opaque elements, for example if a blower-door test has been carried out. As a consequence the standard will not be a market barrier and promote the use of technologies of ventilation in order to improve the energy efficiency of buildings.

ACKNOWLEDGEMENTS

The data about energy certificates and air leakage in existing dwellings were provided by CADEM, the authors would like to acknowledge this contribution.

REFERENCES

[1] Directive 2002/91/EC of the European Parliament and of the Council of 16 December 2002 on the Energy Performance of Buildings. Official Journal of European Communities.

[2] Directive 2010/31/EU of the European Parliament and of the Council of 19 May 2010 on the Energy Performance of Buildings. Official Journal of European Union.

[3] EN 15242:2007. Building ventilation. Methods for the assessment of the air flow rates in buildings, including infiltrations. European Committee for Normalization. 2007.

[4] Building Ventilation, the state of the art, Santamouris M., Wouters P., Earthscan, London, Sterling, VA, 2006. Chapter 2 by Axley J.

[5] EN 13465:2004. Building ventilation. Methods for the assessment of the air flow in dwellings. European Committee for Normalization. 2004.

[6] Royal Decree 314/2006, passing the technical building code (Código Técnico de la Edificación). Spanish Official Journal (BOE) 74. March 2006. http://www.codigotecnico.org/web/recursos/aplicaciones/contenido/texto_0002.html. Basic Document HE-1 Energy needs limitation. Basic Document HS-3 Ventilation and Indoor Air Quality.

[7] C. Dimitroulopoulou, "Ventilation in European dwellings: A review", Building and Environment, vol. 47, pp. 109-125, January 2012.

[8] N. Brelih, O. Sepänen, "Ventilation rates and IAQ in European Standards and national regulations", 32nd AIVC Conference and 1st Tightvent Conference, Brussels, October 2011.

[9] Reglementation Thermique, Guide réglementaire. CSTB. 2010.

[10] Logiciel SIREN. Simulation de renouvellement d'air. CSTB.

[11] Royal Decree 47/2007, approving the Basic Procedure to certify energy efficiency in new-construction buildings. January 2007. http://www.minetur.gob.es/energia/desarrollo/EficienciaEnergetica/CertificacionEnergetica/ProgramaCalener/Paginas/DocumentosReconocidos.aspx

[12] Royal Decree XX/2012, approving the Basic Procedure to certify energy efficiency in existing buildings. Public audience until April 2012. http://www.minetur.gob.es/energia/desarrollo/EficienciaEnergetica/CertificacionEnergetica/propuestaNuevosReconocidos/Paginas/nuevos.aspx

[13] EN 13779:2007 Ventilation for non-residential buildings. Performance requirement for ventilation and room-conditioning systems. CEN, 2007.

[14] NTP 549. Carbon dioxide in the IAQ evaluation, National Institute for Occupational Safety and Health at Work. El dióxido de carbono en la evaluación de la calidad del aire interior. Instituto Nacional de Seguridad e Higiene en el Trabajo.

[15] EN 13829:2002, Erratum 2010 Thermal performance of buildings. Determination of air permeabilityu of buildings. Fan pressurization method. (ISO 9972:1996, modified). CEN, 2002-2010.

[16] W. de Gids, Optimal Air Tightness Levels of Buildings. Session 5B: Philosophy for defining airtightness requirements.TIGHTVENT Conference, Brussels, 2011.

[17] T.R. Nielsen, C. Drivsholm, "Energy efficient demand controlled ventilation in single family houses", Energy and buildings, Vol. 42, pp. 1995-1998, 2010.

[18] J. Laverge, N. van den Bossche, N. Heijmans, A. janssens, "Energy saving potential and repercussions on indoor air quality of demand controlled residential ventilation strategies", Building and environment, Vol. 46, pp. 1497-1503, 2011

[19] D. Johansson, H. Bagge, L. Lindstrii, "Measurements of occupancy levels in multi-family dwellings-Application to demand controlled ventilation", Energy and buildings, Vol. 43, pp. 2449-2455, 2011.

RANS AND LES MODELS COMPARISON FOR A CROSS-SHAPED JET FLOW WITH APPLICATION IN PERSONALIZED VENTILATION

Florin Bode[1], Ilinca Nastase[*2], Amina Meslem[3], and Cristiana Croitoru[4]

1 Technical University of Cluj-Napoca, Mechanical Engineering Department, B-dul Muncii nr.103-105, D03, Cluj-Napoca, Romania CAMBI Research Center, Technical University of Civil Engineering in Bucharest, Building Services Faculty, Avenue Pache Protopescu 66, Bucharest, Romania

2 CAMBI Research Center, Technical University of Civil Engineering in Bucharest, Avenue Pache Protopescu 66, Bucharest, Romania
**Corresponding author: Ilinca.nastase@cambi.ro*

3 LaSIE, University of La Rochelle, Pôle Sciences et Technologie, Avenue Michel Crépeau, 17042 La Rochelle, France

4 CAMBI Research Center, Technical University of Civil Engineering in Bucharest, Avenue Pache Protopescu 66, Bucharest, Romania

ABSTRACT

The need for thermal comfort and clean air for occupants in buildings or vehicles is vital since we spend more than 90% of our time inside these enclosed environments. Worldwide, current directions of the leading powers are oriented towards the reduction of the energy consumptions and HVAC systems make no exception. Personalized Ventilation (PV) applied to buildings may represent a solution to this problem. The main idea of PV is to provide clean air close to the face of each occupant and to improve thermal comfort in his microenvironment.

This study is a part of a larger research work related to the optimization of PV air diffuser for HVAC systems in terms of passive control of jets flows in order to control jet development before it impinges on the occupant's face.

The flow field of an elementary lobed jet from a cross-shaped orifice with straight edges is investigated numerically using Large Eddy Simulation (LES) model and the results are compared with those from seven RANS (Reynolds Averaged Navier-Stokes) turbulence models and with experimental results. The RANS models used are the RNG k-ε turbulence model, k-ε standard, k-ε realizable, the Shear Stress Transport (SST) k-ω, the standard k-ω, the Spalart-Allmaras turbulence models and the Reynolds Stress turbulence Model (RSM). The initial Reynolds number based on the jet centreline exit streamwise velocity and on the equivalent diameter is around 4000. Numerical results are analysed based on PIV measurements performed in the same flow.

The objective is to assess the capability and limitations of the studied viscous models to predict the significant features of the cross-shaped air jet by numerical simulation.

The study revealed that none of the turbulence models was able to predict well all jet characteristics in the same time. For instance, the Reynolds Stress turbulence Model (RSM) predicted better the local jet flow expansion in the longitudinal minor and major planes, whereas global flow expansion and ambient air induction are better predicted by the Shear Stress Transport (SST) k-ω turbulence model. Furthermore the RANS models are not able to capture the flow's temporal dynamics.

Also, the LES model allows obtaining global mean quantities as RANS models but delivers supplementary information about the temporal vortex dynamics.

KEYWORDS

Cross shaped jet, CFD, RANS turbulence models, LES

INTRODUCTION

The need for thermal comfort and clean air for occupants in buildings or vehicles is vital since we spend more than 90% of our time inside these enclosed environments. Worldwide, current directions of the leading countries are oriented towards the reduction of the energy consumptions and HVAC systems make no exception. Personalized Ventilation (PV) for buildings and HVAC systems can be one of the solutions to this problem. The main advantage of PV in comparison to mixing ventilation is to provide clean air close to the face of each occupant and to improve thermal comfort in his microenvironment. In this way we are using energy only to change for each occupant his microenvironment parameters and not to cool the entire building.

A major challenge for personalized ventilation is to be able to control the jet dynamics without causing an increase in energy consumption and occupant discomfort. An innovative concept for optimized air diffusion in buildings using passive control of air jet through lobed diffusers was proposed [1, 2]. Despite the consequent gain in ambient air induction for the lobed perforated panel flow, the streamwise maximum velocities display comparable values in the far field, which signifies comparable throws for the two flows.

Mixing and air entrainment of a lobed jet can be increased or minimized as a function of a set of geometric parameters such as: number of lobes, their heigth and width, the shape of these lobes. Designing an optimal geometry for these lobes only by means of experimental methods the cost both in financial terms and in terms of time required will be huge. Computational Fluid Dynamics (CFD) based methods are an alternative to experimental methods. Large Eddy Simulation (LES) and Reynolds-Averaged Navier–Stokes (RANS) equation solvers are used nowadays in a wide range of engineering and indoor environment modeling. By these method we can extract detailed information on the 3D flow field (wich is not possible nowaday by experiment means), and by changing the boundary conditions and nozzle geometry we can obtain a large parametric study. However the choise of the numerical model to predict well our complex flow at low or moderate Reynolds numbers is still an open question.

The general conclusion of a previous study was that none of the evaluated turbulence models correctly predicted all flow characteristics (the standard k-ε model, the Shear Stress Transport (SST) k-ω model and the Reynolds Stress Model (RSM) were investigated). However, of the seven RANS turbulence models, the SST k-ω appeared capable of reproducing reasonably well flow expansion and ambient air induction when the flow was numerically resolved through a lobed diffuser.

The RNG k–ε turbulence model is generally preferred among all RANS models in indoor airflows modelling [3-6]. Comparative studies [4] showed that the RNG k-ε model is stable and has the best overall performance among all RANS models for enclosed environments.

The RSM is the most elaborate turbulence model among the RANS based models. This turbulence model should be considered whenever non-isotropic effects are important; for example, in flows with strong curvature, swirling flows, and flows with strong acceleration/retardation. Shamami and Birouk [7] showed that the Reynolds Stress Model (RSM) was able to produce much better predictions of the features of the jet in comparison with the eddy viscosity models (the standard k-ε, Renormalization Group (RNG) k-ε, the realizable k-ε, Shear Stress Transport (SST) k-ω were evaluated by the authors). In all RANS turbulence models evaluated, RSM appeared to be the only one capable of reproducing the experimental data reasonably well.

In this study the flow field of an elementary lobed jet from a cross-shaped orifice with straight edges is investigated numerically using Large Eddy Simulation (LES) model and the results are compared with those from seven RANS (Reynolds Averaged Navier-Stokes) turbulence models. Due to the very time step necessary for LES, the corresponding results presented here are only partial. The initial Reynolds number based on the jet centerline exit streamwise velocity and on the equivalent diameter is around 4000. Numerical results are analysed based on PIV measurements performed in the same flow.

JET NUMERICAL SIMULATION

Computational details

The jet air studied in this paper is generated using a cross-shaped orifice with straight edges (Figure 1). The equivalent diameter of the cross-shaped orifice is D_e=10mm.

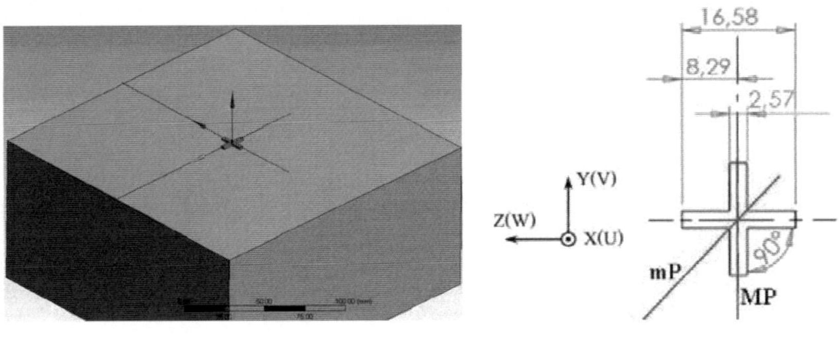

Figure 1. Investigated cross-shaped orifice

The computational domain (Figure 2) is composed of two parts separated by an orifice plate of 1.5 mm (0.15 D_e) thick. The upstream part and the downstream part of the domain has XYZ dimensions of 10×20×20 D_e and 29.85×20×20 D_e, respectively. Owing to the symmetry of the problem, just one fourth of the domain was modelled (so the dimension becomes 10 D_e in the Y and Z directions).

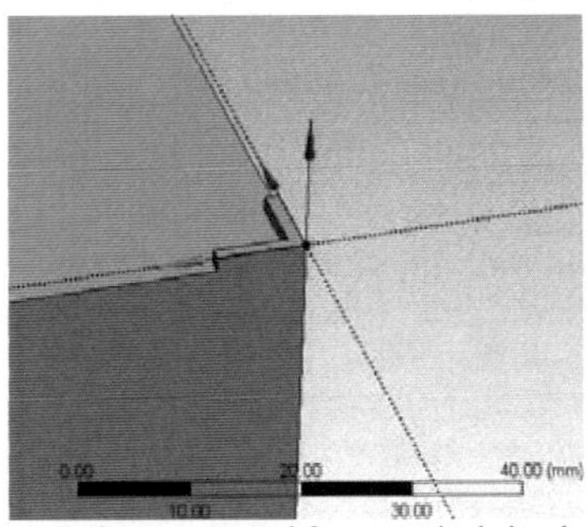

Figure 2. Geometry used for numerical simulation

The numerical analysis was performed using a finite volume based solver Fluent 13.0. Due to the low velocities, the incompressible solver was used. Fluent includes many viscous models, from the simplest laminar model to Large Eddy Simulation (LES), passing by RANS models and hybrid LES-RANS models. A direct full simulation (Direct Numerical Simulation - DNS) of the involved problem is still beyond our computational capabilities at the present time.

The inlet boundary conditions for the numerical simulation were given at the inlet plane of the upstream part of the domain. A uniform velocity normal to boundary of 0.00728 m/s, corresponding to a flow rate of 3.30×10^{-4} m^3/s and turbulence intensity (u' = 2 %) were fixed on this plane.
Given both the interest domain and the cross orifice jet symmetry, the symmetry boundary condition was used in the symmetry planes (*XY*) and (*XZ*) and the wall boundary condition for the orifice plate. There are other references in the literature with non conventional axi-symmetric jets (i.e. lobed jets) for which symmetry planes are used in order to gain computational resources [8-11]. The other boundary conditions applied to the planes which were far enough from the orifice (and are thus not affected by the flow) are atmospheric pressure boundary conditions.
The SIMPLE algorithm was used for pressure-velocity coupling. A second order upwind scheme was used to calculate the convective terms in the equations, integrated with the finite volume method.
Regarding the accuracy of results, the imposed convergence criterion was 10^{-6} for the variables residuals.
Computations were performed on a system Intel Xeon six core dual processor 2.66GHz with 96GBRAM.
In the LES case the numerical simulation was initialized with a time step, matching with a CFL number of 0.06. After solution stabilization the CFL number was raised to 0.2, which corresponds with a time step of 10^{-6} seconds.

Mesh dependency test

From our previous experience [9, 12, 13] we tried to achieve a final computational grid that would try to meet all necessary requirements for a good mesh, such as: the minimum number of cells needed in the critical section (30 cells), the smallest cell size (0.01mm), the largest cell size (2mm), $y+$ (less than 4 and its mean value was 1.3), the rate of cell growth (1.05), skweness of the cells. The resulted grid had a size of 4 million Cartesian non-uniform cells and all the simulation were performed using the same grid. This type of grid has been successfully used in our previous researches and managed to solve the flow field of different type of cross-shaped jets. A Cartesian grid is the best choice for flows with a strong velocity component in one direction such as jet flows.
For low Reynolds models the near-wall region is resolved down to the viscous sub-layer. A fine grid is nedeed in the near wall region which imposes at least one node in the viscous sublayer [14]. Values of y+ close to 1 are most desirable for the near-wall modeling. Wilcox has shown in [15] that the numerical discretization of such a function causes serious numerical errors in the viscous sub layer. He suggested enforcing the analytical solution in all points in the computational grid for which $y+ < 2.5$. The results also show that, in general, the error is the largest when thefirst cell center is located in the region $5 < y+ < 11$, despite the use of correct boundary conditions.
In our case, $y+$ was considered to be satisfactory for values less than 4, knowing that other literature recommends values of $y+$ less than 5 [16-18].
Results were compared with the experimental data of the turbulent cross-shaped jet [19] at moderate Reynolds number and the results of the comparison was satisfactory convincing us that we have a good quality mesh.
However, a mesh dependency study was conducted for five different grids (Figure 3): 0.7, 1.3, 3.3, 4 and 9 million hexaedral cells. In all cases, the initial flow rate was considered to be $Q_0=3.30\times10^{-4}$ m^3/s. From our previous experience [9, 12, 13], we used for the mesh dependency study SST k-ω turbulence model. The results confirmed that the initial grid was indeed a good choice.

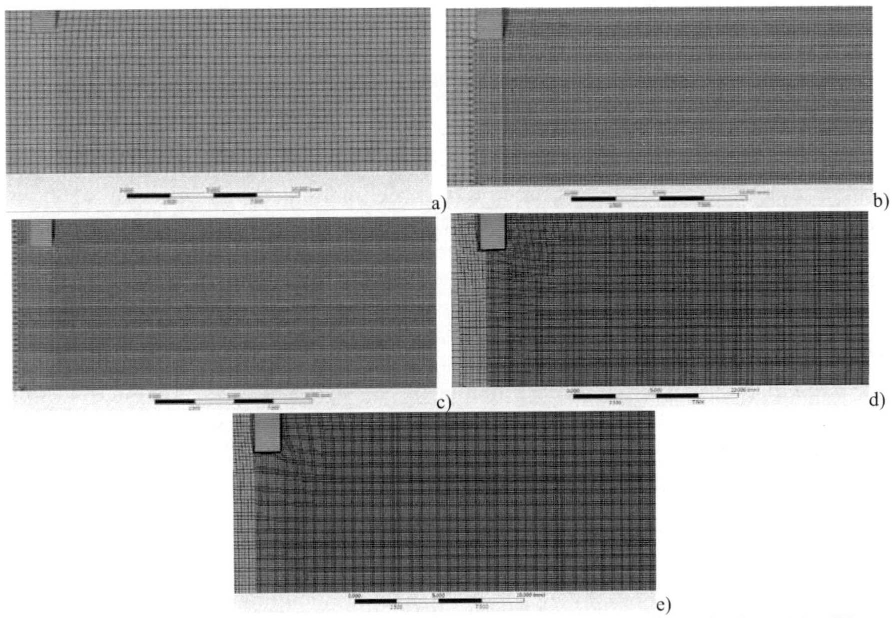

Figure 3. Details of several tested mesh grids: a) 0.7 millions of cells; b) 1.3 millions of cells; c) 3.3 millions of cells; d) 4 millions of cells; e) 9 millions of cells

Results obtained by the use of 4 and 9 million elements meshes are similar (Figure 4) and very close to the experimental results (Particle Image Velocimetry – PIV and Hot Wire Anemometry –HWA). The results obtained with grids of 0.7, 1.3 and 3.3 million elements are far away from experimental results, so grid chosen for numerical simulations was the 4 million elements.

Air entrainment is similar in all studied cases and verify experimental results obtained using PIV (Figure 5). Air entrainment is obtained by integrating streamwise velocity in the cross-planes with a minimum velocity of 0.15 m/s.

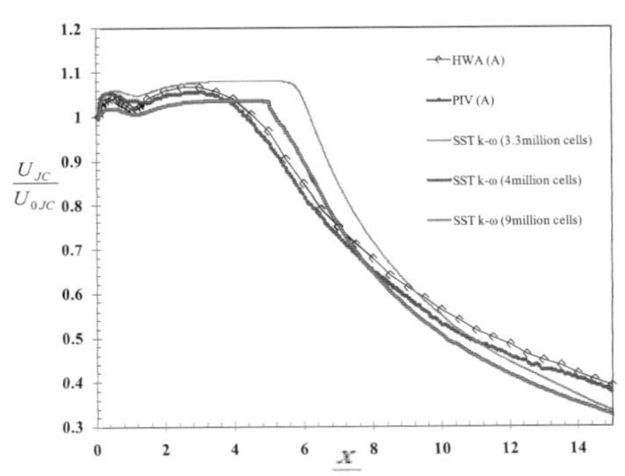

Figure 4. Normalized mean streamwise velocity on the jet centreline for 3.3, 4 and 9 million hexahedral cells compared with experimental results (PIV and HWA)

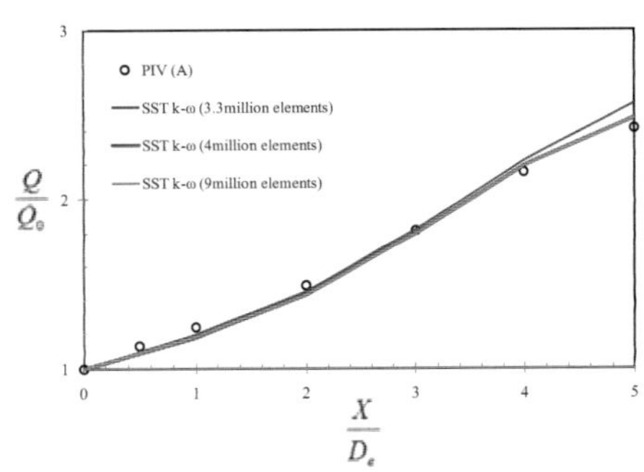

Figure 5. Air entrainment for 3.3, 4 and 9 million hexahedral cells cases compared with PIV experimental results

RESULTS AND DISCUSSION

The Reynolds number based on the Jet Centreline exit velocity (U_{0JC}) and on the equivalent diameter (D_e) is around 4000 for the experimental and numerical cases.

The velocity ratio U_{0JC}/U_{0mean} varies throughout the numerical cases, being closer to the experimental data in the case of SST k-ω and k–ε realizable turbulence models. On the other hand RSM model overestimates and RNG k–ε model underestimates the decay of the jet (Figure 6). k–ε standard, Spalart-Allmaras and k-ω standard are far away from experimental data and further we will not refer to these three turbulence models anymore. SST k-ω and k–ε realizable are the RANS models that predicts fairly well jet decay. Because of very small time step needed for LES model, the solution has not yet stabilized and results presented below for LES are only partial for now.

Ambient air induction (Figure 7) is obtained by integrating streamwise velocity in the cross-planes at 0, 1, 2, 3, 4, and $5D_e$ with a minimum velocity of 0.15 m/s. It is found from inspecting different velocity fields that this value is a good compromise to define the edge of the jet in the present study. As our particular application is directly interested to quantify the mixing between jets generated by HVAC terminal units and their ambience, we considered the 0.15 m/s criterion defining the extinction of the flow from the point of view of the thermal and draft comfort of the occupants [20].

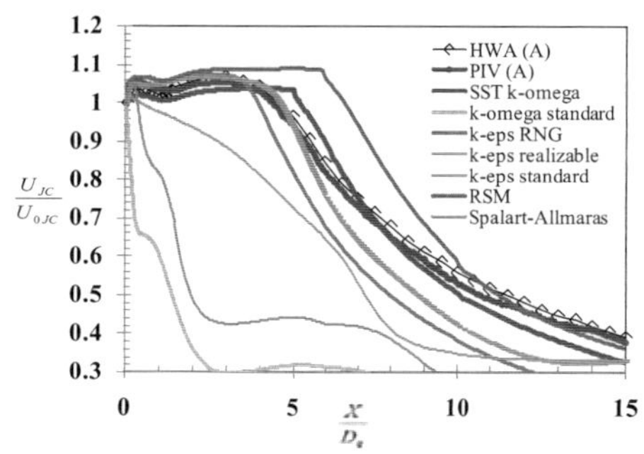

Figure 6. Normalized mean streamwise velocity on the jet centreline for studied numerical cases

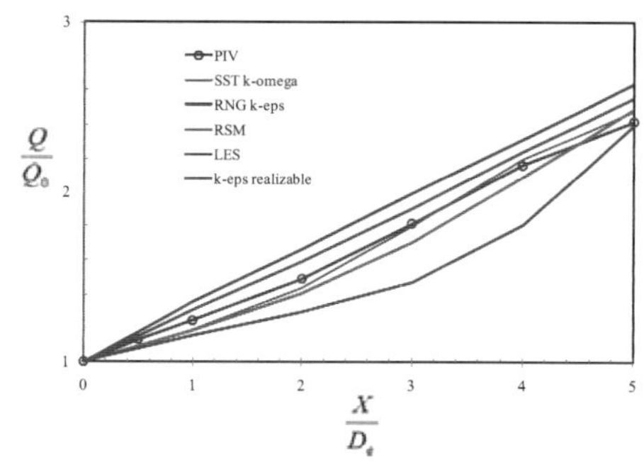

Figure 7. Streamwise evolution of the normalized volumetric flow rates for studied numerical cases

Ambient air induction is closer to the experimental data for SST k-ω and RSM turbulence models.

The difference in the predictions of the axial velocity profiles at 1 D_e in Major Plane (MP) and minor Plane (mP), as is depicted in Figure 1, is clearly visible in Figure 8.

k-ε turbulence models, predict unrealistic inflection points on MP, in the inner jet core region around Y = 0.3 D_e, whereas SST k-ω and RSM turbulence model predict realistic inflection points in the outer jet region, around Z = 0.7 D_e, in RSM case being more pronounced. The inflections are less pronounced than in the experimental situation for both numerical cases.

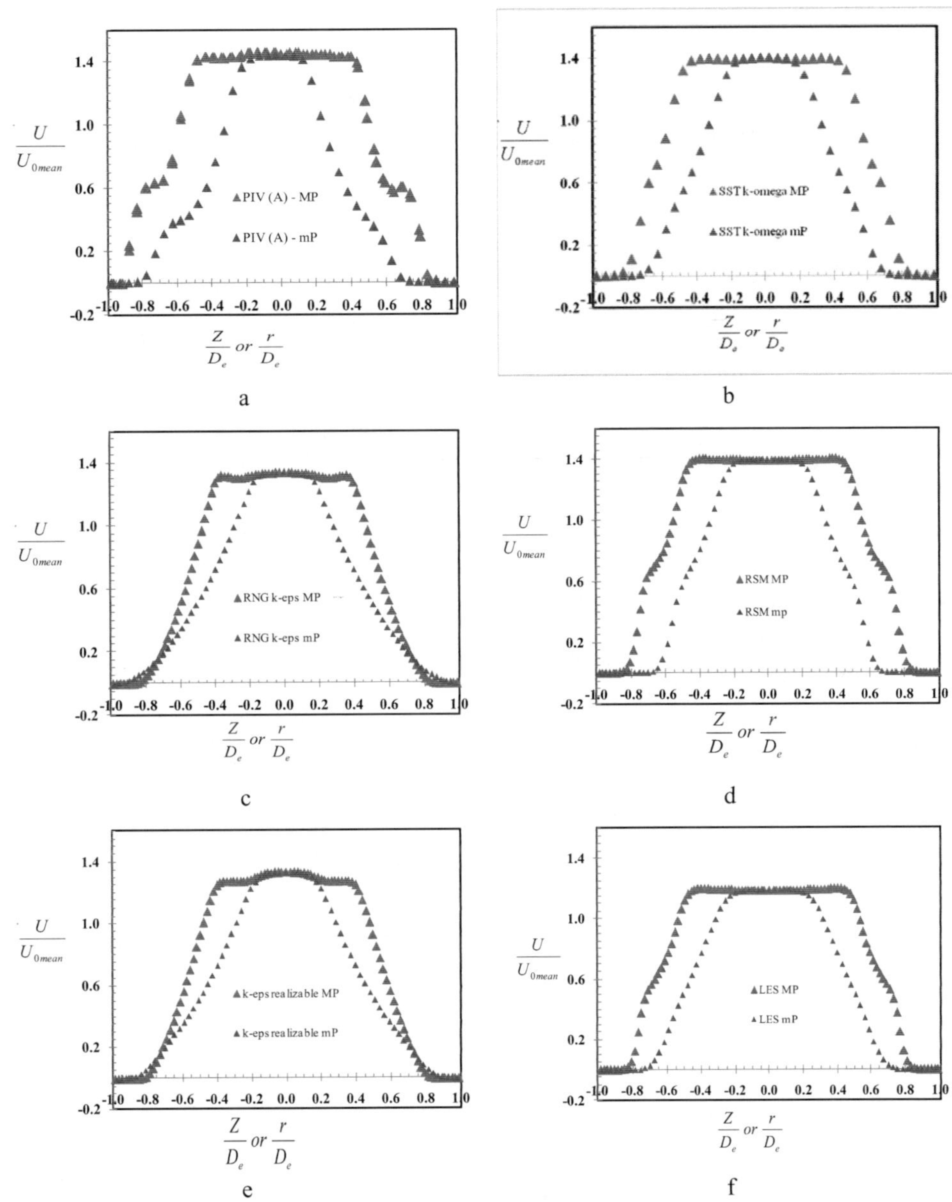

Figure 8. Streamwise velocity profiles in the MP and mP at $X = 1\ D_e$ for a. PIV, b. SST k-ω, c. RNG k–ε, d. RSM, e. k–ε realizable f. LES

The jets completly decays on Z direction at 0.7 D_e on minor Plane and at 0.8 D_e on Major Plane. Among all numerical cases SST k-ω and LES predict realistic decays of the jet both on Major Plane and minor Plane unlike RNG k–ε and k–ε realizable turbulence model.

Although there are only partial results for LES, can be seen in the Figure 9 the strong unsteady nature of the flow. Evolution of the LES results to date and comparison with experimental visualization results show close agreement between LES and experimental method.

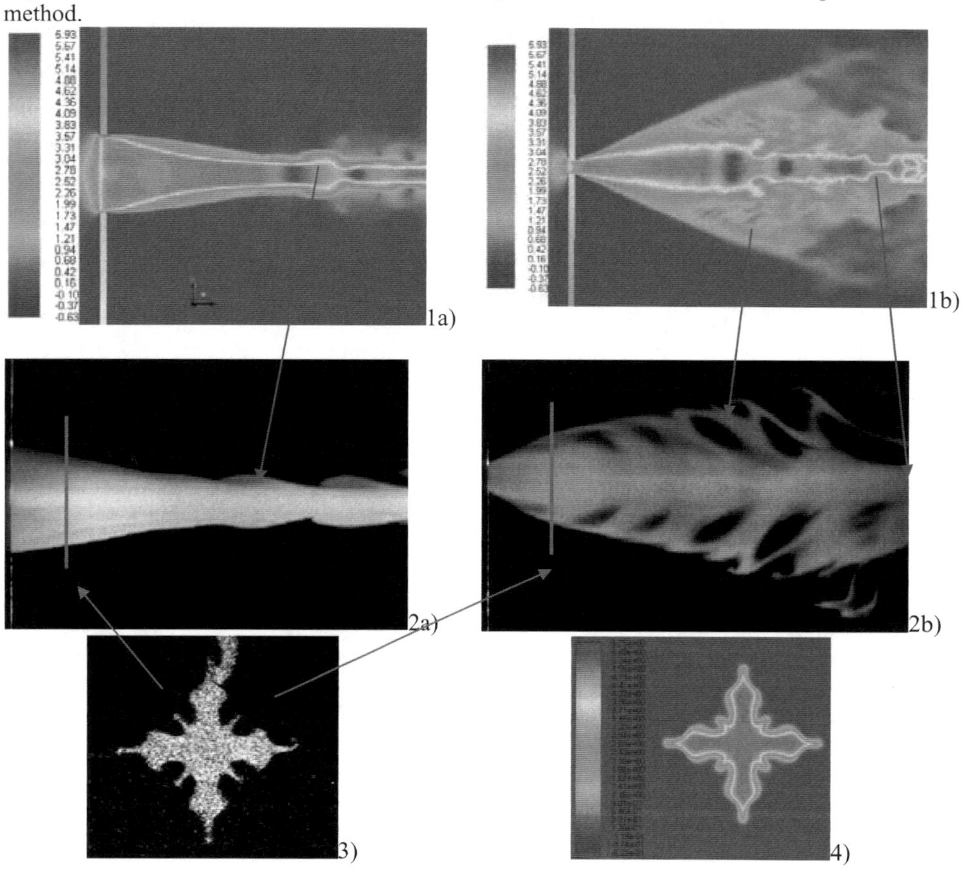

Figure 9. 1) LES isocontours of streamwise velocity, 2) High speed visualizations in the streamwise planes: a) Major Plane, b) minor Plane
3) High speed visualizations in the transverse plane at X=0.5D_e,
4) LES isocontours of streamwise velocity in the transverse plane at X=0.5D_e

CONCLUSION

In this study a three-dimensional flow at moderate Reynolds number through a cross-shaped orifice has been numerically investigated using seven turbulences models: RNG k-ε turbulence model, k-ε standard, k-ε realizable, the Shear Stress Transport (SST) k-ω, the standard k-ω, the Spalart-Allmaras turbulence models and the Reynolds Stress turbulence Model (RSM). The results are then compared with LES and experimental hot-wire anemometry and particle image velocimetry results.

The most accelerated flow at the jet inlet is given by the SST k-ω and RSM turbulence models and the value of the velocity ratio (maximum velocity on jet bulk-velocity) at the jet inlet is close to the experimental data.

From all models the k-ϵ models were not able to predict correct location of the inflection point in centreline streamwise velocity, shown by the experiments in the jet core region at 0.7D_e. Closest results to the experimental data were obtained by RSM and LES.

The study reveals that none of the turbulence models is able to predict well all jet characteristics. The Reynolds Stress turbulence Model (RSM) leads to better agreement between the numerical results and experimental data for the local jet flow expansion in the longitudinal minor and major planes, whereas global flow expansion and ambient air induction are better predicted by the Shear Stress Transport (SST) k-ω turbulence model.

Ambient air induction is closer to the experimental data for SST k-ω and RSM turbulence models.

As stated in [21] we confirm the superiority of the SST k-ω in comparison to all analyzed k–ε turbulence models, k-ω standard, Spalart-Allmaras and RSM, to correctly handle the transition of the flow through the diffuser.

The LES model still needs to be run. Because of very small time step needed for this model, the results presented in this paper are only partial. Evolution of the LES results to date and comparison with experimental visualizations shows very close agreement. Based on these enrouraging first results, LES is expected to well predict the nean flow of the lobed jet and its votex dynamics at moderate Reynolds number.

ACKNOWLEDGEMENTS

„This work was supported by the grants of the Romanian National Authority for Scientific Research, CNCS – UEFISCDI, project numbers: PN-II-PD-PCE-2011- 3-0099, PN-II-ID-PCE-2011-3-0835 and PN-II-PT-PCCA-2011-3.2-1212".

REFERENCES

1. Meslem, A., I. Nastase, and F. Allard, *Passive mixing control for innovative air diffusion terminal devices for buildings.* Building and Environment, 2010. **45**: p. 2679-2688.

2. Nastase, I., et al., *Lobed grilles for high mixing ventilation - An experimental analysis in a full scale model room.* Building and Environment, 2011. **46**(3): p. 547-555.

3. Chen, Q., et al., *Ventilation performance prediction for buildings: Model assessment.* Building and Environment, 2010. **45**: p. 295–303.

4. Zhang, Z., et al., *Evaluation of various turbulence models in predicting airflow and turbulence in enclosed environments by CFD: part 2- comparison with experimental data from literature.* HVAC&R Research, 2007. **13.**(6): p. 871-886.

5. Zhai, Z.J., W. Zhang, and Q. Chen, *Evaluation of various turbulence models in predicting airflow and turbulence in enclosed environments by CFD: part 1 - summary of prevalent turbulence models.* HVAC&R Research, 2007. **13**(6): p. 853 - 870.

6. Haidong Wang, Z.J.Z., *Analyzing grid independency and numerical viscosity of computational fluid dynamics for indoor environment applications.* Building and environment, 2012. **52**: p. 107-118.

7. Shamami, K.K. and M. Birouk, *Three-dimensional numerical simulation of an equilateral triangular turbulent free jet.* International Journal of Numerical Methods for Heat & Fluid Flow, 2009. **19**(7): p. 894 - 912.

8. McGuirk, J.J. and W. Rodi. *The calculation of three-dimensional turbulent free jets.* in *Symposium on Turbulent Shear Flows.* 1977. University Park, Pa ; United States.

9. Meslem, A., et al., *A comparison of three turbulence models for the prediction of parallel lobed jets in perforated panel optimization.* Building and Environment, 2011. **46**(11): p. 2203-2219.

10. Nathan, J.C., M. Parviz, and H. Hu. *Numerical Simulation of the Vortical Structures in a Lobed Jet Mixing Flow.* in *43rd AIAA Aerospace Sciences Meeting and Exhibit.* 2005. Reno, Nevada.

11. Yuan, Y., *Jet Fluid Mixing Control Through Manipulation of Inviscid Flow Structures*. 2000, Virginia Polytechnic Institute and State University: Ph.D. Thesis.
12. Bode, F., I. Nastase, and C. Croitoru, *Mesh dependency study using large eddy simulation of a very low reynolds cross-shaped jet.* Mathematical modelling 2011. **4**: p. 16-23.
13. Bode, F., I. Nastase, and A. Meslem, *Numerical investigation of very low Reynolds cross-shaped orifice jet.* Submitted to International Journal of Heat and Mass Transfer, 2012.
14. Georgi Kalitzin, G.M., Gianluca Iaccarino, Paul Durbin, *Near-wall behavior of RANS turbulence models and implications for wall functions.* Journal of Computational Physics, 2005. **204**: p. 265-291.
15. Wilcox, D.C., *Turbulence modelling for CFD.* La Canada- California: DCW Industries Inc. 536., 1993.
16. N. Martinho, A.L., M. Gameiro da Silva, *Evaluation of errors on the CFD computation of air flow and heat transfer around the human body.* Building and Environment, 2012. **58**: p. 58-69.
17. Tengfei (Tim) Zhang, L.T., Chao-Hsin Lin, Shugang Wang, *Insulation of commercial aircraft with an air stream barrier along fuselage* Building and Environment, 2012. **57**: p. 97-109.
18. Vieser W, E.T., Menter F., *Heat transfer predictions using advanced two-equation turbulence models.* ANSYS CFX Validation Report, 2004(1-62).
19. El-Hassan, M., A. Meslem, and K. Abed-Meraïm, *Experimental investigation of the flow in the near-field of a cross-shaped orifice jet.* Phys. Fluids, 2011. **23**(045101): p. 16 p.
20. Fanger, P.O., *Air turbulence and sensation of draught.* Energy and Buildings, 1988. **12**(1): p. 21-39.
21. El-Behery, S.M. and M.H. Hamed, *A comparative study of turbulence models performance for separating flow in a planar asymmetric diffuser.* Computers & Fluids, 2011. **44**: p. 248–257.

ABSTRACT SUBMISSION FORM FOR THE 2012 AIVC-TIGHVENT CONFERENCE

Send your abstract using this form at info@inive.org

PAPER TITLE

Experimental evaluation of thermal insulation efficiency for the dynamic insulation applied to window frame

MAIN AUTHOR

Shinsuke KATO

OTHER AUTHORS

Sihwan LEE, Daisuke KAWAHARA

ABSTRACT

An efficient thermal insulation of glazing or window frame is important because poor insulating performance usually cause the largest heat loss on residential buildings. As one of the methods decreasing heat loss of residential buildings, we proposed a dynamic insulation system applied to window frame, and its energy saving performance and applicability for residential buliding had been confirmed using numerical simulation in previous study [S. Lee, M. Tanaka, S. Kato, AIVC2010 & AIVC2011]. This system is composed of three parts: a dynamic insulation applied to window frame, a mechanical ventilation system, and a heat-recovery heat pump system.

The aim of this study is to evaluate the thermal insulation efficiency of the proposed system by chamber test using a prototype and field test using experimental model houses among which we mainly focus on the dynamic insulation applied to window frame. First, the prototype of proposed system is designed to ventilate through the dynamic insulation in a window frame based on the fixed window which uses argon gas injected triple glazing. Then, we verified its thermal insulation efficiency in steady state, after designing a new experimental chamber combine with a mechnical ventilator at the hot box method by the International Organization for Standardization 12567-1 : 2000. In addition, to verify its thermal insulation efficiency in transient state, we constructed two experimental model houses located in Sapporo, Japan. The one of the house is installed a normal window frame with argon gas injected triple glazing coated Low-E film and the other one is installed in the proposed window frame. Finally, we evaluated the U-value of the proposed window frame by changing the ventilation rates through the dynamic insulation, after measuring the heat loss differences and the effective air leakage area to compare the airtightness and thermal insulation efficiency of two experimental model houses.

Although it could not measured the same U-value on the proposed window frame compare with the chamber test and field test unfortunately, the thermal insulation efficiency is increased by increasing ventilation rates through the dynamic insulation. The field test results show that the thermal insulation efficiency is increased approximately 86 % (U-value : 1.65 $W/(m^2 \cdot K) \rightarrow 0.23\ W/(m^2 \cdot K)$) at the experimental model house installed in the proposed window frame by increasing ventilation rates ($0.00\ m^3/h \rightarrow 30.00\ m^3/h$). Therefore, the proposed system can be considered as an effective system to increase the thermal insulation efficiency on residential buildings.

SIMPLIFIED HOURLY METHOD TO CALCULATE SUMMER TEMPERATURES IN DWELLINGS

Lone H. Mortensen*[1] and Søren Aggerholm[1]

1 Danish Building Research Institute
Aalborg University
Dr. Neergaards Vej 15
2970 Hørsholm, Denmark
**Corresponding author:lhm@sbi.aau.dk*

ABSTRACT

The objective of this study was to develop a method for hourly calculation of the operating temperature in order to evaluate summer comfort in dwellings to help improve building design.

A simplified method was developed on the basis of the simple hourly method of the standard ISO 13790:2008 but with further simplifications. The method is used for calculating room temperatures for all hours of a reference year. It is essential that the simplified method is able to predict the temperature in the room with the highest heat load. The heat load is influenced by the solar load, internal load, ventilation loss, heat loss and the external temperature. The numbers of hours exceeding 26 °C and 27 °C are summarised in order to compare them with the requirements for summer comfort in the Danish Building Regulations.

The simplified method was qualified by comparison with simulation results from an advanced program for thermal simulations of buildings. The results are based on one year simulations of two cases. The cases were based on a low energy dwelling of 196 m². The transmission loss for the building envelope was 3.3 W/m², not including windows and doors. The dwelling was tested in two cases, a case with an ordinary distribution of windows and a "worst" case where the window area facing south and west was increased by more than 60%.
The simplified method used Danish weather data and only needs information on transmission losses, thermal mass, surface contact, internal load, ventilation scheme and solar load.

The developed method can calculate the number of hours above a given temperature limit. The limits are a prerequisite for the development of the simplified method, and a supplementary maximum temperature limit is suggested to ensure robustness. The setting of the ventilation rate is found to be essential for the fulfilment of summer comfort. Thus it is very important to address both opening areas and ventilation rates.

The developed simplified method makes it possible to test whether or not a building design for a dwelling will prevent excess of the summer comfort limits set by the building regulations.

KEYWORDS

Hourly temperature calculation, summer comfort, simplified method

INTRODUCTION

The Danish Building Regulations 2010 (BR10) [1] set new requirements for the thermal comfort in summer. The Building Regulations applies to buildings in low energy class 2015 (LE2015) or in building class 2020 (BC2020). For larger buildings, the documentation must primarily be based on advanced thermal simulations, and it is up to the developer to specify requirements for temperature conditions in summer. For dwellings, the requirements are that 26 ° C must not be exceeded for more than 100 hours a year and 27 ° C must not be exceeded for more than 25 hours a year.
The purpose of this paper is to present a simplified method for calculation of the temperature conditions in dwellings during summer. The simplified method is developed on the basis of

the simple hourly method of the standard ISO 13790:2008 [2], Annex C but with further simplifications.
BR10 contains a requirement that the energy performance framework of buildings must be verified by the program Be10, which is a method of calculating the energy demands of buildings [3]. In Be10, all energy-related data for a building is provided, but since the program calculates monthly average, it cannot be used directly for evaluation of summer comfort. However, the main idea of developing a simplified method is to connect it to the Be10 program, so that the building data provided for the energy calculation can be reused in the simplified method for evaluation of summer comfort.

Primary, the ability to develop a method that can be applied to the dwelling, as a whole, is preliminarily examined. This is a crucial point for developing a simplified method, which may be attributed to the program Be10 [3].
Additionally, it is investigated whether supplementary temperature requirements for dwellings can enhance robustness.

METHODS AND MODELS

The full set of equations for a simple hourly method given in ISO13790: 2008 [2], Annex C, is the basis of this new simplified method. The method was simplified with assumptions about summer conditions in new low energy dwellings in Denmark.

The preconditions for further simplification of the method for determination of summer temperatures in dwellings were:

- All solar heat gains were assumed allocated in the air, i.e. small error when the sun shines directly on the wall or floor, but with internal blinds or other light surfaces the solar heat will actually be allocated in the air.
- It was primarily the internal room surfaces of the structures that interacted with the room air so the heat transfer from the deep mass to the surface was neglected.
- The operating temperature was a combination of air and surface temperatures but not calculated directly as in ISO: 13790:2008 [2].
- Ventilation with outdoor air but the heat recovery can be by-passed in summer.

The simplified method for calculation of temperature conditions in dwellings during summer was carried out in a spreadsheet by using the full set of equations in ISO 13790. To begin the calculation, the initial and boundary conditions should be given. These included initial air and surface temperatures and a minimum air temperature. In addition, some parameters were retrieved directly from Be10 while others had to be specified. The summer ventilation rate was determined based on the effective opening areas for respectively the dwelling and the most critical room. The ventilation loss was calculated by use of the ventilation rate.

The spreadsheet calculated operating temperature as a combination of air temperature and surface temperatures. The heat gain was found by adding solar load and internal load and the heat loss by adding ventilation loss and transmission loss. By subtracting the heat gain from the heat loss the resulting loss/gain was found. By this the heat transfer between the air and surface temperature was calculated and thus new air and surface temperatures were found and the next hours operating temperature could be calculated.

The monthly solar load was retrieved from Be10 and was adjusted to proportionality of the global radiation for each month.

Test cases

The test cases were based on a dwelling design by "Eurodan huse" where the building envelope has been adapted to comply with energy requirements for both LE2015 and BC2020. The requirement for transmission loss for a BC2020 new construction is that it should not exceed 3.7 W/m² and for the test case the transmission loss for the current building was 3.3 W/m².

The total floor area of the dwelling was 196 m² and a floor plan is shown in Figure 1. The living room had large window areas facing both south and west, and it was considered to be at highest risk of having too high summer temperatures. Therefore, the living room was found to be the most critical room and thus served as a reference room. The living room had a floor area of 29 m² and a room volume of 72 m³. The described dwelling is referred to as Case 0 and is shown in Figure 1.

In this study a "worst case" was constructed by maximising the window areas of the south and west facing facades. This constructed variant is called Case 20 and is shown in Figure 2.

In both cases, the dwelling had balanced ventilation. There was exhaust from; utility room (10 l/s), kitchen (20 l/s) and bathrooms (15 l/s). The ventilation had heat recovery with 85% efficiency, which was bypassed in summer when heat recovery was inappropriate.

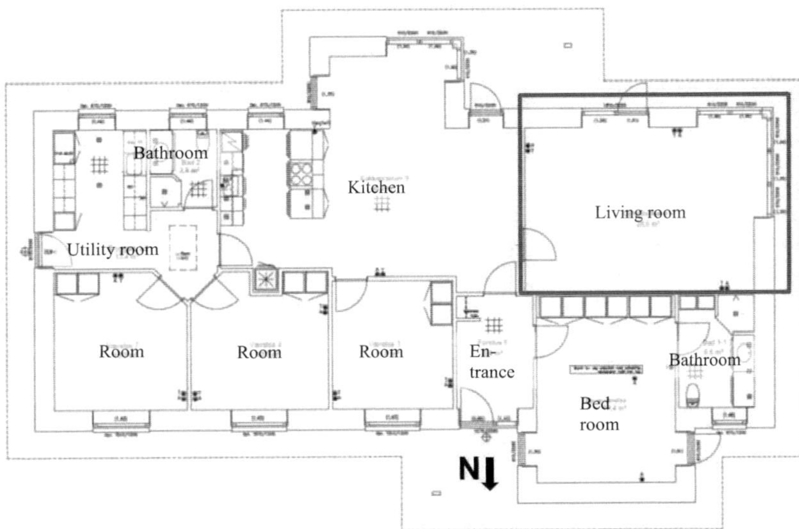

Figure 1. Plan of 196 m² low energy dwelling, corresponding to Case 0. The highlighted area is the living room.

Simulation tool

To test the simplified calculation method, a series of calculations was performed with BSim [4], which is an advanced program for building energy simulation. The program simulates operating conditions and the dynamic interaction between building installations and associated automatic control systems. This makes it suitable for indoor climate analysis and calculation of energy consumption in buildings. In analysis, this program is especially suited for parameter studies, for example of sunlight, use of passive solar protection, window sizes and orientation. In this study, BSim version 6, 11, 6, 20 was used.

Operating time
By definition, the operating hours of a dwelling are 24 hours a day. This means that occupants are always expected to be at home to regulate the indoor climate. In this study simulations were made for a time period of one year.

Venting
In BSim, it is possible to have advanced ventilation control settings. Since the simulations concerned a dwelling, the ventilation was expected to follow the operating time, which is always. However, in order to assess the robustness of the solution a simulation was also performed for a case where the dwelling was vacant and thus had limited ventilation.

Ventilation control

1. When the dwelling was always occupied, it was assumed that the window was opened whenever the air temperature exceeded 23 °C.
2. When the dwelling was vacant, the ventilation was restricted to passive venting or basic ventilation.

Sun protection
The robustness was assessed by simulation where the dwelling was vacant and venting therefore limited. As this easily gave high temperatures, it could be assumed that there were internal solar protection in rooms facing south and west, which was always activated. When the dwelling was vacated, it corresponded to the curtains being drawn.

Glass areas
In the simulations, two cases were investigated. The cases are shown in Figures 1 and 2 and have been described previously. Case 0 corresponds to an ordinary dwelling, while Case 20 is a constructed "worst case" variant of the same dwelling and more likely to experience problems of high temperatures in the summer due to enhanced window areas.

Models
A number of BSim models were created as the example shows for Case 20 in Figure 2, or part of it like Case 50. In the model shown in Figure 2, each room was designed as a separate thermal zone, but there were also models where all the rooms were combined into a single thermal zone. The model variants are described separately below.

Common to all models was that four simulations were performed on each model:

1. A simulation where the dwelling is always occupied and without solar protection
2. A simulation where the dwelling is always occupied and with solar protection
3. A simulation where the dwelling is always vacant and without solar protection
4. A simulation where the dwelling is always vacant and with solar protection

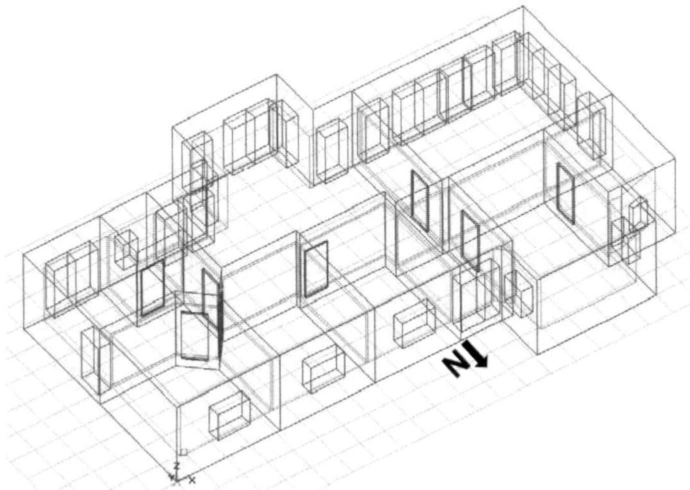

Figure 2. Model of Case 20, a constructed "worst case" variant of Case 0 with substantially increased window area in the South and West façade

Case 0
The geometry of Case 0 is shown in Figure 1. Each room has a separate thermal zone and there is no air mixing between the rooms, which correspond to closed doors between rooms.

Case 0X
In this model the geometry is as in Case 0. Opposite to Case 0 all rooms in the model is associated with the same thermal zone, corresponding to open doors and full air mixing between the rooms.

Case 20
The geometry of Case 20 is shown in Figure 2. The model is similar to Case 0, but with larger window areas to the south and west. Each room has a separate thermal zone and the doors are closed between the rooms.

Case 20X
In this model the geometry is as in Case 20. Opposite to Case 20, all rooms in the model is associated with the same thermal zone, corresponding to open doors between the rooms.

Case 40
This model only contains the living room with adiabatic interior surroundings. The model corresponds to the living room in Case 0 and the room is one single thermal zone.

Case 50
This model only contains the living room with adiabatic interior surroundings. The model corresponds to the living room in Case 20 and the room is one single thermal zone.

Simplified method simulations
By use of the simplified method, calculations were made for Case 0, see Figure 1 and Case 20, see Figure 2. Calculations were performed for every hour throughout the year using the

simplified method and the total number of hours exceeding respectively 26 °C and 27 °C was summarised.

Basically, the simplified method can be used to calculate the temperature condition for every single room in the dwelling. However, if general data for the dwelling is used, as provided in the program Be10 (for energy calculation), the calculation results would correspond to a case with full mixing between zones and open doors between rooms. Thus the results should be comparable with BSim results for cases with just one thermal zone like e.g. Case 0X. Therefore, to adress the possible high temperatures in the living room, a simulation must be performed where the solar load corresponds to the most critical room, which in this case was the living room. The solar load in the most critical room was the maximum solar load.

The highest ventilation rate for the dwelling was proportional to the sum of the basic ventilation and the effective opening areas. As for the solar load, it was nescessary to evaluate a ventilation rate for the entire dwelling as a whole, and one based on the maximum ventilation for the critical room. The lowest of the two ventilation rates per floor area should be used in the simplified method for calculation of summer comfort. When calculating the ventilation rate, it was allowed to asssume cross ventilation for the most critical room through the door if the rest of the dwelling could be cross ventilated. A limit for a maximum air change rate should be provided or possibly converted into a maximum ventilation rate, e.g. 10 $l/s \cdot m^2$.

Additional temperature requirements
In order to ensure the robustness of the dwelling design an additional maximum temperature could be suggested for a vacant dwelling. A supplementary requirement would ensure that pets can stay in vacant dwellings without suffering too high temperatures. A proposal for an additional temperature requirement might be that the temperature in a vacant dwelling should not exceed the maximum outdoor temperature (32.1 °C) for more than a couple of degrees, and thus have a maximum temperature of no more than, e.g. 34 °C.

RESULTS

The results of the simulations are shown in Tables 1 and 2. Table 1 gives the BSim results and Table 2 gives the results of the simplified method.

Temperatures	Case 0		Case 20		Case 0X		Case 20X		Case 40X		Case 50X	
Internal solar protection	+	–	+	–	+	–	+	–	+	–	+	–
Hours above 26 ° C	97	61	247	199	15	6	42	27	119	69	311	191
Hours above 27 ° C	46	23	148	117	3	0	14	6	60	29	205	110
Vacant dwelling Max. operating temperature, °C	*33.6*	*33.0*	*41.9*	*38.3*	*31.3*	*30.2*	*34.3*	*32.6*	*49.3*	*43.2*	*55.8*	*48.0*

Table 1. BSim results of simulations.

BSim results
The results in Table 2 show that Case 0 with internal shading comply with the temperature limits of maximum 100 hours over 26 °C and maximum 25 hours over 27° in the living room. In Case 0X and Case 20X, both with and without shading, the temperature requirements for

thermal comfort were also fulfilled, but these cases did not evaluate the most critical room separately as the entire dwelling was one thermal zone.
For the proposed additional requirement of a maximum temperature of 34 °C when the dwelling is vacant, the results were that Case 0 fulfilled the proposed requirement whereas Case 20 did not. Furthermore, Case 0X with and without solar shading and Case 20X with solar shading were below the proposed minimum temperature.
Case 40X and Case 50X did not fulfil any of the requirements.

Results of the simplified method

Table 2 show results for the calculations with the simplified method. There are results for two different solar loads, one is an average for the dwelling (basic) and one for the most critical room (max). In the case with the basic solar load, a combination with minimum ventilation rate of 0.9 l/s·m² in day and evening hours and 0.6 l/s·m² at night was calculated as described in SBi Guidelines 213 [3], and a similar investigation performed for a doubling of the minimum ventilation rate. The other ventilation rates were based on the effective opening areas of the dwelling and the most critical room, for Case 0 it was respectively 4.3 and 10.6 l/s·m². In the calculations of the ventilation rates it was assumed that all windows could be opened.

Temperatures	**Case 0**						**Case 20**					
Solar load	basic	basic	max	max	max	max	basic	basic	max	max	max	max
Ventilation rate, l/s·m²	0.9	1.8	0.9	1.8	4.3	10.6	0.9	1.8	0.9	1.8	6.1	15.6
Hours above 26 ° C	199	70	908	396	107	41	342	129	1763	798	154	49
Hours above 27 ° C	62	9	611	227	46	20	176	47	1339	531	73	23
Max. operating temperature, °C	*28.1*	*27.3*	*33.8*	*30.8*	*29.1*	*29.1*	*29.5*	*28.0*	*38.4*	*34.1*	*30.0*	*29.6*
With internal solar shading ↓												
Hours above 26 ° C	185	66	825	354	102	37	320	123	1641	744	130	45
Hours above 27 ° C	49	6	547	203	38	17	164	42	1198	483	60	23
Max. operating temperature, °C	*28.0*	*27.3*	*33.4*	*30.5*	*29.0*	*29.0*	*29.3*	*27.9*	*37.6*	*33.5*	*29.9*	*29.6*

Table 2. Results of the simplified method.

The results in Table 2 show that Case 0 both with and without internal shading complied with the temperature limits of maximum 100 hours over 26 °C and maximum 25 hours over 27 °C for the case with basic solar load and a doubling of the minimum ventilation rate and the case with maximum solar load and ventilation rate calculated on basis of the most critical room, living room. For Case 20, the only case to fulfil the requirements, the maximum solar load and ventilation rate was calculated on basis of the most critical room, living room.
According to the results, the influence of internal solar shading was quite small. However, the results differed for the overall image for cases without solar shading by reducing the number of hours above 27 °C.
Case 0 fulfilled the proposed requirement of a maximum temperature of 34 °C, whereas Case 20 did not. However, this was based on the ventilation rates given in Table 2 and not for a vacant dwelling with limited ventilation.

Comparing BSim results and results of the simplified method

For Case 0, the results indicated that it fulfilled the thermal comfort requirements for both the advanced BSim calculation and the simplified calculation method if the ventilation rate was sufficient. For the proposed additional temperature requirement of a maximum temperature of 34 °C, the results were the same that Case 0 fulfilled the requirement for both calculation methods.

For Case 20, the result indicated that it did not fulfil the thermal comfort requirements for the advanced BSim calculation, except in Case 20X. The same result was found for the simplified calculation method except in the case with maximum solar load and ventilation calculated on basis of the most critical room. For the proposed additional temperature requirement of a maximum temperature of 34 °C, the results were similar to the other Case 20 results. The maximum temperature limit for BSim results were exceeded for Case 20 but fulfilled for Case 20X, and for the simplified method the maximum requirement was met for the cases with maximum solar load and ventilation calculated on basis of the entire dwelling and the most critical room.

DISCUSSION

Summer comfort

In BR10 contains requirements for thermal comfort in summer, which apply to buildings in low-energy class 2015 (LE2015) or in building class 2020 (BC2020). Specifically for dwellings, the requirements stipulate that 26 °C must not be exceeded for more than 100 hours a year and 27 °C not for more than 25 hours a year. Two calculation methods were tested on two dwelling cases. Both methods gave results that show that Case 0 complies with the summer comfort temperature requirements. Therefore, it is concluded that Case 0 is an example of a dwelling, which meets the requirements for thermal comfort in summer. Opposite, the results for Case 20 showed that the summer comfort requirements cannot be met. This matches our expectations, since Case 20 is a constructed "worst case" variant of Case 0 with substantially increased window area in the south and west façades.

In general there is agreement between the results of the two simulation methods, the advanced BSim program and the developed simplified method. The results for the influence of internal solar shading are quite small. The effect is that the number of hours above 27 °C is lowered for cases with internal solar shading. This is important, as this temperature limit seems to be the most restrictive requirement for summer comfort.

Supplementary temperature requirements

Occupants have different user behaviours, but in all cases it must be assumed that dwellings will be vacant for longer or shorter periods. This differs from the assumption in these calculations. Therefore, an additional requirement of a maximum temperature, which must be observed when the dwelling is vacant, will help to ensure that pets and flowers do not suffer from extremely high temperatures. The proposed requirement is a maximum temperature of 34 °C, and the results from both calculation methods show that Case 0 complies with the requirement, while Case 20 do not. Thus, an additional requirement of a maximum temperature when the dwelling is vacant will support the other summer comfort requirements for the thermal indoor climate.

Caution with calculated temperatures

As in every other calculation program, both Be10 and BSim give results that are highly dependent on the calculation parameters. Therefore, it should be emphasised that the results in terms of number of hours above 26 °C and 27 °C do not correspond to the number of hours that can be measured in an occupied dwelling. One assumption in the calculation is that dwellings are always occupied and thus that occupants can provide extra ventilation to prevent too high temperatures. However, the requirements for evaluation of summer comfort can help to prevent the construction of dwellings that are vulnerable to solar gain and have limited window openings that are unable to provide sufficient ventilation.

Future work

The simplified method for hourly calculation of the operating temperature in order to evaluate summer comfort in dwellings needs to be tested on other dwelling designs. This should also include block of flats and terraced houses.

CONCLUSION

A simplified method has been developed for hourly based calculation of temperatures in dwellings. The simplified method enables evaluation of the summer comfort in dwellings. The calculation method applies to dwellings that are by definition always occupied and can thus be vented. The highest temperatures will be experienced in the most critical room. Therefore, the calculation is performed on basis of the solar load for that room.

The method ensures that there is adequate ventilation by windows that can be opened in both the most critical room and in the whole dwelling. The simplified method can be implemented in the program Be10, so that program as an additional result can provide the summarised number of hours above respectively, 26 °C and 27 °C, which can be used to evaluate summer comfort.

The calculation method was tested on two dwellings, Case 0 and Case 20. The result was that Case 0 meets the thermal summer comfort requirements, while Case 20 does not. This result was expected, as Case 20 is a constructed "worst case" variant of Case 0 with substantially increased window area in the south and west facade.

Along with the development of the method, a proposal is made for a supplemental maximum temperature requirement for a vacant dwelling. Also in this case the results show that Case 0 can comply with the requirement, while Case 20 cannot. The proposed maximum temperature in a vacant dwelling is 34 °C.

The simplified method summarises the number of hours above 26 °C and 27 °C, but that does not necessarily mean that this number of hours will correspond to the number of hours with high summer temperatures in a real dwelling. The number of hours with high summer temperatures will largely be driven by user behaviour. However, it is expected that the simplified method can help to ensure that there is sufficient opportunities for ventilation of the dwelling and suitable solar protection.

REFERENCES

[1] Building Regulations 2010 (in Danish : Bygningsreglementet 2010), The Danish Ministry of Economic and Business Affairs. Danish Enterprise and Construction Authority.

[2] ISO 13790:2008(E) Energy performance of buildings — Calculation of energy use for space heating and cooling

[3] SBi Guidelines 213, 2nd edition, Energy demands of buildings (in Danish : Bygningers energibehov), Beregningsvejledning, Søren Aggerholm, Karl Grau, Statens Byggeforskningsinstitut, Aalborg Universitet · 2011

[4] Wittchen, K.B. (Eiditor), Sørensen, K.G., Johnsen, K. and Rose, J. 2011. *BSim: User's Guide*. 11th edition. Hørsholm: SBI forlag.

ADDRESSING SUMMER COMFORT IN LOW-ENERGY HOUSINGS USING THE AIR VECTOR: A NUMERICAL AND EXPERIMENTAL STUDY

Axel Cablé*, Ghislain Michaux, and Christian Inard

LaSIE, University of La Rochelle
Pôle Sciences et Technologie
17042 La Rochelle Cedex 1 - France

**Corresponding author: axel.cable@gmail.com*

ABSTRACT

This article deals with summer comfort and room air distribution in low-energy housings. In such buildings, the efficient thermal insulation and air tightness make it crucial to efficiently dispose of the heat released by the internal gains. In this prospect, the comfort in a test room resulting from an integrated cooling and ventilation system is assessed both experimentally and numerically. The air is supplied into the room close to the ceiling through a wall-mounted diffuser of complex geometry composed of 12 lobed nozzles. Experimentally, the air velocity, CO_2 concentration, indoor air, wall and globe temperatures are monitored to assess the indoor environment quality. Numerically, CFD software Star-CCM+ is used to provide valuable information on the airflow patterns in the room. The CFD simulations are run in two steps in order to correctly integrate the complex diffuser's geometry. An excellent indoor environment is obtained in the studied conditions. Furthermore, a parametric study is performed in order to investigate the influence of the heat sources and of the supplying conditions on the airflow and on the resulting comfort.

KEYWORDS

HVAC, air diffusion, summer comfort, experimentations, CFD

INTRODUCTION

Poor indoor air quality and thermal comfort have an important impact on health, well-being and productivity [1]. It is therefore necessary to supply fresh air to the building and dispose of heat, particles, humidity and other pollutants generated in the building. This issue is particularly crucial for low-energy buildings. In fact, they benefit from good thermal insulation and air tightness. Consequently, the heat released by internal gains (*i.e.* the occupants and electronic devices) cannot easily be disposed of. While this reduces to a great extent the energy required for heating in winter conditions, it also makes it particularly difficult to maintain an acceptable indoor temperature in summer conditions. Therefore, the air distribution system should be able to efficiently evacuate the warm, old air, and ensure a homogeneous cooling of the occupied zone. In this prospect, an integrated cooling and ventilation system is considered, with conditioned air blown through a wall-mounted mixing diffuser. The resulting summer comfort in a room is assessed both experimentally and numerically. The experimental method aims at assessing the thermal comfort and ventilation efficiency by measuring the flow parameters in the occupied zone of a test room. At the same time, it provides realistic boundary conditions for the CFD simulations, whose purpose is to give additional information on the airflow patterns inside the room and in the jet region. Additional CFD simulations are then performed for supplying conditions which are not tested experimentally, in order to pinpoint the influence of the supplying conditions on the airflow pattern.

EXPERIMENTAL METHOD

Test room

The experiments are conducted in an airtight test room composed of steel wall panels (see Fig.1a). An air treatment unit is used to blow the inlet air at the specified temperature and flow rate and the wall temperatures are controlled with a thermal guard enclosing the test room. In order to enhance the mixing of the fresh, cold air, with the indoor air, a wall-mounted diffuser is used. It is composed of 12 complex lobed nozzles with a nozzle to nozzle spacing of 8 cm (see Fig.1b) and the first nozzle is located 16 cm below the ceiling. Two black cylindrical manikins of height 1.1 m and diameter 0.5 m are located inside the room to account for occupancy, with a CO_2 emission rate of 18 L/h. Lamps are placed inside the manikins to account for the sensible heat source, corresponding to a metabolism of 1 met.

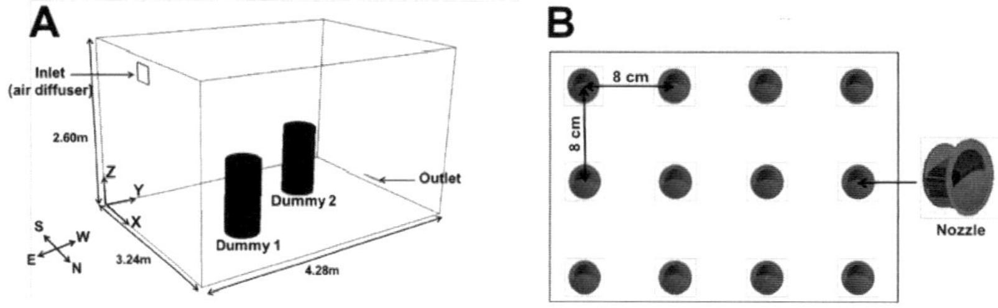

Figure 1. a) Room geometry, b) Diffuser geometry and detail of one nozzle.

Measurements

Measured variables include boundary conditions measurements at the air inlet, outlet and walls, as well as indoor variables measurements in order to assess the comfort. Air and globe temperature, air velocity and CO_2 concentration are monitored on 27 positions inside the occupied zone (Fig.2), corresponding to the ankle level (0.1 m), the neck level of a seated person (1.1 m) and the neck level of a standing person (1.7 m). The results are obtained when steady-state conditions are reached for all the measured values.

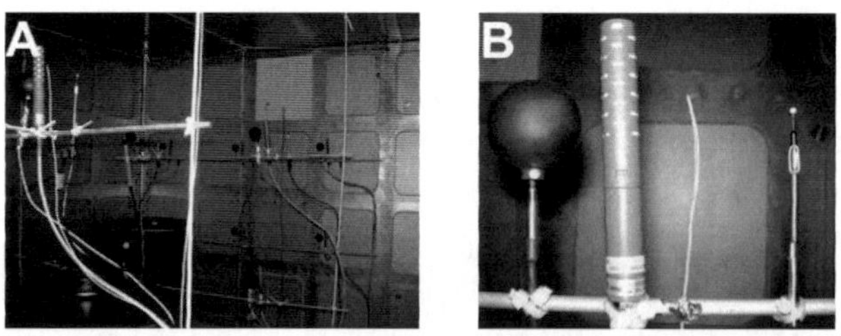

Figure 2. Experimental setup. a) Test room, b) Monitoring position in the occupied zone (globe thermometer, CO_2 probe, type-K thermocouple a nd omnidirectional thermal anemometer)

NUMERICAL METHOD

Numerical parameters

CFD code STAR-CCM+ is used to solve the 3D transport equations for mass, momentum, energy and CO_2 over the computational grid under steady-state conditions. The realizable k-ε turbulence model is chosen, which has been giving good results in room airflow simulations [2]. Two-layer wall functions are employed to model the heat transfer at the walls and buoyancy forces are taken into account through the variation of the density of the fluid considered as an ideal gas. Convergence is ensured by monitoring the residuals and the air and carbon dioxide mass balance evolution.

Diffuser modelling

The air diffuser is responsible for most heat and mass transfers inside the room, it is therefore crucial to correctly implement its geometry in the CFD simulation. However, the complex details of the diffuser make it too costly to directly include its complete geometry in the simulation. Several diffuser modeling methods have been developed to specify simplified boundary conditions either at the level of the diffuser (Momentum method) or on a surface from a distance of the diffuser (Box Method). However, these methods require either precise measurements at the exit of the diffuser or characteristic equations associated with the diffuser geometry [3]. Since this information is not available, another approach is followed. A first simulation of only one nozzle of the diffuser is performed in a reduced computational domain (3M polyhedral cells), whose outlet boundaries are located 20 equivalent diameters away from the nozzle. The velocity components, turbulence quantities and temperature profiles are then collected at the very exit of the nozzle (Fig.3). Prior simulations have been run with several adjacent nozzles in order to ensure that the flow is not affected by the other nozzles at such a close distance. The collected profiles are then set as inlet boundary conditions in the full room simulation for each of the 12 nozzles (Fig.3c). A similar approach had been successfully used by Cehlin and Mosfegh[4] with a perforated displacement diffuser.

The test room calculation domain is then composed of 3M polyhedral cells (Fig.3c), the grid being refined in regions of high gradients (at the exit of the diffuser, close to the walls and to the dummies). The inlet boundary conditions correspond to the collected profiles, and an atmospheric pressure condition is specified at the exhaust. The other boundary conditions used for the CFD simulations were provided by the experiments.

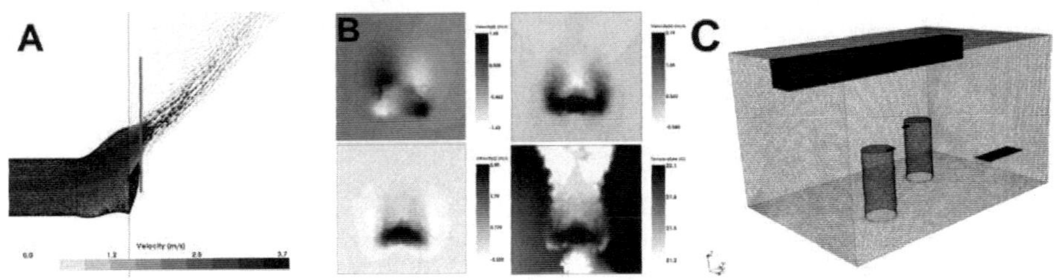

Figure 3. Diffuser modelling approach. a) Single nozzle simulation and profile collecting section, b) Collected velocity components and temperature profiles set as boundary conditions in the full room simulation c) Full room simulation (3M polyhedral cells).

RESULTS AND DISCUSSION

The obtained results are presented below, and analysed in terms of thermal comfort and ventilation efficiency. The airflow patterns are then investigated with the help of the CFD simulations in order to pinpoint the influence of the internal gains and supplying conditions on the airflow. The conditions of the studied cases are provided in Tab. 2, where T_{walls} corresponds to the average surface temperature of the six walls, and P_{inlet} is calculated as follows :

$$P_{inlet} = Q_0 c_p (T_0 - T_{OZ}) \quad (1)$$

Where Q_0 is the inlet air mass flow rate, T_0 the inlet temperature, and T_{OZ} the average temperature in the occupied zone. The air flow rate is the same for all cases (1,6 air change per hour), and is chosen in order to provide the cooling power while maintaining the supplied air temperature high enough. The Reynolds number (Eq.3), which qualifies the nature of the flow, and the Archimedes number (Eq.4), which is the ratio of the buoyancy force to the inertia force, are based on the equivalent diameter D_e of the free area of the 12 nozzles $A_{diffuser}$ composing the diffuser (Eq.2).

$$D_e = 2\sqrt{\frac{A_{diffuser}}{\pi}} \quad (2) \; ; \quad Re_0 = \frac{\rho U_0 D_e}{\mu} \quad (3) \; ; \quad Ar_0 = \frac{g\beta_T |T_0 - T_{OZ}| \, 2\sqrt{A_{diffuser}/\pi}}{{U_0}^2} \quad (4)$$

Where U_0 is the inlet velocity.

	Q_0			T_0	$T_{exhaust}$	T_{guard}	T_{walls}	P_{dummy}	P_0	Q_{CO2}	$T_0 - T_{oz}$	Ar_0	Re_0
	m^3/h	kg/h	ACH	°C	°C	°C	°C	W	W	l/h	°C	-	-
Case 0	55.6	66.7	1.62	21.5	24.9	28.0	25.7	-	69	-	-3.7	9.6E-04	16384
Case 1	57.0	69.4	1.66	17.0	25.3	28.0	25.9	107	162	18.0	-8.3	2.1E-03	17247
Case 2	56.6	69.4	1.65	14.9	24.8	27.2	25.4	213	200	36.0	-10.3	2.7E-03	17356

Table 1. Boundary conditions for the studied cases

Global comfort

The EN 15251 standard [5] was used to evaluate the indoor environment quality (IEQ). It provides a classification of the IEQ depending on the values of the operative temperature, Predicted Mean Vote as defined by Fanger [6], and carbon dioxide concentration level. Category I corresponds to an IEQ level expected in schools or buildings with sensitive persons. Category II is the level expected for new or refurbished buildings, while Category III is the value expected for existing buildings. In addition, the ISO 7730 standard [7] defines the Draught Rate as the percentage of dissatisfied for a given air velocity, air temperature and turbulence intensity.

Category	Maximum operative temperature (cooling season) (°C)	PMV	Maximum CO_2 level above outside concentration (ppm)
I	25.5	< 0.2	300
II	26	< 0.5	500
III	27	< 0.7	800

Table 2. IEQ categories according to the EN 15251 standard[5]

The contaminant removal effectivness (Eq.5) and temperature efficiency (Eq.6) as defined by Sandberg [8] are also considered, and express the ability of the air distribution system to dispose of a pollutant, and to cool the indoor air, respectively :

$$\varepsilon_C = \frac{c_{exhaust} - c_0}{c_{OZ} - c_0} \quad (5)$$

$$\varepsilon_T = \frac{T_{exhaust} - T_0}{T_{OZ} - T_0} \quad (6)$$

The global results obtained for the experimental test cases are provided in Tab.3. The average, minimum and maximum values of the operative temperature, PMV and carbon dioxide concentration are provided, as well as the contaminant removal effectiveness and temperature efficiency.

MIN|AVG|MAX	Top (°C)			PMV			DR (%)			C(ppm)			ε_C	ε_T
Case 0	24.0	25.0	25.6	-0.12	0.20	0.40	0.0	0.2	2.8	-	-	-	-	0.91
Case 1	24.9	25.4	25.8	0.08	0.26	0.41	0.0	0.5	6.7	639	680	719	0.98	1.00
Case 2	24.8	25.3	25.7	-0.03	0.18	0.32	0.0	0.7	5.1	953	1001	1069	1.00	0.97

Table 3. Comfort values and ventilation efficiency obtained from the measurements for the tested cases

Overall, an excellent indoor environment quality is obtained for the studied supplied air conditions. A satisfactory operative temperature is obtained, as well as PMV value under 0.5. Furtermore, the draught risk is limited for all cases, not exceeding a maximum value of 6.7% of dissatisfied. The CO_2 concentration reaches an acceptable mean value of 1001ppm for Case 2, where two occupants are in the room. The contaminant removal effectiveness is close to 1, which means that the mixing strategy is efficient, and that the heat and CO_2 generated by the occupants are efficiently disposed of. The contaminant removal strategy is illustrated on Fig. 4, where the streamlines from the CO_2 injection colored in CO_2 concentration are displayed. The injected CO_2 is immediately entrained into the buoyant plume and brought to the upper part of the room, and is then entrained into the air jet where it is cooled down and falls in the occupied zone with a lower CO_2 concentration. In this case, the polluant sources are associated with the heat sources, the ventilation strategy is hence efficient. However, many indoor air pollutants (VOC, aerosols...) are emitted from passive sources. It would be interesting to determine wether the studied system would be as efficient to remove such kind of polluants.

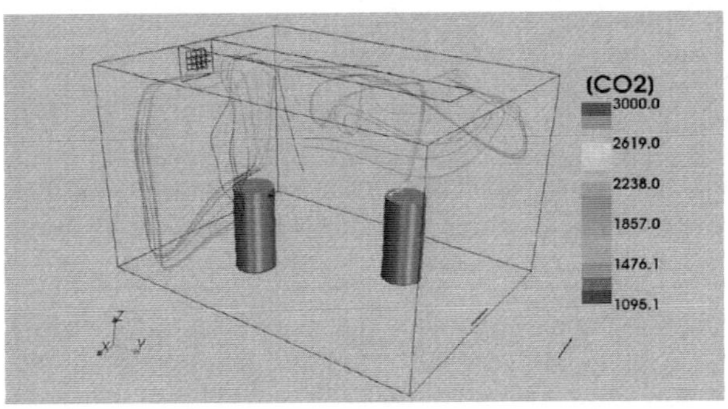

Figure 4. Streamlines colored in CO2 concentration for Case 2

Influence of the heat sources on the airflow pattern

The analysis of the global temperature and velocity values gives valuable information on the influence of the heat sources on the airflow in the room and on the resulting comfort. The mean, median, minimum and maximum values of the air temperature and air velocity for the experimental cases are plotted on Fig.5. All test cases are performed at the same air change rate. It appears that the air velocity in the occupied zone is higher with heat sources (Case 1 and Case 2), due to the entrainment of the ambient air into the buoyant plumes. It also stresses that an increase of the air velocity causes the distribution of the air temperature to be more homogeneous in the occupied zone, which is consistent with a mixing ventilation strategy.

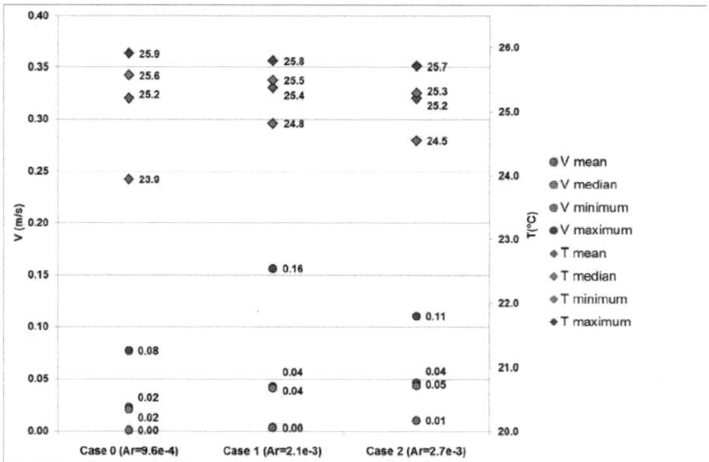

Figure 5. Mean, median, minimum and maximum values of the air temperature and air velocity for the experimental cases

However, the maximum velocity is higher for Case 1 than for Case 2, which indicates that the airflow maybe not be the same for both cases. The velocity streamlines colored in temperature and velocity vectors in the middle plane obtained from the CFD simulations for all three cases are presented in Fig.6 and provide interesting information about the airflow patterns in the room.

In Case 0, the jet is attached to the ceiling thanks to the Coandă effect and goes straight forward, thus reaching the opposite wall. In Case 1 and Case 2, the airflow is tridimensional and follows a clockwise path in the test room. This results from the non-symmetric placement of the dummies. The air jet issued from the diffuser is first deflected to the left by the buoyant plume of the dummy 1. The influence of the heat sources on the airflow is emphasized by the presence of recirculation bubbles above the dummies. The jet then separates from the ceiling and drops slightly in the occupied zone before hitting the opposite wall. It is then deflected backwards toward the occupied zone where it is again entrained in the buoyant plumes and in the air jet. The location and power of the internal gains thus greatly influence the air flow pattern in the room.

Furthermore, it can be seen on the velocity vectors that the jet remains attached to the ceiling longer in Case 0 and Case 1 than in Case 2, which highlights the influence of the supplying conditions on the airflow. However, this influence is hard to pinpoint because of the varying occupancy in the different cases. Hence, a parametric study is performed on the supplying conditions where occupancy is kept constant.

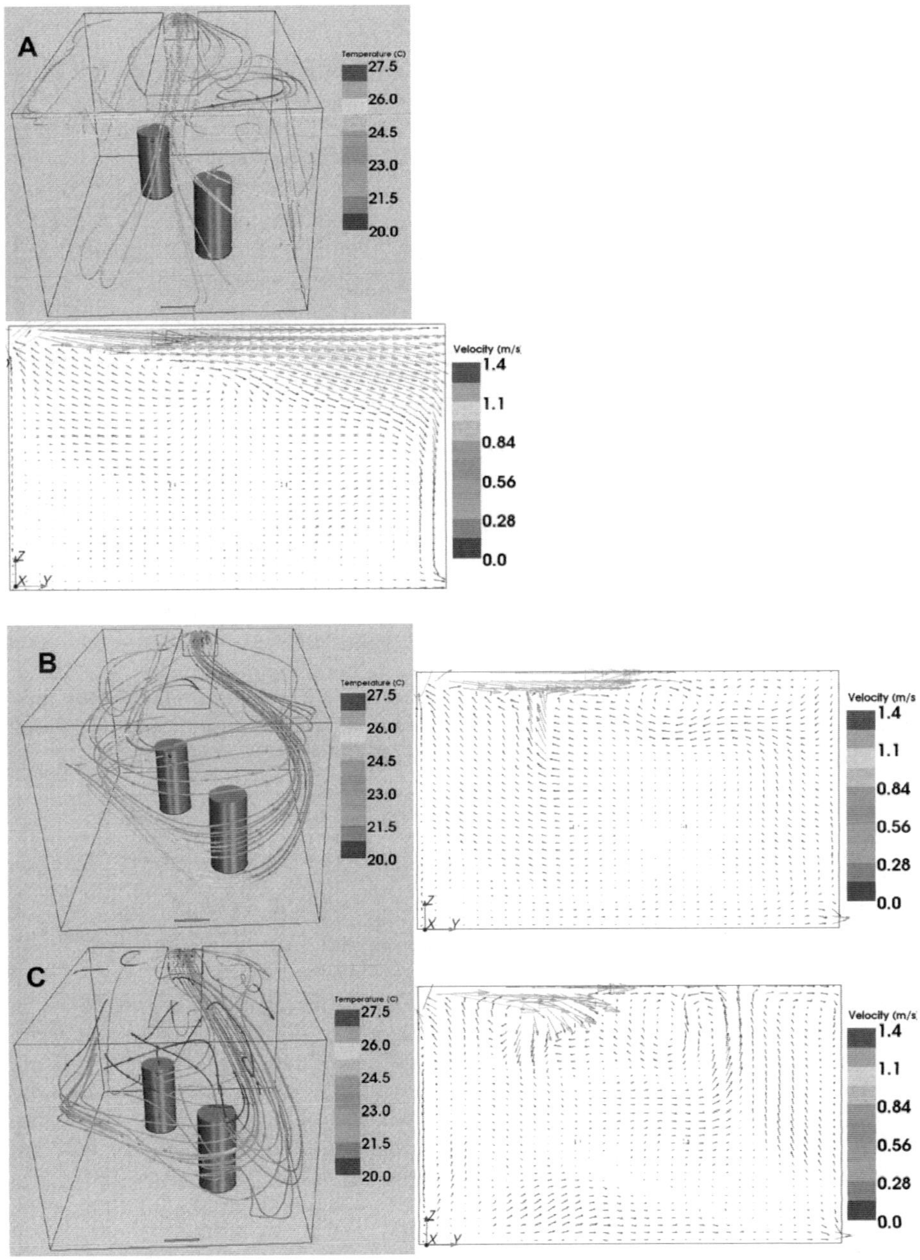

Figure 6. 3D velocity streamlines colored in temperature and velocity vectors in the middle plane. a) Case 0, b) Case 1, c) Case 2.

Parametric study on the supplying conditions

The choice of the inlet air flow rate and air temperature is crucial during the room air distribution system design, in order to reach the desired IEQ level. Therefore, additional simulations are performed for cases that were not tested experimentally, in order to evaluate the influence of the supplying conditions on the airflow pattern and on the resulting IEQ. The cooling power brought by the supply, the internal gains, the wall temperatures and the CO_2 source are kept constant (corresponding to Case 1), while different inlet air flow rates and supply temperatures are considered. The conditions of the tested cases are listed in Tab.4.

	Qinlet		Tinlet	Toutlet	Toz-Tinlet	Ar0	Re0
	m3/h	ACH	°C	°C	°C	-	-
Case 1	57.0	1.66	17.0	25.3	8.3	2.1E-03	17247
Case 1⁻	34.3	1.00	11.7	25.3	14.1	1.0E-02	10722
Case 1⁺	85.7	2.50	19.7	25.8	6.2	6.8E-04	25497

Table 4. Supplying conditions for the parametric study

The air flow rate should be high enough to maintain an acceptable air quality, but low enough to not cause any draughts in the occupied zone. A way to define the maximum air flow rate is to choose a maximum velocity allowed in the occupied zone, or at the upper limit of the occupied zone (1,8 m), for instance 0.15 m/s [9]. Buoyant plumes may be responsible for draught as well, but the high air velocity resulting from the dummy is not taken into account in the following maximum air velocity values, the studied parameter being the supplied air conditions.

The contaminant removal effectiveness and temperature efficiency for the three tested cases are displayed in Tab. 5. Compared to Case 1, both Case 1^- and Case 1^+ display a better mixing in the occupied zone, which is translated into a better temperature efficiency.

	ε_C	ε_T	V_{mean}	$V_{max\_1.8m}$	V_{max_OZ}
Case 1	0.98	0.94	0.04	0.14	0.14
Case 1⁻	=	+ 0.02	+ 0.01	+ 0.13	+ 0.13
Case 1⁺	+ 0.05	+ 0.02	+ 0.04	- 0.01	+ 0.06

Table 5. Contaminant removal effectiveness, temperature efficiency, mean, maximum and maximum velocity et Z = 1.8m in the OZ, compared to Case 1.

However, a different phenomenon is responsible for the increased mixing in both cases. In Case 1^-, the cold air jet seperates from the ceiling because of the high Archimedes number and drops into the occupied zone, as shown on Fig.7a. Consequently, the mixing is increased, but so is the draught risk with a maximum air velocity of 0.27 m/s in the occupied zone. In Case 1^+, the jet reaches the opposite wall thanks to the higher air flow rate. The latter also explains the increased mixing, caused by an higher entrainment of the ambiant air into the cold air jet. But this is then responsible for a maximum air velocity at the ankle level of the occupied zone of 0.20 m/s, which results from the jet hitting the opposing wall and being deflected backwards. Therefore, it appears that a low enough Archimedes number which would ensure that the jet would not fall into the occupied zone is not a sufficient condition to guarantee that a high draught risk may not occur, caused by locally high values of air velocity.

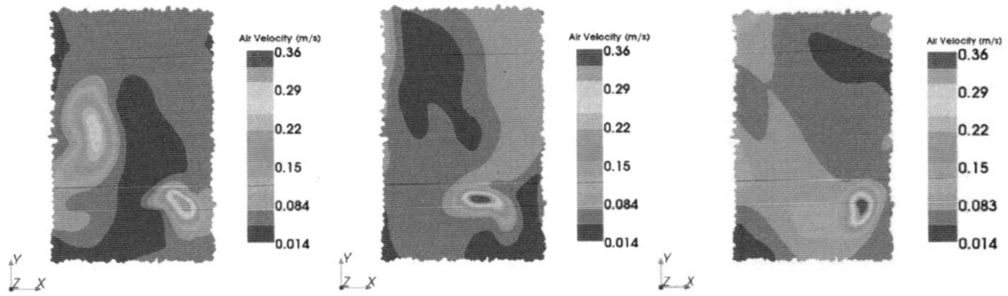

Figure 7. Velocity contours at the upper limite of the occupied zone : Z = 1,8 m. a) Case 1^- , Ar0 = 1.0E-2 b) Case 1, Ar0 = 2.1E-3 c) Case 1^+, Ar = 6.8E-4

CONCLUSION

The summer thermal comfort and the ventilation efficiency have been assessed in a test room using a wall-mounted mixing diffuser. Both experimental measurement and CFD simulations have been used. An excellent indoor environment quality has been obtained in the tested conditions. It has been found that the location of the heat sources in respect to the location of the air diffuser plays a major role on the airflow in the room and on the resulting IEQ. This emphasises that the location and power of the heat sources should be taken into account during the design of an air distribution system. Furthermore, it appeared that predicting the throw of the cold air jet to ensure that it does not fall in the occupied zone, or predicting the average air velocity in the occupied zone with the Archimedes number may not ensure that there would not be locally high values of air velocity responsible for local discomfort by draught. It would therefore be advised to perform new CFD simulations to predict the airflow whenever a new room geometry is considered.

ACKNOWLEDGEMENTS

This study is part of the HABISOL-VABAT programme and is supported by the French National Agency of Research (ANR), which is gratefully acknowledged.

REFERENCES

[1] Lan, L., Wargocki, P., Wyon, D.P., and Lian, Z. 2011. *Effects of thermal discomfort in an office on perceived air quality, SBS symptoms, physiological responses, and human performance*, Indoor Air 21, 376–390.

[2] Costa, J.J., Oliveira, L.A., and Blay, D. 1999. *Test of several versions for the k-e type turbulence modeling of internal mixed convection flows*, Int. Journal of Heat and Mass Transfer 42, 4391-4409.

[3] Fontaine, J.R., Rapp, P., Koskela, H., and Niemelä, R. 2005. *Evaluation of air diffuser flow modeling methods experiments and computational fluid dynamics simulations,* Building and Environment 40, 377-389.

[4] Cehlin, M., and Moshfegh, B. 2010. *Numerical modeling of a complex diffuser in a room with displacement ventilation*, Building and Environment 45, 2240-2252.

[5] EN 15251. 2007. *Indoor environmental input parameters for design and assessment of energy performance of buildings- addressing indoor air quality, thermal environment, lighting and acoustics*, CEN, Brussels.

[6] Fanger, P.O. 1970. *Thermal Comfort,* Copenhagen: Danish Technical Press.

[7] EN 7730. 2005. *Analytical determination and interpretation of thermal comfort using calculation of the PMV and PPD indices and local thermal comfort criteria.*
[8] Sandberg, M. 1981. *What is Ventilation Efficiency?*, Building and Environment 16, No. 2, 123-135.
[9] Nielsen, P.V. 2007. *Analysis and design of room air distribution systems*, HVAC&R Research 13, 987-997.

FRENCH POLICY FOR SHELTER-IN-PLACE: AIRTIGHTNESS MEASUREMENTS ON INDOOR ROOMS

Gaëlle Guyot[*1], Daniel Limoges[1], François-Rémi Carrié[2]

1 CETE de Lyon
46 rue Saint Théobald. F – 38081 L'Isle d'Abeau Cedex, France,
**Corresponding author: Gaelle.Guyot@developpement-durable.gouv.fr*

2 ICEE,
93 rue Molière, 69003 Lyon, France

ABSTRACT

Accidental dispersion of toxic gas clouds may occur around industrial platforms or during hazardous materials transportation. In case of such a toxic risk, the best protection strategy is to remain inside a building and seek refuge in an airtight room identified as "shelter" until the toxic cloud has finally been swept off. This strategy called "passive shelter-in-place" also includes obstructing all external openings and turning off all mechanical ventilation systems

Following the AZF chemical accident (Toulouse, 2001, 31 deaths), a French law was adopted in 2003 that can compel public and private building owners to adopt such a shelter-in-place strategy. To prove that the shelter airtightness is sufficient and that the occupants will not be exposed to irreversible effects, the shelter's air leakage measurement is compulsory for buildings owners. Envelope leakage does not need to be measured.

This paper gives an overview and first analysis of collected airtightness measurements for these indoor shelters. More than 100 results have been collected, with information on the building use (one-family dwelling / multi-family dwelling / non residential), the required airtightness level, the volume, the floor area, the year of construction. The final goal of this database is to give a picture of the vulnerability of housing stock around industrial platforms.

The aim is to help local decision makers with information related to the cost and the extent of works to be done on buildings in order to protect people against toxic risk, e.g. to reach the expected airtightness requirement, regarding some criteria like building use, geometric characteristics of the shelter, year of construction.

These experimental data can also be used as inputs in multi-zone airflow and pollutant transfer model, when data on internal airtightness are needed to study inter-zone airflows.

KEYWORDS

Air infiltration, air leakage, shelter-in-place, vulnerability, toxic risk, land-use, indoor air transfer, airtightness measurement

INTRODUCTION

Accidental dispersion of toxic gas clouds can occur around industrial platforms or during hazardous materials transportation. In case of such a toxic risk, two strategies can be implemented to protect people: shelter-in-place or evacuation [1]. In France, like in other countries, passive shelter-in-place has been found the best protection strategy. It consists in having people remain inside a building and seek refuge in an airtight room identified as 'shelter' until the toxic cloud has finally been swept off. Following the AZF chemical accident (Toulouse, 2001, 31 deaths), a French law adopted in 2003 established a land-use tool around all SEVESO II (high level) classified establishments [2]: the technological risk

prevention plan (PPRT) [3]. Such a plan specifies protective construction works for future and existing buildings in case of toxic risk in the plant, which consist into the implementation of a shelter-in-place system against toxic risk.
On 14th August 2012, 182 PPRT have been established, 8 PPRT have not begun and 212 are under development.

DESCRIPTION OF SHELTER-IN-PLACE REQUIREMENTS ON BUILDINGS

Shelter-in-place requirements are detailed in a guide we wrote up for the French Ministry in charge of PPRT plans development [4]. It is compulsory for a shelter-in-place system to achieve the protection of people during 2 hours against irreversible effects caused by a toxic cloud.

Firstly, a shelter-in-place system includes general constraints on the whole building and on a room used as shelter. These constraints do not depend neither on the toxicity of the products, nor on the intensity of the toxic cloud.
For instance, each building has to be equipped with a system that quickly stops all voluntary airflows, which supposes an emergency circuit breaker on ventilation systems and devices to close rapidly the air inlets and outlets. The room used as shelter must respect a minimum size per occupant (1 m^2, 2.5 m^3). The heating system must be adjustable from the room. Toilets are compulsory in the shelter for non-residential buildings, but not for dwellings.

Secondly, the shelter's airtightness level must guarantee that the concentration in the shelter remains lower than the irreversible effects threshold (SEI) during 2 hours, for the considered toxic cloud.
During the elaboration of the PPRT, different zones are defined along with the severity of the effects (irreversible, lethal 1%, or lethal 5% effects) and types of pollutants. For each zone, a conventional toxic cloud (60 min duration) can also be defined.
Then, the maximum attenuation rate on concentrations A (Eq.1) is calculated, defined as the ratio between the threshold in the shelter and the concentration of the conventional outdoor toxic cloud. As a result, the maximum attenuation rate depends on the toxicity of the products, and on the severity of the effects caused by the toxic cloud. In case of several toxic products, the lowest attenuation rate is selected.

$$A \quad (\%) \frac{S)\ E\ 2I\ I}{C_{o\,u\,t\,d\,o\,o}(\,1} \qquad (1)$$

With this, it is possible to calculate the airtightness level of the shelter that will be able to guarantee this maximum attenuation rate. For shelter-in-place issues, we use as an indicator the air change rate at 50 Pa: n_{50} (Eq.2, [5]). Pressure codes like CONFINE can be used, under conditions described in the guide [4].

Since 2005, we have developed CONFINE ([6],[7]), a software that calculates the minimum airtightness level required for a shelter in order to maintain the internal concentration under a given limit. With CONFINE, we assume that any building can be modeled as a 3-zones building with a default envelope airtightness level: Q_{4Pa_surf}, the airtightness indicator in French Thermal regulation (Eq.3, [8]).

$$n_{50} = \frac{q_{50}}{V} \quad (2)$$

$$Q_{4Pa\_Surf} = \frac{q_4}{A_{Tbat}} \quad (3)$$

$q_{\Delta P}$: volumetric airflow through envelope leakage defaults with an induced pressure difference ΔP, between indoor and outdoor ($m^3.h^{-1}$)
V: internal volume of the tested zone (m^3)
A_{Tbat}: total envelope area of the building, excepted ground floor area, according to the French thermal regulation (m^2)

SHELTER-IN-PLACE AIRTIGHTNESS REQUIREMENTS FOR DWELLINGS

Airtightness requirement on an internal room envelope

The French Ministry for Ecology wished to avoid an airtightness calculation for each dwelling, which would result in an additional cost for individual owners. In this goal, we used CONFINE software to generate abacus, using "standard dwellings", as presented hereafter. These abacuses deliver the shelter airtightness requirement (n_{50}) depending on the maximum concentration attenuation rate (Eq.1). They are includes in the guide [4] used by State departments responsible to design PPRT. As a result, PPRT-plans include, for each zone, airtightness requirements for dwellings, and not only the maximum attenuation rate, which is not directly applicable. Contrarily to non-residential buildings, there is no need to use modeling software such as CONFINE to define the shelter airtightness level of dwellings.

The "standard single-family dwelling" (Figure 1) has been considered in the abacus as a single level house, with a 98 m² ground floor and whose envelope airtightness level is estimated as the 95th percentile of the CETE airtightness database[1]: Q_{4Pa_surf} = 2 $m^3/h/m^2$ (n_{50}=7.7 h^{-1}, considering V/A_{Tbat}=1.4 m).
The "standard multi-family building" (Figure 2) has been considered as a four-stories building, with an envelope airtightness level estimated following the 95th percentile of the CETE airtightness database on multi-family dwellings[2]: Q_{4Pa_surf}, = 3 $m^3/h/m^2$ (n_{50}=6.5 h^{-1}, considering V/A_{Tbat}=2.5 m).

For both types of buildings, two configurations were studied depending on whether the shelter is down- or upwind.
Lastly, three wind velocities have been considered: 3-5-10 m/s.
As a result, we computed 12 abacuses (Figure 3), which are used by State services during technological risk prevention (PPRT) plans design.

1 217 single-family dwellings in 2007
2 190 multi-family dwellings in 2007

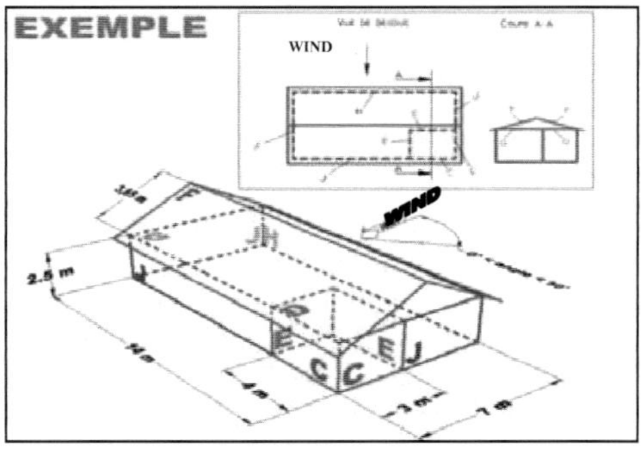

$V_{shelter}$ (m^3)	27
V_{attic} (m^3)	98
$V_{rest\ of\ the\ building}$ (m^3)	251
$H_{building}$ (m)	4.2
Slope of the roof (°)	25
$Q_{4Pa\ Surf,attic}$ ($m^3/h/m^2$)	30
$Q_{4Pa\ Surf,building}$ ($m^3/h/m^2$)	2

Figure 1 and Table 1: Characteristics of the 3-zones "standard single-family dwelling", with downwind shelter

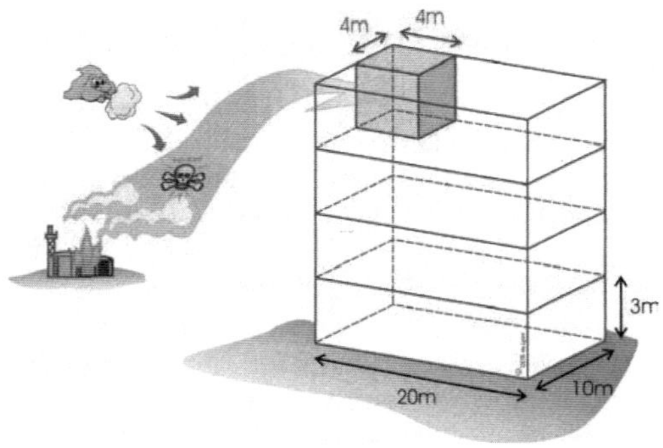

$V_{shelter}$ (m^3)	48
V_{attic} (m^3)	0
$V_{rest\ of\ the\ building}$ (m^3)	2352
$H_{building}$ (m)	12
Slope of the roof (°)	0
$Q_{4Pa\ Surf,attic}$ ($m^3/h/m^2$)	/
$Q_{4Pa\ Surf,building}$ ($m^3/h/m^2$)	3

Figure 2 and Table 2: Characteristics of the 2-zones "standard multi-family building", with upwind shelter

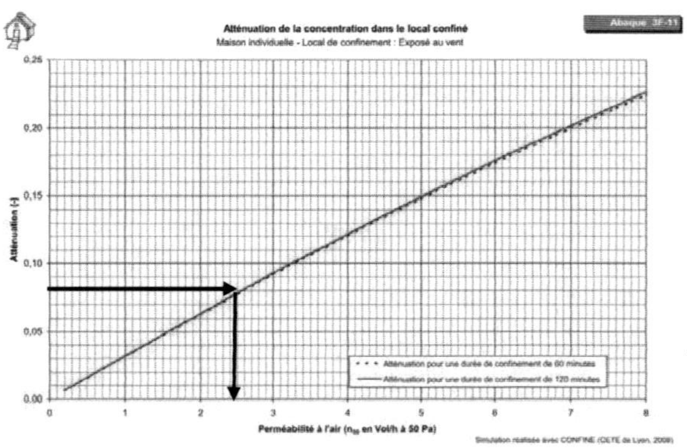

A =	**8 %**
	n_{50}
Upwind shelter, v=3m/s	2.5 h^{-1}
Downwind shelter, v=3m/s	> 8 h^{-1}
Upwind shelter, v=5m/s	1.3 h^{-1}
Upwind shelter, v=10m/s	0.5 h^{-1}

Figure 3 and Table 3: Example of use of "standard single-family dwelling" abacus for shelter airtightness requirement establishing

Measurement requirements

For every building, air leakage level of the shelter must also be measured after constructive works have been implemented, including works on ventilation systems.
As it was shown in Rolfsmeier et al.'s paper [9], there can be misinterpretations of the measurement protocol and analysis, with consecutive errors in the estimations of derived quantities that are used in the calculation method. As a consequence, the French Ministry for Ecology decided that the air leakage measurers will have to be authorized to perform such measurements, and has streamlined a procedure in this goal. This procedure described in a paper [10] concerns measurements in the field of low-energy labels and of the new French thermal regulation (RT2012). On September 2012, around 400 persons have been authorized.
In the PPRT plan, buildings owners are encouraged to work with those authorized measurers. A special measurement protocol has been developed and published [11].
In order to accompany the market transformation in this field, we have conducted a free training program for authorized measurers, including information on the PPRT context and the shelter-in-place strategies, and works to be realized on buildings. On September 2012, around 80 persons have been trained. The list of the trained professionals is maintained on a website [12] and largely distributed to State organizations and local authorities.

COLLECTED DATA

Context

During the working out of each PPRT, shelter-in-place studies may be implemented by local State organizations, in order to get information on the vulnerability of the territory, and have an idea on the financial impact of the PPRT.
For selected dwellings and with their owners' agreement, a free-of-charge vulnerability diagnostic may be realized, supported by the Ministry for Ecology, including an air leakage measurement. In those cases, measurements are performed before any constructive work has been done. Thanks to these diagnostics we were able to collect data and to generate a small database.

Description of the database

In September 2012, data from 140 measurements performed between 2008 and 2012 on 95 single-family dwellings and 45 multi-family dwellings were collected.
For each dwelling, the database includes the location, the type of dwelling (single-family or multi-family), ground floor area and volume of the shelter, required and measured airtightness of the shelter. Year of construction and envelope airtightness level are sometimes given.

First analysis

Airtightness measurements on internal rooms give results from n_{50}=0.7 h^{-1} to 30.7 h^{-1}, with a median value of 6.1 h^{-1} and a mean value of 8.0 h^{-1}. A first analysis in terms of cumulative frequency shows that 95% of the tested rooms have an air leakage level under n_{50}=22 h^{-1}. Figure 4 shows that it is much higher for single-family dwellings (n_{50}=22 h^{-1}) than for multi-family dwellings (n_{50}=17 h^{-1}). For 6 cases, we were able to compare the airtightness of the shelter to the envelope airtightness of the dwelling (Table 3). For one case only, shelter envelope is tighter than dwelling envelope. Internal rooms are rarely designed to be tight because there is rarely an energy issue, even if acoustics or IAQ problems could contribute to design airtight rooms. On the field, we often observe that high air leakage is due to a leaky internal wall: for instance wood intermediate floor without concrete slab.

We observe also that year of building's construction, volume and ground floor area of the shelter have no influence on its airtightness level.

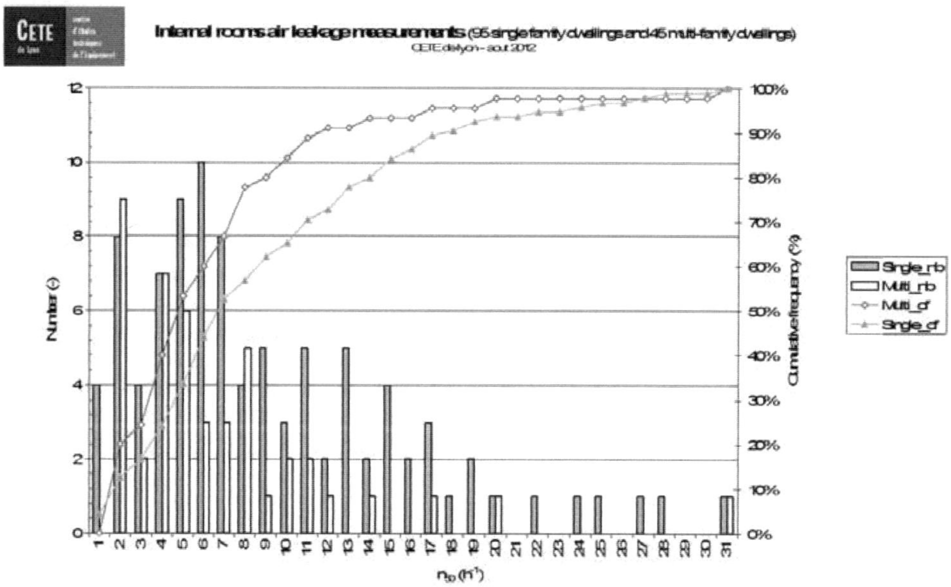

Figure 4: Internal rooms air leakage measurements on 140 dwellings. Cumulative frequencies.

Type of dwelling	n_{50} room (h^{-1})	n_{50} enveloppe (h^{-1})
Single-family	5.5	8.2
Single-family	6.9	5.0
Single-family	13.3	10.3
Single-family	15.8	8.0
Single-family	20	9.0
Multi-family	3.1	2.6

Table 4: Comparison between internal and envelope airtightness levels

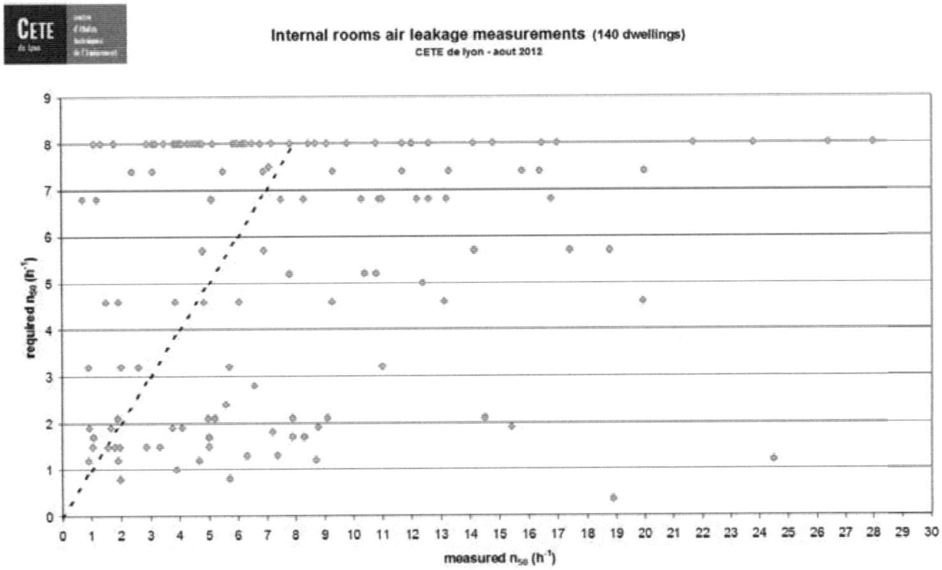

Figure 5: Internal rooms air leakage measurements on 140 dwellings. Difference between airtightness requirement and measurement

As a result, 60 % of the tested shelters have lower performance than expected (Figure 5). In those cases, private individuals have to perform constructive works in the room in order to achieve an airtightness level, which will guaranty their protection. On the other size, a significant number of shelters are tight enough (40%). In those cases, buildings owners would just have to do works to respect general constraints on the whole building and on the room used as a shelter (e.g. a system to quickly stop all voluntary airflows).

CONCLUSION

When all PPRT plans will be promulgated, we expect to have a database of about 1000 airtightness measurement of shelters. Later on, it will be more difficult to collect these data because each dwelling's owner will order its own measurement.
Analysis of this database allows us to estimate the territory vulnerability around Seveso facilities in France and overall cost consequences of this public policy.
This database is also a good opportunity to collect precise information about internal air leakage in dwellings. These experimental data can be used as inputs in multi-zone airflow and pollutant transfer model, when data on internal airtightness are needed to study inter-zone airflows.

ACKNOWLEDGEMENTS

The authors are very grateful to the French ministry for ecology, sustainable development and energy (MEDDE), and in particular to the DGPR department, for its support throughout this work. The sole responsibility for the content of this publication lies with the authors. It does not necessarily reflect the opinion of the Ministry.

REFERENCES

[1] Montoya, M.I, Guyot, G., Planas, E., *French and Catalan approaches to assess shelter in place effectiveness in the event of a toxic gas release*, 32nd AIVC Conference, 1st Tightevent, 6p., 2011

[2] European Commission : Environment. *Chemical Accidents (Seveso II) - Prevention, Preparedness and Response.* Available at: http://ec.europa.eu/environment/seveso/index.htm

[3] French ministry of Ecology and Sustainable Development, *Technological risk prevention plan (PPRT)-Acting together to control risks*, 3p., 2007

[4] CETE de Lyon, CERTU, INERIS, *Guide PPRT-Complément technique relatif à l'effet toxique*, 108 p., 2008

[5] EN 13829: February 2001 – *Thermal performance of buildings - Determination of air permeability of buildings - Fan pressurization method* (ISO 9972:1996, modified)

[6] Guyot, G., Carrié, F-R., CETE de Lyon, *Modélisation des transferts aérauliques en situation de confinement. Bases théoriques et éléments de validation*, 38 p., 2009

[7] Guyot, G., Gentilhomme, O., Carrié, F-R., *Shelter-in-place strategy : CONFINE, an airtightness level calculation tool to protect people against accidental toxic releases*, 32nd AIVC Conference, 1st Tightevent, 6p., 2011

[8] *Arrêté du 24 mai 2006 relatif aux caractéristiques thermiques des bâtiments nouveaux et des parties nouvelles de bâtiments*, JO 25 mai 2006

[9] Rolfsmeier, S., Vogel, K., Bolender, T., 2010. *Ringversuche zu Luftdurchlässigkeitsmessungen vom Fachverband Luftdichtheit im Bauwesen e.V.*, 5 th International Symposium on Building and Ductwork Air-tightness, October 21-22, 2010, Copenhagen/Lyngby, Denmark

[10] Leprince, V., Carrié, F-R., Olivier, M., *The quality framework for Air-tightness measurers in France: assessment after 3 years of operation*, 32nd AIVC Conference, 1st Tightevent, 6p., 2011

[11] CETE de Lyon, *Risque toxique : Protection des personnes dans les PPRT, Précisions sur les études de confinement*, 7p., 2011, http://www.cete-lyon.developpement-durable.gouv.fr/pprt-guides-et-rapports-a165.html

[12] CETE de Lyon, http://www.cete-lyon.developpement-durable.gouv.fr/formations-r131.html

ABSTRACT SUBMISSION FORM FOR THE 2012 AIVC-TIGHVENT CONFERENCE

Send your abstract using this form at info@inive.org

PAPER TITLE
Effect of Measurement Location of Air Air-tightness Performance on Apartment Units in Korea

MAIN AUTHOR
Shin Cheol Woong

OTHER AUTHORS
Lee Yun Gyu

ABSTRACT

The purpose of this study is to evaluate the effects of measurement location of air-tightness performance in Flat-type and Tower-type apartments. Air-tightness performance was measured at the front door using Blower Door system in accordance with CAN/CGSB 149 and at the windows using Airtightness Measuring in accordance with JIS A 2201. The air-tightness test was performed with apartment completed in 2011 a month before its final construction. The air-tightness test results on location were converted into ACH50 for comparison. The result on the windows was higher compared to the result of the front door. According to the results, the front door has more effect on the air-tightness performance than the window which is less air-tightness. In conclusion, it is appropriate to take measurement of air-tightness performance on the windows of apartments in korea

OPTIMAL SIZING RULES FOR NATURAL, SIMPLE EXHAUST AND MECHANICAL RESIDENTIAL VENTILATION SYSTEMS

Jelle Laverge* and Arnold Janssens

Ghent University
Jozef Plateaustraat 22
9000 Gent, Belgium
**Corresponding author: jelle.laverge@ugent.be*

ABSTRACT

Sizing rules in residential ventilation standards lack uniformity in both methodology and resulting design flow rates. In order to investigate the best achievable performance of natural ventilation, exhaust and fully mechanical ventilation systems, this paper presents a multi-zone simulation based optimization study for both a detached dwelling.
Total ventilation heat loss, including mechanical ventilation, adventitious ventilation and infiltration as well as occupant exposure to carbon dioxide are used as criteria. The results show that the relative optimal performance of all three system approaches can be very different, when average exposure or peak exposure is considered.

KEYWORDS

Optimization, Exhaust Ventilation, Residential, Simulation, IAQ

INTRODUCTION

The first wave of energy conservation interventions in buildings, ushered into existence by the 1970's oil crisis, considerably reduced the amount of fresh air infiltration through improved airtightness of newly built dwellings and intensive weatherisation campaigns. As an unintended consequence of this, the incidence of indoor mould problems peaked and reports on high prevalence of occupants complaining of a wide variety of symptoms or physical discomfort, baptised 'sick building syndrome'[1-5], emerged.
The continued scientific interest in these emerging problems was the basis for the indoor environmental science field, growing fast over the last decades. The state of the art within this field has demonstrated positive correlations between indoor air pollution and human health [6, 7], comfort [8-10] and productivity [11, 12]. As people spend about 90% of the time indoors [13, 14], the minimization of these effects is essential. The issue has been prioritized by WHO [15]. Although source control is the most effective and straightforward way to reduce exposure to harmful pollutants, some emissions are related to the very function of the building, such as housing the occupants in residences. The sources related to these essential functions can't be eliminated. Therefore, the pollutant concentrations are diluted by ventilation. For occupant health, fresh air flow rates below 25 l/s per person for offices or building air change rates lower than 0.5 are associated with higher prevalence of symptoms of sick building syndrome and allergies respectively [16]. With respect to comfort, flow rates

below 7 l/s per person are considered to result in unacceptably poor perceived air quality in European ventilation standards [17, 18].
Political action in the aftermath of the reports on the consequences of poor indoor air quality saw to the introduction of ventilation requirements in building codes all over the western countries. The air flow rates required in these standards, as well as the sizing required for non-mechanical system components such as trickle ventilators and transfer grilles [19], can vary quite considerably, with whole building air change rates ranging from 0.3 to 1 [20] for dwellings.
In addition, due to differences in climate, boundary conditions, occupant behaviour and consumer preference, 3 main types of residential ventilation systems are dominant in the Northern, Western and Southern part of Europe respectively. Heat recovery ventilation is the most common system in Scandinavia [21]. In the moderate climate region, simple exhaust systems are more popular [22], while natural ventilation is widespread in Southern Europe.
The performance of the different systems [23-25] and approaches to sizing of their components that are put forward [26] is the object of relentless debate. Presenting the results from a multi zone simulation based optimization study of residential ventilation design flow rates and sizing of the system components, this paper aims to provide a benchmark for achievable performance for the different systems for moderate climate regions (eg. Western Europe), as well as point to possible sizing strategies for future standards.

SIMULATION MODEL

The results presented in this paper are based on airflow simulations. These were executed in the multi-zone airflow simulation package Contam [27], which takes effects of buoyancy, wind and fan pressure into account and is used in numerous ventilation studies [eg. 28, 29].

Building Model

The geometry used in the model is based on a detached house that is statistically representative for the average Belgian dwelling. It has been designed for and used in several previous research projects [30-34] and is currently used to assess the performance of residential ventilation systems in the EPBD framework in Belgium [35]. Table 1. lists the dimensions (m^2) of the spaces in the building model. The airflow in the dwelling has been modelled taking into account both the ventilation system and leakage. Overall leakage, characterized by the v_{50} value, is modelled by means of cracks in the roof and wall surface. The v_{50} value is the ratio of the air leakage rate at 50 Pa pressure difference and the building envelope heat loss area. According to observations by Bossaer [36], the specific leakage rate through roof and walls has a 2/3 ratio, which has been implemented in the model. Each wall is fitted with two cracks, one at 1/4 of its height and the second one at 3/4. The internal doors are simulated with additional cracks in the walls. For the internal walls, a fixed specific leakage value is assumed. This methodology is in agreement with guidelines given in EN 15242 [37]. In the results presented, a specific airleakage (v_{50}) of 1 is used. This represents the 'best practise' for current construction in Belgium and other european countries [38]. This high performance level for building leakage was selected to assess the performance of the ventilation system as such, unbiased by the effect of leakage on both heat loss and indoor air quality.
The production of CO_2 within the model is only related to the occupants' metabolism and corresponds to their whereabouts. A constant outdoor background concentration of 350 ppm is assumed. A different occupancy scheme is used for each day of the simulated week. This makes sure that the reported exposure is insensitive to the specific occupancy schedule and the promotion of 'tailor made' sizing rules is prevented. The mean occupancy for the whole week, 3.4 persons, corresponds to the average occupancy of a 3 bedroom dwelling in

Belgium. Humidity production, like carbon dioxide production, is linked to the metabolism of the occupants with additional production linked to activities such as cooking, bathing and drying clothes. An effective moisture penetration depth model is used to take buffering into account. An odour tracer is produced in sync with the use of the toilets.

Ground Floor	Area	Supply	Exhaust
Living room	35.7	128.4	
Office	8	28.9	
Kitchen	10.2		50
Service room	7.7		50
Toilet	1.7		25
Hallway	28.1		
1st Floor	**Area**	**Supply**	**Exhaust**
Bedroom 1	17	61.1	
Bedroom 2	18.2	65.6	
Bedroom 3	18.3	65.8	
Bathroom	8		50
Hallway	28.1		

Table 1. Geometrical characteristics of reference dwelling.

Ventilation system design and model

The ventilation scheme used in both dwelling geometries is based on the sizing rules put forward in the Belgian residential ventilation standard [39]. This standard is chosen because it contains clear and simple sizing rules for natural, exhaust and mechanical ventilation.

The Belgian standard requires a design flow rate of 1 l/s*m² for each occupied space. For kitchens, bathrooms and service rooms, a minimum design flow rate of 14 l/s should be taken into account. The design flow rate for a toilet is 7 l/s. The design flow rates for the reference dwelling according to these sizing rules are listed in Table 1.

The occupied spaces and the wet spaces should be connected to each other or via circulation spaces by transfer grilles sized at 7 l/s at 2 Pa pressure difference, which corresponds to 70 cm², except for the kitchen, in which the transfer grille should be sized twice as large. For natural and exhaust ventilation systems, supply trickle ventilators and exhaust grilles should be sized at the design flow rate for that space at 2 Pa pressure difference.

The design flow rates for supply, transfer and exhaust as mentioned above are varied from 10% to 200% in 10% steps in order to assess the optimal performance of 3 ventilation system approaches. Since the flow rates proposed in the Belgian standard are moderate to high in comparison to other residential ventilation standards [19, 20], this covers a broad range of sizing options that are both realistic and within ranges used in existing standards.

All mechanical exhaust vents were modelled as constant volume flow rate components in the respective zone node, while transfer grilles and trickle ventilators were modelled with single direction power law flow components with a flow exponent of 0.5 [40]. All systems were modelled with windows and internal doors closed, in order to simulate the performance of the systems as such, without user interaction.

Assessment parameters

Through the correlation between excess CO_2 concentration and mean percentage of dissatisfied [17] and Fanger's Perceived Air Quality approach [10], excess CO_2 concentration is now widely accepted as a proxy for perceived indoor air quality [18], especially if the main pollution sources are related to the human metabolism. In contrast to the basic model, steady state conditions are rarely applicable to real ventilated environments. CO_2 concentrations are inherently transient, due to changes in environmental boundary conditions. Additionally, the relevant CO_2 sources tend to constantly move around in the multi-spaced dwelling, introducing discontinuous sources and further increasing the transient character of the indoor air quality. There is no consensus in literature about the way transient concentrations have to be interpreted. This lack of agreement is reflected in the disparate list of performance criteria provided in EN 15665 [41]. From the suggested parameters in this standard, 2 were selected for use in this paper, namely the total dose of CO_2 for an occupant over the total heating season and the dose of CO_2 over 1000 ppm excess CO_2. Exposure to concentrations in excess of 1000 excess CO_2 is considered to correspond to poor perceived indoor air quality [18] and is therefore a relevant parameter for peak exposure.

The total, heating season averaged, convective heat loss through the combination of intended ventilation, adventitious ventilation and infiltration is used to assess the energy performance of the different sizing options. Fan power was not taken into account because it is very system specific.

RESULTS AND DISCUSSION

As was discussed above, the sizing of supply, transfer and exhaust components was, each in 20 steps, varied within a realistic range. This amounts to $8*10^3$ cases for every system option. From these sizing options, the achievable performance is assessed based on the proposed criteria. Ventilation is always faced with a trade-off between indoor air quality and associated ventilation heat loss. Therefore, the achievable performance is defined as the set of pareto optimal sizing cases for a specific indoor air quality criterion. Pareto optimal cases are cases where none of the other cases achieve better results on both indoor air quality and heat loss.

Figures 1-2 show the pareto optimal cases for the airtight detached house with natural ventilation, simple exhaust ventilation and mechanical ventilation and using occupant exposure to carbon dioxide as indoor air quality criterion. As was mentioned in the methods section, two separate parameters were initially selected to characterize this exposure: the average concentration to which occupants are exposed and the total dose over 1000 ppm excess carbon dioxide. When the former is considered, the pareto optimal solutions for the different installation concepts are rather similar (figure 1.). Considering the latter, however, two striking aspects are observed: in a large number of cases, the occupants are not exposed to excess carbon dioxide concentrations higher than 1000 ppm, regardless of the system concept while the optimal cases off the different system concepts demonstrate larger discrepancies (Figure 2.). The exposure to high carbon dioxide concentrations for an optimal case of the natural ventilation system is on average 7.5 times higher than that for an optimal case of the mechanical ventilation system at the same level of heat loss, while this increase was only about 50% when considering average exposure. The average heat loss for an optimal case of the natural ventilation system is 16, and 22 % higher than the heat loss for an optimal case of the mechanical ventilation system at equal exposure, considering average exposure and dose over 1000 ppm excess carbon dioxide concentration respectively. The differences between the spread in optimal performance found with the different criteria is readily explained by the higher variability of the air flow in the natural system. This increases the exposure to peak concentrations, while the average is less affected.

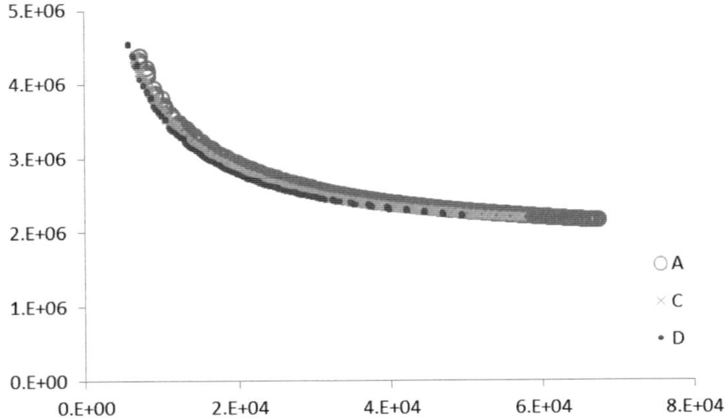

Figure 1. Heating season averaged ventilation heat loss traded-off against total exposure to carbon dioxide concentration to which occupants are exposed during the heating season for all pareto optimal cases for natural (A), simple exhaust (C) and mechanical (D) residential ventilation.

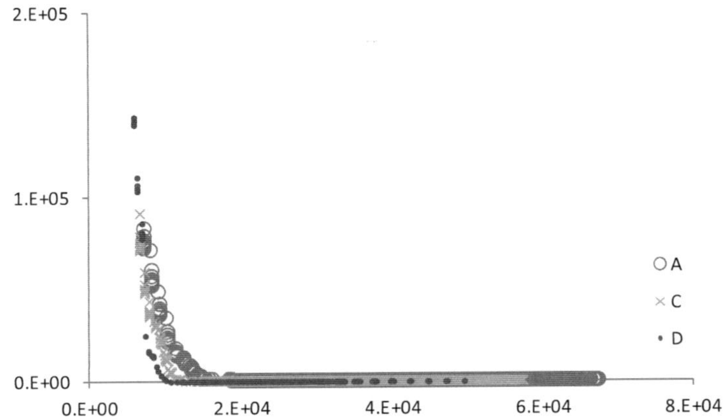

Figure 2. Heating season averaged ventilation heat loss traded-off against exposure to carbon dioxide concentration in excess of 1000 ppm above the outdoor concentration to which occupants are exposed during the heating season for all pareto optimal cases for natural (A), simple exhaust (C) and mechanical (D) residential ventilation.

CONCLUSION

In this paper, the differences in optimal performance of the natural ventilation, the simple exhaust ventilation and the mechanical ventilation concept was assessed with numerical simulations. Optimal performance was defined as the best achievable indoor air quality for a given ventilations heat loss or vice versa. Total and peak exposure to carbon dioxide was used as indoor air quality assessment parameters. Heat loss through mechanical flow rate, adventitious ventilation and infiltration were considered part of the total ventilation heat loss. Considering total exposure to carbon dioxide, only slightly better performance of the mechanical ventilation concept compared to simple exhaust ventilation was observed, while the latter demonstrated slightly better performance than natural ventilation. The spread in optimal performance increased when exposure to peak concentrations was considered instead of average exposure. Nevertheless, the differences remained moderate.

REFERENCES

[1] W.J. Fisk, D. Black, G. Brunner, Changing ventilation rates in U.S. offices: Implications for health, work performance, energy, and associated economics, Building and Environment, 47 (0) (2012) 368-372.
[2] K. Engvall, P. Wickman, D. Norbäck, Sick building syndrome and perceived indoor environment in relation to energy saving by reduced ventilation flow during heating season: a 1 year intervention study in dwellings, Indoor Air, 15 (2) (2005) 120-126.
[3] O.A. Seppänen, W.J. Fisk, M.J. Mendell, Association of Ventilation Rates and CO2 Concentrations with Health andOther Responses in Commercial and Institutional Buildings, Indoor Air, 9 (4) (1999) 226-252.
[4] W.J. Fisk, A.G. Mirer, M.J. Mendell, Quantitative relationship of sick building syndrome symptoms with ventilation rates, Indoor Air, 19 (2) (2009) 159-165.
[5] R. Ruotsalainen, J.J.K. Jaakkola, R. Rönnberg, A. Majanen, O. Seppänen, Symptoms and Perceived Indoor Air Quality among Occupants of Houses and Apartments with Different Ventilation Systems, Indoor Air, 1 (4) (1991) 428-438.
[6] C.G. Bornehag, J. Sundell, T. Sigsgaard, Dampness in buildings and health (DBH): Report from an ongoing epidemiological investigation on the association between indoor environmental factors and health effects among children in Sweden, Indoor Air, 14 (2004) 59-66.
[7] J.M. Daisey, W.J. Angell, M.G. Apte, Indoor air quality, ventilation and health symptoms in schools: an analysis of existing information, Indoor Air, 13 (1) (2003) 53-64.
[8] L. Fang, G. Clausen, P.O. Fanger, Impact of temperature and humidity on the perception of indoor air quality, Indoor Air-International Journal of Indoor Air Quality and Climate, 8 (2) (1998) 80-90.
[9] H.N. Knudsen, G. Clausen, P.O. Fanger, Sensory characterization of emissions from materials, Indoor Air-International Journal of Indoor Air Quality and Climate, 7 (2) (1997) 107-115.
[10] P.O. Fanger, Introduction of the olf and decipol unit to quantify air-pollution perceived by humans indoors and outdoors, Energy and Buildings, 12 (1) (1988) 1-6.
[11] P. Wargocki, L. Lagercrantz, T. Witterseh, J. Sundell, D.P. Wyon, P.O. Fanger, Subjective perceptions, symptom intensity and performance: a comparison of two independent studies, both changing similarly the pollution load in an office, Indoor Air, 12 (2) (2002) 74-80.
[12] U. Haverinen-Shaughnessy, D.J. Moschandreas, R.J. Shaughnessy, Association between substandard classroom ventilation rates and students' academic achievement, 21 (2011) 121-131.
[13] I.M. Glorieux, J., Belgisch tijdsbudgetonderzoek, in, Brussels, 2008.
[14] M. Basner, K.M. Fomberstein, F.M. Razavi, S. Banks, J.H. William, R.R. Rosa, D.F. Dinges, American time use survey: Sleep time and its relationship to waking activities, Sleep, 30 (9) (2007) 1085-1095.
[15] M. Braubach, D.E. Jacobs, Environmental burden of disease associated with inadequate housing. Methods for quantifying health impacts of selected housing risks in the WHO European Region. Summary report., WHO, 2011.
[16] J. Sundell, H. Levin, W.W. Nazaroff, W.S. Cain, W.J. Fisk, D.T. Grimsrud, F. Gyntelberg, Y. Li, A.K. Persily, A.C. Pickering, J.M. Samet, J.D. Spengler, S.T. Taylor, C.J. Weschler, Ventilation rates and health: multidisciplinary review of the scientific literature, Indoor Air, 21 (3) (2011) 191-204.

[17] CEN, Ventilation for buildings - Design criteria for the indoor environment, in, Brussels, 1998.
[18] CEN, Ventilation for non-residential builidings - performance requirements for ventilation and room-conditioning systems, in, Brussels, 2007.
[19] J. Laverge, X. Pattyn, A. Janssens, Performance assessment of residential mechanical exhaust ventilation systems dimensioned in accordance with Belgian, British, Dutch, French and ASHRAE standards, Building and Environment, (0).
[20] C. Dimitroulopoulou, Ventilation in European dwellings: A review, Building and Environment, 47 (0) (2012) 109-125.
[21] J. Kurnitski, O.A. Seppanen, Trends and drivers in the Finnisch ventilation and AC market, in: AIVC (Ed.) Ventilation Information Papers, AIVC, Brussels, 2008.
[22] F. Durier, Trends in the French building ventilation market and drivers for changes, in: AIVC (Ed.) Ventilation Information Papers, 2008.
[23] J. Laverge, A. Janssens, Heat recovery ventilation operation traded off against natural and simple exhaust ventilation in Europe by primary energy factor, carbon dioxide emission, household consumer price and exergy, Energy and Buildings, 50 (0) (2012) 315-323.
[24] J. Laverge, A. Janssens, IAQ exposure of sleeping occupants under different residential ventilation configurations, in: Roomvent 2011, Trondheim, 2011.
[25] J.-L. Savin, J. Laverge, Demand-controlled ventilation: an outline of assessment methods and simulations tools, in: AIVC & TIGHTVENT 2011, Brussels, 2011.
[26] J. Laverge, Comparison of the use of trickle ventilators in european residential ventilation standards, in: Roomvent 2011, Trondheim, 2011.
[27] W.S. Dols, A tool for modeling airflow & contaminant transport, Ashrae Journal, 43 (3) (2001) 35-+.
[28] G. Nirvan, F. Haghighat, L. Wang, H. Akbari, Contaminant transport through the garage – House interface leakage, Building and Environment, 56 (0) (2012) 176-183.
[29] Y.L. Chen, J. Wen, The selection of the most appropriate airflow model for designing indoor air sensor systems, Building and Environment, 50 (0) (2012) 34-43.
[30] G. Verbeeck, Life cycle optimization of extremely low energy dwellings, J. Build Phys., 31 (2) (2007) 143-177.
[31] G. Verbeeck, H. Hens, Energy savings in retrofitted dwellings: economically viable?, Energy and Buildings, 37 (7) (2005) 747-754.
[32] G. Verbeeck, H. Hens, Life cycle inventory of buildings: A calculation method, Building and Environment, 45 (4) 1037-1041.
[33] G. Verbeeck, H. Hens, Life cycle inventory of buildings: A contribution analysis, Building and Environment, 45 (4) 964-967.
[34] A. Janssens, L. Willems, J. Laverge, Performance evaluation of residential ventilation systems based on multi-zone ventilation models., in: International Building Physics Conference, Istanbul, 2008.
[35] N. Van Den Bossche, A. Janssens, N. Heijmans, P. Wouters, Performance evaluation of humidity controlled ventilation strategies in residential buildings, in: Thermal performance of the exterior envelopes of whole buildings X, Clearwater, 2007, pp. 7.
[36] A. Bossaer, J. Demeester, P. Wouters, B. Vandermarke, W. Vangroenweghe, Airtightness performances in new Belgian dwellings, in: 19th AIVC conference: ventilation technologies in urban areas, AIVC, Oslo, 1998, pp. 77-84.
[37] CEN, Ventilation for buildings - Calculation methods for the determination of air flow rates in buildings including infiltration, in, Brussels, 2007.
[38] J. Laverge, S. Vandevelde, T. Debrauwere, M. Delghust, A. Janssens, Airtightness assessment of newly build single family houses in Belgium, in: Buildair 2010, Copenhagen, 2010.
[39] BIN, Ventilatievoorzieningen in woongebouwen, in, Brussels, 1991.

[40] P. Karava, T. Stathopoulos, A.K. Athienitis, Investigation of the performance of trickle ventilators, Building and Environment, 38 (8) (2003) 981-993.
[41] CEN, Ventilation in buildings - Determining performance criteria for design of residential ventlilation systems, in, Brussels, 2009.

CHECKING "FABRIC FIRST" REALLY WORKS: IN-CONSTRUCTION TESTS USING THERMOGRAPHY

Tim Taylor*[1], John Counsell[1], Andrew Geens[2], Steve Gill[1] and Gerraint Oakley[3]

1 Cardiff Metropolitan University
Western Avenue
Cardiff CF5 2YB, UK
**Corresponding author: tijtaylor@cardiffmet.ac.uk*

2 Chartered Institute of Building Services Engineers
222Balham High Road
London SW12 9BS, UK

3 Coastal Housing Group
220 High Street
Swansea SA1 1NW, UK

ABSTRACT

The UK Government strategy for all new homes to be built to zero carbon standards by 2016 is based upon a "fabric first" approach to design. This means prioritising energy efficiency improvements to the building envelope through: increasing overall levels of insulation; reducing thermal bridging; and making buildings more airtight. However, recent research has raised concerns about the standards that are actually achieved in the construction of new housing. More robust quality assurance procedures for construction work may be required to ensure that energy efficiency targets are met in practice. One potential approach is the use of thermal imaging (thermography) to inspect new buildings at different stages during the construction process. The effectiveness of this technique has been tested during the construction of two affordable housing projects in Swansea, UK. Thermal performance issues were identified at both of the schemes, including infiltration through the building envelope and poor insulation of ductwork for mechanical ventilation systems. The results of these two case studies illustrate some practical considerations for the application of the thermography technique and also shortcomings in the current approach to determining compliance with energy performance requirements in UK Building Regulations. This research topic will be of interest to housing developers, built environment professionals, thermographers and researchers interested in methods of investigating the thermal performance of new housing.

KEYWORDS

Thermography, performance testing, construction process, low carbon housing, Building Regulations

INTRODUCTION

In the 'Building a Greener Future' policy statement of 2007, the UK Government announced proposals for all new homes to be built to "zero carbon" standards by 2016 [1]. These standards are to be based upon a "fabric first" approach to design, which means prioritising energy efficiency improvements to the building envelope through: increasing overall levels of insulation; reducing thermal bridging; and making buildings more airtight [2]. As the UK construction industry moves towards full implementation of the 2016 zero carbon target, a series of small-scale research studies have raised concerns that significant discrepancies can exist between the predicted energy performance of a new home as calculated at the design stage compared to the actual performance of the completed building – with evidence of significant under-performance in some cases [3]. This phenomenon is widely referred to within the industry as the "performance gap". The extent of concern is such that, in a recent consultation on changes to Building Regulations in England, the Government acknowledged that *"the risk of wider scale underperformance cannot be ignored and that the potential performance gap could be very significant"* [4].

The main focus of this paper is the relationship between construction quality and performance testing in the delivery of low carbon homes in the UK. Specific consideration is given to the use of thermography as a quality control test for fabric energy efficiency and quality of workmanship during the construction of new housing. It is proposed that conducting tests at appropriate points during the construction process (or 'in-construction testing') will help support the management of construction quality and increase confidence that design targets for thermal performance will be achieved in practice [5, 6]. Moreover, it is advantageous if defects can be identified through testing within a reasonable timescale prior to completion, since remedial work can become increasingly costly and disruptive once a building is occupied. The practical experience of the authors has shown that some specific considerations apply to conducting thermographic surveys on a construction site. However, a literature review has identified a lack of detailed guidance on the effective application of thermography in this context (e.g. [7–12]). Having identified this gap in existing knowledge, an approach has been developed for in-construction tests using thermography. The main elements of the testing approach are shown in Figure 1 below.

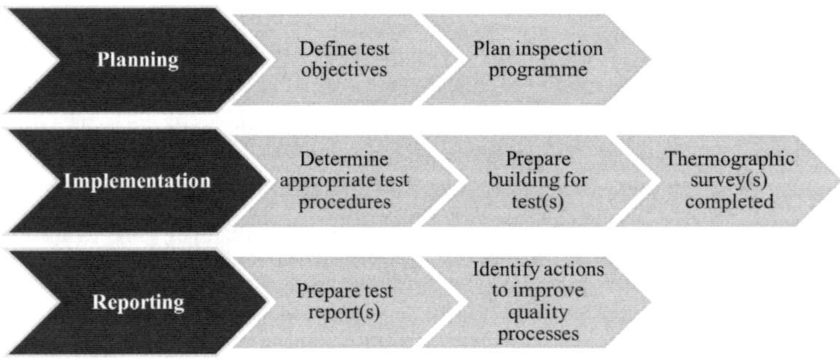

Figure 1. Main elements of the testing approach.

The content of the paper is organised into three main sections as follows:

1. Performance testing and UK Building Regulations
2. Examples of construction defects detected using thermography
3. Introduction to testing approach

PERFORMANCE TESTING AND UK BUILDING REGULATIONS

Levels of compliance with energy efficiency requirements are reportedly a "weaker area" of UK Building Regulations [13]. A report commissioned by the Department for Communities and Local Government 'Performance Testing of Buildings' [14] reviewed the scope for additional performance tests to check compliance with the requirements of the Regulations. The report concluded that: *"To be useful, pre-completion performance tests must be quick and inexpensive. They must not delay occupancy, or have to be carried out after occupancy when they may become impracticable"*. In a previous publication, Taylor et al. [5] argued that the 2016 zero carbon target will likely result in a significant shift in the procedures and practices of Building Control[1] and, furthermore, that new approaches to testing in-situ performance

[1] In the UK, Building Control Bodies have the responsibility for checking compliance with Building Regulations.

would need to be developed. In the next section of the paper, this argument is developed further with reference to two case studies where in-construction thermography tests revealed performance defects in low carbon housing projects. The tests conducted at these case studies illustrate the limitations of pressurisation testing as a means of verifying that construction quality is consistent with the predicted energy performance of a building.

EXAMPLES OF DEFECTS DETECTED USING THERMOGRAPHY

Case Study A test results

The environmental design strategy at Case Study A (a block of 69 flats in timber frame construction) was developed with an assumption that the building would be constructed to high standards of airtightness (achieving an air permeability of 3.0 $m^3/h.m^2$ at 50Pa). To determine if this level of performance was likely to be achieved in practice, one of the flats in the development was brought to a more advanced stage of completion so that a pressurisation test could be carried out at an early stage of the construction process. The result of this test showed a measured air permeability of 1.09 $m^3/h.m^2$ at 50Pa had been achieved – a significant improvement on the design target. However, masking tape was used extensively to seal around openings and sockets prior to the pressurisation test. The use of temporary seals in this way is not permitted according to the testing protocol that is specified in the Building Regulations: *"All external doors and windows should be closed (but not additionally sealed). This includes door thresholds"* [15]. A thermographic survey of the flat 14 months later showed extensive air leakage around the balcony doors as shown in Figure 2 below.

(a) Photograph taken of balcony doors in living room of test flat at time of thermographic survey

(b) Thermal image of balcony doors corresponding to photograph (a)

Figure 2. Thermal images of flat in Case Study B.

The masking tape applied around the balcony doors before the pressurisation test effectively concealed these locations of air leakage[2]. On this basis, the pressurisation test result is unrepresentative of actual performance and higher levels of infiltration may mean that the energy performance of the flat is less than predicted. The wind pressures that caused the air leakage observed in Figure 2 can also result in other types of heat loss. In the same flat, a thermographic survey indicated a localised surface temperature decrease in the ceiling of one of the bedrooms. The area of the ceiling where this was observed corresponds with the location of an extract duct for the mechanical ventilation unit. It follows that a possible

[2] It should be further noted that the pressurisation test was not performed as part of mandatory testing and the test result was not used to determine compliance with Building Regulations.

explanation of this thermal pattern could be wind penetration above ceiling level around a poorly-sealed or damaged duct air terminal. This type of defect would contribute to heat loss but does not constitute an infiltration mechanism. It would therefore not be detected by a pressurisation test (this is also the case for thermal bypass caused by "wind-washing" [16]).

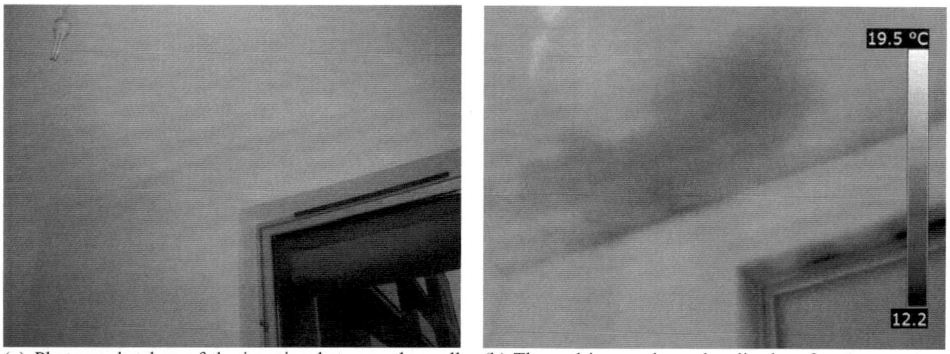

(a) Photograph taken of the junction between the wall and ceiling above a window in one of the bedrooms of the test flat

(b) Thermal image shows localised surface temperature decrease in ceiling area corresponding with the position of the extract duct for the mechanical ventilation unit

Figure 3. Wind cooling effect observed in ceiling below mechanical ventilation extract duct.

Case Study B test results

At Case Study B (a block of 32 flats, part new build and part refurbishment) the ventilation strategy utilised mechanical ventilation heat recovery (MVHR). A thermographic survey in one of the top floor flats indicated that the inlet duct for the MVHR unit was not correctly insulated as shown in Figure 4 below. Poor installation of the MVHR ductwork will reduce the energy performance of the dwelling.

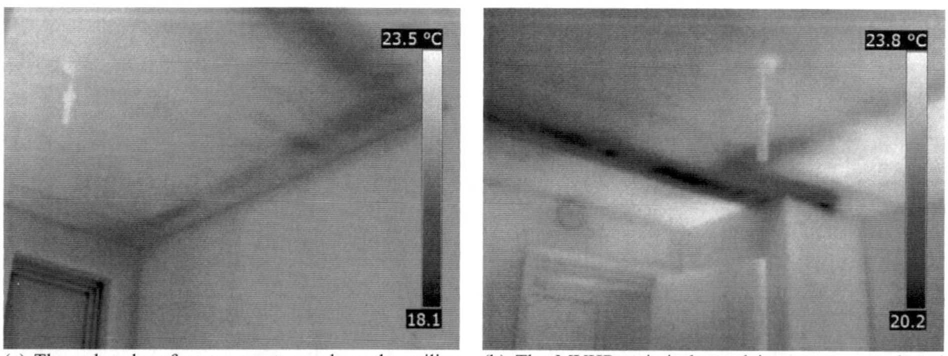

(a) The reduced surface temperatures along the ceiling follow a linear pattern corresponding with the position of the MVHR unit ductwork

(b) The MVHR unit is located in a storage cupboard adjacent to this corner of the room

Figure 4. Insulation not continuous around MVHR ductwork.

Summary of case study results

The case study results illustrate how thermography can be used to identify: air leakage around window and door openings; wind penetration through the external leaf, or "wind-washing"; and poor insulation of ductwork. In these cases, low standards of workmanship and poorly performing building components were not identified by pressurisation testing. At Case Study A the effect of air leakage around window and door openings would not have been measured by pressurisation testing because the preparation of the building deviated from standard test protocols. Although one example cannot be considered representative of wider industry practices, it does indicate the sensitivity of pressurisation test results to correct site test procedures. The risk of making unrealistic assumptions about the airtightness of the building envelope based upon a pressurisation test result is not only important in the context of demonstrating compliance with Building Regulations – it also has implications for the design of heating and ventilation systems. This emphasises the importance of a holistic approach to environmental design; encompassing design, construction and maintenance practices. For designs that include provision for passive means of ventilation, current approaches to performance testing could be extended by carrying out pressurisation testing with vents sealed and then repeating the test with vents open to calculate an in-situ 'equivalent area' of the open vents. This approach could be use to verify design assumptions for background ventilation rates [14].

IN-CONSTRUCTION TESTS USING THERMOGRAPHY

The testing approach follows a process illustrated in Figure 1 comprising three main stages: planning, implementation and reporting. The main purpose of in-construction testing using thermography is to assess the continuity of insulation and identify air leakage paths in the test building. The testing approach is generally consistent with the requirements of BS EN ISO 13187:1998 for simplified testing with an infra-red (IR) camera [10] and is intended to be applicable to all dwelling types. However, practical experience has shown that some specific considerations apply to conducting tests effectively during the construction process and this is reflected in the test procedures outlined in the following sections of the paper. Firstly, it is useful to outline some general principles which help to determine the most appropriate approach to testing:

- **Location of insulation layer(s) within the building structure**

For assessing the continuity of insulation, it will be advantageous if the insulation layer is positioned close to the surface of the construction that is being inspected. Defects within the insulation layer will have two effects when the building is heated: increased heat loss and reduced internal surface temperatures [17]. This localised decrease in surface temperature is the basis on which defects can be identified using thermography. A defect appears more obvious during a thermographic survey when the insulation layer is located closer to the inspected surface because the contrast between the thermal pattern of the defect and the surrounding structure is enhanced. To verify that the insulation layer has been correctly installed, if follows that the optimal time to survey the building is once the insulation layer has been fixed and covered over, but before any finishes have been applied. Surveying the building at a later stage of the construction may involve additional re-work if partial deconstruction is required to repair a defect.

- **Completion of airtightness layer during the construction process**

The airtightness of the building envelope is ideally tested early once the build has been completed up to the airtightness layer and windows and doors are in place (or at least can be temporarily sealed) [18]. Thermography can be used to identify the location of leaks within the envelope if testing is carried out in conjunction with a fan pressurisation test. It may also be useful to conduct tests when the building is exposed to strong prevailing winds (without mechanically pressurising the building). In this case, wind pressures will cause air movement through any leaks in the envelope and it may also be possible to observe "wind-washing" effects. Testing under these conditions may enable the identification of defects that are more significant in terms of the actual performance of the building in use.

Planning

Certain forms of construction will be more amenable to thermographic testing. This is illustrated with reference to three external wall construction types given in Table 3 below. These forms of construction were selected on the basis that they have been developed to reflect good practice in fabric energy efficiency and represent the main construction types often specified in UK housing.

Construction type	Type code	Description
Timber frame	TF01	140mm fully filled timber frame, sheeted externally, air barrier/vapour control layer and insulated lining internally. Service void and plasterboard. Clear cavity with brick outer leaf.
Cavity masonry	MV01	100mm block inner leaf internally plastered. 150mm fully filled insulated cavity. Brick outer leaf.
Light steel frame	SF01	70mm fully filled light steel frame, sheeted both sides, air barrier/vapour control layer. Service void and plasterboard. Partially filled insulated cavity with brick outer leaf.

Table 1. Energy Saving Trust Enhanced Construction Details [19].

- **Timber frame external walls (TF01):** An initial thermographic survey to check the continuity of insulation would ideally be carried out once the insulated lining is fixed to the timber frame (enclosing the insulation between the studwork). The purpose of this initial survey would be to confirm that the insulation between the studwork has been correctly installed before the insulated lining is covered over with plasterboard. Either an internal or external survey would be effective for this first test. However, an internal survey is likely to be more appropriate since it allows greater flexibility with respect to the timing of the test[3] and the external envelope may also be obscured by scaffolding. At this stage of the construction process the building heating systems would not have been installed and commissioned and so an alternative method of heating the building would need to be adopted for the test. A second survey would usefully be carried out in conjunction with a pressurisation test to identify locations of air leakage through the building envelope once the heating, plumbing and electrical services have been installed (to check the effectiveness of sealing around the building services).
- **Masonry Cavity external walls (MV01):** An initial thermographic survey to check the continuity of insulation within the cavity is ideally carried out once the building is weathertight but before a parge coat is applied to the internal face of the inner leaf. Either an internal or external survey would be effective for this first test. If any defective areas of insulation are identified then repair work may necessitate partial deconstruction of the inner or outer leaf. A longer heating period would be required in comparison to the testing of timber frame structures because of the thermal mass of the block inner leaf. Once the plasterwork has been applied to the internal leaf (this effectively acts as the airtightness layer), a second survey would be usefully carried out in conjunction with a pressurisation test to identify air leakage.
- **Light steel frame (SF01):** As with the timber frame external wall type, an initial survey would ideally be carried out to check the continuity of insulation between the studwork once the sheeting has been fixed to the internal face of the steel frame (before it is covered over with plasterboard). Thermography will be less effective for checking the continuity

[3] The best conditions for external surveys are found during the night, sometime after sunset, when the effects of direct solar radiation on the surface temperature distribution of the external envelope can be discounted.

of the insulation within the cavity since any discontinuities in this insulation layer will be difficult to detect from either an internal or external survey. This is because the intermediate layers of the external wall structure will reduce the effect the defect has on the internal and external surface temperatures. In this case, supervision of the construction process becomes increasingly important to ensure the insulation is securely fixed back to the inner leaf to prevent air from circulating around the insulation. A second survey would usefully be carried out in conjunction with a pressurisation test to identify air leakage.

Implementation

The interpretation and reliability of thermographic testing is facilitated by a stable pattern of heat flow through the building envelope and a sufficiently large difference between internal and external temperatures so that surface temperature variations are detectable. Pearson [8] recommends a minimum temperature difference of 10°C between internal and external temperatures for thermal performance surveys. Wahlgren & Sikander [18] state that a temperature difference of at least 5°C is acceptable for surveys to identify air leakage. Prior to testing, the building may be heated using either electrical fan heaters, radiant heaters or the building heating system (if this has been installed and commissioned). A decision tree for selecting the most appropriate approach is given in Figure 5 below. However, experience indicates that a useful daytime temperature difference can be obtained through solar gain alone for internal surveys and thus in some circumstances it may be possible to identify defects in the building envelope without providing supplementary heating.

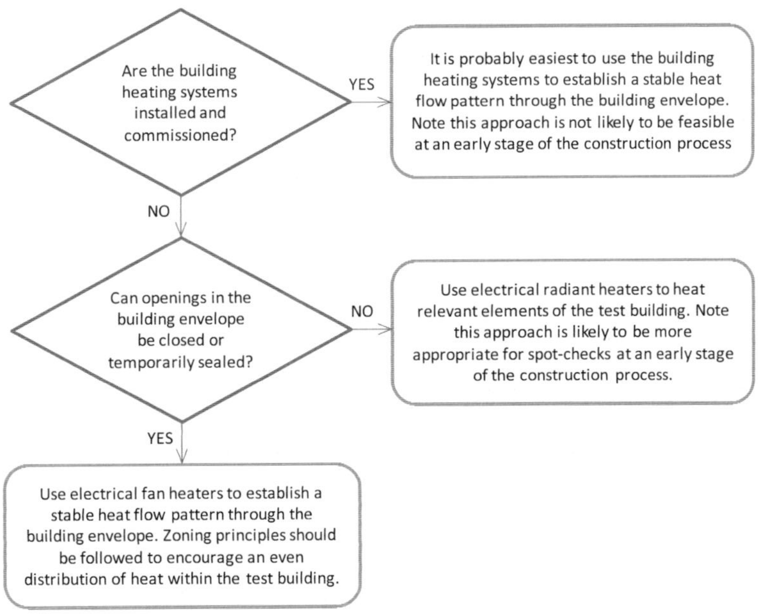

Figure 5. Decision tree for selection of heating method.

In outline, testing consists of two stages as follows:

- **Pre-test requirements:** to prepare the building for testing, including a walkthrough of the test building and the installation of heaters and other equipment (if required).

- **Site test procedure:** the process of examining thermal patterns on the internal and/or external surfaces of the test building.

The **pre-test requirements** are as follows:

1. Select a heating approach using the decision tree in Figure 5.
2. It is preferable to commence heating of the test building at least 24 hours before the inspection. However, a shorter heating period may be adopted if it is not possible to obtain access or permission to operate the heaters outside of normal site working hours. In this case, the number and/or power output of heaters may need to be adjusted to compensate for the reduced heating period.
   If using electrical fan heaters:
   - Install 110V electrical fan heaters in the test building[4]. The power output and number of heaters required will depend upon the configuration of the building. The placement of circulation fans in appropriate locations to encourage air movement may assist with achieving a more even temperature distribution.

   If using electrical radiant heaters:
   - Install 110V electrical radiant heaters in the test building. It may be necessary to adjust the distance of the radiant heater from the target building element and also the angle of inclination of the heater element to the building surface to achieve an even heating profile. The IR camera can be used to assist with this process. Care should be taken not to point the IR camera directly at the radiant heater when it is switched on as this may damage the detector.

   If using building heating systems:
   - Adjust the heating controls in the test building according to the instructions provided by the manufacturer.
3. Prior to switching on the heaters, all external doors, windows and trickle vents should be closed. Internal doors should be fully opened and restrained (if necessary) to encourage an even distribution of heat within the test building.
4. If the inspection personnel are on site before the heaters are to be switched on then this may be an appropriate point at which to conduct a walkthrough of the test building. The walkthrough presents an opportunity to record visual images, taking note of any factors that may influence heat flow through the building envelope (e.g. service penetrations), and review health and safety issues with the site manager and/or other responsible person(s).
5. If a meteorological station is located in close proximity to the test building then this may be a convenient way of noting the local weather conditions during the 24 hours preceding the survey. Temperature and humidity sensors may also be installed in appropriate internal and external locations if required. All surfaces to be inspected during the survey must be dry and therefore any precipitation in the 24 hours preceding the survey is likely interfere with surveys of the external facade of the test building.

The **site test procedure** is as follows:

1. The external air temperature, external relative humidity (RH) and wind speed should be recorded at the start of the survey using a suitably calibrated environment meter (with thermometer, hygrometer and anemometer functions). The air temperature and relative

[4] Note the Health and Safety Executive (HSE) recommends a reduced low voltage 110V supply system for all portable electrical equipment used on construction sites in the UK. Further information is provided in BS 7375:2010 Distribution of electricity on construction and demolition sites – Code of practice.

humidity inside the test building should also be recorded. Ideally, these measurements should be repeated at the end of the survey.

2. Thermal patterns should be examined using the IR camera on the internal surfaces of the test building and/or all aspects of the external facade (unless radiant heaters are used, in which case only the relevant element of the building envelope need be inspected). Particular note should be taken of windows and any joints in the construction (e.g. wall-ceiling junctions). Any areas of special interest and any thermal irregularities should be studied in detail. Written or audio notes should be taken to accompany the thermal images recorded during the inspection to aid the interpretation of results.

Reporting

The results of the survey should be presented in a report including a description and interpretation of the thermal images recorded during the survey, and preferably accompanied with corresponding visual images. Recommendations for the detailed content of the report are given in Pearson [8] and BS EN 13187:1999 [10].

CONCLUSIONS

The UK Government expects carbon savings to be delivered by increasing the energy efficiency of new housing to zero carbon standards. However, a growing body of evidence for a potential "performance gap" suggests that planned carbon savings may not be delivered in practice. Underperformance poses a reputational risk to the UK construction industry, as Government carbon reduction targets may be undermined and householders may not benefit from the expected savings in their energy bills. This paper has developed an argument for extending current industry practices for in-situ performance testing of new housing to help address these risks. A testing approach using thermography to check the continuity of insulation and locate air leakage in the building envelope is outlined in the paper. Existing literature on thermography does not provide detailed guidance for the effective implementation of testing during the construction process. The testing approach, which is being developed as part of a PhD research programme at Cardiff Metropolitan University, seeks to address this gap in existing knowledge. The main benefit of 'in-construction testing' is that defects can be identified at an early stage of the construction process when it is likely to be easier and less costly to carry out any remedial work that may be required. Therefore, thermography is potentially a useful complement to pressurisation testing, and the use of both techniques together could provide a more representative assessment of fabric energy efficiency and quality of workmanship in residential construction projects.

ACKNOWLEDGEMENTS

The PhD project is funded by Cardiff Metropolitan University, Coastal Housing Group, and the European Social Fund through the European Union's Convergence programme administered by the Welsh Government.

REFERENCES

[1] Department for Communities and Local Government, 2007. *Building a Greener Future: policy statement*. London, UK: Department for Communities and Local Government.

[2] Zero Carbon Hub, 2009. *Defining a Fabric Energy Efficiency Standard for Zero Carbon Homes: Task Group Recommendations November 2009*. London, UK: Zero Carbon Hub.

[3] Zero Carbon Hub, 2010. *Carbon Compliance for Tomorrow's New Homes - Topic 4: Closing the gap between designed and built performance*. London, UK: Zero Carbon Hub.

[4] Department for Communities and Local Government, 2012. *2012 consultation on changes to the Building Regulations in England. Section two - Part L (Conservation of fuel and power). Proposed changes to technical guidance*. London, UK: Department for Communities and Local Government.

[5] Taylor, T., Littlewood, J., Goodhew, S., Geens, A., Counsell, J., Hopper, J., et al., 2011. *In-construction testing of the thermal performance of dwellings using thermography*. In: Sustainability in Energy and Buildings, SEB11, Proceedings of Third Annual Conference. 1-3 June 2011, Marseilles, France: Springer.

[6] Littlewood, J., Taylor, T., Goodhew, S., Evans, N., Counsell, J., Wilgeroth, P., et al., 2011. *Development of a thermography protocol for the in-construction testing of the thermal performance of low carbon dwellings*. In: CIBSE Technical Symposium, 6-7 September 2011, Leicester, UK.

[7] Hart, J.M., 1991. *BR 176 A practical guide to infra-red thermography for building surveys*. Watford, UK: BRE.

[8] Pearson, C., 2002. *BSRIA TN 9/2002 Thermal Imaging of Building Fabric: A best practice guide for continuous insulation*. Bracknell, UK: BSRIA.

[9] Lo, T.Y., Choi, K.T.W., 2004. *Building defects diagnosis by infrared thermography*. Structural Survey. Vol. 22, No. 5,pp.259–263.

[10] British Standards Institution, 1999. *BS EN 13187:1999 Thermal performance of buildings - Qualitative detection of thermal irregularities in building envelopes - Infrared method (ISO 6781:1983 modified)*. London, UK: BSI Group.

[11] Walker, N.J., 2004. *Infrared Thermography Volume 1: Principles and Practice*. Northampton, UK: The British Institute of Non-Destructive Testing.

[12] Nowicki, A.N., 2004. *Infrared Thermography Volume 2: Applications*. Northampton, UK: The British Institute of Non-Destructive Testing.

[13] Cox, C., 2006. *Compliance with Part L1 of the 2002 Building regulations (An investigation into the reasons for poor compliance) Final Report 3.5.06*. Didcot, UK: Future Energy Solutions.

[14] Department for Communities and Local Government, 2008. *Performance Testing of Buildings BD2535*. London, UK: Department for Communities and Local Government.

[15] ATTMA, 2010. *Technical Standard L1. Measuring Air Permeability of Building Envelopes (Dwellings). October 2010 Issue. [Internet]*. Available at: http://www.attma.org/downloads/ATTMA%20TSL1%20Issue%201.pdf [Accessed 6 July 2011].

[16] Siddall, M., 2009. *The Impact of Thermal Bypass [Internet]*. Green Building magazine. Vol. 19, No. 1, Available at: http://www.aecb.net/PDFs/Impact_of_thermal_bypass.pdf [Accessed 4 June 2012].

[17] Ward, T.I., 2006. *Assessing the effects of thermal bridging at junctions and around openings IP 1/06*. Watford, UK: BRE.

[18] Wahlgren, P.S.M., Sikander, E., 2010. *Methods and Materials for an Airtight Building*. In: Thermal Performance of the Exterior Envelopes of Whole Buildings XI International Conference, 5-9 December 2010, Florida, USA.

[19] Energy Saving Trust, 2008. *CE297 Enhanced Construction Details: introduction and use*. London, UK: Energy Saving Trust.

ENERGY RETROFIT OF THE EXISTING HOUSING STOCK IN ENGLAND

Raniera Barbisan[*1] and Hasim Altan[1]

1 Building Environments Analysis Unit (BEAU), The University of Sheffield
ICoSS, 219 Portobello, Sheffield S1 4DP, United Kingdom
**Corresponding author: r.barbisan@sheffield.ac.uk (h.altan@sheffield.ac.uk)*

ABSTRACT

The housing stock in the United Kingdom is the oldest in Europe and great part of its energy consumption is due by space and water heating. The high costs of energy are a national matter not only for their economic and environmental implications, but also because they contribute largely to a social problem, known as *fuel poverty*. The cost of heating the housing stock is rather high for different reasons, one of each being the heat loss through the building envelope. The thermal performance of existing buildings can be increased in two ways: by adding insulation to external fabric, and by reducing the unintended air leaks of the envelope. This study focuses on this second method, which can lower heat waste of about 20%.
Typical pre-1970 English buildings are characterised by an evident degree of permeability, emphasised either in the construction techniques or in the materials used. They are also pretty leaky: they have cracks and holes in the fabric, unused open chimneys and fireplaces, gaps around windows and door frames. Unwanted air leakage allows the waste of heat toward the outside and can cause interstitial condensation. As a result, there can be a decrease in the performance of thermal insulation (up to 70%) and fabric damages. Air-proof improvements increase the energy and cost efficiency of a building, raise the level of internal comfort and lower the risk of thermal bypass. They can be very cost-effective and easy to do, but their effects are underestimated. After draught-proofing, less heat will be wasted through the envelope (this can save on average £55 per year) and thus less heat will be required to have a comfortable temperature in the inside (this can save another £60 per year). The Energy Saving Trust (EST) evaluates that if every dwelling in the UK was draught-proofed at its best, £190 million would be saved every year and the unloosen energy would be sufficient to heat nearly 400,000 houses. The benefits from the economic, energy-efficiency, environmental and social point of view would be remarkable and worthy. After reviewing the literature on the topic, this work presents a case study, in which the most common air-leaking points are found in an existing traditional masonry solid wall house. Some easy-to-do and affordable recommendations are suggested to fix those points. Reducing air leaks is a sustainable way to improve thermal performance in dwellings; it is a relatively low-cost solution, with no big impact or negative effects on the environment, and it facilitates further strategies to lowering energy consumption and CO_2 emissions.

KEYWORDS

Airtightness, energy efficiency, building envelope, existing housing stock, eco-retrofit.

INTRODUCTION

The United Kingdom has the oldest housing stock in Europe, counting almost nine million buildings older than 60 years [1]. In March 2012 the UK Department of Energy and Climate Change (DECC) [2] highlighted that the energy utilised for space and hot water heating in households is nearly 80%. DECC reports that the heat demand in the housing sector has

enhanced in the last 40 years, despite the noticeable improvements in the buildings energy efficiency and the more temperate winters; such a growth is due primarily to the raise of either the average internal temperature in homes or the number of dwellings.
The Department for Communities and Local Government (DCLG) [3] informs that the existing housing inventory is somewhat weak from the energy efficiency point of view: more than 40% of dwellings is in band D and more than 30% is in band E. Retrofitting the current buildings to take them to a higher band (hopefully at least to band B) is loosely acknowledged to be a smart goal.
Apart from the economic and environmental aspects of the matter, reducing the amount and costs of heat demand is worthy also to improve the social downside infamously known as *fuel poverty*. In 2009 nearly 20% of households in England were listed as being in fuel poverty and most of them live in private rented dwellings [4]. In other words, four million households cannot afford to heat the place they live in.
As shown in Figure 1, the public sector holds just a small amount of the dwelling stock, which is on average more energy efficient than the private one; this finding is not surprising, as local authorities and housing associations can access more easily to funding and grants, and can get bulk rebate from contractors. Nevertheless, in England approximately 10% of inhabitants in the social housing sector live in fuel poverty [5].

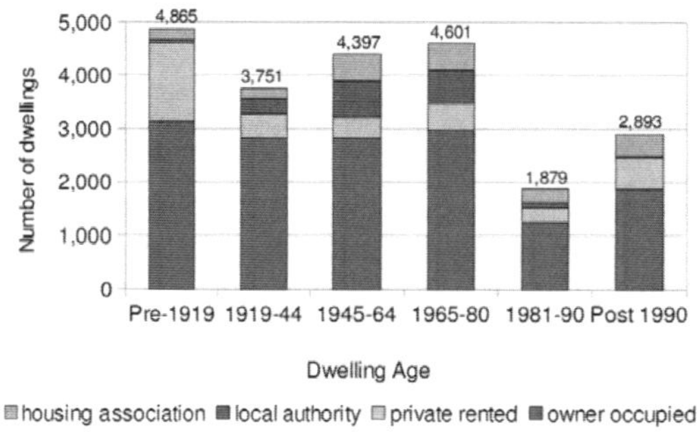

Figure 1. English dwelling stock profile in 2010, in numbers [3].

Most of the heat generated in the UK is wasted through inefficiently insulated envelopes, heats vacant rooms or warms too much lived spaces, causing loss of money and increment of CO_2 emissions. Lowering heat demand is the key to prevent short-term peaks and to encourage the use of energy from low carbon sources; it would also push for low-cost warmth and cut down fuel poverty [2]. The thermal performance of the current dwellings can be improved in two ways: by adding insulation to the envelope, and by reducing the unintended air leaks of the envelope [2, 6]. This study focuses on this second method, which can reduce heat waste of about 20%; moreover, an air-proof envelope can enhance the performance of the insulation and limit the damages to it. A traditional masonry solid wall building is analysed as case study.

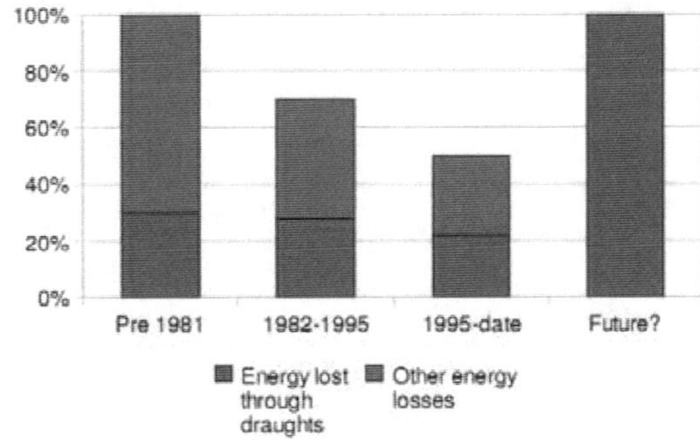

Figure 1. Heat losses [7, 8].

AIMS AND OBJECTIVES OF THE STUDY

This work does not aim at suggesting improvements to bring the existing English housing stock to the highest level of energy performance. A lot of literature is already been published on the topic and it is well known that, without funds from the government, very few owners can afford such an enhancement to their house. The main goal of this study is to suggest recommendations for lowering the heat losses through the building envelope due to air leakage that each owner can do by her/himself in a cost-effective manner. Thus the solutions suggested are affordable and feasible; practical, quick and simple, easy to find in the current market, and easily replicable in different buildings. Pushing for high-energy-performing retrofits is today very costly and it is feasible only if governments intervene with public funds or grants. There are some very efficient examples of retrofits, but they still are not frequent.

The rationale behind the approach proposed in this study is that by promoting feasible and affordable improvements, a larger audience can be involved in the energy-retrofit field. Walking baby-steps toward improving the efficiency of a house is easier and cheaper than performing massive works to insulate it, and more people can afford doing this. Once the existing housing stock reaches its best level of draught-proof, all other improvements (insulation, high performing systems) will be much more effective. The expected results on the long term are the ones suggested by the Energy Saving Trust such as reducing the level of energy needed to heat buildings, and as a result, cutting down energy bills. As explained in The Poverty Site, four million of households in England cannot afford to warm their houses. It is not conceivable that they commit in hugely demanding energy retrofit works. The approach proposed in this study has a surplus value due to the significant social impact it may have; this is another reason why it may be worth pushing for it.

Sealing against draughts is a cost-effective and efficient way to save money and energy in any building, as draught-proof solutions help warm air staying within the building. Householders can save lot money on their bills by simply reducing air leakages. As explained by the EST in the following:

"Full draught-proofing will save you on average £55 per year. Draught-free homes are comfortable at lower temperatures – so you'll be able to turn down your thermostat. This could save you another £60 per year. If every household in the UK used the best possible draught proofing, every year we would save £190 million, and enough energy to heat nearly 400,000 homes." [9]

LITERATURE REVIEW

Traditional Building Construction

Traditional buildings in England have masonry solid thick walls (usually 50-60 cm) made up of stones laid in lime or earth mortar. These walls are inhomogeneous, composed by an external stone leaf, an internal stone leaf and the gap between the two filled with mortar and rubble (a cheap local material); the three layers are glued together by the mortar. The most common internal wall finishes are as follows: plastered on the hard; plaster on laths; dry-lining; or timber lining. All these finishes, apart from the plaster on the hard, are attached to the wall through flat wooden strips (battens) and studs. The distance from the wall to finish could vary from 25 mm (for plaster on laths) to 50 mm for dry-lining and the resulting cavity was usually air-filled [10, 11] (see Figure 3).

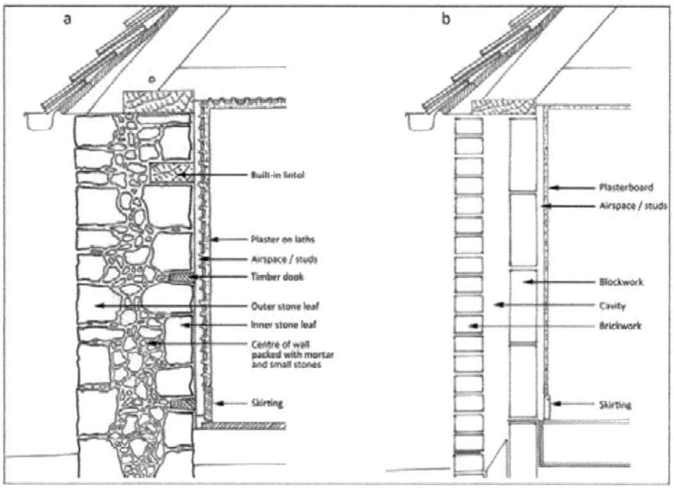

Figure 3. Comparison of (a) traditional solid masonry wall construction and (b) non-traditional cavity wall construction [10].

Air Leakage in Traditional Buildings

Air leakage occurs when air passes through the building envelope through cracks and gaps. The amount of air infiltration depends on the design and quality of the construction. The wind blowing against the construction may also induce air infiltration, as it causes a difference in pressure between the inside of the house and the outside: on the windward wall air gets into the house through gaps (infiltration), while in the leeward face air gets out (exfiltration) [12]. Despite their permeability, pre-1900 houses are frequently more airtight than some modern buildings; this might be due to the fact that the skilled labour who built traditional constructions was more precise than the modern one [13, 14, 15].

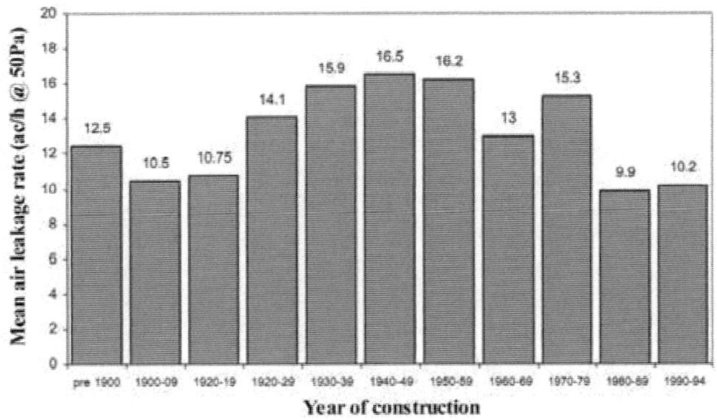

Figure 4. Relationship between dwelling age and air leakage [13].

Antretter et al. [15] highlight also that houses built after 1980 are tighter than previously built ones. Such a difference is explained by the use of upgraded and new materials, and improved construction and design techniques. The evolution of building codes has also played a key role in promoting the enhancement of airtightness. The same research found that bigger houses tend to have more air leakage and that seasonal changes have no effects on building air leakage.

Sedlak et al. [16] found that 1950s houses are more airtight than modern and refurbished ones, and this probably depends on the construction method (solid walls with wet plastering).

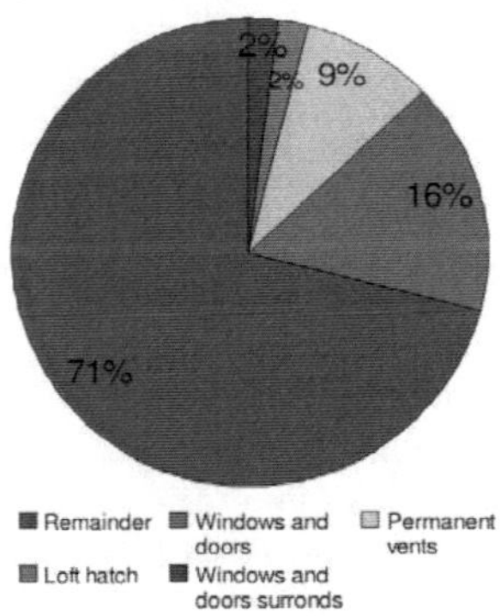

Figure 5. Air leakage paths [17].

As the above figure highlights (see Figure 5), the bigger slice (71%) is composed by the remainder, namely [17]:

- *Plasterboard dry lining on dabs or battens;*
- *Cracks, gaps and joints in the structure;*
- *Joist penetrations of external walls;*
- *Timber floors (under skirting and between boards);*
- *Junctions between internal stud walls and with floors and ceilings;*
- *Electrical components such as sockets, switches and light fittings;*
- *Service entries and ducts;*

• *Areas of unplastered wall between intermediate floors, behind baths etc.*

The Holistic Approach

Adopting a holistic approach in energy retrofitting a traditional building is crucial. Before making any changes to improve energy efficiency, it is fundamental to understand how traditional buildings were designed to work; how tick walls perform in term of thermal mass; how moisture moves within the structure; and what is the actual thermal performance of the house. Several completed retrofits are based on modern methods of insulation and internal climate control which has been deleterious for buildings [6, 10, 11, 18].
As explained previously, traditional constructions are made of soft and pervious materials, and therefore the building envelope is permeable to moisture and water vapour. Traditional buildings behave in a quite different way from modern ones, and this must be taken into consideration when working on energy retrofit.

Traditional masonry wall	Modern masonry wall
Massive construction, thick walls	Relatively slender construction
Porous, highly permeable construction	Porous materials but low permeability finishes
Large volume of absorbent materials (porous stone and mortar)	Limited ability to absorb moisture
Moisture can penetrate into and evaporate easily from the wall. This helps stabilise moisture levels in rooms	Moisture within the wall not easily evaporated
Modern levels of insulation may not be achievable	Insulation materials may be adversely affected by moisture
Construction can absorb small thermal and moisture movements	Well insulated construction Prone to cracking due to hard, brittle materials Cracks in external finishes permit water penetration into construction
Air movement required behind dry lining to prevent raised moisture content of timbers	Ventilation of cavities not required
Damp-proof course not normally installed	Damp-proof course essential
No vapour checks or barriers	Vapour checks integrated into construction

Table 1. Comparison of traditional masonry wall and modern masonry wall constructions [11].

As Urquhart suggests, new additions should never be stronger or denser than the existing elements, as they could deteriorate the original structure, which he also encourages the use of material of natural origin. Furthermore, adding too different materials to a traditional solid wall could affect its moisture equilibrium and risk to damage the timber elements in touch with the wall.
The traditional construction techniques exploit the higher thermal mass of thick walls to regulate the temperature and comfort within the building, cooling it in the summer and warming it in the winter. Despite the potential energy efficiency of traditional building envelopes, they often over-leak heat, mainly through cracks and holes in the fabric, unused open chimneys and fireplaces, gaps around windows and door frames [6, 10, 11, 18].
Traditional masonry constructions usually do not have damp-proof courses or damp-proof membrane under the ground floors. They use the mass of the walls and the movement of the air to balance the moisture transfer coming from the ground; porous walls exploit capillarity forces to drag moisture in their inside and let it evaporate on their external surface. It is therefore important to keep the porosity of the walls, in order to allow air to pass and draw moisture. The experience has shown that inserting modern moisture barriers may not be an effective solution, and it may be better to add perimeter drainage to regulate the effects of raising damp [11].

CASE STUDY: J&M'S HOUSE IN NETHER EDGE, SHEFFIELD, UK

Nether Edge is a Conservation Area located few miles south west of Sheffield, South Yorkshire, UK. The area is a leafy suburb, characterised by fine Victorian and Edwardian buildings. J&M's house is a 19th century Victorian semi-detached house located in this neighbourhood.

Air Leakage Points

The main air leakage points in J&M's house have been identified [19]. Like in thermal insulation, in correcting airtightness it is crucial to have continuity. All gaps and cracks in the structure allow air to get into and out the building, so they all should be sealed at the same time. The table below lists the most common problems and the suggested improvements (see Table 2).

Most common problems	Suggested improvements
Gaps in cellar's walls and ceilings	• All gaps, holes and breaks in the walls of the cellar need to be repaired with materials similar to the existing ones. • Top hat elements or collars may be used to seal around services pipes. • Small cracks around pipes can also be sealed with silicon fillers, while larger gaps can be filled with expanding polyurethane foam.
Gaps in and around suspended timber floors	• All damages and gaps in the timber floor should be repaired. •The junction between suspended timber floors and the skirting should be sealed with a flexible sealant. • Joints in timber floor should be sealed with suitable glue and any angular edge joints in the decking to the joists should be fully supported and fixed. • If there is an air barrier, all penetration through it should be filled up with a flexible sealant. • Gaps around service pipes passing through suspended timber floors should be adequately sealed.
Gaps around windows and doors	Draught-proofing <u>timber windows</u> limit air leakage and improve the internal comfort. • It is not a good practice to use foaming gap-filling adhesives, as they tend to contract and break the seal. Internally silicone is an appropriate sealant, while for the exterior it is better to use moisture-resistant strips of EPS between the window frame and the reveals. • Window closing devices need to be checked to secure a tight closure. • The wall-to-frame junctions need to be air-proof and can be sealed, especially at sills. • It is better to avoid draught-stripping existing windows and external doors in kitchens and bathrooms unless there is an extract fan [14]. Sealing <u>outside doors</u> is very cost-effective as it is cheap but can save a lot of heat. • Four main elements have to be considered: 1. the keyhole has to be covered by a metal disk that falls down when the key is not in; 2. the letterbox should have a flap or a brush; 3. the gap at the bottom should be closed with a brush or hinged flap draught excluder; 4. gaps around the edges should be filled up with fit foam, brush or wiper strips like the ones used for windows. The <u>doors in the inside</u> have to be draught-proofing, especially if they connect unheated to heated spaces. • Gaps around these doors should be sealed with a draught excluder that can be made by used plastic bags or pieces of spare material [9].

Gaps at the ceiling-to-wall joint at the eaves	• The ceiling junction has to be continuous. If it is not, an airtightness membrane or tape should be installed; as alternative, a flexible sealant should be applied between ceiling board and wall. • The ceiling should be repaired first (if necessary), and then all gaps between ceiling and the masonry wall should be sealed with flexible or adhesive sealant.
Open chimneys	• It is appropriate to close fireplaces and chimneys, and this can be done in two ways: by adding a vent in the fireplace or a chimney balloon (an inflatable cushion that blocks the chimney); and by capping the chimney-pot with a hood (it is recommended to have this work done by a professional) [9, 14].
General air leakage through walls	• Collars should be applied around the pipes, in order to better airtight them; otherwise a weather-resistant sealant should be applied. • All gaps and cracks in the wall should be repaired by using materials similar to the existing ones.
Bathroom and kitchen wall vent or extract fan	• Gaps around the extractor fans and cooker hoods have to be properly sealed. • If in the wall there are old fan outlets, these should be filled with material similar to the existing wall and sealed either in the inside or the outside [9].
Gaps around bathroom and kitchen waste pipes	• Gaps around pipes should be sealed with top hat elements or collars. •All services passing through the wall (water, drainage, gas pipes, boiler flues and electrical cables) that have some gaps around the pipes should be sealed with an appropriate sealant. • Sealants around boiler flues have to be heat resistant [14].
Gaps around loft hatches	• The loft hatch has to be sealed with an air-proof tape. • Having the hatch thermally insulated, as well as the rest of the ceiling, is a plus [14].
Gaps around service pipes	• Gaps around service pipes, cables and ducts should be blocked with top hat elements or collars and be sealed with an appropriate sealant. • Sealant around hot pipes has to be heat resistant.
Ceiling roses and recessed ceiling lights	• Holes around light fittings and pull cords in the ceiling should be appropriately sealed and, if possible, an airtight box over the light fitting in the ceiling void should be installed. • When changing the light fittings, it is good practice to choose airtight ones [14].

Table 2. List of most common problems and suggested improvements [19]

CONCLUSION

Old houses have modified over time to adjust to the changing needs of the owners and to the environment. However, the variations they had not always benefit the energy performance of the building, as shown by the case study.

J&M's house has very special features that make it very characteristic and appealing: beautiful chimneys almost in every room, high ceilings, large windows, spacious cellar. The downside of these features is that they ask for a lot of energy in order to keep rooms warm and comfortable and that they favour the waste of heat.

The house perfectly adapted to the changing needs of the owners: as bedrooms were rented to different people, new bathrooms were added and the new pipes were installed on the outside walls. While making the house more liveable, these improvements required to pass the pipes through the external walls, leaving gaps around them where heat is lost. Also the age of the house does not help much in term energy performance, as the settlements of the ground over time caused cracks to the ceiling-to-wall joints and to the floor-to-wall joints.

The issues found in J&M's house are common in pre-1900 English buildings, but they can be easily fixed with simple and cost-effective improvements. The benefits householders can have from such enhancements are often underestimated, but they may be worth the work and the cost of doing them. Raising the airtightness of old buildings limits the amount of heat wasted through the envelope, with the resulting less heat required to have a comfortable temperature in the rooms. Home owners can save up to more than £100 per year on their energy bills (mainly heating). Furthermore, bringing the draught proof of the dwelling to its best will lift considerably the effectiveness of the insulation, of the high performance systems, and of all other enhancements done to improve the energy performance of the house.

The literature suggests that up to 20% of energy can be saved by bringing the air-proofing of existing buildings to their best. Such an amount is significant and commonly undervalued in the energy retrofit processes. Energy performance software usually focuses on U-values and do not consider the raise in thermal performance due to the improved airtightness. For practical purposes this is tricky.

Let's suppose the thermal performance of a building is actually improved by 20% only by limiting its air leakage. No software will corroborate a shift to a higher energy efficiency rating band; but in the reality, this could happen, above all for buildings laying at the highest limit of a band (a dwelling in band E, rating 53 or 54, could raise to band D).

ACKNOWLEDGEMENTS

I would like to thank my supervisor, Prof. Cristina Benedetti, Faculty of Science and Technology, Free University of Bolzano, Italy, for pushing me into independent thinking and practising. I am also grateful for the precious feedback and encouragement from my co-supervisor, Dr. Hasim Altan, who hosted me at the Building Environments Analysis Unit (BEAU), University of Sheffield, UK.

Ms Jenny Fortune and Ms Beatrice Greenfield at the South Yorkshire Energy Centre and Mr Nick Parsons, Sustainable Building's consultant, gave me valuable hints, information and literature. I have really appreciated their willingness to share their knowledge and experience with me. Special thanks to Ms Joyce and Mr Mark Higton for their kindness, enthusiasm and patience in letting me examining their house.

REFERENCES

[1] Energy Saving Trust. (2010b). *Sustainable Refurbishment*. London.

[2] Department of Energy and Climate Change. (2012). *The Future of Heating: A Strategic Framework for Low Carbon Heat in the UK*. London.

[3] Department for Communities and Local Government. (February 2012). *English Housing Survey. Headline Report 2010-11*. London. ISBN: 978-1-4098-3314.

[4] The Poverty Site. (2012). *Fuel Poverty. Key Points*. (Accessed on April 2012): http://www.poverty.org.uk/.

[5] Energy Saving Trust. (2010a). *Reducing Emissions from Social Housing*. London.

[6] Jenkins, M. (2011). *Improving Energy Efficiency in Traditional Buildings*. Historic Scotland, Longmore House, Salisbury Place, Edinburgh.

[7] Jennings, P. (2001) *Airtightness in Buildings. (Parts 1 & 2)*. Building for a Future, Winter 00/01, Spring 01.

[8] Morgan, C. (2006). *Design and Detailing for Airtightness*. SEDA Design Guide for Scotland No. 2. SEDA.

[9] Energy Saving Trust. (Accessed April 2012): http://www.energysavingtrust.org.uk/.

[10] Baker, P. (2011). *Historic Scotland Technical Paper 10*. Historic Scotland.

[11] Urquhart, D. (2007). *Conversion of Traditional Buildings. Application of the Scottish Building Standards. Part 1. Principles and Practice*. Historic Scotland, Edinburgh.

[12] Jaggs, M. and Scivyer, C. (2009). *A Practical Guide to Building Airtight Dwellings*. Published by IHS BRE Press on behalf of the NHBC Foundation.

[13] Johnston, D., Wingfield, J., Miles-Shenton, D. and Bell, M. (2004). *Airtightness of UK Dwellings: Some Recent Measurements*. Paper presented at COBRA 2004 conference.

[14] GreenSpec (Accessed April 2012): http://www.greenspec.co.uk/.
[15] Antretter, F., Karagiozis, A.N., TenWolde, A. and Holm, A. "Effects of Air Leakage of Residential Buildings in Mixed and Cold Climates." in *Thermal Performance of the Exterior Envelopes of Buildings X*, Proceedings of ASHRAE THERM X, Clearwater, FL, Dec. 2007.
[16] Sedlak, P. and Sheward, N. (2008). *Advanced Ventilation Approaches for Social Housing (AVASH) - Irish Sampling & Survey Report*. DWEcoCo and Cluid Housing Association.
[17] Trainenergy (Accessed April 2012): http://www.trainenergy-iee.eu/.
[18] Pickles, D., Brocklebank, I. and Wood, C. (2011). Energy Efficiency and Historic Buildings. English Heritage. SEDA.
[19] Barbisan, R. (2012). *Energy Retrofit of the Existing Housing Stock in England.* Master Thesis (Master Klimahaus). Faculty of Science and Technology, Free University of Bolzano, Italy.

Lessons learnt from the regulatory quality management scheme in France

Sarah Juricic*[1], Sandrine Charrier[1], Florent Boithias[1] and Joris Biaunier[1]

1 Centre d'Études Techniques de l'Équipement de Lyon

Boulevard Bernard Giberstein

71400 AUTUN Cedex, France

**Corresponding author: sarah.juricic@developpement-durable.gouv.fr*

ABSTRACT

From January 1st 2013 on, the French energy performance regulation will demand that the airtightness level is justified and that airtightness of a building should be below 0,6m³/h/m² at 4Pa for single family housing and 1m³/h/m² for multi-family dwellings, resulting into an important growth in the airtightness market. It is the role of the State to accompany this market evolution and to supervise the quality of airtightness measurements used for the EP calculation. This is why it has been decided that there are two possibilities to justify the airtightness level of a building. Either the constructor makes a systematic measurement of their building or the constructor proves they have a quality management approach so that more than 85% of their production reaches the wanted airtightness.

In order to ensure the quality of the quality management schemes for airtightness, a specific committee has been created. Its goal is to authorize constructors to justify an airtightness level by a quality management scheme. The CETE de Lyon is in charge of this committee.

This paper deals with the role and results of the committee and discusses the advantages and issues raised by such authority, thanks to the experience gained by the CETE de Lyon on these matters. Results show an improvement in the airtightness levels reached by authorized constructors in comparison to levels reached without any quality management approach. Flaws in the control process and biased tests show several possibilities for the State to improve the frame of this authorization.

KEYWORDS

Envelope airtightness, quality management

INTRODUCTION

With the future obligation to prove a certain level of compliance with the French Energy Performance Regulations, airtightness has got a key role in the construction field. Indeed, the application of the 2012 EP regulation demands that buildings comply with an airtightness level below 0,6m^3/h/m^2 at 4Pa for single family housing and 1m^3/h/m^2 for multi-family dwellings. To prove the compliance, a constructor has two choices. Either they make a systematic measurement of their buildings or they prove by hand of a quality management scheme for airtightness that more than 85% of their production has the wanted airtightness.

This paper deals with the role of this committee and discusses the advantages and issues raised by such authorities. This paper also presents the results of a state driven control campaign. This paper will hence try to give some answer to the question: is it worth it to implement such a procedure for quality management schemes?

REGULATORY QUALITY MANAGEMENT SCHEME

Context

As described in Leprince 2011, quality management process for airtightness of buildings has been set up in order to improve air tightness treatment during all design and construction stages and in order to spread good practice among professionals.

The French 2005 energy performance regulation introduced the possibility to use an airtightness value lower than the default value in the EP-calculation. This possibility is given only if a measurement proves the lower airtightness value or if the constructor follows a State authorized quality management procedure for airtightness, without systematically performing a test.

Soon, the 2012 energy performance regulation, applicable from January 1st 2013 for housing, makes the airtightness test compulsory. The quality management scheme gives the applicants the possibility to reduce the amount of compulsory tests at commissioning since only a minimum of 5% of the production has to be tested. It gives also the possibility to make energy performance calculations with an airtightness factor lower than the regulatory 0,6 m^3/h/m^2.

Requirements

Applications are sent to a specific committee dealing with the quality management procedure in airtightness. Any application has to include basic requirements linked to quality management approach, tests on a sample of the production and training documents focusing on airtightness destined to co-workers and craftsmen. Furthermore, some documents have to be submitted to the committee, among others:

-Identification of the chain of liabilities: who does what and when

-Description of the approach applied to the company

-Description of the design characteristics of the buildings on which the quality management approach applies

-Results of tests on a sample of the buildings production proving that more than 85% of the tests are below the target airtightness value

The 2012 quality management process will also require all documents produced in the frame of the quality approach for randomly selected buildings.

Self declared results obtained by approved companies in 2011

So far, the committee received follow ups of a dozen of applicants implementing a 2005 quality approach. The follow ups included bar charts of all measurements performed internally.

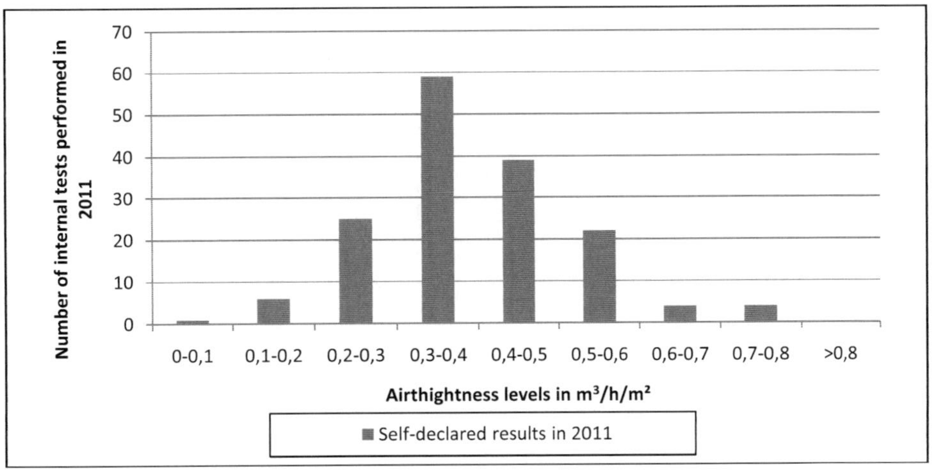

Chart 1: Bar chart of all self-declared results (follow-up 2011) N=160

Chart 1 presents a sum up of all self-declared test results made in 2011 by all constructors that had been authorized in 2010. Obviously, the results in Chart 1 show that every single building tested by these 14 constructers scored below the Q_{4Pa_surf} target of 0,8 $m^3/h/m^2$. The bar chart also shows a normal distribution.

Controls by state technicians

The results presented in Chart 1 are based on measurements performed by State authorized testers. These testers however are not necessary independent of the applicant. Indeed, applicants get advice from ISO9001 bodies working in the field of airtightness that audit the applicants and most likely test the production of the applicant. The independence of the measure is therefore not guaranteed.

To avoid such a bias, the committee started in 2011 a control campaign. Every year, each applicant is asked to hand in a list of all buildings expected to be delivered in the coming year, including date of commissioning, name and address of the client. If the applicant is reluctant to give the demanded information, the applicant might see his agreement suspended.

Then a state technician performs control tests on randomly selected buildings. The amount of buildings tested is supposed to cover more than 5% of all buildings delivered. As of September 2012, 74 control measurements have been performed, whereas 99 had been planned. It represents so far 3,7% of the yearly production of all constructors. Further tests are still expected.

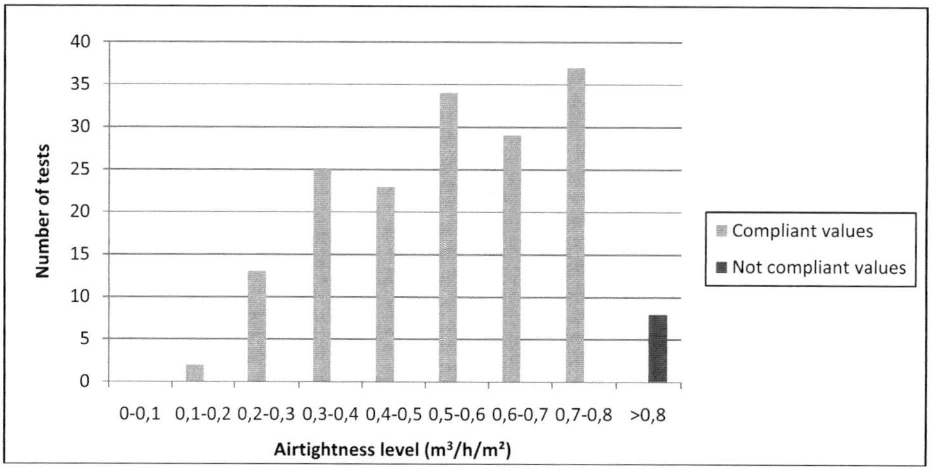

Chart 2: Bar chart of airtightness levels from the control campaign, compliance to the target level

Chart 2 shows a bar chart of all airtightness values measured. From Chart 2 can be inferred that if most of the tests show a result lower than the target airtightness level, a few are above the wished Q_{4Pa_surf} of $0{,}8m^3/h/m^2$. In the dwellings showing a higher airtightness measure, the leaks are mainly located around water and gas ducts, around boxes integrating roller shutters and window frames. Other leaks are due to a misunderstanding of the constructor of the moment of commissioning. Indeed, some constructors leave the possibility to the client to do a part of the building works themselves, for example installing toilets or a wood-burning stove. So when the dwelling is handed in to the client, these elements are not installed, but they do have an influence on airtightness, which explains part of the high airtightness results obtained.

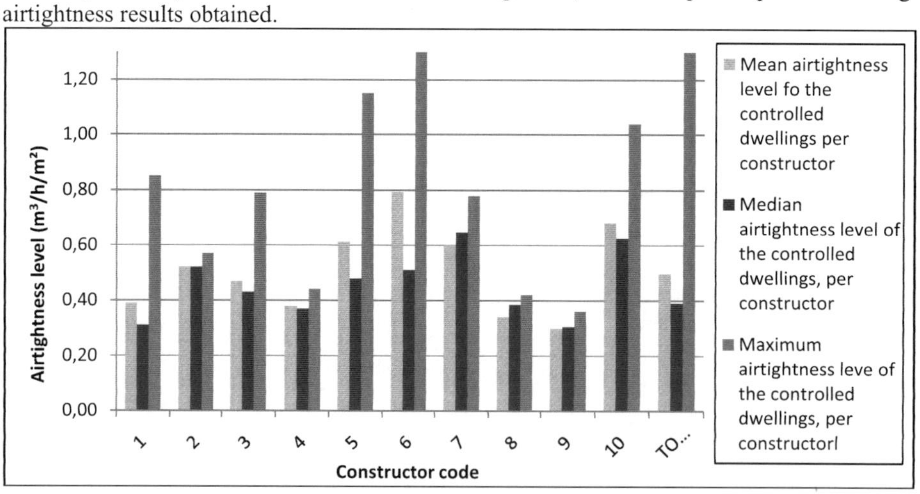

Chart 3: Bar chart of the mean, median and maximum airtightness values from the control campaign, per constructor [Maximum value of constructor 6 is above $1m^3/h/m^2$ and is out of the scale of the chart]

Mean and median values showed in Chart 3 corroborate the results implied from Chart 2. There are all under the 0.8 $m^3/h/m^2$ target level. However, this means that some constructors have been controlled with frequent too high airtightness results, whereas other constructors comply at 100% to the target level. A particularly problematic point in this chart is that one of the constructors showed control results above $1.3m^3/h/m^2$, which is above the default value of the Energy Performance Regulation and which is a specific requirement to be respected.

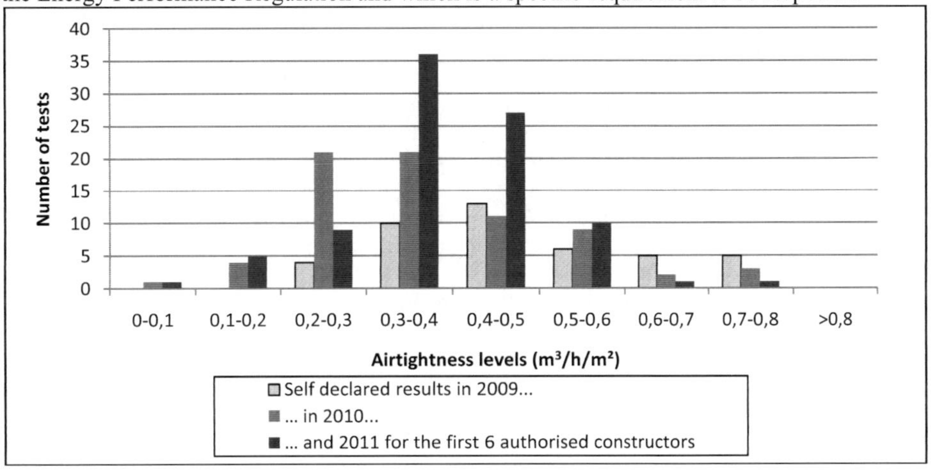

Chart 4: Bar chart of the self-declared results for constructors that have had more than two follow-ups (2009/2010/2011 or 2010/2011)

Chart 4 shows the self-declared results of the constructors that have had more than two follow-ups since beginning. It can be implied that results of 2010 and 2011 are better than results of 2009, but that the latest follow-ups show higher results than in 2010. This means that there is a certain improvement in the general airtightness level, but also that the efforts are probably not being pursued when the target level, $0.8m^3/h/m^2$ is reached.

Discussion

As already mentioned above, buildings are not always completely finished when the keys are handed to the owner, for example clients take in charge bathrooms or chimney. As a consequence, testers should not seal the holes left because they have to comply to the norm NF EN 13829 and its implementation guide, which demand to leave the holes open, hence there are probably some improper measurements done internally, which gives a bias in the results showed by the constructer.

The committee discussed this point and decided that it is still the liability of the constructor to justify the level of airtightness at commissioning, even when holes are left open. The committee will therefore expect the following requirements to be fulfilled. The first possibility is to reach an airtightness level low enough even if the building is not yet finished. If not and/or if works are to be done in the house by the client, the constructor has to prove that those works are not a threat to the airtightness, and a test is performed after the works by the client. On the contrary, if the works are a threat, the test will still be done after finishing the works. Hence the constructor is expected to give a specific training about air permeability to the client so that they will not deteriorate the airtightness.

As a consequence, the committee advises the constructors to inform in early stages their clients that their house has had a specific airtightness treatment and that there have to be precautious if they do not want to ruin the work done.

Another bias seen in the control tests performed by the state technician is that the controller is given name and address of clients with approximate date of commissioning by the constructor. The controller randomly selects buildings to test, but still relies on the constructor to visit the construction site. It has been seen that some controlled buildings have been “prepared” for the venue of the controller, with among others fresh foam material filling in vacant spaces for toilets. The test is done in the conditions the building has been delivered, but the real final airtightness value will be higher than what is measured, since the foam material is not meant to stay.

To improve the efficiency of the controls, it has been suggested that they should focus on buildings with sensible spots. We identified among others wooden intermediate floors or mechanical ventilation as quite difficult to apprehend from an airtightness point of view. If the focus is on buildings presenting that type of characteristics, it is to expect that the rest of the buildings production complies with the target airtightness level. Plus, the committee witnesses a growth in the number of applicants and with the application of the 2012 energy performance regulation. In only a few months, the number of applicants for the 2012 version has already exceeded the number of 2005 applicants over three years. It will then be difficult for control testers to measure more than 5% of the production of all these constructors. It is then all the more understandable to focus on sensible construction types.

Seeing that constructors having a quality management process succeed more easily to reach a target airtightness value raises an issue concerning other constructors. Every building will soon have to comply with the Q_{4Pa_surf} of $0,6m^3/h/m^2$ but it is feared that without proper preparation especially in early design stage, it might be difficult for average constructor to obtain such airtightness results.

Finally, let us note that controls are informative. But what if in the future, controls show more applicants that do not comply with their own target? There are still questions here: will the company lose immediately its agreement, will they be warned for a year, or will they have to hand in more documents? The balance between understanding and harsh decisions is yet to be found.

CONCLUSION

With the January 1st 2013 deadline approaching, it is of the greatest importance to prepare the market for lowered requirements in airtightness of buildings.

With the increase of applications the committees receives, it is to be understood that more and more constructors see the importance of treating airtightness by hand of a quality management scheme, which is in a way a success knowing the initial purpose of this authority. Self-declared tests as well as control tests show that in general, constructors gain advantage of such a scheme, for they reach satisfying airtightness levels, even for the 2012 version of the quality management requirements.

At the same time, it is feared that companies that have their authorization for long do not make any effort anymore to continuously improve their scheme, which is the opposite of what was hoped for. Plus, knowing the difficulty of testing the building at the exact moment of commissioning makes the committee doubt about the good faith of certain self-declared tests and makes it a necessity to communicate to all authorized constructors about what is testing at commissioning.

ACKNOWLEDGMENTS

The sole responsibility for the content of this publication lies with the authors. It does not necessarily reflect the opinion of the Ministry.

BIBLIOGRAPHY

V. Leprince, J. Biaunier, R. Carrié and M. Olivier, Quality management approach to improve buildings airtightness, requirements and verification, Tightvent conference, October 2011

S. Juricic, F. Boithias, J. Biaunier, Lessons learnt from their qualification of the airtightness testers and regulatory quality management scheme in France, Brussels, 2012

ABSTRACT SUBMISSION FORM FOR THE 2012

AIVC-TIGHTVENT CONFERENCE

Send your abstract using this form at info@inive.org

PAPER TITLE

Build Large Buildings right or suffer a life sentence of Energy, Comfort & IAQ Problems

MAIN AUTHOR

Name of the 1st Author: Colin Genge

Institution / company / authority, etc.: Retrotec Energy Innovations Ltd.

OTHER AUTHORS

None

ABSTRACT

Build Large Buildings right or suffer a life sentence of Energy, Comfort & IAQ Problems

Large buildings needlessly suffer from continual air quality complaints, excessive mechanical system maintenance and moisture problems that can lead to structural damage and unhealthy building syndrome. Although we have gotten much better at patching them up since so many have problems, we can now build new structures to last at about the same price that have none of these problems. Learn how.

All high rise buildings suffer from one or more of the following problems:

- air quality complaints – cigarette and cooking odors
- higher energy bills than anticipated
- excessive mechanical system maintenance
- vehicle exhaust fumes coming up from garages
- powerful drafts in cold weather
- moisture problems; condensation, mold, unhealthy humidity levels

We will show how building engineers regularly fix these problems in existing buildings after occupancy and alternatively, how they can be fixed for one tenth to one thousandth the cost during design and construction.

Once completed, almost all high rise buildings sentenced to a life of high energy costs, indoor air pollution, smoke risk and potential structural (or moisture) problems, without much hope of doing anything but patching up the problem.

We will present testing results on several high rise buildings that have undergone compartment by compartment air leakage tests that would have identified these problems. LEED only touches on the issue; there is much more that must be done to modernize our high rise buildings. We will show how problems can be fixed easily prior to occupancy by proper commissioning and how to designs can be improved to avoid problems in the first place.

POSTULATE FOR AIRTIGHTNESS LIMITS IN LARGE BUILDINGS

Paul Simons and Stefanie Rolfsmeier

BlowerDoor GmbH
MessSysteme für Luftdichtheit
Zum Energie- und Umweltzentrum 1, D-31832 Springe-Eldagsen, Germany
Tel. +49 (0)5044/975-40, E-Mail: info@blowerdoor.de

ABSTRACT

DIN 4108-7 requires a limit of $q_{50} \leq 3.0$ m³/m²h for the air permeability of large buildings. Even stricter limits with respect to q_{50} can be found at DGNB [German Sustainable Building Council] and in the Swiss MINERGIE Standard.

It is the objective of this presentation to develop awareness of this topic in the audience and to give recommendations as to which limits can be applied to new building projects.

Theoretical considerations and experience from measurements lead to the conclusion that a volume-based limit of n_{50} is not a suitable target value for large buildings. Because of the changing surface-area-to-volume ratios (SA:V-ratio) in large buildings, it makes sense to require an envelope-based limit, especially since there are requirements for limiting air permeability for building component joints and service apertures. Existing limits and results from airtightness measurements are presented. The presentation will also outline the main points of the approach to achieving airtightness as planned.

KEY WORDS

Air permeability q_{50}, limits (table), roll-up doors, loading bridges, smoke extraction for elevators

INTRODUCTION

Airtightness tests of large buildings such as office buildings, schools, homes for the elderly, warehouses, and production halls are fortunately becoming increasingly common in Germany. They are frequently performed in order to meet the requirements or exploit the benefits of building airtightness as defined in the German Energy Savings Regulation or conducted because of increased public awareness as to preventing waste of energy. Another reason is the ever higher number of quality certificates required.

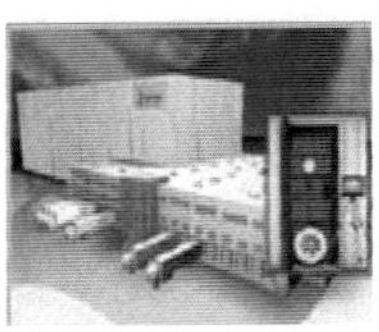

Photos:
Left: www.hammersen.de
Right: www.sanidetectif.be

Figure 1: Two buildings with an internal building volume of 200,000 m^3 each. The building on the left with a V_{50} = 2,600 m^3/h is very airtight. The building on the right has a V_{50} = 86,500 m^3/h.

LIMITS AND MEASURED VALUES

Air change rate n_{50} at 50 Pascal

The German Energy Savings Regulation limits the air change rate n_{50} of a building to the following values when conducting an airtightness test according to European Standard EN 13829:

$n_{50} \leq 3.0\ h^{-1}$ for buildings without a ventilation system and
$n_{50} \leq 1.5\ h^{-1}$ for buildings with a ventilation system

According to the German Energy Savings Regulation 2007, the energy balance for non-residential buildings is calculated according to the series of German Industrial Standards DIN V 18599. Based on the project experience of Mr. Moritz Wagner, Dipl.-Ing., of Büro IFB Sorge (Nuremberg), the following can be stated with the DIN V 18599 assessment:

- Considering an airtightness test usually has a positive effect on the annual primary energy requirement.
- For common types of buildings, the reduction comes to 10-15%.

The German Industrial Standard DIN V 18599 allows for applying the measured n_{50}-value as a rated value. The standard rated value according to DIN V 18599 for buildings without ventilation systems is $n_{50} \leq 2.0\ h^{-1}$ and for buildings with ventilation systems is $n_{50} \leq 1.0\ h^{-1}$. Figure 2 shows that the real measured values often amount to $n_{50} \leq 0.5\ h^{-1}$. By using the real measured n_{50}-values, improvements in the energy balance beyond the standard rated values can be expected.

It is important to determine this value according to Method A in German Industrial and European Standard DIN EN 13829.

The experience from testing large buildings has shown that the limits of the German Energy Savings Regulation and German Industrial Standard DIN V 18599 are usually met and to some extent the measured values remain far below them. The following diagram shows a compilation of the air change rates at 50 Pascal (depressurization tests) of 82 buildings measured by a series of testing teams. The smallest building has an internal volume of approximately 1,300 m^3, the largest one of approx. 520,000 m^3.

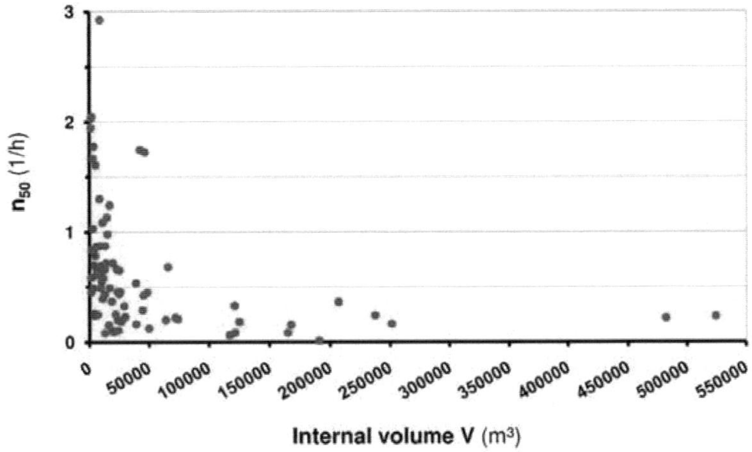

Figure 2: Air change rates n_{50} (82 depressurization tests) of large buildings

The air change rates of all buildings are below 3.0 h^{-1}. Almost 90% of the air change rates are even lower than 1.5 h^{-1}.

What is the reason for these seemingly excellent results for the air change rates at 50 Pascal? Is the quality of the building envelope of large buildings so much better than that of single-family homes? Or are there other reasons?
The air change rate n_{50} is a volume-based indicator. It is calculated by dividing V_{50}, the leakage flow determined at 50 Pa, by the internal building volume V:

$$n_{50} = V_{50} / V$$

This results in large buildings achieving better (lower) air change rates than single-family homes because they have a smaller SA:V-ratio (surface-area-to-volume ratio), which means that a "large" volume is enclosed by a relatively small building enveloping area containing the leakages.

Examples for SA:V-ratios:

Type of building	SA:V-ratio(1/m)
high-rise building	from 0.2
apartment building/multiple family home (MFH) (3 to 4 floors)	approx. 0.3 to approx. 0.6
center row house (2 to 3 floors)	approx. 0.5 to approx. 0.7
single-family home (SFH)	from 0.8

In this context, the air change rate n_{50} does not yet provide any information on the quality of the building envelope. An evaluation can only be conducted when the air change rates of the same quality of the airtight layer are related to the SA:V-ratios.

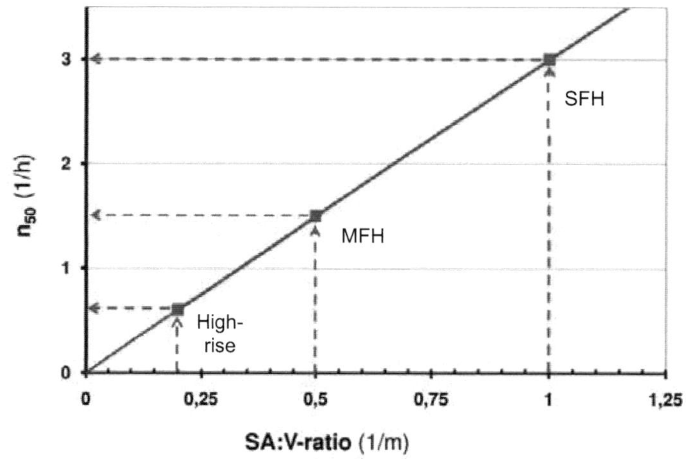

Figure 3: Air change rates of the same quality of the airtight layer related to the SA:V-ratio

The diagram shows the example of three buildings: a single-family home (SFH) with a SA:V-ratio of approx. 1 m^{-1}, an apartment building/multiple-family home (MFH) with a SA:V-ratio of approx. 0.5 m^{-1}, and a high-rise building with an SA:V-ratio of 0.2 m^{-1}. The single-family home at 50 Pascal is supposed to have a maximum air change rate of $n_{50} = 3.0\ h^{-1}$. Assuming that the multiple-family home and the high-rise building feature just as many leakages per square meter of enveloping area, the airtightness test of the multiple-family home would determine an air change rate of n_{50} of 1.5 h^{-1} at 50 Pascal and that of the high-rise building a rate of 0.6 h^{-1}.

CONCLUSION

The air change rate of large buildings should always be evaluated in relation to the SA:V-ratio of the building.

Air permeability q_{50} at 50 Pascal

To better compare the quality of the building envelope of different buildings, an additional indicator can be used: the air permeability q_{50}. German Industrial Standard DIN 4108-7 in its version of January 2001 also requires limiting the air permeability for buildings with an internal volume of > 1.500 m^3 to $q_{50} \leq 3.0\ m^3/(h \cdot m^2)$.

The air permeability q_{50} is calculated by dividing the leakage flow V_{50} at 50 Pascal by the respective building enveloping area A_E:

$$q_{50} = V_{50} / A_E$$

It indicates how many cubic meters of air per hour at a building pressure differential of 50 Pascal flow over one square meter of building enveloping area.

The following diagram shows the air permeability q_{50} of 42 depressurization tests.

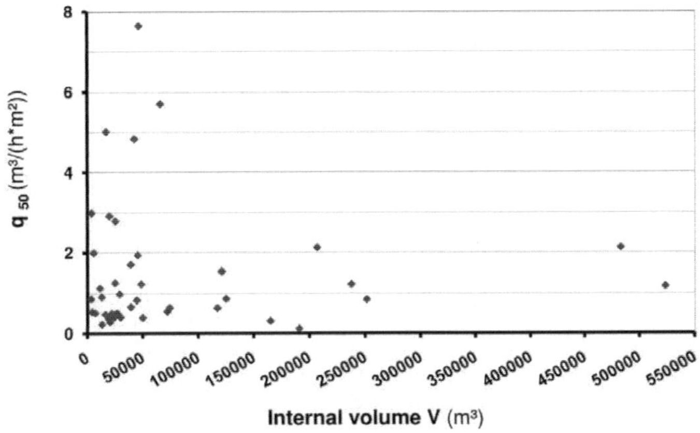

Figure 4: Air permeability q_{50} (42 depressurization tests) of large buildings

90% of the buildings meet a $q_{50} \leq 3.0$ m³/(h*m²). 70% remain below a $q_{50} = 1.5$ m³/(h*m²).

At an international level, limits for buildings larger than 1,500 m^3 have already been formulated:

- Minimum standard 4108-7 $q_{50} \leq 3.0$ $m^3/(h \cdot m^2)$
- Minimum standard DGNB $q_{50} \leq 2.5$ $m^3/(h \cdot m^2)$
  (German Sustainable Building Council)
- Improved standard DGNB $q_{50} \leq 2.0$ $m^3/(h \cdot m^2)$ rule of technology
  (German Sustainable Building Council)
- Swiss MINERGIE Standard $q_{50} \leq 1.25$ $m^3/(h \cdot m^2)$ soon rule of technology
- Optimum standard $q_{50} \leq 0.6$ $m^3/(h \cdot m^2)$ state of the art

Rule of technology means that these values are already met today by applying the generally used techniques and working methods. Since awareness in practice has been increasing, the authors estimate that the rule of technology will very soon shift towards a $q_{50} \leq 1.25$ $m^3/(h \cdot m^2)$.

State of the art means that it is possible to achieve these values by applying special diligence. This usually implies quality assurance during the construction phase. The authority for public buildings in Luxemburg already applies a limit of $q_{50} \leq 0.6$ $m^3/(h \cdot m^2)$ for new school or office buildings. For halls, the q_{50} is adjusted depending on the quality of the roll-up doors.

SUGGESTIONS FOR QUALITY IMPROVEMENTS

To purposefully achieve good quality in the airtight building envelope, an airtightness concept for the building should be developed as early as the planning phase, as is the case for single-family homes. The airtight layer as well as the thermal building envelope have to completely enclose the entire heatable volume. Selecting sufficiently airtight materials, planning details diligently and avoiding unnecessary penetrations are requirements for successful implementation later.

Based on the testing experience to date, improvement is needed for, for example, post-and-rail façade structures, smoke extraction in elevators, roll-up doors and movable loading bridges.

Post-and-rail façade structures

Figure 5 gives an example of early airtightness testing of a post-and-rail façade structure.

Figure 5: A sample façade with connected casing allowed the airtightness of the façade component and the connecting joint to be tested before construction. In this case, improvements and a second airtightness test were necessary.

Smoke extraction in elevators

Smoke extraction in elevators is intended for cases of fire. It is mostly an aperture at the elevator head. In case of fire, these apertures serve as smoke extractors from the shaft. Elevator doors in many cases are only authorized if smoke extraction apertures exist. If these apertures remain open all year, they cause ventilation heat loss or, in air-conditioned buildings, ventilation cold losses in summer. Flap valves that will only open as needed are now available on the market. In some cases, the smoke extraction apertures also serve to cool the elevator drive motor. Should this be the case, the function of the smoke extractor shutter can be combined with switch-on/switch-off temperature for cooling the motor.

Installation shafts frequently also feature smoke extraction, and thus also have to be equipped with flap valves.

Roll-up doors

Roll-up doors are used in many larger projects, e.g., warehouses. Table 1 shows the airtightness of roll-up doors: "Airtightness classes 0 to 5 for roll-up doors according to German Industrial and European Standard Din EN 12426." The indicated values correspond to the q_{50}-value in German Industrial Standard DIN 4108-7. A roll-up door of the airtightness class 4 with an air permeability of 3 $m^3/(h \cdot m^2)$ corresponds to the limit stipulated in DIN 4108-7.

Class	Air permeability(AP) at a pressure of 50 Pa $m^3/(h \cdot m^2)$	Value defined
0		No value defined
1	24	
2	12	
3	6	
4	3	
5	1.5	

Table 1: Airtightness classes 0 to 5 for roll-up doors according to [DIN EN 12426]

Movable loading bridges

Different loading-bridge systems are in use for loading and unloading trucks.

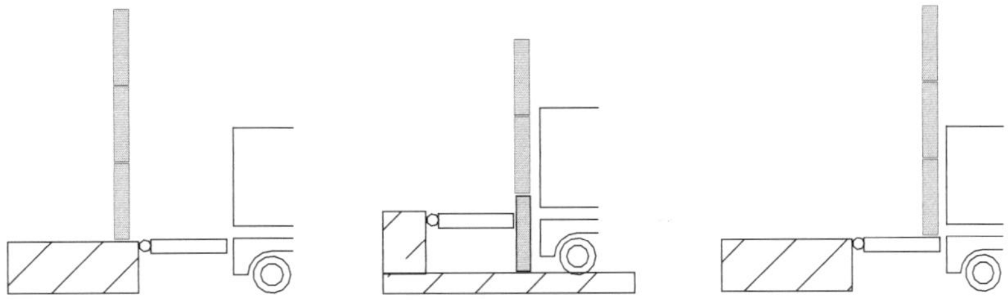

Figure 6: The loading bridge on the right has an effect on the airtightness since it forms part of the building envelope.

For loading bridges that form part of the external building envelope, the two-centimeter joint between the loading bridge and the floor has a critical effect on airtightness. Attention must be paid to sealing this joint (Figure 7). The authors are not aware of any airtightness targets for movable loading bridges.

Figure 7: Detail, movable loading bridge with integrated sealing. A clearly visible air leakage only remains at door level.
(Source: Bauphysikkalender 2012/Calendar of Building Physics 2012)

TEST EXAMPLE

Figure 8: New school building in Luxemburg, cafeteria building "Public" with integrated BlowerDoor MultipleFan measuring system.

Building envelope = 15,000 m^2

Internal building volume = 45,000 m^3

Target value: $q_{50} \leq 1.25\ m^3/(h \cdot m^2)$

Test results: $V_{50} = 7{,}000\ m^3/h$, $q_{50} = 0.5\ m^3/(h \cdot m^2)$, $n_{50} = 0.15\ h^{-1}$

Figure 9: During the BlowerDoor test in the building "Public"

Conclusion: The authors recommend discussing the target values for large buildings as pertains to setting a target value for newly planned large buildings of $q_{50} < 2.0\ m^3/(h \cdot m^2)$ in calls for tender and stricter requirements, e.g., determining a target value of $q_{50} \leq 1.25\ m^3/(h \cdot m^2)$ for office buildings.

REFERENCES/LITERATURE

Rolfsmeier, S., Simons, P., Dorschky, S.: *Luftdichtheit in Planung, Ausführung und Messung*. Bauphysikkalender 2012, Ernst & Sohn, Berlin 2012 [Rolfsmeier, S., Simons, P., Dorschky, S.: *Airtightness in Planning, Implementation and Testing*. Calendar of Building Physics 2012, Ernst & Sohn, Berlin 2012

[DIN EN 13829] *Wärmetechnisches Verhalten von Gebäuden – Bestimmung der Luftdurchlässigkeit von Gebäuden – Differenzdruckverfahren* (ISO 9972:1996, modifiziert), Deutsche Fassung EN 13829: 2000. Beuth Verlag Berlin, Februar 2001. [DIN EN 13829 – *Thermal-Technical Behavior of Buildings. – Determining Air Permeability of Buildings – Differential Pressure Method* (ISO 9972:1996, modified), German Versions EN 13829: 2000. Beuth Verlag Berlin, February 2001.]

[DGNB-Standard] DGNB = Deutsche Gesellschaft für nachhaltiges Bauen. www.dgnb.de [DGNB-Standard. DGNB = Deutsche Gesellschaft für nachhaltiges Bauen/German Sustainable Building Council. *www.dgnb.de*]

[Minergie] *MINERGIE®-Anforderungen* (Norm SIA 380/1:2009) www.minergie.ch [Minergie. *MINERGIE®-Requirements* (Standard SIA 380/1:2009) *www.minergie.ch*]

[DIN EN 12426] *Tore – Luftdurchlässigkeit – Klassifizierung*. Beuth-Verlag Berlin, 2000-11. [DIN EN 12426. Doors – *Air Permeability – Classification*. Beuth-Verlag Berlin, 2000-11.]

[DIN V 18599-2] *Energetische Bewertung von Gebäuden – Berechnung des Nutz-, End- und Primärenergiebedarfs für Heizung, Kühlung, Lüftung, Trinkwarmwasser und Beleuchtung. Teil 2: Nutzenergiebedarf für Heizen und Kühlen von Gebäudezonen*. Beuth-Verlag Berlin, Februar 2007. [DIN V 18599-2. *Energetic Evaluation of Buildings – Calculating the Useful, Secondary, and Primary Energy Need for Heating, Cooling, Ventilation, Potable and Hot Water, and Lighting. Part 2: Useful Energy Need for Heating and Cooling in Building Sections*. Beuth-Verlag Berlin, February 2007.]

[www.passiv.de] Passivhaus-Institut: *Kriterien für Passivhäuser mit Wohnnutzung – Zertifizierung als „Qualitätsgeprüftes Passivhaus"*. Veröffentlicht unter www.passiv.de [www.passiv.de. Passive House Institute: *Criteria for Passive Houses with Residential Use – Certification as a "Quality-Approved Passive House"*. Published at www.passiv.de]

[EnEV 2007] *Verordnung über energiesparenden Wärmeschutz und energiesparende Anlagentechnik bei Gebäuden* (Energieeinsparverordnung – EnEV, Inkrafttreten 1. Oktober 2007). Bundesgesetzblatt vom 26. Juli 2007. Berlin. [EnEV 2007/German Energy Savings Regulation 2007. *Regulation on Energy-Saving Thermal Insulation and Energy-Saving Building Services Equipment*.(German Energy Savings Regulation – EnEV, Effective as of 1 October, 2007). Federal Gazette of 26 July, 2007. Berlin.]

TREMCO
illbruck

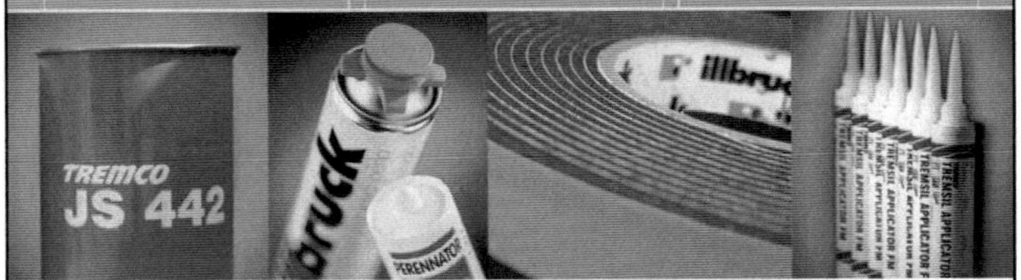

Best Practice Window installation in Sweden
Copenhagen
Stefan Tenbuß

An RPM Company

TREMCO
illbruck

Tremco illbruck – At a Glance

Business Area:
Sealing and Bonding for Construction and Industry

Strategic Segments:
Windows, Facades, Waterproofing, Fire Protection, Distribution, Industry

International Presence:
1,000 Employees
25 Sites in Europe, Africa & Middle East

Turnover:
290 million €

Structure:
Part of RPM International Inc., USA

TREMCO illbruck

Requirement for the Connection Joint

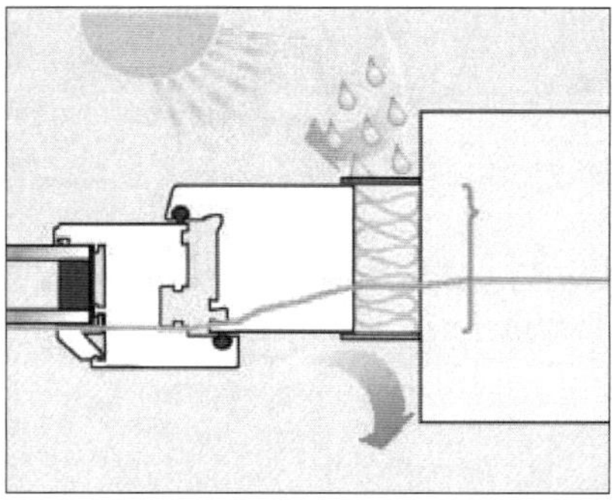

Outside:
- Driving rain tight
- No damage by weathering
- No cold wind pass through
- Water vapour open

Insulation:
- Sound protection
- Thermal insulation
- No condensation

Inside:
- No warm/humid air pass through
- „inside more water vapour tight than outside

3

TREMCO illbruck

Solution: Multifunctional Tape = 3 level seal in one

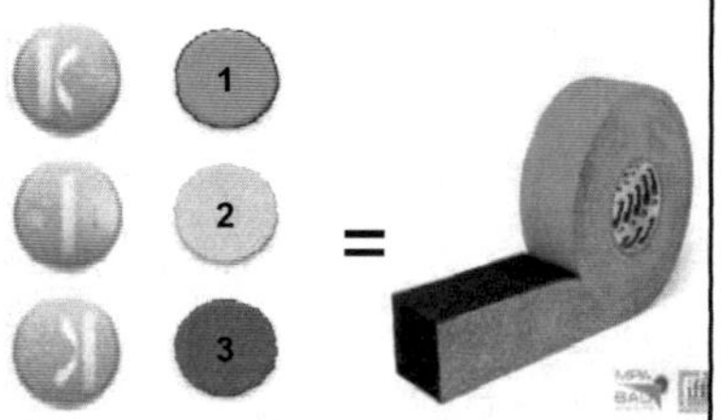

Training and Guidelines

Illmod Trio

Don't forget to put the corners down into the angle. NOTE: The wihite side must show to the inside

Fold down the overlap of the tape during the installation.
NOTE: Fold up after fixing the window to insure the airtightness in the corner

Quality Control by the installer

Egenkontrollplan /illmod Trio

Projektnummer:	Projektnamn:	Lokalisering:	Del:	datum:	upprättad av:

Kontroll	Signatur	Datum	Avvikelse/anmärkningar
Kontrollera att rätt dimension används			
Kontrollera Hörnen			
Vita sidan inåt			
övermått i hörnen med 10mm			
förvara bandet i +20grader vintertid			
sommartid undvik för varm förvaring.			

[illegible]	
[illegible]	
[illegible]	
[illegible]	

☐ Arbetet är helt klart

☐ Arbetet klart med undantag av:

Datum:

Namn:

TREMCO illbruck

Airtightness Requirements

	Air tightness parameter	Limit value by 50 Pa
Passive Energy Building, Finland	n_{50}	0,6 h^{-1}
Passive house by FEBY, Sweden	q_{50}	0,3 l/sm^2 [a]
Passive house by Norwegian Standard	n_{50}	0,6 h^{-1}
Low Energy Building Class 1, Denmark	q_{50}	1,5 l/sm^2 [b]
International Passive House	n_{50}	0,6 h^{-1}

a) area related to $A_{envelope}$

b) area related to A_{floor}

From: Northpass, Important design aspects of a very low energy building

TREMCO illbruck

Skanska, Munksjöpromenaden, Jönköping

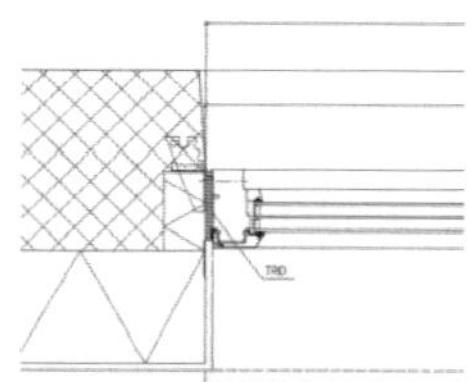

TREMCO
illbruck

Test Result

SKANSKA

Skanska Sverige AB 2 (8)
Teknik och Projekteringsledning

Uppdragsansvarig
Patrik Bengtsson

Datum
2011-08-26

SAMMANFATTNING

Lägenheterna har ett medelläckage om 0,18 l/s·m² (omslutande area) och överskred aldrig 0,24 l/s·m². Det är mindre än vad som beställaren har angivet som maximalt luftläckage (0,6 l/s·m² (omslutande area)).

De större luftläckagen som kunde konstateras var internt i balkongdörren, då mellan dörrblad och karmstycke men även runt dörrkarm i och med att fogning saknades.

q_{50}=0,18 l/sm²

TREMCO
illbruck

Office Building - Stenungsund Sweden

MATS & ARNE ARKITEKTKONTOR AB

q_{50}=0,29 l/sm²

Conclusion

+ Proper planning for the requirements of the installation
+ Getting sufficient support
+ Use easy to install products

= Airtightness

Thank you very much for your attention!

DEMAND SPECIFYING VARIABLES AND CURRENT VENTILATION RATE REQUIREMENTS WITH RESPECT TO THE FUTURE USE OF VOC SENSING FOR DCV CONTROL

Jakub Kolarik[1*]

International Centre for Indoor Environment and Energy, Technical University of Denmark
Nils Koppels Alle, b. 402
2800 Kgs. Lyngby, Denmark
**Corresponding author: jakol@byg.dtu.dk*

ABSTRACT

Demand Controlled Ventilation (DCV) is a well established principle to provide a certain indoor environmental quality, defined both in the terms of air quality and thermal comfort. This is accomplished by adjusting the supplied airflow rate according to a certain demand indicator, which conventionally has been the temperature or the CO_2-concentration. When compared to schedule driven ventilation, application of DCV can lead to substantial energy savings. However, CO_2 is the pollutant related to human occupancy and it does not provide any indication of so called building-related pollution. Building itself as well as its furnishing and equipment together with different human activities happening in them, are significant sources of different chemicals that may aggravate comfort and in some cases even negatively affect the health of the occupants. That is why emissions of those compounds should be also taken into account in the ventilation control. Recent development in gas sensing technology resulted in a new generation of relatively cheap and practically applicable sensors that can offer measurements of some of the pollutants mentioned above – mainly Volatile Organic Compounds (VOC). This seems to bring a new dimension into the control of DCV systems. This paper is a contribution to the workshop on utilization of VOC sensing technology used for DCV control. The aim of the paper is to provide a short review of different types of demand variables used to control DCV systems and summarize ventilation rate requirements contained in current standards and guidelines with respect to the future potential of VOC sensing.

KEYWORDS
Demand Controlled Ventilation, Volatile Organic Compounds, ventilation rate, air quality

INTRODUCTION
Demand Controlled Ventilation (DCV) is a well established principle to provide a certain indoor environmental quality, defined both in the terms of air quality and thermal comfort. This is accomplished by adjusting the supplied airflow rate according to a certain demand indicator, which conventionally has been the temperature or the CO_2-concentration. When compared to schedule driven ventilation, application of DCV can lead to substantial energy savings [1]. The main principle of DCV is that the building is ventilated only in the case that there is an appropriate need for that. In periods of time, where need for fresh outdoor air is low, the amount of supplied air is substantially decreased. However, DCV is not suitable for all types of buildings, since several conditions need to be fulfilled to select DCV. DCV is suitable for buildings with unpredictable variation of occupancy and buildings where heating or cooling is required during all year. DCV is most effective in buildings, which are characterized by well defined dominant pollutant and, at the same time, have low emission of other non-dominant or non-occupant related pollutants. Human body odours, so called human bioeffluents are considered as a dominant pollutant in non-industrial premises like offices, residences, schools or public buildings. It is a state-of-the-art of DCV systems to use the

concentration of carbon dioxide (CO_2) as an indicator of the demand for fresh outdoor air. However, not only human bioeffluents are polluting the air in today's buildings. Buildings themselves as well as their furnishing and equipment together with different human activities happening in them, are significant sources of different chemicals that may aggravate comfort and in some cases even negatively affect the health of the occupants. Emissions of those compounds will not be detected by a CO2 sensor and thus the ventilation rate will not be adjusted to dilute them. Recent development in gas sensing technology resulted in a new generation of relatively cheap and practically applicable sensors that can offer measurements of some of the pollutants mentioned above – mainly Volatile Organic Compounds (VOC). This seems to bring a new dimension into the control of DCV systems. This paper is a contribution to the workshop on utilization of VOC sensing technology used for DCV control. The aim of the paper is to provide a short review of different types of demand variables used to control DCV systems and summarize ventilation rate requirements contained in current standards and guidelines with respect to the future potential of VOC sensing.

DCV STRATEGIES – DEMAND INDICATING PARAMETERS

Currently three main parameters are used to control DCV systems. Those are occupancy, indoor air humidity and the concentration of carbon dioxide (CO_2) indoors. Temperature can also be considered as one of the main demand indicators; however, due to the need to decrease energy consumption it becomes a general trend to separate climate conditioning and ventilation. Therefore this option will not be further considered in the present paper. As the goal of ventilation is to provide fresh outdoor air to ensure health and comfort of the occupants, the role of all the above mentioned parameters is to give an estimate of the air flow, needed to meet the goal.

Occupancy detection

The simplest solution to optimize outdoor airflow according to the current demand is to use the occupancy as the control variable. Occupancy is indirectly related to air quality, as people are important sources of odorous gases, so called human bioeffluents. Occupancy sensors are relatively cheap and represent an effective solution in spaces with rather steady occupation patterns [2]. In some cases, utilization of occupancy sensors together with sensors for relative humidity can be beneficial, because of the poor short term correlation between real concentrations of pollutants and the presence of people [3].

Humidity of the air

Measurement of air humidity is an attractive way of the demand specification for residential buildings. Moisture is generated not only by the occupants of apartments and houses, but also by their activities (showering, cooking, drying clothes, etc.). In the work of ECBCS Annex 18 is moisture considered a dominant pollutant in residential settings [4]. Absolute humidity tends to correlate with the CO2 concentration, and thus with the concentration of occupant related pollution in the space, much better than relative humidity. Therefore it is recommended to use absolute humidity as a control variable. Another factor that has to be taken into account in humidity control is the effect of hygroscopic buffering in building constructions. Humidity levels indoors tend to have less fluctuation when indoor materials have high ability to absorb moisture. Aspects of hygroscopic buffering were investigated as part of the IEA Annex 41 [5]. The study of Mortensen et al. [6] showed the importance of accounting also for the water content of the outside air. Its seasonal increase can lead to higher indoor relative humidity levels. If the DCV controller works with a fixed set point this leads to unnecessarily high airflow rates.

Carbon Dioxide - CO_2

Most of the current sensor based DCV systems use CO_2 concentration as the control input. CO_2 is a good indicator of human occupancy and can also be used, to some extent, as an indicator of the emission of VOCs or particulate matter related to human activities (using photo copiers, printers etc.) [2]. Use of CO_2 as a control variable is only appropriate at sufficiently high levels of occupancy, which has unpredictable variation patterns [2, 7].

It is important to note that CO_2 is not an indicator of perceived air quality *per se*, as it gives no indication of the emission of pollutants not related to the occupants (emitted from building materials, furniture, etc.). Moreover, building ventilation also needs to take into account pollutants that are not sensory irritants and thus cannot be identified by human perception, but they may be a risk for human health - radon, CO or NO_x.

Measurement of VOC and other chemicals

An obvious step towards better control of DCV seems to be utilization of sensors that can detect not only CO2 to account for pollution related to occupancy, but also other typical indoor pollutants (VOC, aldehydes, radon). This would give a possibility to account also for background pollution related to the building. Herberger et al. [8] describes the development of a metal oxide semiconductor sensor (MOS) that integrates measurement of human emitted VOCs [9] and several other typical indoor pollutants. The sensor expresses its measurement in so called equivalent CO2 concentration, which is quite similar to the Total Volatile Organic Compound (TVOC) approach [10, 11]. Moreover, another very important possibility for application of chemical sensors could also to use them to warn building occupants in the case of sudden releases of highly toxic pollutants in the space [12].

DEMAND SPECIFICATION

Required ventilation rates from current guidelines and standards

The main aim of ventilation is to exchange indoor air polluted by emissions from various sources like people, their activities, building materials etc. with fresh unpolluted air from outdoors. Primarily, ventilation was rather question of comfort than health, as it was realized that people emit rather smelly more than noxious compounds [13, 14]. This approach to ventilation as a way of eliminating body odours, updated with the task of maintaining thermal comfort was used in standards until the '80s. The concentration of CO_2 was used as the primary indicator of human bioeffluents, which was considered as the only problematic pollution source in non-industrial buildings. Fanger and Berg-Munch [15] showed that in a space where humans performing sedentary work were the only pollution source, an outdoor air supply rate of 8 L/s person would lead to about 80% of unadapted visitors to be satisfied with air quality. This leads to an indoor concentration of CO_2 of 660 ppm above outdoors and thus absolute concentration (assuming a CO_2 production rate of 19 L/h person and outdoor levels of CO_2 in the range between 300 and 500 ppm) of about 1000 ppm. This value was recommended already by Pettenkofer (1858) [16] in the 19th century. The requirement of a steady-state CO_2 concentration below 800 ppm above outdoors has been adopted in all current ventilation standards [17, 18, 19, 20].

Later studies showed that occupants were not the only source of pollution in buildings. There are thousands of other compounds than human bioeffluents originating from building materials, office equipment etc. Some of these compounds, like for example formaldehyde or radon [21, 22, 23] can mean significant health threat for people, other can be strong sensory irritants and cause degradation of the perceived air quality [24]. This implies that there is a need to ventilate not only for the pollution load originating from people, but other sources need to be also included.

World Health Organization (WHO) drafted 1983 a document on the possible risks of indoor air quality problems [25]. The publication was later followed by guidelines regarding the most important indoor pollutants [26, 27].

In 1988 Fanger [28] introduced a method to account not only for human odours, but also for other pollutants that may aggravate perceived air quality (PAQ). This approach to ventilate for "acceptable air quality" was consequently adopted by ventilation standards ASHRAE 62.1 [20] and EN 13779 [19] as well as international ventilation guidelines CEN Report 1752 [17] and European guidelines for ventilation requirements in buildings – ECA [29].

Requirements concerning health have been implemented for well-known noxious pollutants specifying Threshold Limit Values (TLV), which should not be exceeded. Inclusion of building related sources increases the required minimum ventilation rates that are specified based on air quality expressed in terms of percent of dissatisfied occupants (PD) with the PAQ. The typical range of PD is from 10% to 30%. The European guideline ECA [29] established three categories of buildings according to the achieved indoor air quality corresponding to PD values of 10%, 20% and 30%. CEN Report 1752 [17] increased the percentage in the first category to 15%. This division of categories was later adopted by the European standard EN 15251 [18], where also a fourth category for PD > 30% was added. The American standard ASHRAE 62.1 [20] works with a single PD value of 20%.

Table 1 gives an overview of recommended ventilation rates for non-residential buildings where smoking is not allowed according to CEN Report 1752 [17], EN 15 251 [18] and ASHRAE 62.1 [20]. It can be seen that although both the EN and ASHRAE standards use the principle of calculating the total outdoor air flow rate as a combination of the occupant related element (to remove pollution related to human occupancy) and the building related element (to remove pollution related the building) the resulting air flow rates differ significantly. The reason is that the US standard uses the PD of adapted occupants as a criterion for acceptable air quality, whereas both CEN Report 1752 and EN 15 251 use PD of unadapted visitors. As human olfactory senses adapt rather quickly to the sensory pollution load in the room [30], utilization of adapted occupants leads requires less ventilation air to provide the same level of satisfaction.

In addition to the method based on the person and building components, both CEN Report 1752 [17] and EN 15 251 [18] states recommended CO_2 concentrations above the outdoor concentration to be used when DCV systems are designed. The recommended values differ slightly in the two standards being: 460 ppm, 660 ppm and 1190 ppm in [17] and 350 ppm, 500 ppm and 800 ppm in [18] for the first three building categories. It is also worth noting that outdoor air flow rates calculated according to these target concentrations do not include the building-related pollution component.

Ventilation requirements can also be set to avoid high levels of relative humidity. This is of special importance in residential buildings where there are, besides people, many other sources of moisture (cooking, washing, drying of clothes or bathing/showering). Relative air humidity levels of 30% to 70% are considered as normal, but the supply air flow rate should be designed so that the relative humidity indoors does not exceed 45%. This value is considered as a limit to prevent growth of house dust mites during winter.

For dwellings, ASHRAE 62.1 [20] recommends ventilation rates for acceptable air quality as a function of the area of the residence and the number of bedrooms. EN 15 251 [18] specifies air change rates per m^2 of the dwelling as well as air flows per person for lining rooms and bedrooms. The largest value expresses the total ventilation rate for the residence. The standard specifies also exhaust air rates for bedrooms and kitchens. These should be adjusted according to the supply flows from the living room and bedrooms.

DISCUSSION

Although many VOC sensors are commercially available, there is no general agreement regarding their suitability for use in indoor air quality applications. Won and Schleibinger [31] conducted a review of commercially available sensors for formaldehyde, radon and VOC. To allow for comparison and to ensure their meaningful use in ventilation systems, the study proposes performance requirements (including selectivity, cross-sensitivity, resolution, response rate, price etc.) for such sensors. For formaldehyde and radon, several sensors meeting the proposed requirements were identified, but no commercial VOC sensors meeting were found. The main drawback factors were: high price for sensor arrays, low resolution (~100 ppb at best vs. ~5 ppb requirement) and the fact that sensors do not measure the concentration of individual chemicals (in the case of photo ionization detector sensors – PID and metal oxide semiconductor sensors – MOS). Due to this drawback, the authors concluded that no VOC commercial sensors were identified that would be sensitive and specific enough to be used for ventilation control. Therefore it seems that sensors measuring VOC may need future development.

Also more field- and laboratory tests seem to be needed to bring the technology to the level where it will be fully comparable with the current state of the art CO_2 based control. The main argument for the adoption of VOC sensors is that they are able to detect also diverse odorous events taking place in the space. This may certainly be an advantage, however the performance of VOC sensors was not, on a long term basis, compared to current CO_2 driven DCV. So far only comparisons to schedule based ventilation are available in the literature [32]. In the future, VOC/chemical sensors can maybe find their use as detectors with particular sensitivity to specific noxious contaminants. Their signal would become a control variable in the case of release of a toxic pollutant, while normal comfort ventilation would be based on for example CO_2 concentration measurement.

Ventilation rates recommended by current standards and guidelines are based on target levels of indoor air quality estimated while assuming emissions of typical indoor pollutants (human bioeffluents, background pollution from building). This so called prescriptive approach is mostly used, because most often no precise information about pollutants in a building is available. It is a simple and practical solution for a design engineer who has no possibility of making qualified judgement of occupants' exposure too different contaminants.

Another way to determine the amount of ventilation air is the so called performance-oriented/performance-based approach. This approach specifies requirements in terms of pollutant concentrations that have to be maintained in a given space and/or limit values of an exposure to a given pollutant – "a dose" that cannot be exceeded [33]. ASHRAE 62.1 [20] includes such design approach and states that contaminants of concern of the design shall be identified in terms of their indoor and outdoor sources and their strength. The standard also provides some concentration guidelines. However regarding concentration of VOCs the guideline states that predicted (or measured) acceptability of air quality of more that 80% of occupants or visitors should be used as an indicator regarding odours, while regarding TVOC concentration no precise guideline value can be given. EN 15 251 [18] does not include the performance-based approach, but another European standard EN 13779 [19] deals with this approach providing room mass balance equations (both steady state and time dependent) to calculate the airflow needed to reduce emission of other than human related pollutants. The steady state equations are also mentioned in CR 1752 [17].

The performance oriented approach creates an alternative to the prescriptive approach and opens a possibility for innovative technologies to improve both air quality and energy-efficiency. The study of Mortensen et al. [34] used occupant exposure to pollutants integrated over time as the metric to evaluate the effectiveness and air quality implications of demand controlled ventilation in residences. The study compared indoor air quality at two situations: the exposure to pollutants over a long term (typically used in ventilation standards) and the peak exposures associated with time variations in ventilation rates and pollutant generation (background building-related rate and occupant related rate). The DCV system operated at a low and high air flow rate when the residence was unoccupied and occupied, respectively. The authors used analytical solutions to the continuity equation to determine the ventilation effectiveness and the long-term chronic dose and peak acute exposure Logue et al. [35]. The study showed that the peak changes in pollutant concentration were a significant consequence of the dose based design for DCV system. The authors used acute to chronic ratio [36] as a metric to evaluate the possible effects of short term concentration peaks on the occupants. Formaldehyde with an acute to chronic ratio of 4.7 (1h average) was used as the limiting case. The results of the study showed that it was possible to optimize the DCV air flow rates to reduce the quantity of air used for ventilation without introducing problematic acute conditions as the increase in acute to chronic exposure was well below the acute to chronic exposures of concern derived from health standards.

With current knowledge there is still not enough scientific evidence to say how much fresh outdoor air is needed to ensure a healthy indoor environment. A recent review of the scientific literature on ventilation rates and health [37] showed that higher ventilation rates in offices (up to 25 L/s) are associated with lower prevalence of sick building syndrome symptoms (SBS) while ventilation rates higher than 0.5 1/h in residences are associated with reduced risk of allergy among children in a Nordic climate. The panel of scientists concluded that there is still a substantial lack of knowledge on the effect of climate, outdoor pollution and non-office buildings. The project HealthVent [38] funded by the Executive Agency for Health and Consumers of the European Union (EAHC) represents efforts to develop health-based ventilation guidelines. The project aims to develop such guidelines while reconciling the health of the occupants and low energy use in offices, homes and public buildings such as schools, nurseries and day-care centres.

DCV systems controlled by sensors that are sensitive to one or several well-defined pollutants seems to be an appealing alternative to current, mostly schedule or CO_2 controlled systems. However, building occupants are exposed to hundreds of different chemical compounds and even if some representatives are chosen from a so called "universe" of Organic Compounds in Indoor Air (OCIA) introduced by Wolkoff and Nielsen [39], control of ventilation according to them does not necessarily ensure good indoor air quality. Moreover, the chemical universe in buildings is under constant development. According to Weschler [40], concentrations of several indoor pollutants like formaldehyde or PCB have increased and then decreased again. At the same time, concentrations of other chemicals have increased and still remain high (e.g., phthalate esters or brominated flame-retardants). Despite the fact that the amount of scientific literature on the topic of indoor air quality has increased dramatically during last 20 years [41], we still have little knowledge about the compounds that are the most important ones for "the good indoor air quality". This means a substantial limitation and thus a huge challenge to the control of ventilation systems by chemical sensors.

CONCLUSIONS

- Occupancy detection, measurement of air humidity (relative or absolute) and measurement of the CO_2 concentration or their combinations represent the state-of-the-art of control variables for DCV systems.
- VOC sensing seems to be a promising approach to improve the control of DCV systems. However more research is needed to clarify whether VOC sensors should be used as a complementary feature to current systems or if they can be used as standalone sources of a signal for DCV controllers.
- Current ventilation standards and guidelines apply mostly a prescriptive approach to specify required air flow rates depending on the type of premises, occupancy and level of building related pollution.
- The performance based approach specifying ventilation requirements in terms of limit concentrations for specific pollutants is now part of both US and European standards and can be used to design DCV systems.

ACKNOWLEDGEMENTS

The work on this paper was funded the European Community's Seventh Framework Programme (FP7/2007-2013); grant agreement n° 211948 as a part of the project Clear-up.

REFERENCES

[1] Mysen, M., Rydocka, J.P., Tjelflaat, P.O. 2003. Demand controlled ventilation for office cubicles-can it be profitable? Energy and Buildings, 35, 657–662

[2] W. J. Fisk and A. T. D. Almeida. Energy and Buildings, 29(1):35, 1999.

[3] V. Pavlovas. Demand controlled ventilation - A case study for existing Swedish multifamily buildings, 2003. ISBN 1400-9544. Licentiate thesis, Chalmers University of Technology

[4] L.-G. Mansson, S. A. Svennberg, and M. W. Liddament. Technical Synthesis Report. A Summary of IEA Annex 18. Demand Controlled Ventilating Systems. Technical report, International Energy Agency, 1997. ISBN 0 946075 99 9

[5] IEA. Annex 41: Whole Building Heat, Air and Moisture Response. Closing seminar. Technical report, International Energy Agency, 2008

[6] Mortensen, D. K., Nielsen, T. R., Topp, C. 2008. Comparison of a Constant Air Volume (CAV) and a Demand Controlled Ventilation (DCV) System in a Residential Building. In Proceedings of the 8th Symposium on Building Physics in the Nordic Countries, 1, pages 191–197

[7] Seppanen, O., Goeders, G., van Steenberghe, T. 2010. Sustainable Energy Use in Buildings- REHVA Workshops at Clima 2010. Technical Report 3, REHVA – Federation of European Heating, Ventilation and Air-Conditioning Associations, ISBN 978-2-930521-04-6

[8] Herberger, S., Herold, M., Ulmer, H., Burdack-Freitag, A., Mayer, F. 2010. Detection of human effluents by a MOS gas sensor in correlation to VOC quantification by GC/MS. Building and Environment 45, 2430-2439

[9] Burdack-Freitag A, Rampf R, Mayer F, Breuer K. 2009. Identification of anthropogenic volatile organic compounds correlating with bad indoor air quality. In: Proceedings of the 9th International Conference and Exhibition Healthy Buildings 2009, Syracuse, NY

[10] Mølhave, L. (1990) "Volatile organic compounds, indoor air quality and health". Ln: Walkinshaw, D.S. (ed.) Proceedings of Indoor Air '90, Ottawa, Canada Mortgage and Housing Corporation, Vol. 1, pp. 15-33

[11] Seifert, B. (1990) "Regulating indoor air". In: Walkinshaw, D.S. (ed.) Proceedings of Indoor Air '90, Ottawa, Canada Mortgage and Housing Corporation, Vol. 5, pp. 35-49.

[12] Sreedharan, P., Sohn, M.D., Gadgil, A.J., Nazaroff, W.W. 2006. Systems approach to evaluating sesor characteristics for real-time monitoring of high-risk indoor contaminant releases. Atmospheric Environment 40 (19), 3490-3502
[13] Brown-Séguard, C. E., d'Arsonval, A. 1887. Demonstration de la pussiance toxique des exhalations pulmonaries provenant de l'hombre et du chien. Compt. redn. Soc. de Biol., 39, 814
[14] Yaglou, C. R., Riley, E. C., Coggins, D. I. 1936. Ventilation requirements. ASHRAE Transactions, 42, 133-162
[15] Fanger, P. O., Berg-Munch, B. 1983 Ventilation and Body Odour. In: Proceedings of an engineering foundation Conference on Management of Atmospheres in Tightly Enclosed spaces, ASHRAE, Atlanta, pp. 45-50
[16] Pettenkofer , M. V. 1858. Űber den Luftwechsel in Wohngebräuden. Munich, Litterarisch-Artistiche Anstalt der J. G. Cotta'schen Buchhandlung
[17] CR 1752. 1998. CEN Report Ventilation for Buildings: Design Criteria for the Indoor Environment, CEN/TC 156/WG 6, European Committee for Standardization, Brussels, Belgium
[18] EN 2007. EN 15251: Indoor environmental input parameters for design and assessment of energy performance of buildings addressing indoor air quality, thermal environment, lighting and acoustics. European Committee for Standardization, Brussels, Belgium
[19] EN 2004. EN 13779: Ventilation for non-residential buildings – performance requirements for ventilation and room-conditioning systems. European Committee for Standardization, Brussels, Belgium
[20] ASHRAE. 2010. ANSI/ASHRAE Standard 62.1-2010 - Ventilation for acceptable indoor air quality, American Society of Heating, Refrigerating and Air Conditioning Engineers, Atlanta, USA
[21] Andersen, I. 1979. Formaldehyde in the indoor environment – health implications and the setting of standards. In: Fanger, P. O. and Valbjørn, O. (eds.) Indoor Climate, Copenhagen, Danish Building Research Institute, 67 – 87
[22] Korsgaad, J. 1979. The effect of the indoor environment on the house dust mite. In: Fanger, P. O. and Valbjørn, O. (eds.) Indoor Climate, Copenhagen, Danish Building Research Institute, 187 – 205
[23] Swedjemark, G. A., 1979. Ventilation requirements in relation to the emanation of radon from building materials. In: Fanger, P. O. and Valbjørn, O. (eds.) Indoor Climate, Copenhagen, Danish Building Research Institute, 405 – 421
[24] Wargocki, P., Wyon, P. D., Baik, Y. K., Clasusen, G. and Fanger, P. O. 1999. Perceived air quality, Sick Building Syndrome (SBS) symptoms and productivity in an office with two different pollution loads. Indoor Air, 9(3), 165 – 179
[25] WHO 1983. Indoor air pollutants: Exposure and health effects. Copenhagen WHO Regional Office for Europe, (EURO Reports and Studies, No. 78)
[26] WHO. 2000. Air quality guidelines for europe. Technical Report WHO Regional Publications, European Series, No. 91
[27] WHO. 2010. WHO guidelines for indoor air quality: selected pollutants. Technical report, ISBN 978 92 890 0213 4
[28] Fanger, P. O. 1988. Introduction to the olf and decipol units to quantify air pollution perceived by humans indoors and outdoors. Energy and Buildings, 12, 1-6
[29] ECA 1992. European Concerted Action "Indoor Air Quality and its Impact on Man" Guidelines for ventilation requirements in buildings. Report no. 11, EUR 14449 EN, Luxembourg: Office for Publications of European Communities
[30] Gunnaresen, L., Fanger, P. O. 1992. Adaptation to indoor air-pollution. Environment International, 18 (1), 43-54

[31] Won, D. Y., Schleibinger, H. 2011. Commercial IAQ Sensors and their Performance Requirements for Demand-Controlled Ventilation. Report no. IRC-RR-323, National Research Council Canada

[32] Herberger, S., Ulmer, H. 2012. Indoor Air Quality Monitoring Improving Air Quality Perception. Clean-Soil Air Water, 40 (6), 578-585

[33] Sherman, M. H., Mortensen, D. K., Walker, I. S. 2011. Derivation of equivalent continuous dilution for cyclic, unsteady driving forces. International Journal of Heat and Mass Transfer, 54, 2696–2702

[34] Mortensen, D. K., Walker, I. S., Sherman, M. H. 2011. Optimization of Occupancy Based Demand Controlled Ventilation in Residences. INTERNATIONAL JOURNAL OF VENTILATION, 10 (1), 49-60

[35] Logue, J. M., McKone, T. E., Sherman, M. H., Singer, B. C. 2010. Hazard assesment of chemical air contaminants measured in residences. Indoor air, 21(2), 92–109

[36] Sherman, M. H., Logue, J. M., Singer, B. C. 2011 Infiltration Effects on Residential Pollutant Concentrations for Continuous and Intermittent Mechanical Ventilation Approaches. ASHRAE HVAC&R Research Journal, 17(2), 159–173

[37] Sundell, J.; Levin, H.; Nazaroff, W. W.; et al. 2011. Ventilation rates and health: multidisciplinary review of the scientific literature. INDOOR AIR, 21 (3), 191-204

[38] HeatlhVent. 2010. Health-Based Ventilation Guidelines for Europe. http://www.healthvent.byg.dtu.dk/, visited Sept. 2012

[39] Wolkoff, P., Nielsen, G. D. 2001. Organic compounds in indoor air - their relevance for perceived air quality? Atmospheric Environment, 35, 4407-4417

[40] Weschler, Ch. J. 2009. Changes in indoor pollutants since the 1950s. Atmospheric Environment. 43 (1), 153-169

[41] Clausen, G.; Beko, G.; Corsi, R. L.; et al. 2011. Reflections on the state of research: indoor environmental quality. Indoor Air, 21 (3), 219-230

Table 1 – Comparison of recommended ventilation rates for non-residential buildings (non-smoking) according to CR 1752 [17], EN 15 251 [18] and ASHRAE 62.1 [20]

Building type	Occupancy	Category CEN	Category EN	Minimum ventilation rate (occupants=only source of pollution), L/s person		Additional ventilation rate (pollution from building) L/s m2				Total air change rate L/s m2		
				CEN*	ASHRAE	EN Very low-polluting	CEN* Low-polluting	CEN* Not low polluting	ASHRAE	EN Very low-polluting	CEN* Low-polluting	ASHRAE
Single office	0.1	A	I	10	2.5	0.5	1.0	2.0	0.3	1.5	2.0	0.55
		B	II	7		0.3	0.7	1.4		1.0	1.4	
		C	III	4		0.2	0.4	0.8		0.6	0.8	
Landscaped office	0,07	A	I	10	2.5	0.5	1.0	2.0	0.3	1.2	1.7	0.48
		B	II	7		0.3	0.7	1.4		0.8	1.2	
		C	III	4		0.2	0.4	0.8		0.5	0.7	
Conference room	0,5	A	I	10	2.5	0.5	1.0	2.0	0.3	5.5	6	1.55
		B	II	7		0.3	0.7	1.4		3.8	4.2	
		C	III	4		0.2	0.4	0.8		2.2	2.4	
Auditorium	1,5	A	I	10	3.8	0.5	1.0	2.0	0.3	15.5	16	6
		B	II	7		0.3	0.7	1.4		10.8	11.2	
		C	III	4		0.2	0.4	0.8		6.2	6.4	
Cafeteria/restaurant	0,7	A	I	10	3.8	0.5	1.0	2.0	0.9	7.5	8	1.17
		B	II	7		0.3	0.7	1.4		5.2	5.6	
		C	III	4		0.2	0.4	0.8		3.0	3.2	
Classroom	0,5	A	I	10	3.8	0.5	1.0	2.0	0.3	5.5	6	2.2
		B	II	7		0.3	0.7	1.4		3.8	4.2	
		C	III	4		0.2	0.4	0.8		2.2	2.4	
Kindergarten	0,5	A	I	12	5.0	0.5	1.0	2.0	0.9	6.5	7	3.4
		B	II	8.4		0.3	0.7	1.4		4.5	4.9	
		C	III	4.8		0.2	0.4	0.8		2.6	2.8	
Department store	0,15	A	I	14.7	3.8	1	2.0	3.0	0.6	3.1	4.1	1.17
		B	II	10		0.7	1.4	2.1		2.2	2.9	
		C	III	6		0.4	0.8	1.2		1.3	1.7	

*Requirements of CEN 1752 [17] and EN 15 251 [18] are equal

NEW DEVELOPMENTS IN VOC SENSING FOR DCV

Heiko Ulmer[1], Simone Herberger[*1]

1 AppliedSensor GmbH
Gerhard-Kindler-Str.8
72770 Reutlingen, Germany

**Corresponding author: simone.herberger@appliedsensor.com*

ABSTRACT

The paper summarizes the activites undertaken by AppliedSensor within the European Clear-up project with respect to new developments in volatile organic compound sensing for demand controlled ventilation. State-of-the-art is to use non-dispersive infrared sensor technology for indoor carbon dioxide detection. Carbon dioxide so far serves as indicator for bad indoor air quality and required ventilation rates. Indoor events coming along with complex mixtures of gaseous compounds, mainly volatile organic compounds, released by human bio-effluents, cooking odours, outdoor pollutants, cleaning supplies as well as building material and furniture emissions, play a pronounced role for human air quality perception and health, but cannot be dectected with carbon dioxide as indicator. Within the European Clear-up project, AppliedSensor worked on the development of metal oxide semiconductor based indoor air quality monitoring modules spezialized for demand controlled ventilation, closing the gap between pure CO_2 and volatile organic compound only detection, their implementation in test facilities and enhancement for wireless application.

KEYWORDS

IAQ, VOCs, MOS gas sensor, DCV, Energy savings

INTRODUCTION

Buildings have to operate over a wide range of environmental and load conditions and contribute to a large part of the world's energy consumption and carbon emission. Almost continuous adjustment of many operating parameters in buildings is required in order to achieve minimum energy consumption while maintaining acceptable comfort levels. Intelligent sensors for safety, comfort and control appliances are considered as the keys to energy-efficiency and a high standard of living. Improvements in low power sensor systems and wireless communication continuously increase the application of wireless sensor networks for monitoring and control systems in the building sector, especially when there is no infrastructure available [1].

Energy efficient ventilation relating to good indoor air quality (IAQ) is a major task for building performance according to the requirements set by the European Energy Performance of Buildings Directive (EPBD) in 2010 [2]. Today's ventilation schemes are mainly based on time-scheduled ventilation, leading to a waste of energy and often do not account for the right ventilation demand. More advanced systems apply demand controlled ventilation (DCV) using sensors for IAQ control. DCV ensures that fresh air is supplied to interior spaces whenever necessary, saving energy and lowering building operation costs. Sensors alter the climate control system to increase ventilation when defined indoor pollution threshold limits are exceeded. Decreasing pollution levels result in ventilation rate reduction. State-of-the-art is to use carbon dioxide (CO_2), whose production is proportional to the human metabolic rate, as indicator for bad IAQ [3]. Ventilation standards and guidelines in Europe specify ventilation rates for different IAQ categories based on CO_2 concentrations above outdoors, related to the percentage of dissatisfied people with the IAQ [4,5]. The leading sensor technology for IAQ monitoring based on CO_2 quantification is non-dispersive infrared adsorption (NDIR) [6].

Clear-up [7] focuses on sensor technologies and strategies for DCV taking into account events coming along with more complex mixtures of gaseous compounds, mainly volatile organic compounds (VOCs), released by human bio-effluents, cooking odours, outdoor pollutants, cleaning supplies as well as building material and furniture emissions, that cannot be detected with carbon dioxide as indicator, but play a pronounced role for human air quality perception and health [8].

AppliedSensor developed IAQ modules specialized for DCV, closing the gap between pure CO_2 and VOC only detection. The detection of VOCs and potentially harmful substances (e.g. CO) is facilitated by a micromachined metal oxide semiconductor (MOS) sensing element optimized for the detection of anthropogenic VOCs. Sensor data of various indoor spaces (meeting rooms, offices, kitchens etc.) has been correlated with analytical measurement data and user perception. The use of NDIR-CO_2 reference sensors led to the development of an empirical algorithm correlating proportionality of human CO_2 production rate and other bio-effluent (VOC) generation. This way, the sensors are able to give measures of both, human related CO_2 events as well as VOC related events at the same time. Based on VOC detection the sensor module provides a standardized output signal in CO_2 equivalents [ppm] that can be linked to specific IAQ levels requested by HVAC planners and established ventilation standards referring DCV settings to CO_2 concentrations above outdoors [9].

The sensor modules have been installed in different test buildings for long-term stability measurements and investigations on the energy saving potential using the sensor module for DCV. Further development led to power reduced IAQ sensors suitable for battery operated devices.

FIELD STUDIES - VOC vs. CO_2 APPROACH

Fig. 1 shows the developed IAQ-module used for the field studies within the European clear-up project.

Figure 1. Sensor module for IAQ monitoring - micromachined MOS gas sensor.

The benefit of VOC-based IAQ control compared to CO_2-based DCV becomes obvious in indoor spaces where changes of CO_2 are too small for ventilation control but indoor air is dominated by odorous events, affecting perceived air quality by a large extent (e.g. kitchens, restrooms, smoking areas).

The developed IAQ-module allows capture of CO_2 concentrations and odorous events at the same time. The main signals caused by cooking activities, that can be attributed to VOCs are only detected by the developed IAQ-module in the background of CO_2 (Fig. 2, right hand side: cooking event). The NDIR reference sensor however, is only able to measure the CO_2 production linked to the grade of occupancy. Implementation of the empirical algorithm for CO_2 prediction based on anthropogenic VOC detection allows reliable correlation of predicted and measured CO_2 concentrations in indoor spaces where no appreciable human activity takes place. The perfect correlation of predicted CO_2 equivalents calculated from the detected VOC level compared to the measured CO_2 concentration using NDIR technology is shown by means of a meeting room on the left hand side of Fig. 2.

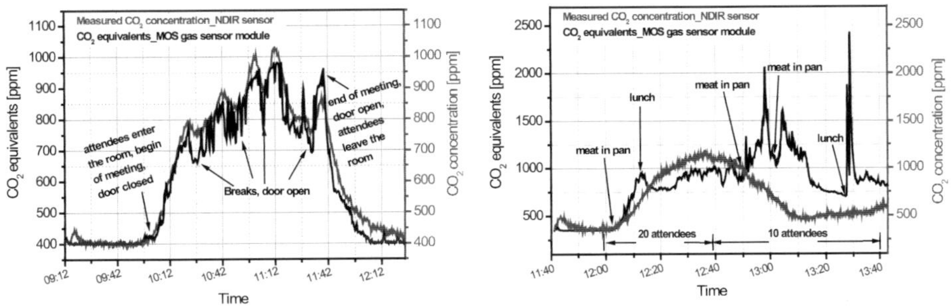

Figure 2. VOC vs. CO_2 approach: predicted and measured CO_2 concentrations (left: meeting room/right: kitchen).

DCV vs. TIME-SCHEDULED VENTILATION

DCV – office

The developed IAQ-module has been tested for DCV compared to natural and time-scheduled ventilation with the main focus on energy demand and resulting IAQ.

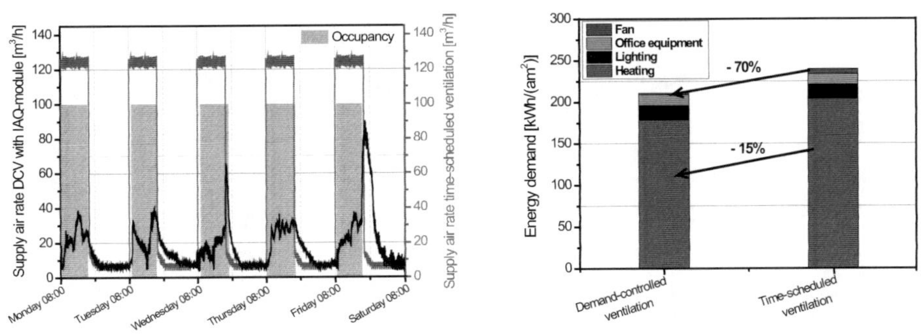

Figure 3. Left: Supply air rate [m^3/h] - DCV compared to time-scheduled ventilation. Right: Energy demand during winter period 2009/2010 for DCV and time-scheduled ventilation.

The results when triggering the supply air rate with the IAQ-module instead of applying time-scheduled ventilation are very promising: In a 80m^3 office with two employees, the supply air rate could be reduced by 60% compared to time-scheduled ventilation, set up according to DIN EN 15251 (Fig. 3, left hand side). Good air quality was maintained, measured CO_2 concentrations did not exceed 1000 ppm for the design value of 2 people in the office.

Natural, demand-controlled and time-controlled ventilation were carried out reiteraiting weekly for 4 months during winter time. The ventilation settings can be taken from Table 1.

Ventilation Strategy	Description
Natural ventilation	Window opening is up to people working in the office
Time-scheduled ventilation	Ventilation according to EN 15251 (assuming IAQ category II, low polluting building material and equipment, 2 attendees) • Total air flow rate: 125 m^3/h (Mo-Fr 08:00 a.m. to 06:00 p.m.) • 6 m^3/h otherwise
DCV using iAQ-module	Sensor output, 450-2000 ppm CO_2 equivalents serves to trigger the ventilation rate in linear scale from 6 m^3/h to 125 m^3/h

Table 1. Ventilation settings - office.

60% less supply air resulted in 70% less power consumption for the fan and 15% less heating energy demand during heating period 2009/2010 (Fig.3, right hand side).

DCV – meeting room

In a highly frequented meeting room designed for 8 people and time-scheduled centralized ventilation of 70 l/s from 08:00 a.m. to 6 p.m., 50% supply air rate could be saved within one week applying a 3-step DCV scheme as shown in Fig. 4, left hand side. Step 1: 35 l/s, corresponding to sensor output > 700 ppm, step 2: 50 l/s, corresponding to sensor output > 900 ppm and step 3: 70 l/s, corresponding to sensor output > 1300 ppm.

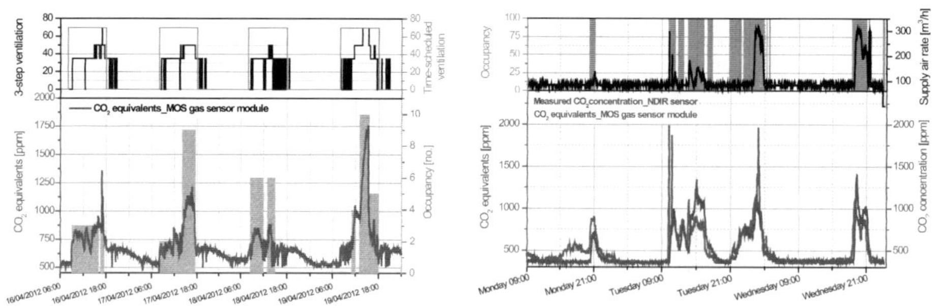

Figure 4. Left: DCV – meeting room. Right: DCV – fitness center.

DCV – fitness center

Installation in an air handling unit in a fitness center resulted in 24% less operating time, which translates to 60% energy cost saving with the overall IAQ rated "good" (Fig. 4, right hand side).

LONG-TERM STABILITY

Long-term stability of sensors in the building sector is essential because of maintenance reasons. Field-tests comparing permanently operated IAQ-modules with new modules showed, that perfect functionality in the field is guaranteed even for more than 50.000 operating hours (Fig. 5). As result of the standardized output signal, the IAQ-module is not affected by typical chemical sensor drift effects.

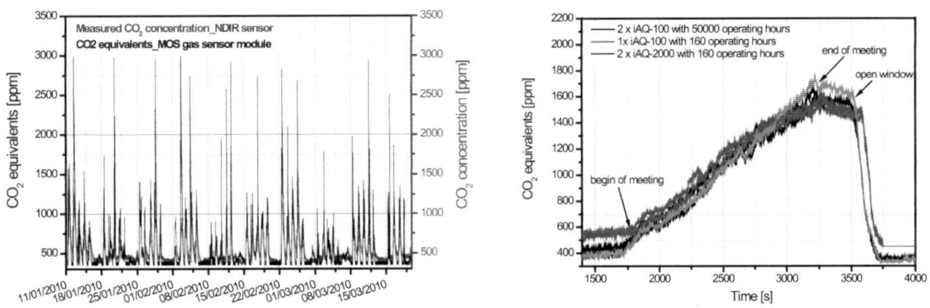

Figure 5. Long-term stability: Left: 10 weeks sensor data permanently operated IAQ-module. Right: Real-life test after 50.000 operating hours.

POWER REDUCED IAQ-MODULES

Wireless sensor networks require low power consuming sensors. Constant power operated IAQ-modules currently consume ~200 mW, too much for desired battery operation at an arguable maintenance level. Temperature modulation of the sensing element is seen as the way forward to reduce power consumption drastically whereat performance of the sensor has to remain unaffected.

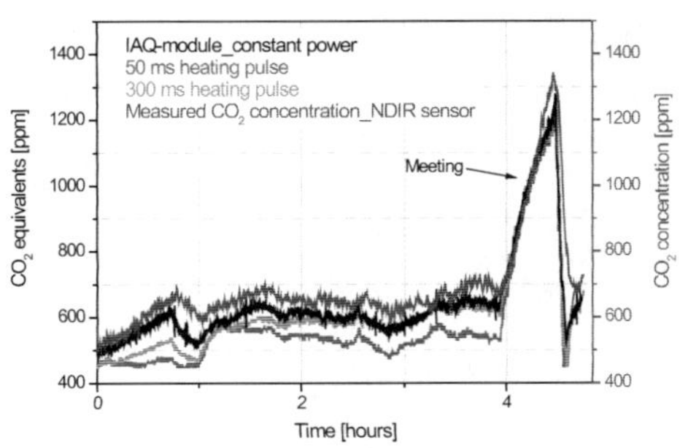

Figure 6. Predicted CO_2 concentrations for constant powered IAQ-module, temperature modulated modules in comparison to measured CO_2 concentration.

Different temperature profiles have been applied and tested in real-life in comparison to the constant power operated IAQ-module and NDIR-CO_2 reference sensor. 50 ms and 300 ms heating pulse of 3 V showed comparable performance to the constant powered IAQ-module while lowering the power consumption between ~50-90%. Fig. 6 shows output signals in CO_2 equivalents applying AppliedSensor's evaluation algorithm to the temperature modulated sensor raw data in comparison to the constant operated IAQ-module and measured CO_2 concentration, as obtained for an office scenario where a meeting of 6 people took place.

BUILDING MONITORING SYSTEM

Together with sensors for humidity, temperature, iluminance level and CO_2, the developed IAQ-modules build the core of environmental condition monitoring within the Clear-up project. In the test bulding in Prague and in the demonstrator building in Cadiz, the IAQ-modules serve to trigger ventilation. Monitoring and evaluation of data is supported by an installed web-based monitoring system that provides real-time web-access to the sensor data (Fig. 7).

The system enables the setup of thresholds for temperature, relative humidity and IAQ that provides the chance to immediately interfere upon sensor/system failures or significant events by text notification. Floor plans of the buildings can be used to display the information about each analyzed room on a monitor. A detailed view of time series for the last day up to the last month of the monitored parameters can be selected by the user.

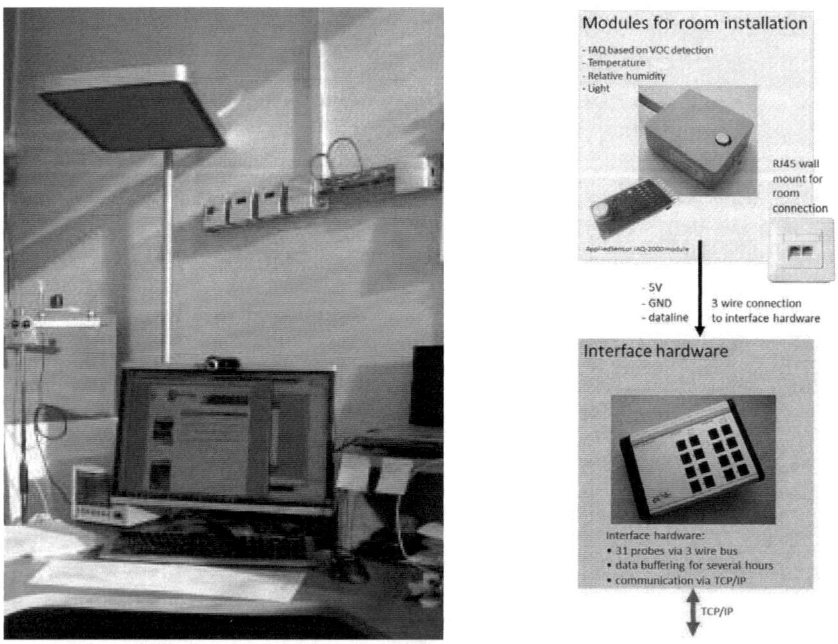

Figure 7. Left: Wall-mounted IAQ-modules for ventilation control and dimmable lighting at the test building in Prague. Right: Web-based monitoring system with room probes for IAQ, temperature, relative humidity and illuminance measurement using the normal UTP infrastructure of the building. The FlexProbe serves as interface hardware between room probes and software.

CONCLUSION

The energy findings accentuate the need for DCV regarding actual VOC load conditions in buildings using the developed IAQ-module. Human adaption often prevents air quality perception and results in high indoor pollution loads. Time-scheduled ventilation needs air flow rates to be adjusted before start-up depending on the average grade of occupancy or activity in the respective room which often results in a lack of overlap of operation time, ventilation demand, occupancy profiles and indoor pollution load. IAQ control with the developed IAQ-modules however ensures that fresh air is supplied to rooms whenever necessary minimizing the amount of supply air and hence the energy consumption compared to time-scheduled ventilation (Table 2).

Case study	Supply air rate reduction
Office	• 60%
Meeting room	• 50%
Fitness center	• 24%

Table 2. Case studies DCV - Supply air rate reduction.

Remote building monitoring and control systems are seen as a way forward for intelligent buildings, providing very meaningful information about the metabolism of the building, occupancy profiles, long-term data, comfort conditions and applied ventilation strategies. Implementation of low power consuming IAQ-modules in wireless monitoring and control systems will even lead to new applications in the building sector.

Clear-up test building measurements have just started and we are looking forward to gain new experiences with the developed IAQ-module.

REFERENCES

[1] Gassmann, O., Meixner, H. (Editors) 2001. *Sensors in Intelligent Buildings, Sensors Applications*, Volume 2, Wiley-VCH Verlag GmbH.

[2] EPBD, Directive 2010/31/EU of the European Parliament and of the Council of 19 May 2010 on the energy performance of buildings, Official Journal of the European Union, L 153, Volume 53, 18 June 2010, 13-23.

[3] Pettenkofer M. 1858. *Über den Luftwechsel in Wohngebäuden (Air change in residential housing)*, Munich: Cottasche Buchhandlung [German].

[4] CR 1752 1998. *CEN Report Ventilation for Buildings: Design Criteria for the Indoor Environment, CEN/TC 156/WG 6*, European Committee for Standardization, Brussels, Belgium.

[5] EN 15251 2007. *Indoor environmental input parameters for design and assessment of energy performance of buildings addressing indoor air quality, thermal environment, lighting and acoustics*, European Committee for Standardization, Brussels, Belgium.

[6] Won, D. and Yang, W. 2005. *The State of-the-art in Sensor Technology for Demand-Controlled Ventilation*, National Research Council Canada, October 2005.

[7] Clear-up 2012. www.clear-up.eu

[8] Burdack-Freitag, A., Rampf, R., Mayer, F. and Breuer, K. 2009. *Identification of anthropogenic volatile organic compounds correlating with bad indoor air quality*, Proceedings of the 9th International Conference and Exhibition Healthy Buildings, 2009, Sept. 13-17, Syracuse, NY, paper 645.
[9] Herberger, S., Herold, M., Ulmer, H., Burdack-Freitag, A. and Mayer, F. 2010. *Detection of human effluents by a MOS gas sensor in correlation to VOC quantification by GC/MS*, Building and Environment, 45 (11), 2430-2439.
[10] Herberger, S. and Ulmer, H. 2012. *Indoor Air Quality Monitoring Improving Air Quality Perception*, CLEAN – Soil, Air, Water, 40 (6), 578-585.

RESIDENTIAL VENTILATION SYSTEM PERFORMANCE: OUTCOMES OF A FIELD STUDY IN THE NETHERLANDS

Atze Boerstra[a,*], Jaap Balvers[a], Rik Bogers[b], Rob Jongeneel[b] and Froukje van Dijken[a]
[a] BBA Indoor Environmental Consultancy, Rotterdam, The Netherlands
[b] Center for Environmental Health, National Institute for Public Health and the Environment, Bilthoven, The Netherlands

* *Corresponding author; email: ab-bba@binnenmilieu.nl*

ABSTRACT
This paper describes the results of a Dutch national study into performance of mechanical ventilation systems and its effect on the self-reported health and perceived indoor environmental quality of occupants.
Ventilation systems with natural supply and mechanical exhaust ventilation (MEV) and balanced mechanical supply and exhaust systems with heat recovery (MVHR) were investigated. Surveys were performed in 299 homes, which included visual inspections and measurements of ventilation rates per room and installation noise levels. Furthermore, dwellers were questioned regarding perceived indoor air quality and self-reported health. Results show that shortcomings are common in many homes and for both MVHR and MEV. Shortcomings include insufficient ventilation rate, high noise levels, unclean systems and insufficient maintenance. The indoor environmental quality was perceived more positive in homes with MEV when considering air quality, dryness of air, noise and control options. There was no clear relationship between self-reported health and shortcomings of the ventilation.

KEYWORDS
Mechanical ventilation, balanced ventilation with heat recovery, self-reported health, indoor air quality, houses and homes.

INTRODUCTION
In recent years, nearly all newly built dwellings in the Netherlands are equipped with mechanical ventilation systems. The two most common systems are mechanical air supply with mechanical exhaust and heat recovery ('balanced mechanical ventilation with heat recovery' or 'MVHR') and natural air supply with mechanical exhaust ('mechanical exhaust ventilation' or 'MEV'). See figure 1. MEV works with natural ventilation grilles above windows and is in use since the 1970s. MVHR is only widely used since the past 10 years in The Netherlands. All MVHR systems in The Netherlands use plate air-to-air heat exchangers.

European directive EPBD (Energy Performance of Buildings Directive, EU, 2002) is the driving-force for national regulations to reduce energy consumption in buildings. The directive "lays down requirements [...] on the energy performance of buildings". But the EPBD has a broader scope and also states that "requirements shall take account of general indoor climate conditions, in order to avoid possible negative effects *such as inadequate ventilation*" and that "... measures [...] should take into account climatic and local conditions *as well as indoor climate environment* and cost-effectiveness".

Newly built dwellings in The Netherlands are required to have a certain degree of energy-efficiency (an 'energy performance index'), which over the past 5 years has become more and more demanding. Balanced mechanical ventilation has a significant impact on the energy performance index and can (in theory) reduce energy consumption. Therefore, many new homes in The Netherlands have been equipped with balanced mechanical ventilation over the past years.

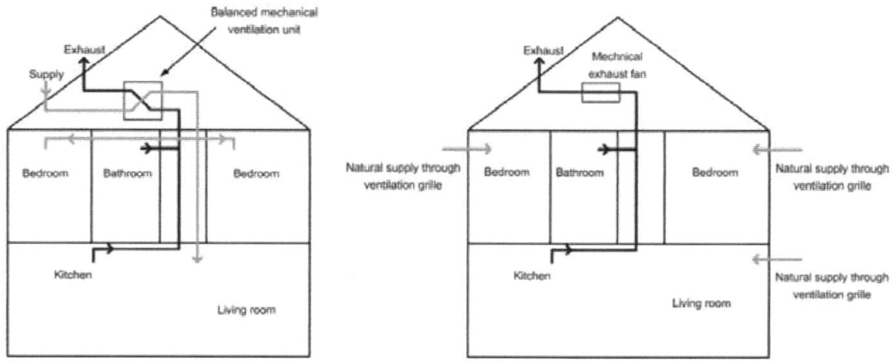

Figure 1 Left: Balanced mechanical ventilation with heat recovery; right: Natural supply with mechanical exhaust ventilation.

However, many occupants of newly built Dutch homes seem to be bothered by noise, draught, heat, poor indoor air quality and a lack of personal control caused by the ventilation system (Duijm, 2006). There has been considerable media-coverage in The Netherlands on societal concerns about the supposed relationship between health-effects and balanced mechanical ventilation in homes. This included reports on public television linking energy-efficient houses with balanced ventilation to negative health effects. A local study by the municipal health services (Duijm et al., 2007) and a nationwide study (Leidelmeijer et al., 2009) showed that health problems were indeed related to the presence of balanced ventilation systems. However, the reason why health symptoms are more prevalent in dwellings equipped with these systems could not be determined. One possible explanation is that mechanical ventilation systems, in particular balanced ventilation systems, produce relatively high levels of noise. Noise annoyance results in occupants setting the system in a low setting or turning it completely off, leading to insufficient ventilation. Other shortcomings of the ventilation systems may also play a role, but the technical performance of the systems was not determined in these studies.

The Dutch Ministry of Housing, Spatial Planning and the Environment commissioned a study into the performance of mechanical ventilation systems in Dutch homes and the relationship with perceived indoor environment and self-reported health of occupants. The overall aim was to inventory the most prevalent technical shortcomings of the two most common mechanical ventilation systems with respect to design, construction, maintenance and use. Moreover, a study was performed to explore the relationship between performance of the systems and the perceived indoor environment and self-reported health, and if occupants' report of indoor environment and health differed between the two types of ventilation systems.

The 4 research questions were as follows:

- What are the most common shortcomings of mechanical ventilation systems?
- What is the relationship between shortcomings of mechanical ventilation systems and self-reported health?

- What is the relationship between shortcomings of mechanical ventilation systems and the perceived indoor environmental quality?
- Is there a difference in self-reported health and the perceived indoor environmental quality between occupants of homes with balanced mechanical ventilation and occupants of homes with mechanical exhaust ventilation?

METHODS

The study focused on ventilation systems in single-family houses that were completed between June 2006 and January 2008. Only terraced houses, semi detached houses and detached houses were included in this study (apartment buildings and other multi-family houses were excluded). Selected dwellings were fairly equally distributed across the Netherlands. A number of 150 dwellings with balanced mechanical ventilation or (MVHR) and 149 dwellings with mechanical exhaust or (MEV) were inspected between December 2009 and June 2010.

Data acquisition consisted of two parts: (1) technical building surveys combining visual inspections and measurements and (2) questionnaires for building occupants. Part 1 aimed to determine the performance and common shortcomings of mechanical ventilation systems. In part 2, the relationship between the performance (from part 1) and perceived indoor environment and self-reported health (from the questionnaires) was explored. Results were also used to compare performance of the two different ventilation systems.

Surveys were performed in all homes, which included measurements of ventilation rates per room and installation noise levels. Mechanical ventilation rates were determined per room using a pressure-compensating air flow meter. Installation noise was determined using an integrating sound level meter. Additional technical inspections were carried out in order to qualify risk factors for the aspects indoor air quality, thermal comfort and noise. This included visual inspections of risk factors such as the hygienic condition of the ventilation unit, filters and ducts, the position of the outdoor air inlet, commissioning and the presence of a bypass on the heat recovery.

Performance of ventilation systems was compared with the requirements in the Dutch Building Code (BRIS, 2003) and the Dutch GIW/ISSO guidelines for installations in dwellings set by the Dutch Guarantee Institute for House-building (ISSO, 2008). The basic assumption in this study was that a proper ventilation system matches all requirements of both the Building Code and the GIW/ISSO guidelines.

On the day of the survey, one occupant per home completed a questionnaire on perceived indoor environment and self-reported health. The questionnaire contained mostly standardized questions on subjective health, nonspecific health symptoms, indoor environment related health symptoms and perceived indoor environment (MM-040 (Andersson, Stridh et al., 1993 & Andersson, 2010)), sleep disturbance, and perceived controllability of the ventilation system.

The indoor environment related health symptoms were not determined by the complete MM-040 scale (for which data is available) but by a selection of five items comparable to Burge's et al. (1993) Building Symptom Index (BSI 5). These were i. fatigue; ii. headache; iii. burning, itchy or irritated eyes; iv. irritated, stuffy or runny nose; v. hoarse, dry throat.

The statistical analysis was done in three steps. First, all 95 associations between technical aspects and performance of the ventilation systems and perceived health and indoor environment were tested in a univariate way. Non-parametrical tests were used, i.e., Spearman's correlation coefficient, Kruskal-Wallis, Fisher exact and Chi square tests. Second, to investigate confounding, all associations with a statistical significance level of p<0.10 were stratified according to categories of potential confounding variables such as smoking and educational level.

RESULTS

Ventilation capacity

The most common problems (where >30% of the homes did not comply with the reference level) related to the ventilation capacity are shown in Table 1. Dutch ventilation systems usually have 3 user controlled settings, where 1 is the lowest and 3 highest. The actual air supply and exhaust rates in the highest setting were insufficient in most homes. In 48% of the dwellings with balanced mechanical ventilation the total air supply rate was insufficient (< 0,7 $l/s/m^2$), while the air supply rate in 85% of dwellings was insufficient in one or more rooms in comparison with the Dutch Building Code (Figure 2). Total air exhaust rates were insufficient in 55% of the dwellings with balanced ventilation and 69% of the dwellings with mechanical exhaust. The exhaust rates in one or more rooms did not comply with the standards in 80% (MVHR) and 76% (MEV) of the dwellings (Figure 3).

Table 1 Common shortcomings related to the ventilation capacity for balanced mechanical ventilation (MVHR) and mechanical exhaust ventilation (MEV). (n/a = not applicable)

	% dwellings with shortcomings	
	MVHR	MEV
Performance		
The measured air supply rates are insufficient (Figure 2).	48%	n/a
Calculated (theoretical) air supply rate of natural ventilation grilles in bedrooms is insufficient.	n/a	36%
The measured air exhaust rates are insufficient (Figure 3).	55%	69%
Construction		
Ventilation controls settings are not properly adjusted. *The differences between the control switch steps should be >10% and >30% in total in order to provide sufficient control options.*	34%	27%
Setting of the air supply and exhaust valves is not locked or marked. *The settings of the valves should be locked or marked after commissioning in order to secure the proper settings after e.g. cleaning of the valves.*	61%	35%
Location of the air supply and exhaust valves is not marked. *The location or room of the valves should be marked after commissioning in order to secure the proper settings after e.g. cleaning of the valves.*	99%	96%
No commissioning reports about the ventilation system available.	87%	94%
Ductwork is not properly installed. *Non-smooth, bendy or tight air ducts cause extra air resistance. Leaky joints result in the loss of supply or exhaust air.*	48%	40%
The capacity of air transfer devices is insufficient. *Air transfer devices (such as a slit under a door) should be present in order to balance air supply and exhaust.*	40%	32%

Indoor Air Quality

The most common shortcomings related to the indoor air quality (where >30% of the homes did not comply with the reference level) are listed in Table 2. A surprising finding is that the air supply ducts were contaminated with dust and dirt in 77% of all homes with MVHR, while all homes were completed only several years ago. See figure 4 for a typical example of a polluted air supply duct as found during the field study. Annual maintenance is not common: 66% of homes with MVHR and 82% of homes with MEV have a longer maintenance interval. Air filters were visibly not clean (e.g. dirty enough to warrant filter replacement) in 43% of homes with MVHR, which could also negatively affect the ventilation rate. Almost all occupants (96% in both cases) do not use their ventilation system as prescribed by manufacturers.

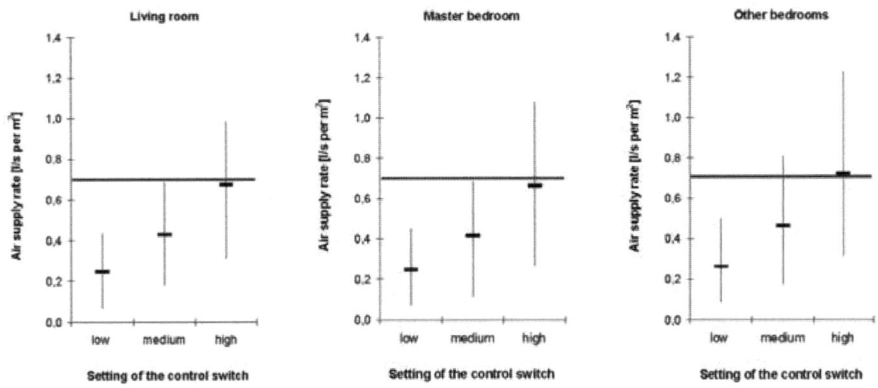

Figure 2 Air supply rates (average, P10 and P90) in the living room, master bedroom and other bedrooms in dwellings with balanced mechanical ventilation. The horizontal line gives the reference (minimum) level according to the Dutch Building Code (0,7 l/s/m^2).

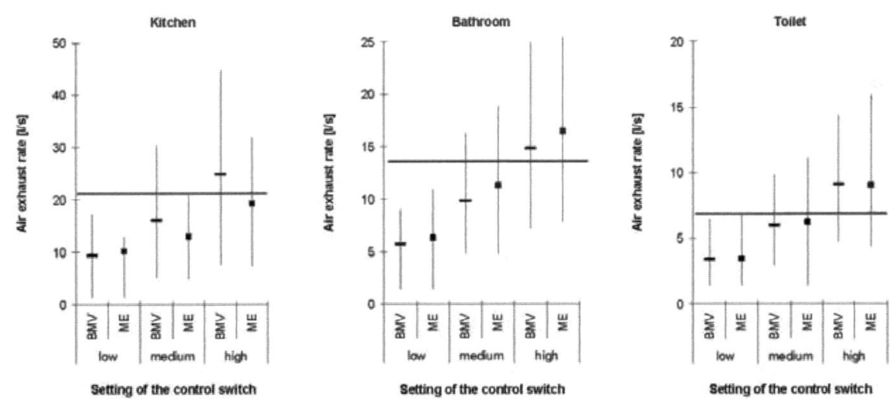

Figure 3 Air exhaust rates (average, P10 and P90) in the kitchen, bathroom and toilet in dwellings with balanced mechanical ventilation (MVHR) and mechanical exhaust (MEV). The horizontal line gives the reference (minimum) level according to the Dutch Building Code (21, 14 and 7 l/s respectively).

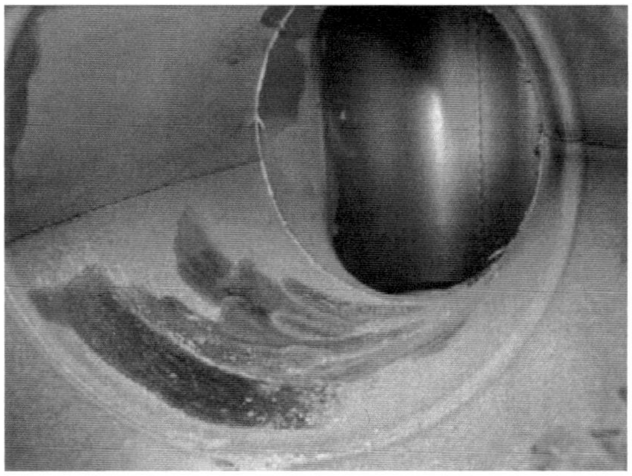

Figure 4 Example of dust accumulation in an air supply duct of a MVHR system.

Table 2 Common shortcomings related to indoor air quality for balanced mechanical ventilation (MVHR) and mechanical exhaust (MEV). (n/a = not applicable)

	% dwellings with shortcomings	
	MVHR	MEV
Design		
Insufficient control options. *A control switch should be present in the kitchen as well as the bathroom.*	81%	70%
Supply and exhaust valves in one room are situated too close together (<1 m). *As a result, parts of the room cannot be properly ventilated.*	53%	n/a
Construction		
Visible dirt in air supply ducts. *Ducts were soiled with dirt introduced during construction works, like cement or plaster.*	67%	n/a
Recirculation of exhaust air. *Qualitative experiments with smoke and particulate measurements (Balvers et al. 2011a) show that exhaust air is partly recirculated while no recirculation section is present.*	59%	n/a
Maintenance		
Filters are visibly not clean.	43%	n/a
Visible dust and dirt in air supply ducts. *Ducts were soiled with dist due to poor maintenance as well as dirt introduced during construction works, like cement or plaster.*	77%	n/a
Filters are changed less then 2x per year.	47%	n/a
No annual inspection of the overall functioning of the ventilation unit.	66%	82%
Usage		
Improper use of control switches. *The control switch is mostly used in a lower setting than recommended for proper ventilation.*	96%	96%
No oral instruction is given about the operation and functioning of the ventilation system (according to the dwellers).	42%	43%

Thermal Comfort

The most common shortcomings related to thermal comfort (where >30% of the homes did not comply with the reference level) are shown in Table 3. Heat recovery in a balanced ventilation system is unbeneficial when indoor temperatures are high (e.g. >24 °C), while the outdoor temperature is lower. This will mostly occur in summer months. A balanced system can be equipped with a bypass, which will conduct some or all air around the heat exchanger. This prevents heat from indoors being transferred to cooler outdoor air and will prevent unnecessary overheating of homes. In this study, about half of all MVHR systems were equipped with a bypass.

Table 3 Common shortcomings related to thermal comfort for balanced mechanical ventilation (MVHR) and mechanical exhaust (MEV). (n/a = not applicable)

	% dwellings with shortcomings	
	MVHR	MEV
Design		
No bypass on heat recovery.	49%	n/a
Air supply valves without air flow deflector too close to a ceiling or wall. *If supply air valves are positioned too close to a ceiling or wall (less than 30 cm) the valve should be equipped with an air flow deflector in order to deflect the air over a 180° sector instead of over 360° to avoid draughts caused by air that collides with the nearby surface and moves down.*	31%	n/a

Noise

The most common shortcomings related to high noise levels (where >30% of the homes did not comply with the reference level) are listed in Table 4. Noise levels are higher than 30 dB(A) in one or more bedrooms in 86% of homes with MVHR in the setting in which the ventilation system is providing a sufficient ventilation rate (>0,7 l/s) in many dwellings (or in the highest setting if the ventilation rate is insufficient). This was mainly a problem in dwellings with balanced mechanical ventilation. High noise levels are in part caused by incorrect placement of (flexible) ductwork, e.g. sharp bends (see Figure 5 for an example). At the moment of this study, maximum sound levels were not required by law, but a maximum level will be required in the next version of the Dutch Building Code (in effect as of January 2012). In most cases, the ventilation unit is placed on a position that will increase the chance of ventilation noise (53% of MVHR and 67% of MEV). Results from noise level measurements are shown in more detail in Figure 6. The differences between balanced mechanical ventilation and mechanical exhaust were the largest in the bedrooms.

Figure 5 | Example of incorrectly placed flexible ductwork near the ventilation unit.

Table 4 Common shortcomings related to noise of the ventilation system for balanced mechanical ventilation (MVHR) and mechanical exhaust (MEV). (n/a = not applicable)

	% dwellings with shortcomings	
	MVHR	MEV
Performance		
Measured noise levels in the living room are too high (>30 dB(A)).	72%	54%
Measured noise levels in one or more bedrooms are too high (>30 dB(A)).	86%	21%
Design		
Ventilation unit is mounted at a position that is susceptible to noise. *Ventilation units installed e.g. in a built-in cupboard in the bedroom or on light-weight walls without proper vibration control provide a high noise risk.*	53%	67%
Construction		
Silencers are not properly installed on either air supply or exhaust duct.	66%	n/a
Ductwork is not properly installed.	48%	40%

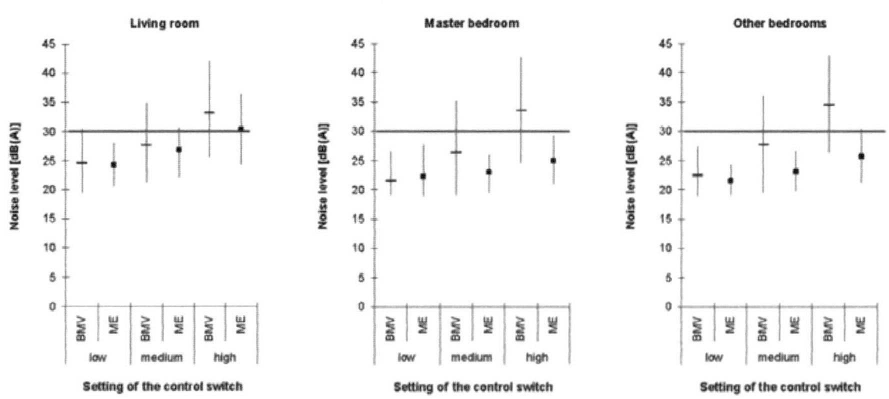

Figure 6 Measured noise levels from mechanical ventilation (average, P10 and P90) in the living room, master bedroom and other bedrooms in dwellings with balanced mechanical ventilation (MVHR) and mechanical exhaust (MEV). The horizontal line gives the reference level (30 dB(A)) (GIW/ISSO, 2008).

Perceived indoor environment and health

Results from the occupant questionnaires are presented below. Table 5 shows mean scores for the questions on self-reported health and sleep disturbance. The number of occupants with increased (i.e. unfavorable) scores was relatively low for all measures of self-reported health but somewhat higher for sleep disturbance.

Occupants of dwellings with balanced mechanical ventilation perceived the indoor environment as less favorable than occupants of dwellings with mechanical exhaust (Table 6). This was the case for perceived air quality, perceived dryness of the air, noise annoyance due to the ventilation system, and perceived personal control.

Table 5 Subjective health and sleep disturbance for occupants of dwellings with MVHR and MEV.

	Balanced mechanical ventilation	Mechanical exhaust	p-value
Nonspecific health symptoms (4-DKL sum score; range 0-32)[b]	2.7 ± 3.7	2.8 ± 4.5	0.81
Low (%)	91	92	
Moderately increased (%)	5	6	
Strongly increased (%)	1	1	
Perceived health status (VOEG-13 sum score; range 0-13)[c]	2.4 ± 2.7	2.1 ± 2.8	0.14
Increased (%)	14	12	
Indoor environment specific health symptoms (MM-040 BSI sum score; range 0-5)[d]	0.6 ± 1.1	0.5 ± 0.9	0.51
Sleep disturbance score (GSK; range 0-10)[e]	1.7 ± 2.4	1.7 ± 2.3	0.58
Increased (%)	17	18	

[a] Mean ± SD, unless otherwise specified. Higher scores are less favorable
[b] Terluin and Duijsens, 2006
[c] Sonsbeek van, 1990
[d] Andersson, Stridh et al. 1993
[e] Meijman, de Vries-Griever et al. 1988

Table 6 | Perceived indoor environment for occupants of dwellings with MVHR and MEV.

	Balanced mechanical ventilation	Mechanical exhaust	p-value
Perceived temperature			0.048[a]
Very good (%)	12	13	
Good (%)	72	68	
Acceptable (%)	15	12	
Not good (%)	1	7	
Bad (%)	0	0	
Perceived air quality			0.01[a]
Very good (%)	5	11	
Good (%)	61	68	

Acceptable (%)	24	18	
Not good (%)	10	3	
Bad (%)	1	0	
Perceived draft (% often (every week))	8	15	0.06[a]
Perceived stale air (%often (every week))	1	0	n.a.[b]
Perceived dry air (% often (every week))	20	7	0.001[a]
Perceived noise annoyance due to the ventilation system (mean ± SD; range 0-10)	1.9 ± 2,4	1.3 ± 2,1	0.003[c]
Severely annoyed by the ventilation system's noise (%)	5	2	0.34[a]
Perceived controllability (% insufficient)	32	18	0.01[a]

[a] tested with Fisher's exact or Chi^2-test
[b] not calculated because only one person complained of dry air
[c] tested with Kruskal-Wallis test

In addition, an analysis was made on the correlation between technical performance and perceived comfort and health. Three of the technical features of the ventilation systems showed a statistically significant association with perceived indoor environment or subjective health. These were:

1. A negative association between (in the case of balanced mechanical ventilation) the flow of supplied air in the most frequently used position in the bedroom during the night and nonspecific health symptoms (Spearman's r=-0.25, p=0.01);
2. A less positive perceived air quality in the case of a short-circuit between supplied fresh air and exhausted air in balanced mechanical ventilation systems (48% vs. 70% 'good' air quality, p=0.01) (Balvers et al., 2011a);
3. A positive association between the flow of exhausted air in the most frequently used position in the whole dwelling during the night and noise annoyance due to the ventilation system (in both systems; Spearman's r=0.16, p=0.01).

The statistically significant associations between performance/quality of the ventilation system and perceived health and perceived indoor environment were weak, but in line with the expectations (perception less favorable in homes with MVHR).

DISCUSSION & CONCLUSION

The overall conclusion from this study is that many things go wrong in standard Dutch housing projects with mechanical ventilation (both balanced mechanical ventilation and mechanical exhaust). The most common shortcomings are:

- Insufficient ventilation capacity. About half of all homes have at least one room with insufficient ventilation.
- Noise from ventilation systems. Noise exceeds the reference value in more than half of all homes with MVHR, especially in bedrooms.
- No operable windows available. 5% of all homes had at least one living room or bedroom without any operable windows.
- No bypass on the heat exchanger. There was no bypass on the heat exchanger in half of the homes with MVHR.
- Unclean ventilation systems. Many ventilation systems were (internally) dirty or dusty. Dust and dirt was found in the air supply ducts of about 66% of homes with MVHR. The ventilation-unit was internally dirty in a third of the homes with MVHR. Air filters were dirty in almost half of the homes with MVHR.
- Incorrect design or installation. Air ducts were not properly installed (e.g. with unnecessary bends) and air supply grilles were not placed at the optimal location in a number of homes.

- Incorrect use. Most users do not control ventilation systems as recommended. The highest setting, for example, remains mostly unused because of high noise levels. Moreover, occupants often don't know what the recommended use is, due to insufficient user information.
- Insufficient maintenance. In most cases, there was no maintenance contract for the ventilation system.
- The supply of (part of the) exhaust air. Part of the exhaust air is recirculated in more than half of all homes with MVHR. Further study is necessary to determine the exact cause and amount of recirculation.

There was no obvious relationship found between shortcomings of the ventilation systems and self-reported health. The only weak relationship found, was one between a higher nightly ventilation rate in the bedroom and a decrease in unspecific health problems.

Two shortcomings in mechanical ventilation systems were found to have a relationship with perceived indoor environmental quality. First, indoor air quality was rated less positive when recirculation of exhaust air was found in a MVHR system. Second, occupants reported more complaints about noise in situations where the ventilation system had a higher exhaust air rate at night.

No difference was found in the self-reported health of occupants with balanced ventilation or with mechanical extract ventilation. There were however some differences in perceived indoor environmental quality between homes with the two different ventilation systems:

- Occupants rate the indoor air quality higher in homes with MEV;
- Occupants rate the air less often as 'dry' in homes with MEV;
- Occupants have less complaints about noise in homes with MEV;
- Occupants rate the control options better in homes with MEV.

For more detailed information about the study and the study outcomes, see Bogers et al (2011), Balvers et al (2012) and van Dijken & Boerstra (2011) (in Dutch).

The original research reports (in Dutch) can be downloaded at http://www.binnenmilieu.nl/v2/bba/?id=141.

ACKNOWLEDGEMENTS

This study was commissioned and financed by the the Dutch Ministry of Housing, Spatial Planning and the Environment (currently: the Ministry of Infrastructure and the Environment). The authors would like to acknowledge the help and support during this study by Ellen Koudijs, Irene van Kamp, Erik Lebret, Tim Beuker en Mark Verlinde.

LITERATURE

1. Andersson, K., 2010, *The MM Questionnaires*, Retrieved 9 December 2010, from http://www.mmquestionnaire.se/index.html.
2. Andersson, K., G. Stridh, et al., 1993, *The MM-questionnaires – A tool when solving indoor climate problems*, Örebro, Sweden, Department of Occupational and Environmental Medicine, Örebro University Hospital.
3. Balvers J.R., Bogers R., Jongeneel R., van Kamp I., Boerstra A.C. and van Dijken F., 2012, *Mechanical ventilation in recently built Dutch homes: technical shortcomings, possibilities for improvement, perceived indoor environment and health effects*, Architectural Science Review, 55:1, 4-14.
4. Balvers J.R, Boerstra A.C, and van Dijken F., 2011a, *An investigation into re-entrainment of exhaust air in mechanically ventilated dwellings*, In: Proceedings of the 12th International Conference on Indoor Air Quality and Climate – Indoor Air 2011, Austin.

5. Bogers R, Jongeneel R, van Kamp I, and Koudijs E., 2011, *Health and wellbeing in relation to the quality of ventilation systems in newly built dwellings in the Netherlands*, In: Proceedings of the 12th International Conference on Indoor Air Quality and Climate - Indoor Air 2011, Austin.
6. BRIS, 2003, *Bouwbesluit 2003* (Dutch Building Code), available online through www.bouwbesluitonline.nl (in Dutch).
7. Burge PS, Robertson AS, Hedge A., 1993, *The development of a questionnaire suitable for the surveillance of office buildings to assess the building symptom index, A measure of the sick building syndrome*. Indoor Air conference 1993.
8. Duijm F., 2006, *Balansventilatie en gezondheid; waarom klagen bewoners?* ISSO ThemaTech, 9:20-22 (in Dutch).
9. Duijm F, Hady M, van Ginkel J, ten Bolscher GH., 2007, *Gezondheid en ventilatie in woningen in Vathorst; onderzoek naar de relatie tussen gezondheidsklachten, binnenmilieukwaliteit en woningkenmerken*. Amersfoort: GGD Eemland.
10. European Union, 2002, *Directive 2002/91/EC of the European Parliament and of the Council of 16 December 2002 on the energy performance of buildings*, available online through http://eur-lex.europa.eu.
11. ISSO, 2008, *GIW/ISSO publicatie 2008 Ontwerp- en montageadviezen*. ISBN: 978-90-5044-161-2, Stichting ISSO en Stichting GIW, Rotterdam, juli 2008
12. Leidelmeijer K, Menkveld M, Cozijnsen E, Heemskerk H., 2009, *Mechanische ventilatie in nieuwbouwwoningen: ervaringen en oordelen van bewoners over de kwaliteit van ventilatie en de eigen gezondheid*, reportnr. P10700. Amsterdam: RIGO Research en Advies BV.
13. van Dijken, F., J. R. Balvers, and Boerstra A.C., 2011, *The quality of mechanical ventilation systems in newly built Dutch dwellings*, Proceedings of the 12th International Conference on Indoor Air Quality and Climate – Indoor Air 2011, Austin.
14. van Dijken F. and Boerstra A.C., *Onderzoek naar de kwaliteit van ventilatiesystemen in nieuwbouw eengezinswoningen*, Report commissioned by the Dutch ministry of infrastructure and the environment, report date: February 23, 2011 (In Dutch), BBA Binnenmilieu, Rotterdam, 2011.

PERFORMANCES OF VENTILATION SYSTEMS: ON SITE MEASUREMENTS RELATED TO ENERGY EFFICIENCY, COMFORT AND HEALTH

PAPER TITLE
Performances of ventilation systems: on site measurements related to energy efficiency, comfort and health

MAIN AUTHOR
Caillou Samuel

OTHER AUTHORS
Van den Bossche Paul
Dinne Karla
Vandaele Luk

ABSTRACT

The real performance of ventilation systems is a recurrent point of discussion: too low ventilation flow rates, leading possibly to too low indoor air quality; electric consumption much higher than expected; acoustic complaints from the occupants; effective heat recovery lower than expected; etc. Moreover, health risks due to microbial growth in the filters or even in the ductworks are more and more proclaimed, without any evidences in terms of measurements or analyses on site.
By evaluating the on-site real performances of ventilation systems, this study aims to identify the most critical problems of such systems but above all to propose solutions for a better design and installation, improving comfort and energy efficiency.

More than 40 ventilation systems in dwellings have been, at least partially, monitored. Most of them are installed in recent buildings but some are in use since several years, up to 16 years. They are mainly balanced ventilation systems, but also some extract only ventilation systems with natural supply. The following characteristics have been evaluated: air flow rates, electrical consumption of the fans, temperatures of the air flows (for heat recovery), noise levels in the rooms, micro-organisms (moulds and bacteria) in the air (outdoor, indoor and supply air), at the duct surface and in the filters.

The most important outcomes are summarized as follows.

The total air flow rate in the dwelling is in general sufficient but the repartition of the flow rate in the different rooms is usually very poor, showing possible improvements thanks to a correct adjustment of the ventilation system.
The electrical consumption in the samples vary for a very large extent from 0.17 to 0.90 $W/(m^3/h)$, demonstrating the huge potential of energy savings by the choice of duct type and by the correct dimensioning of the ductwork.
The real heat recovery efficiency have been evaluated continuously in some systems, showing in some cases as high efficiencies as obtained in laboratory tests, but also revealing some critical points for heat recovery in practice, such as flow balance, control of the by-pass, etc.
The measured noise levels are usually too high, but the comparison of the measured levels with the solutions used for the design and installation should allow to identify the most important attention points for a better acoustical comfort.
Finally the microbiological results demonstrated that well-designed and maintained ventilation systems present no additional microbiological risk compared to outdoor air. Ventilation systems with supply air filtration are even able to significantly decrease the amount of micro-organisms from the outdoor air. The results also highlight the critical point to limit the microbiological risk of ventilation systems: maintenance of the filters (for example replacement once a year), position of air intake to avoid recirculation of polluted air, etc.

AIRTIGHTNESS AND VENTILATION OF NEW ESTONIAN APARTMENTS CONSTRUCTED 2001-2010

Leena Paap[1*], Alo Mikola[2], Teet-Andrus Kõiv[2], Targo Kalamees[1]

Chair of Building Physics and Architecture
Tallinn University of Technology
Ehitajate tee 5
19086 Tallinn, Estonia
**Corresponding author: leena.paap@ttu.ee*

Chair of Heating and Ventilation
Tallinn University of Technology
Ehitajate tee 5
19086 Tallinn, Estonia

ABSTRACT

The performance of ventilation and airtightness of the building envelope was studied in field measurements in recently constructed Estonian apartment buildings. The buildings were selected with different building envelopes and ventilation systems. The mean air leakage rate at the pressure difference of 50 Pa in the database was 1.7 $m^3/(h \cdot m^2)$. The mean air change rate at the pressure difference of 50 Pa from the database was 2.3 h^{-1}. Ventilation airflows in apartments were low in general, resulting in bad indoor air quality. Only in a few apartments did the general airflow correspond to the requirements of indoor climate category II. Together with increasing air tightness of the building envelope, more attention should be paid to the performance of ventilation. The capacity of the ventilation system is not the only concern. Only increasing the ventilation airflow, without proper design (noise reduction, avoiding draft, energy performance, etc.), may not guarantee good results.

KEYWORDS
Airtightness of building envelope; Performance of ventilation;

INTRODUCTION

Energy performance of buildings and indoor air quality are becoming more important in many European countries – especially when the latest version of EPBD (European Energy Performance of Buildings Directive) calls for all new buildings to be nearly-zero energy by the end of 2020. This sets higher requirements on performance of ventilation and airtightness of the building envelope.

Air tightness of building envelopes directly affects indoor quality (health, thermal comfort), hygrothermal performance, noise and fire resistance, and energy consumption of the building. Jokisalo et al. showed that in Finnish cold climate infiltration causes about 15–30% of the energy use of space-heating, including ventilation, in a two-story detached house, when the building leakage rate is typically (n_{50} = 3.9ach), while the corresponding proportion is about 30–50% in a leaky house (10ach). Because the correlation between the airtightness of the building envelope and the infiltration rate is almost linear, heating energy use of the houses also increases almost linearly at the same time. Therefore, the preceding correlation reduces into a simple rule of thumb: a one unit change in n_{50} corresponds to a 7 % change in the energy use of space heating, including ventilation. At the same time, the change in total heat energy use is about 4%. In the studied cases, these increment percentages vary from 4 % to 12 % regarding space heating, or from 2 % to 7 % regarding total energy use. The variation of these percentages mainly results from different wind conditions [1].

Therefore airtightness has become a single requirement in low energy buildings. In certification of new passive houses, the Passive House Institute requires air leakage rates below 0.6 air changes per hour at 50 Pascals pressure difference.

A good principle regarding the balance between airtightness and ventilation is: « *build tight — ventilate right* ». To guarantee indoor air quality, the performance of ventilation in airtight buildings is an important issue. If buildings become more airtight, leakage airflow is smaller and ventilation airflow should be larger. This is partly the reason, why in many countries the ventilation airflows are increased. But just increasing the ventilation airflow may not guarantee good results. It is not only a question of what the capacity of ventilation system is, but also the way how inhabitant uses the ventilation, how the ventilation is designed and built (noise reduction, avoiding draft, energy performance, etc.). Because there are many unknowns in the final performance of ventilation and airtightness of the building envelope, field measurements can give a good overview of the situation.

To give an overview of the final performance of the ventilation and airtightness of the building envelope in Estonian modern apartment buildings, field measurements were carried out in 28 buildings built between 2000 and 2010. The study is a part of the research project about decreasing environmental impact of buildings by improving energy performance of buildings in Estonia, and collecting a database of airtightness in Estonian apartment buildings.

METHODS

Studied buildings

63 apartments from 28 buildings were investigated in a cross-sectional study of the technical condition of recently built apartment buildings. The airtightness of the building envelope was measured in 26 apartments in 23 buildings. Ventilation airflows were measured in 30 apartments.

Buildings were selected with different external wall structures (Figure 1, left) and with different ventilation systems (Figure 1, right). The selection should represent an average of recently built Estonian apartment buildings.

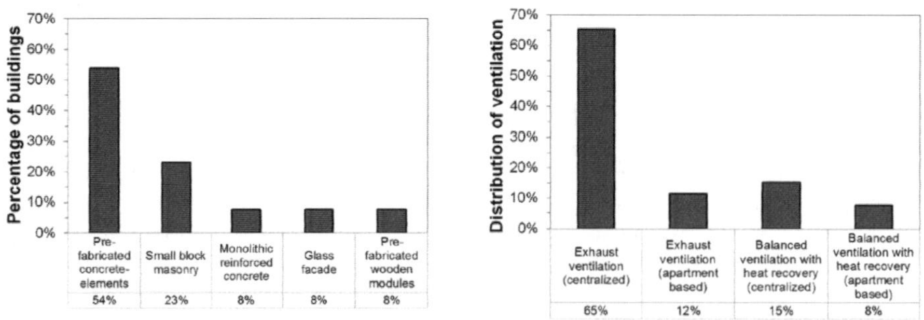

Figure 1. Distribution of studied apartments according to external wall type /left) and ventilation system (right).

Measurement methods

The air tightness of each apartment was measured with the standardized fan pressurization method [2], using "Minneapolis Blower Door Model 4" equipment (flow range at 50 Pa 25-7 800 m^3/h, accuracy ±3 %). To determine the air tightness of the building envelope, depressurizing and pressurizing tests were conducted with closed exterior openings, windows and doors and sealed ventilation ducts. To compare air leakage of different apartments, the air flow at pressure difference 50 Pa was divided by the apartment's internal envelope area (including intermediate walls) resulting air leakage rate at 50 Pa q_{50}, $m^3/(h \cdot m^2)$ or by the internal volume of the building, resulting in an air leakage rate at 50 Pa n_{50}, h^{-1}.

To determine typical air leakage locations and their distribution during the winter period, an infrared image camera FLIR Systems E320 was used (accuracy 2% or 2°C, measurement range: -20…+500°C). The temperature difference between the indoor and outdoor air was at least 20°C. Thermography investigations were conducted twice: first, to determine the normal conditions, the surface temperature measurements were performed without any additional pressure difference, and then to determine the main air leakage locations, the 50 Pa negative pressure under the envelope was set with fan pressurization equipment. After the infiltration airflow had cooled the inner surface (~30…45 min) of the envelope, the surface temperatures were measured with the infrared image camera from the inside of the building.

Ventilation airflows were measured with an anemometer (SwemaFlow 233 (accuracy ±4 % read value, minimum 1 l/s, measurement range 2 to 65 l/s)). The supply air flow rates were measured with a manometer Alnor/TSI AXD610 Digital Differential Micromanometer.

Assessment criteria

Requirements on airtightness of the building envelope is different country by country and in different standards [3][4]. In Estonia, the first requirement on the envelope's air tightness for apartment buildings was set in 1995: air leakage rate should be $q_{50} < 3.0\ m^3/(h \cdot m^2)$ [5]. The minimum requirements on energy performance of buildings [6] suggest that the general air leakage rate could be $q_{50} < 1.0\ m^3/(h \cdot m^2)$, and to avoid problems due to moisture convection critical joints should be made airtight.

Requirements on indoor climate are set in the standard [7]. The indoor air quality is expressed as the required level of ventilation. The general ventilation airflow in new apartments (indoor climate category II) should be at least 0.42 $l/(s \cdot m^2)$ or 0.6 h^{-1} and airflows in living rooms and bedrooms should be at least 1.0 $l/(s \cdot m^2)$ or 7 l/(s·person).

From measurement results, the reference value of air tightness for different types of buildings was calculated. The reference value $q_{50,delc}$ (Eqn. 1) represents median value (50% fractal) with a confidence level of 90% for air tightness. The reference value of air tightness is applicable for energy calculations, when air tightness is not measured or the air tightness base value given in energy performance regulation is not suitable to use (too large or too small).

$$q_{50,decl} = \overline{q_{50}} + k \cdot \sigma_{q50},\ m^3/(h \cdot m^2) \quad (1)$$

where: $\overline{q_{50}}$ is the mean value of air tightness of this building type, $m^3/(h \cdot m^2)$; k is the factor which takes into account the median value with a confidence level of 90% (Eqn. 2), and σ_{q50} is standard deviation of air tightness measurement results for this building type.

$$k = \frac{1.645}{\sqrt{n}} \quad (2)$$

where: n is the number of measurements

RESULTS

Airtightness of building envelope

The mean air leakage rate at the pressure difference of 50 Pa in the database was 1.7 $m^3/(h \cdot m^2)$, the minimum being 0.8 $m^3/(h \cdot m^2)$ and the maximum 4.6 $m^3/(h \cdot m^2)$. The mean air change rate at the pressure difference of 50 Pa from all the data was 2.3 h^{-1} (minimum being 0.9 h^{-1} and the maximum 6.6 h^{-1}). The average values of air leakage rates and air change rates at 50Pa pressure difference of all measured apartments are shown in Figure 2.

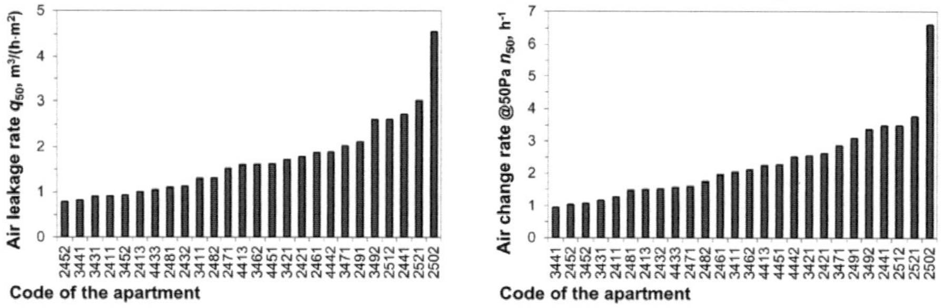

Figure 2. Air leakage rate q_{50} (left) and air change rate n_{50} (right) of all apartments.

Airtightness measurements show only a small difference between different building types (q_{50}=1.5…2.2 $m^3/(h \cdot m^2)$ and n_{50}=2.2…2.7 h^{-1}), Figure 3. It shows that it is possible to build airtight building envelopes within all types of structures. A larger deviation within the same building type shows that the quality of construction work can be a stronger influence.

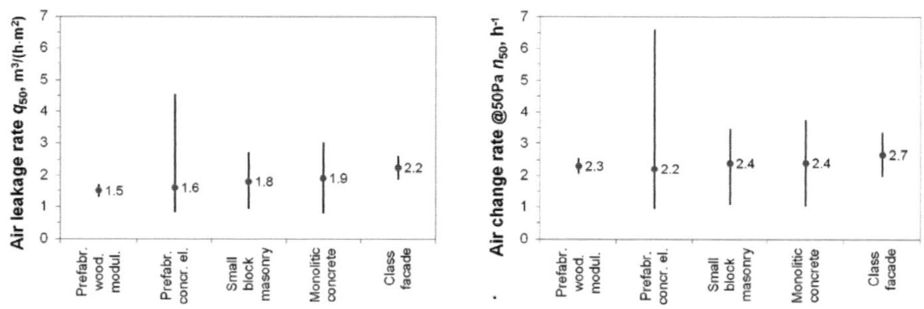

Figure 3. Air leakage rate q_{50}, $m^3/(h \cdot m^2)$ (left) and air change rate n_{50}, h^{-1} (right) of different building types.

By comparing current results with previous airtightness measurement results done in Estonia it shows that modern building envelopes are tighter, see Table 1. Older apartment buildings built with concrete and brick show similar airtightness, while old wooden apartment buildings are the leakiest.

Airtightness of the building envelope influences the heating energy consumption. Energy audits of buildings are commonly done with limited resources without airtightness measurements. Nevertheless, estimated values should be sufficiently on the safe side, to avoid too optimistic economic estimations. Reference values of airtightness that represent median values with a confidence level of 90% can then be used, see Table 1.

Type of the apartment building	Air leakage rate q_{50}, $m^3/(h \cdot m^2)$		Air change rate @50Pa n_{50}, h^{-1}	
	Average	Reference value	Average	Reference value
Modern buildings built 2000-2010 [current study] 26 apartments	1.7	2.0	2.3	2.7
Prefabricated concrete elements, 1960-1990 [8] 19 apartmnets	4.2	4.7	6.0	6.8
Brick walls, 1960-1990 [9] 30 apartments	4.0	4.4	5.7	6.4
Wooden structures, 1900-1940 [10] 35 apartments	10	11	13	14

Table 1. Comparison of airtightness of apartment buildings with different structures in Estonia.

Typical air leakage places in modern apartment buildings were:

- Leakages around the windows (Figure 4);
- Junction of the roof/floor with the external wall;
- Junction of the ceiling with the external wall;
- Junction of the separating walls with the external wall (Figure 5);
- Penetrations of pipes through the external wall;
- Surroundings of the fresh air valves in the external wall.

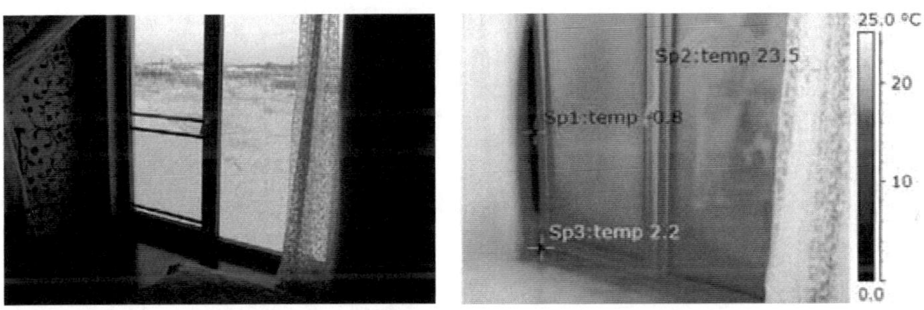

Figure 4. Leakages around the windows.

Figure 5. Air leakage between separating walls and the external wall.

Performance of ventilation

The performance of ventilation was assessed on the apartment level and on the bedroom level. Indoor climate category II (EN 15251: normal level of expectation and should be used for new buildings and renovations) was selected as reference.

Ventilation airflows in apartments were low in general (Figure 6) resulting in bad indoor air quality (Figure 7). Only in a few apartments did the general airflow correspond to the requirements of the indoor climate category II (>0.42 l/(s·m^2)). Even average general airflow (0.3 l/(s·m^2)) was below the indoor climate category II target value (>0.35 l/(s·m^2)).

Based on measurements of indoor CO_2 levels and estimated CO_2 (as tracer gas) emissions from residences during the night (≈20:00…8:00), the air change in bedrooms was estimated [11], Figure 8. As measurements were done in the main bedroom, the required airflow there should be at least 14 l/s. This average airflow was guaranteed only in 16 % of bedrooms during winter. Probably due to window airing during summer, this airflow was in 44 % of apartments.

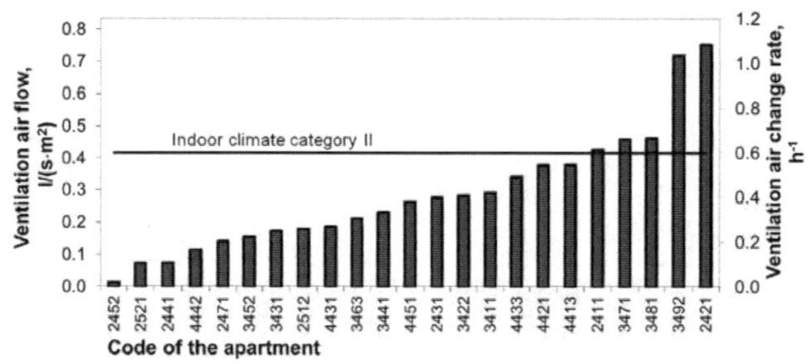

Figure 6. Ventilation airflow in apartments.

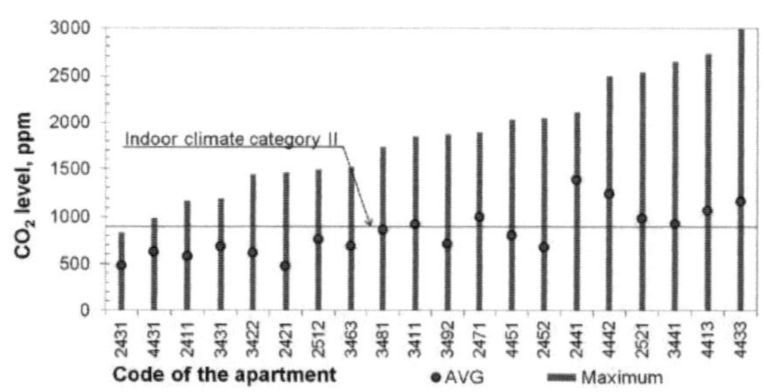

Figure 7. CO_2 levels in apartments.

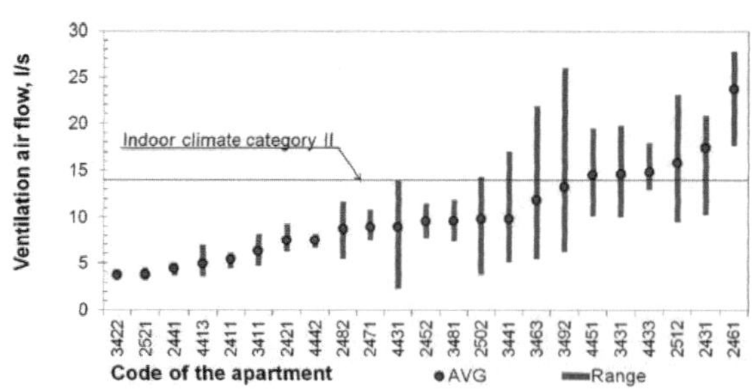

Figure 8. Ventilation airflows in bedrooms during winter.

DISCUSSION

The results of current study show that studied buildings are substantially more airtight compared to buildings from the period 1960-1990. About 92% of studied buildings satisfied minimum requirements for airtightness in Estonia (q_{50} < 3.0 $m^3/(h \cdot m^2)$). According to international standards on ventilation [14] and heating energy consumption [15], studied apartment buildings are buildings with low air leakage rates.

However, given the fact that according to the EPBD all new buildings must meet nearly-zero energy building's requirements by the end of 2020, the Estonian building sector has a lot of improvement to do if the airtightness requirements will be changed, for example to level q_{50} < 1.0 $m^3/(h \cdot m^2)$)).

Airtightness measurements showed small variations between newly built buildings with different structures, and large variations within similar structural solutions. For example air leakage rates of buildings made of prefabricated concrete elements were between q_{50} 0.82…4.55 $m^3/(h \cdot m^2)$, which clearly shows the impact of varied workmanship quality. Also typical air leakage distribution indicates that poor workmanship quality is behind the reason for low airtightness performance not low-grade building products. Airtight materials and good workmanship play important role in order to achieve high airtightness of building envelopes.

Due to larger air pressure differences over the building envelope in airtight buildings [12] and due to considerable moisture convection [13], special attention should be paid to the correct performance and balance of ventilation systems for ensuring a good indoor environment.

The performance of ventilation was not good in studied apartments. There was a very low correspondence for target values of indoor climate category II. The bad performance of ventilation is due to the extensive use of exhaust ventilation systems. In cold climates, taking outdoor air through the external wall without preheating does not provide thermal comfort (low temperatures, draft). If heat recovery is not used, it results in larger energy bills. These are the main two reasons why people decrease the ventilation airflows to the lower speed. If the exhaust fan is located in an apartment (bathroom, toilet, kitchen), then the noise may prevent the use of ventilation in a proper way.

CONCLUSIONS

The mean air leakage rate at the pressure difference of 50 Pa in the database was 1.7 $m^3/(h \cdot m^2)$, the minimum being 0.8 $m^3/(h \cdot m^2)$ and the maximum 4.6 $m^3/(h \cdot m^2)$. The mean air change rate at the pressure difference of 50 Pa from all the data was 2.3 h^{-1} (minimum being 0.9 h^{-1} and the maximum 6.6 h^{-1}). Based on the results it can be said that with all structural types it is possible to build airtight buildings, and quality of workmanship plays an important role in reaching a low leakage rate level. Future airtightness requirements may need improvement of current constructional style.

Together with the increase of the air tightness of building envelopes more attention should be paid to the performance of ventilation. The capacity of the ventilation system is not the only concern. Only increasing the ventilation airflow without proper design (noise reduction, avoiding draft, energy performance, etc.) may not guarantee good results.

AKNOWLEDGEMENTS

The research is done in project “Reducing the environmental impact of buildings through improvements of energy performance, AR12059” and has been financed by SA Archimedes. The study utilizes the measuring data of the national research projects financed by Estonian Credit and Export Guarantee Fund and by Ministry of Economic Affairs and Communications Research team of Chair of Building Physics and Architecture of Tallinn University of Technology is gratefully acknowledged for helping to carry out the measurements.

REFERENCES

[1] Jokisalo, J.; Kurnitski, K.; Korpi, M.; Kalamees, T.; Vinha, J. (2009). *Building leakage, infiltration, and energy performance analyses for Finnish detached houses*. Building and Environment, 44(2), 377 - 387.

[2] EN 13829. Thermal performance of buildings - Determination of air permeability of buildings – Fan pressurization method. 2001. European Committee for Standardization.

[3] C3. 2002. *Thermal insulation in a building,*. National building code of Finland.

[4] NBN D50-001. 1991.*Ventilatie voorzieningen in woongebouwen*. Brussels: Belgian Institute for Standardisation; (in Dutch).

[5] EVS 837-1:2003. Piirdetarindid. Osa 1: Üldnõuded [Enclosing structures. Part 1: General requirements]. Estonian Centre for Standardisation; 2004 (in Estonian).

[6] Estonian Government Ordinance No. 258. *Energiatõhususe miinimumnõuded.* (*Minimum requirements for energy performance of buildings*); RT I 2007,72, 445, 2007.

[7] EN 15251:2007/AC: 2012 *Indoor environmental input parameters for design and assessment of energy performance of buildings addressing indoor air quality, thermal environment, lighting and acoustics*. European Committee for Standardization,

[8] Kalamees, T; Õiger, K; Kõiv, T-A; Liias, R; Kallavus, U; Mikli, L; Lehtla, A; Kodi, G; Luman, A; Arumägi, E; Mironova, J; Peetrimägi, J; Korpen, M; Männiste, L; Murman, P; Hamburg, A; Tali, M; Seinre, E. (2009). *Technical condition and service life of Estonian apartment buildings, built with prefabricated concrete elements*. Tallinn University of Technology (in Estonian).

[9] Kalamees, T; Kõiv, T-A; Liias, R; Õiger, K; Kallavus, U; Mikli, L; Ilomets, S; Kuusk, K; Maivel, M; Mikola, A; Klõšeiko, P; Agasild, T; Arumägi, E; Liho, E; Ojang, T; Tuisk, T; Raado, L-M; Jõesaar, T. (2010). *Technical condition and service life of Estonian brick apartment buildings*. Tallinn University of Technology (in Estonian).

[10] Kalamees, T.; Arumägi, E.; Just, A.; Kallavus, U.; Mikli, L.; Thalfeldt, M.; Klõšeiko, P.; Agasild, T.; Liho, E.; Haug, P.; Tuurmann, K.; Liias, R.; Õiger, K.; Langeproon, P.; Orro, O.; Välja, L.; Suits, M.; Kodi, G.; Ilomets, S.; Alev, Ü.; Kurik, L. (2011). *Technical condition and service life of Estonian wooden apartment buildings.* Tallinn University of Technology (in Estonian).

[11] CIBSE, *Applications Manual AM 10, Natural Ventilation in Non-domestic Buildings*, Chartered Institution of Building Services Engineers (CIBSE), London, 2005.

[12] Kalamees, T; Kurnitski, J; Jokisalo, J; Eskola, L; Jokiranta, K, Vinha, J. (2010). *Measured and simulated air pressure conditions in Finnish residential buildings. Building Services Engineering Research & Technology*, 31(2), 177 - 190.

[13] Kalamees, T.; Kurnitski, J. (2010). Moisture Convection Performance of External Walls and Roofs . Journal of Building Physics, 33(3), 225 - 247.

[14] EN 15242:2007 *Ventilation for buildings - Calculation methods for the determination of air flow rates in buildings including infiltration.* European Committee for Standardization

[15] ISO 13789:2007 *Thermal performance of buildings -- Transmission and ventilation heat transfer coefficients -- Calculation method.*

THERMAL COMFORT ANALYSES IN CLASSROOMS

David Venus[1], Armin Knotzer[1]

1 AEE – Institute for Sustainable Technologies (AEE INTEC)
Feldgasse 19
A-8200 Gleisdorf, Austria

ABSTRACT

Thermal comfort in classrooms seems to be the most important parameter in the sensation of pupils, for both warm and cold season. It could be more important than IAQ (CO_2-concentration) or daylight-conditions of the classroom. Focus in schools will be the comfort conditions of transition and summer period which increases importance due to all-day and summer schools, changing climate conditions and technical equipment of pupils and teaching.

KEYWORDS

Investigations and simulations of the ERACOBUILD-project "school*vent*cool" could show that in the middle European climate classrooms are at high risk for overheating, not only during summer time, also during transition period. It is believed that the high temperatures decrease the ability to concentrate during lessons and is as important as the IAQ in classrooms.

INTRODUCTION

As there is very low energy demand for heating in high performance modernized and new passive house standard school (and similar) buildings the importance shifts towards the cooling demand of these buildings. First of all this is significant for buildings where technical equipment is the main factor for internal heat gains like schools and office buildings.

On that point, following Figure 1 shows the measured average hourly room temperatures of a school building which was retrofitted to passive house standard. On the x-axis the outside temperatures and on the y-axis the room temperatures were plotted.
There are a large number of hours with high room temperatures outside the defined comfort zone (red mark). Overheating is therefore a present issue. Passive cooling concepts have to include measures like "intelligent" shading systems to avoid this overheating.

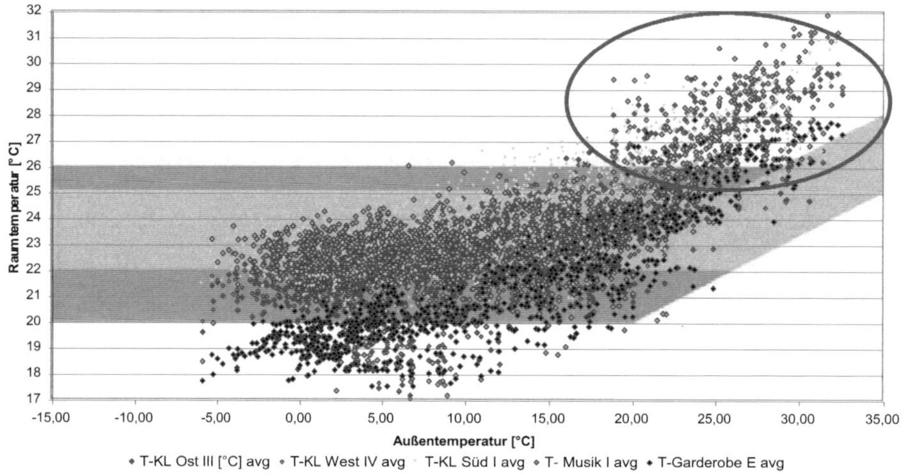

Figure 1: Temperatures in five different sections of an Austrian school building renovated to passive house standard (Source: AEE INTEC)

The existing school building stock is mostly equipped with big window areas without acceptable thermal and shading standards. High performance renovation and high new building standards lead to well insulated and airtight building envelope coupled with good indoor air quality (IAQ) like passive house standard offers and normally equipped with outside shading. But is this enough to ensure good summer comfort?

In the present ERACOBUILD-project "school*vent*cool", the Austrian partner AEE INTEC investigates the different reasons for and the amount of overheating in existing and high performance renovated schools and solutions for protection from heat and glare during the warm season. There are made calculations (PHPP, iDbuild, TRNSYS,...), measurements and interviews, analysing classroom comfort conditions before and after shading solutions, and before and after renovation, including the use of daylight. The comfort situation of the pupils is a central point in all evaluations and solutions.

THE INVESTIGATED SCHOOL BUILDING

All results of the calculations and the interviews shown on the following pages are an outcome of the detailed investigation of the vocational school Gleinstätten. Here is some information about the investigated school (see Table 1):

Parameter	Value
Years of construction	1974 – 1977
Numbers of floors	Basement and 3 floors
Number of classrooms	10
GFA (school building)	6.253 m²
Energy performance (calculated)	100 kWh/m²a
Heating supply	Central heating system fed by biomass-solar thermal district heating
Ventilation	Natural ventilation by windows

Table 1: Characteristic values of the existing school building (Source: AEE INTEC)

Figure 2: Views of the existing school building in Gleinstätten/Austria (Source: AEE INTEC)

Figure 3 shows the ground plan of the school with the investigated classrooms (marked in colour). Classroom E007 is mainly oriented to the north, classroom E004 to the west and classroom E001 to the south. The selection of these three rooms enables also information of all other classrooms in this storey because (nearly) all orientations are depicted by these three rooms. In the existing building classroom E007 is used as a computer room.

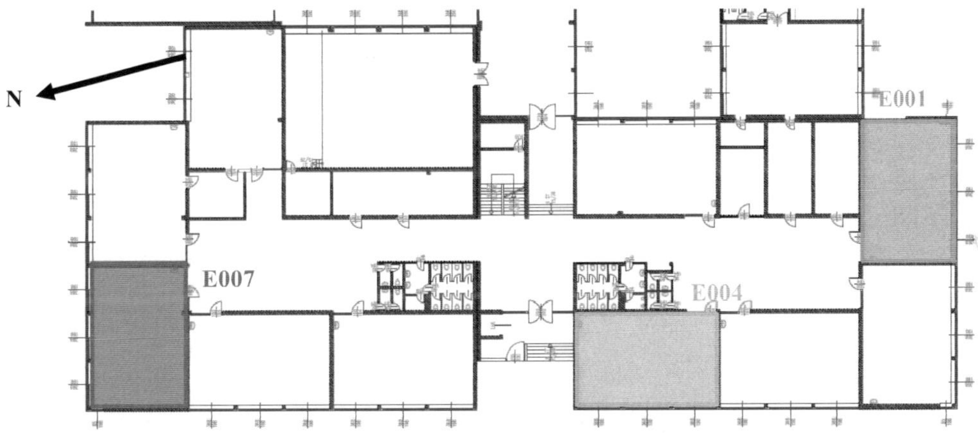

Figure 3: Ground plan of the existing school building with the three investigated rooms (marks) (Source: LIG)

The calculations of the school building were performed for the existing building and the retrofitted building whereas for the retrofitted building assumptions were made concerning the insulation of the building parts, the mechanical ventilation with heat recovery and other measures based on the passive house standard.

RESULTS

Calculations

Figure 4 shows the assorted room temperatures of classroom E001 (large south façade) plotted to the hours of the school year in %. In this case the calculation was performed with the internal gains of 18 pupils and one computer in the classroom.

The analysis of this figure shows that in the existing building the room temperature is about 88 hours (or just above 5% of the total school year) higher than 26°C. In the retrofitted building the room temperature is about 242 hours (or 14.5% of the total school year) above this limit. This is equal to the 2.75-fold of the value of the existing building! Another calculation (with hybrid ventilation, 18 pupils, each pupil equipped with a computer) resulted in even 637 hours (or 38.2% of the total school year) higher 26°C.

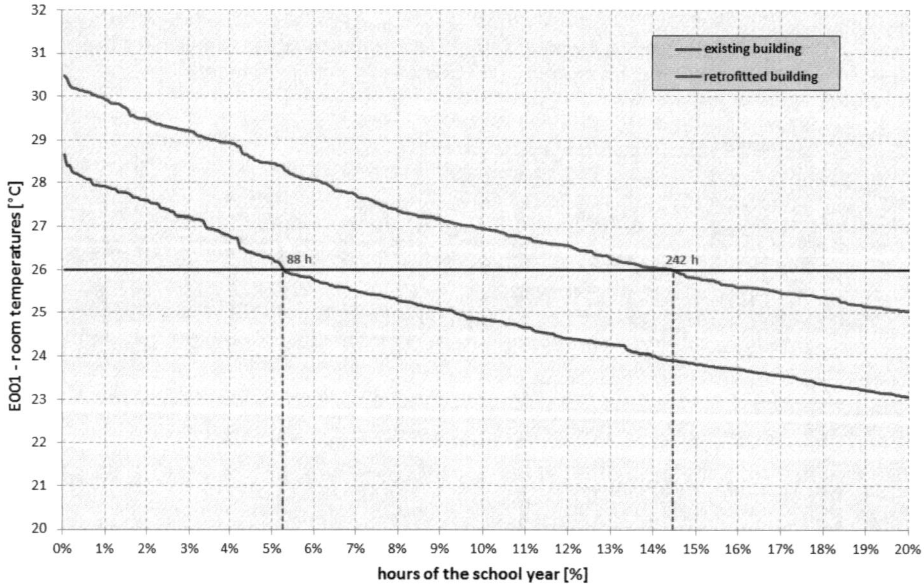

Figure 4: Sorted room temperatures for room E001 – comparison of existing and retrofitted building (Source: AEE INTEC)

In another calculation the influence on the high room temperatures was investigated. Figure 5 shows four different scenarios. Case 1 (red bars) represents the retrofitted building with 18 resp. 11 pupils inside the classroom, an automatic control of the shading and mechanical ventilation (including one calculation with an additional night ventilation). Case 2 (blue bars) represents case 1 plus the assumption that every pupil is equipped with a computer. The comparison of this two scenarios shows that the technical equipment (in this case computers) has more influence on the overheating in the classroom than the number of pupils inside.
This statement can confirmed by the comparison of the cases 3 (green bars) and 4 (orange bars). These two scenarios also represent the retrofitted building with 18 resp. 11 pupils inside the classroom and an automatic control of the shading but in this time executed with an optimized hybrid ventilation. With an increased number of computers in case 4 the school

days with temperatures above 26°C in the classroom also increase (from 51 to 86 and from 47 to 68) more than they do because of the higher number of pupils in the classroom.

Figure 5 also shows the "reduction potential" of an optimized ventilation system. By the implementation of a hybrid ventilation system the school days with room temperatures above 26°C can be reduced from 67 to 51 resp. 64 to 47 school days per year. An additional night ventilation can even reduce the school days with room temperatures above 26°C from 67 to 11 days per year.

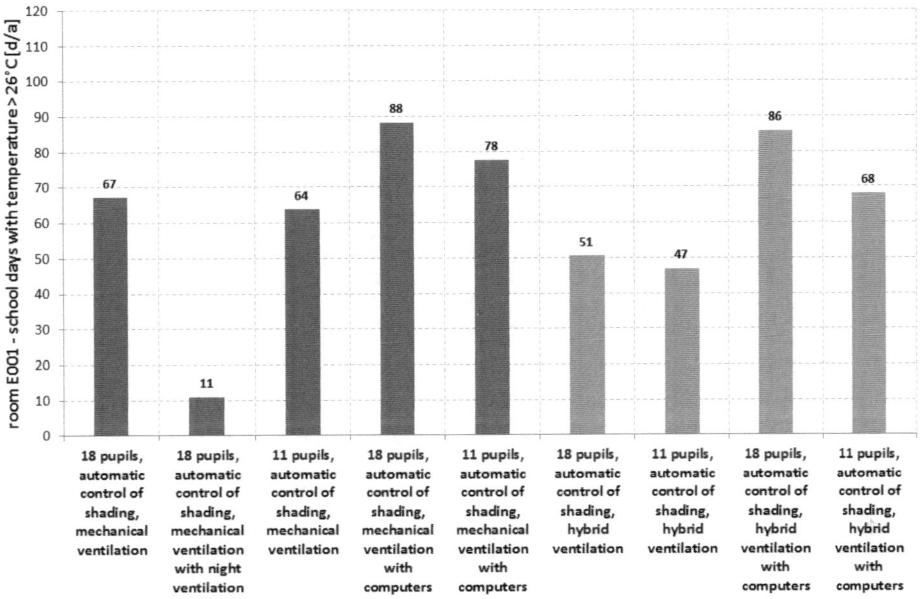

Figure 5: Influence on the overheating in classrooms – comparison of different varieties of the retrofitted building (Source: AEE INTEC)

Besides the room temperature other comfort parameters in the classroom have to be regarded as well. Pooled can this comfort parameters for instance in the characteristic value of the air quality according to EN 15251. This value was evaluated for the three classrooms, both existing and retrofitted building.

Figure 6 shows the results of the calculation of the air quality according to EN 15251 of room E001. Thereby the number of pupils, the solar shading and the ventilation of the building was varied. The best result was achieved by the scenario with more pupils, no shading and with mechanical ventilation (green bar). The results show in general that the two scenarios with no shading devices (green and purple bar) achieve the best values (30% resp. 25% in class I of EN 15251).

The further analysis shows that the hybrid ventilation system performs worse than the scenarios with mechanical ventilation systems. The use of a hybrid ventilation system reduces the characteristic value of the air quality in this case from 23% in class I to 16%.
Astonishingly the number of pupils in the classroom has no influence on the air quality in this calculation.

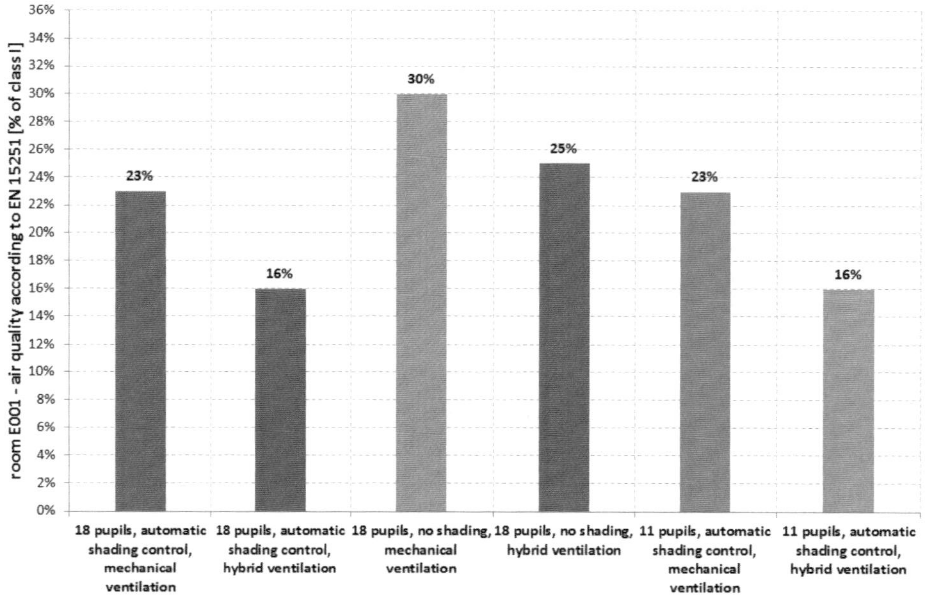

Figure 6: Air Quality in room E001 - comparison of different varieties of the retrofitted building (Source: AEE INTEC)

For the visual comfort and the health of the pupils the daylight use is of importance. For different scenarios the percentage of possible daylight use and the resulting primary energy demand for the lighting was calculated. For the simulation the illumination was set to 500 lux in the middle of the classroom. Again all three classrooms were investigated and analysed. Figure 7 shows the results of room E001. The blue bars characterize the existing building scenarios, the green bars the retrofitted building scenarios.

The result of the calculation shows that the percentage of daylight use in the retrofitted building scenarios is fairly high while the existing building scenarios perform worse. This indicates that there is a lot of potential to optimise the daylight use in the retrofit. Coincident not only the daylight use can be increased, also the primary energy demand for the lighting in the classroom can be lowered with an optimized shading solution (automatic control) in the retrofit.

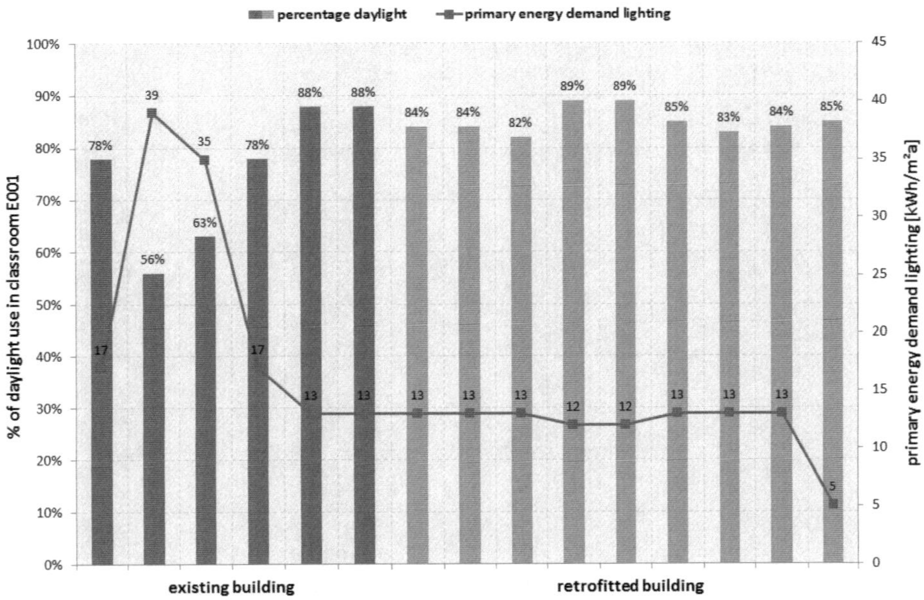

Figure 7: Daylight use and primary energy demand for lighting – comparison of different varieties of the existing and the retrofitted building (Source: AEE INTEC)

Interviews

73 pupils plus one teacher were asked in personal interviews to answer questions about the classroom comfort situation, written down in a specially developed questionnaire. Parameters like noise, IAQ, smell, temperatures, draught, interior design, acoustic and daylight conditions were asked to be assessed by the interviewee.

The interviews showed that pupils have the greatest sensation for the thermal situation in classrooms (cold and draught in winter, hot conditions in summer). Further results indicate that parameters like noise, IAQ, daylight conditions are much more linked to the special situation of the classrooms, individuals and their constitution. For example only 32% of the pupils perceive the smell in the classrooms as an annoying thing, but the amount of 79% of the pupils say, that the classrooms are too cold or too hot (see Figure 8).

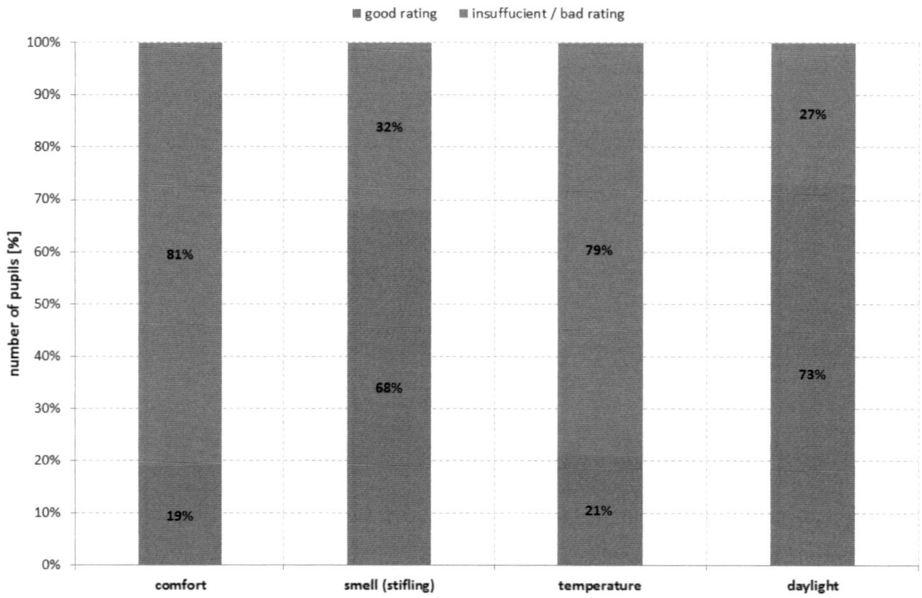

Figure 8: Simplified illustration of the results regarding the interviews about comfort sensation with the pupils and the teacher (Source: AEE INTEC)

CONCLUSION

Within the program "building of tomorrow" a lot of big-volume passive houses and down to lowest energy consumption renovated buildings were implemented and scientifically analysed, most of them by AEE INTEC. One of the results was the fact that a number of very energy efficient buildings exceeded thermal summer comfort (temperatures, relative humidity) limits in a quantity which was not expected [1]. In two more Austrian research projects there were found indications for the assumption that classrooms in schools built to nearly zero energy standard are at high risk for overheating, not only during summer time, and due to the high indoor temperatures the ability to concentrate during lessons decreases [2][3]. Analyses of the project "school*vent*cool" again could show that overheating is a serious problem and to be recognised in the very early planning stage of a nearly zero energy building. Suitable thermal comfort for pupils and teachers without active cooling is only possible with a range of passive cooling measures like outside shading and night ventilation in middle European climate.

ACKNOWLEDGEMENTS

The work in "school*vent*cool" was highly supported by the Austrian Solar Shading Organisation, J. Gerstmann, and the Styrian "Landesimmobilien-Gesellschaft mbH" (LIG), H. Reichl, A. Scharl. Many thanks also go to the headmistress of the vocational school Gleinstätten, M. Vörös-Achleitner, and the school caretaker, J. Strohmeier.

REFERENCES

[1] Wagner, W., Spörk-Dür, M., Lechner, R., Suschek-Berger, J. 2011: *Leitfaden: Ergebnisse der messtechnischen Begleituntersuchungen von "Haus der Zukunft"-Demonstrationsbauten*; edited in the frame of the project „Energy relevant and ecological monitoring studies of demonstration buildings, which were built in the framework of the program 'building of tomorrow'"; Editor: Austrian Federal Ministry of Transport, Innovation and Technology (bmvit)

[2] Luttenberger, C., Knotzer, A., Venus, D.: *Energetischer Maßanzug - Modul „Kommunale Sanierung"*, project financed by the local government of Styria/Austria 2011-2012

[3] Schwarzl, I. 2011: *Excessive heat in classrooms and impact on Student's concentration abilities, results of measurments in different schools;* presentation at "ökosan'11" conference, Graz

EXPERIMENTAL STUDY OF DIFFUSE CEILING VENTILATION IN A CLASSROOM

Christian Anker Hviid[1], Søren Terkildsen[*1]

1 Tech.Univ. of Denmark, Dept. of Civil Engineering
Brovej, building 118
DK-2800 Kgs. Lyngby, Denmark
**Corresponding author: sterk@byg.dtu.dk*

ABSTRACT

Diffuse ceiling ventilation is a novel air distribution device that combines the suspended acoustic ceiling with ventilation supply. A diffuse ceiling distributes the supply air above the acoustic tiles and has proven performance in laboratory experiments. To study the performance in real conditions a classroom was retrofitted with mechanical ventilation and a diffuse ceiling. The employed ceiling comprises active panels penetrable to air and impenetrable passive panels. The performance was studied with regard to air movements, temperatures and air change efficiency at two different air changes. The experiments were carried out during class to obtain realistic conditions. At both airflows did the ceiling perform satisfactorily with air movements and temperatures within the requirements of indoor climate standards. The air change efficiency is comparable to conventional mixing ventilation.

KEYWORDS
Diffuse ceiling ventilation, ventilation efficiency, ventilation, draught rating, tracer gas

INTRODUCTION
School classrooms are characterized by high occupancy and high thermal load. Consequently high ventilation rates are required which can be difficult to fulfil with conventional inlet diffusers without causing draught

Development of new concepts to ventilate school classrooms is therefore relevant and one promising solution is diffuse ventilation air inlet. The concept is commonly used in live stock buildings [1] and in the clean room industry [2], yet it is increasingly employed in comfort ventilation. The principle of diffuse air inlet in comfort ventilation is to inject the supply air into a pressure chamber above a standard suspended acoustic ceiling. The air is distributed to the room below through cracks and perforations. The flow velocity is very small and irregular, hence the term diffuse.

The reported research in this area mostly relies on laboratory experiments where results have been promising. In [3] diffuse ceiling ventilation outperformed five conventional air distribution systems in a laboratory office environment. The findings were supported by [4] who carried out experiments in a test facility resembling a small classroom. Tracer gas experiments showed that diffuse ventilation inlet provided perfect mixing, while air temperature and air velocity measurements did not disclose any local discomfort in the occupied zone over a broad range of flow rates and inlet temperatures. Similar draught assessments in a class room laboratory setup is reported by [5], who, in the same paper, also report a pilot study in a real classroom. However, the measurements only included overall CO2-concentration and a questionnaire about perceived air quality and not quantifiable measurements of air temperature and velocity or air quality.

To investigate the performance of diffuse ceiling inlet under real conditions a diffuse ceiling was installed in a classroom at Vallensbæk primary school. The investigation encompassed several elements to document the thermal comfort and to map the air distribution in a class room with diffuse ventilation air inlet, including: local comfort by air temperature and air velocity, air change efficiency, and radiant asymmetry. The objective was to validate how diffuse ceiling inlet performs in practise and to document the applicability of diffuse ceiling ventilation in classrooms.

METHODS

Room description

A classrooms at Vallensbæk primary school was refurbished with a new mechanical ventilation system including a new suspended acoustic ceiling functioning as diffuse ceiling inlets. Vallensbæk School has outer masonry walls and partitions of gypsum. The floor is a concrete deck with linoleum covering and the ceiling is a wooden roof structure, see Figure 1. The ceiling was suspended 0.2 m from the roof structure to create a plenum to distribute the air and resulted in a floor height of 2.5 m in the main part of the room.

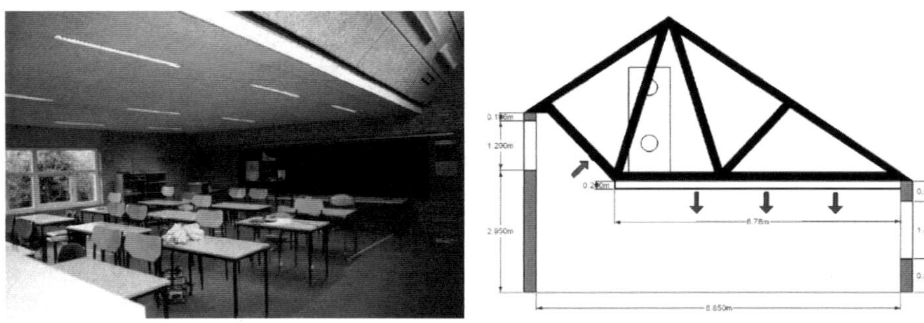

Figure 1. Left: Picture of classroom where measurements were performed, Right: Schematic of classroom with dimensions and principle of air distribution (blue arrows inlet, red arrows exhaust).

The floor area was 9.15x8.85 m, see Figure 3 and the windows on the main facade is oriented North-West, the solar gain during the occupied hours is therefore limited and only fabric curtains are installed as shading device. The classroom is usually occupied by 24 pupils in 6th grade (age 11-12).

Diffuse ceiling layout

The ventilation ceiling consists of a metal frame suspension system to create the plenum on which cement bonded wood-wool panels were mounted. This is a commonly used product in Danish classrooms for its acoustic properties. For use as a diffuse ventilation inlet, two types of panels are needed : active and passive. The active panels are permeable so the supply air can penetrate and passive panels has 20 mm of hard painted mineral wool glued to the backside making them non-permeable, see Figure 2. The mineral wool improves the acoustic properties and permits control of the supply air distribution in the room. A small overpressure is created in plenum above the panels that ensures the supply air is equally distributed through the active panels. In the room 6 active panels were used and placed as shown in Figure 3 equalling an area of 8.6 m^2. The panels were not uniformly placed as intended but this was discovered after the measurements were performed.

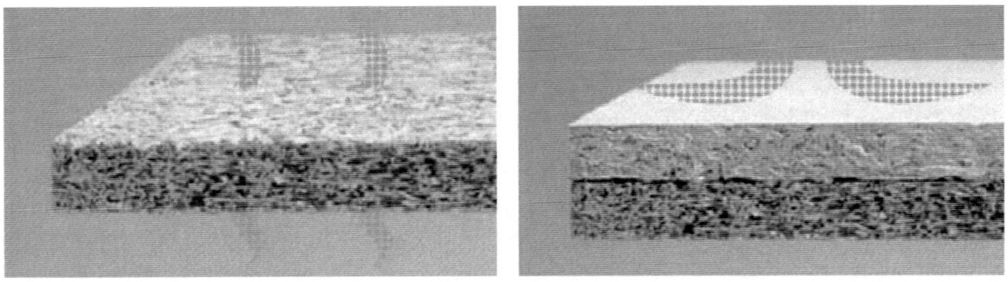

Figure 2. Left: Active plate of cement bonded wood-wool, Right: Passive plate has a layer of impenetrable mineral wool. Source: www.troldtekt.dk

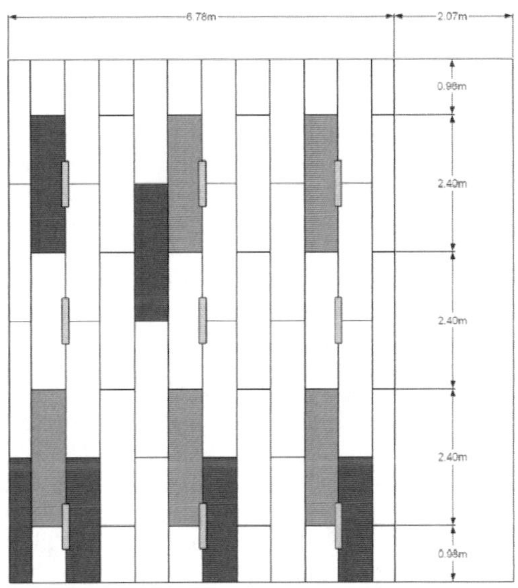

Figure 3. Placements of active cement bonded wood wool panels (red) and passive panels (white), green indicate the intended placement of the active panels.

The fresh air inlet was placed slightly off-centre of the suspended ceiling, blowing vertically down against a passive panels thereby distributing the air uniformly in the pressure chamber. The outlet air was extracted via existing diffusers located on the sloped part of the ceiling, see Figure 1. The supply and extract flow rate was balanced and controlled by dampers. The lighting fixtures were integrated in the ceiling for design reasons and to avoid vandalism, see Figure 1. The embedding of lighting fixtures caused leaks in the ceiling in addition to the active supply panels.

MEASUREMENTS

Four scenarios of measurements were performed at airflow rates of 500 m^3/h and 1000 m^3/h and inlet temperatures of 10 and 17 ^0C as stated in Table 1. The airflows correspond approximately to indoor environment category 3 and 1 in the European Standard 15251.

During the measurements 21 out of 24 pupils were present and one radiator below the windows where on and had a temperature of 55 °C corresponding to a heat load of approximately 1100 W. The only other heat load in the room was the lighting system 9x2x18 W.

	Supply flow	Air change	Air flow per pupil	Supply air temperature	Measurements
Scenario 1	500 m^3/h	2.45 h^{-1}	5.8 l/s	10 °C	Velocity, temperature
Scenario 2	500 m^3/h	2.45 h^{-1}	5.8 l/s	17 °C	Vel., temp., air change efficiency
Scenario 3	1000 m^3/h	4.90 h^{-1}	11.6 l/s	10 °C	Velocity, temperature
Scenario 4	1000 m^3/h	4.90 h^{-1}	11.6 l/s	17 °C	Vel., temp., air change efficiency

Table 1. Investigated ventilation scenarios.

Local air velocity and air temperature

Mixing ventilation is achieved by inducing air at relatively high velocity above the occupant zone. The entrainment of the jet causes supply air to be mixed with room air. Efficient mixing is therefore an indicator of turbulent flow patterns and discomfort by draught. In diffuse ceiling ventilation the supply air is induced into the room with relatively low air velocities limiting the risk of draught caused by the ventilation air inlet. Mixing is achived by the convective plumes of the occupants present. However, in [6] CFD simulations indicated that the thermal plumes force the supply air to regions with no heat sources, leading to cold downdraft and risk of draught in those regions.

Therefore air velocity and air temperature measurements were carried out with two Brüel og kjær Indoor Climate Analyzer model 1213, to examine the risk of discomfort due to air movements in the room. The accuracy of the temperature sensor is ±0.2 ^0C and the velocity sensor has an accuracy of ±0.1 m/s. The measurements were performed in 8 points evenly distributed and placed close to tables in the occupied zone of the room, marked with triangles in Figure 3.

At each point the air velocity and temperature was measured at heights of 0.1 m and 1.1 m corresponding to ankle and neck level of a sitting person as specified in CR1752 [7] and each measurement was averaged over 3 minutes.

According to (CR1752) maximum air velocities of 0.12, 0.15 and 0.18 m/s are allowed for indoor environment category A, B and C, respectively, and category B is used as reference to evaluate the measurements. The air temperature must be between 20-24 °C for winter conditions in category B which is applicable for the conditions under which the measurements was performed.

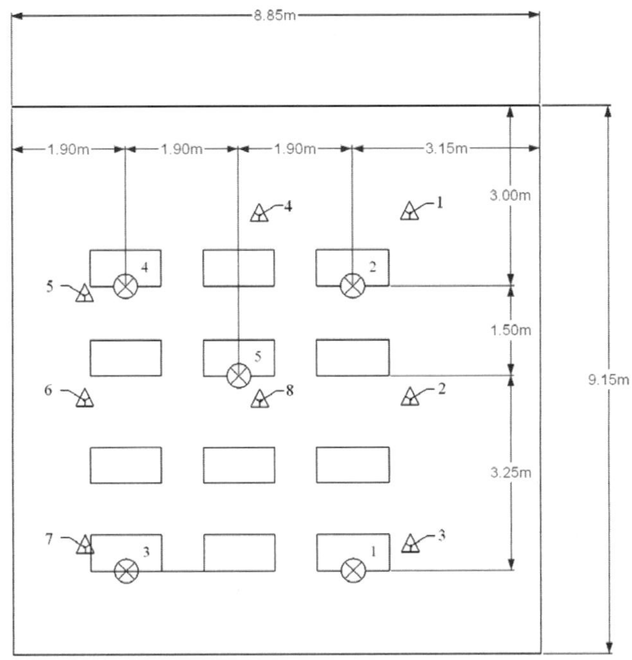

Figure 4. Location of measuring points in classrooms, triangles represent air temperature and –velocity measuring points and circles represent tracer gas measuring points.

Air change efficiency and local air change index

Air change efficiency ε^a is a measure of the average time it takes to replace the air in the room compared to the shortest possible air change time. It is the rati of the nominal time constant τ_n, and the actual air change time $\bar{\tau}_r$. The latter can be derived from the mean age of air concept [8].

$$\epsilon^a = \frac{\tau_n}{\bar{\tau}_r} \text{ x } 100\%$$

If the mixing of the supply air in a room is optimal the actual air change is twice the shortest possible time. Therefore full mixing has a maximum air change efficiency of 50 %. Lower efficiencies indicate incomplete mixing or short-circuited flow and higher efficiencies indicate flow regimes transitioning into displacement ventilation.

Local air change index represents the ratio of mean age in the exhaust and local mean age of air at the point of interest. At full mixing the age of air in the exhaust is the same as the age of air in the room and the ratio is equal 100 %.

Tracer gas

The concentration-decay method was used to determine the mean age of air. The equipment used was a photo-acoustic gas monitor model 1312 from Innova and a multipoint sampler and doser from Brüel & Kjær model 1303. The gas used was Freon R-134a that has a density of 4.25 kg/m^3, about 3.5 times higher than air. The multipoint sampler had 6 tube connections that were divided with five in the room and one in the extract duct. The five points chosen in the room were at each corner table in the occupied zone and at a table in the middle, all at a height of 1.1 m similar to the height of a sitting person. The points are marked with circles in Figure 4 and were chosen to give a representative picture of the mixing efficiency in the occupied zone and identify any stagnant zones. In the experiments the tracer gas was injected into the room until a constant concentration was reached and two swivel fans ensured fully mixing in the room. Depending on low and high airflow rate, a constant concentration of approximately 5 or 10 ppm was achieved, respectively, and the swivel fans were turned off and the following decay at the 6 sampling points was recorded. From this the local mean age of air in each point and the air change efficiency of the room were determined.

RESULTS AND DISCUSSION

Air velocities

The air velocity was measured at two heights, two flow rates and two supply temperatures. In Figure 5 and Figure 6 are shown the results for the supply temperature of 17 °C and 10 °C respectively. In the figures the mean air velocity over the sampling period of three minutes is shown as columns with an error bar displaying the standard deviation. The standard deviation shows that negative (physically invalid) air velocities has been recorded, especially for very low mean values. The standard deviation assumes Gaussian data distribution but this can be disturbed by outliers in the sampled data. These outliers are probably generated from turbulent flows caused by seated, yet active children during the measurements.

At both inlet temperatures the comfort threshold of 0.15 m/s is fulfilled except at point 3 and 7. Point 3 is located close to the leaky ceiling hatch and point 7 is located in the corner where, erroneously, two and not one active ventilation plate were installed.

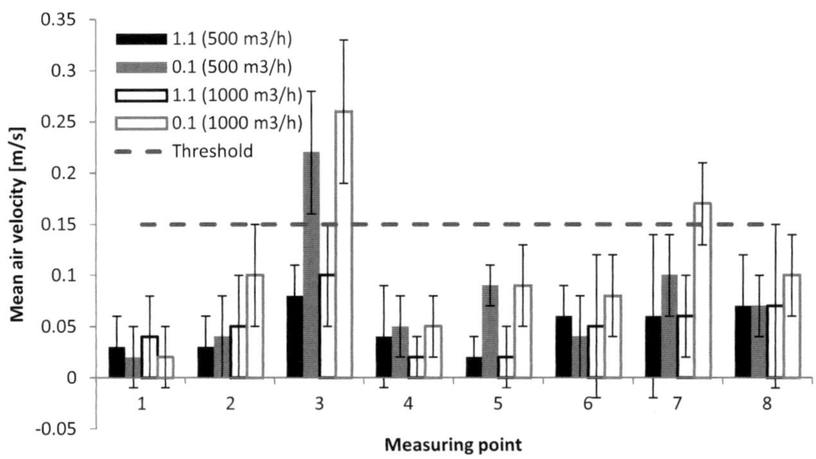

Figure 5. Measured air velocities at 0.1 and 1.1 m above the floor and flow rates of 500 and 1000 m^3/h with a supply temperature of 17 °C, scenario 2 and 4.

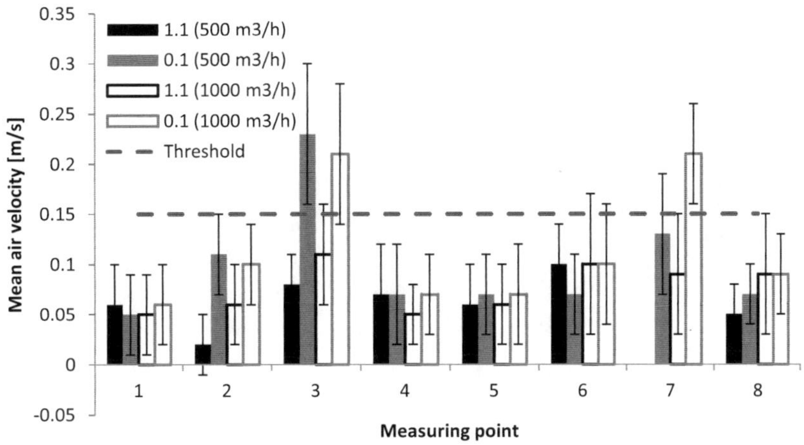

Figure 6. Measured air velocities at 0.1 and 1.1 m above the floor and flow rates of 500 and 1000 m^3/h with a supply temperature of 10 °C, scenario 1 and 3.

The measured air temperatures are shown in Figure 7 for an inlet temperature of 17 °C and Figure 8 for an inlet temperature of 10 °C. The results show that the air temperatures were quite uniform across the room, maximum 1 degree difference between the points when comparing the respective heights and flow rates measurements.

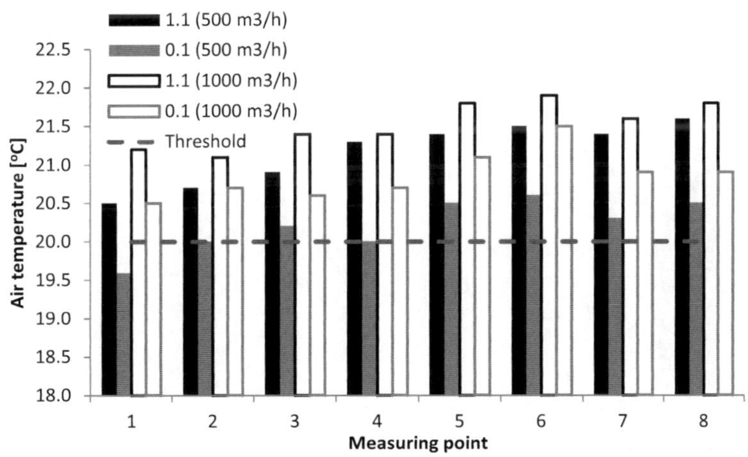

Figure 7. Measured air temperatures at 0.1 and 1.1 m above the floor and flow rates of 500 and 1000 m^3/h with a supply temperature of 17 degrees, scenario 2 and 4.

At the inlet temperature of 10 °C the air temperature measurements were below 20 °C at all points under the high flow rate and many points under the low flow rate. One radiator was malfunctioning and therefore the heating system was not able to compensate.

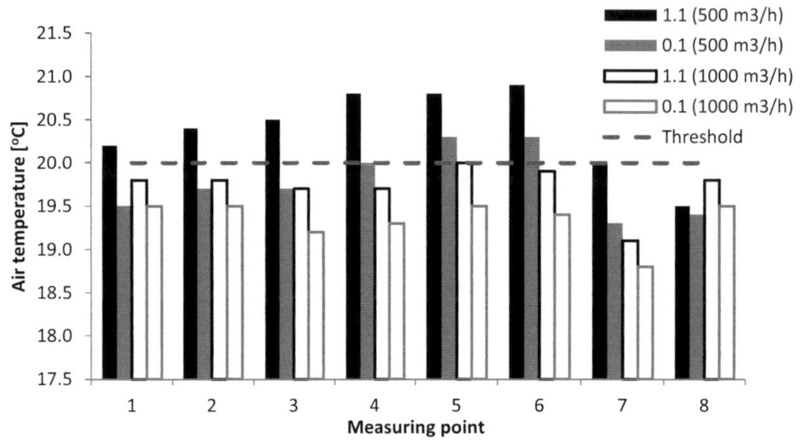

Figure 8. Measured air temperatures at 0.1 and 1.1 m above floor and flow rates of 500 and 1000 m^3/h with a supply temperature of 10 degrees, scenario 1 and 3.

Air change efficiency

In Figure 9 (left) is the local air change index of 5 sampling points in the classroom shown. The values are close to 100 % meaning similar air quality in all of the sampled points. The air change efficiency shown in Figure 9 (right) support the conclusion on room level, however, higher flow rates indicate transitional flow regimes into displacement ventilation which increases the risk of downdrafts.

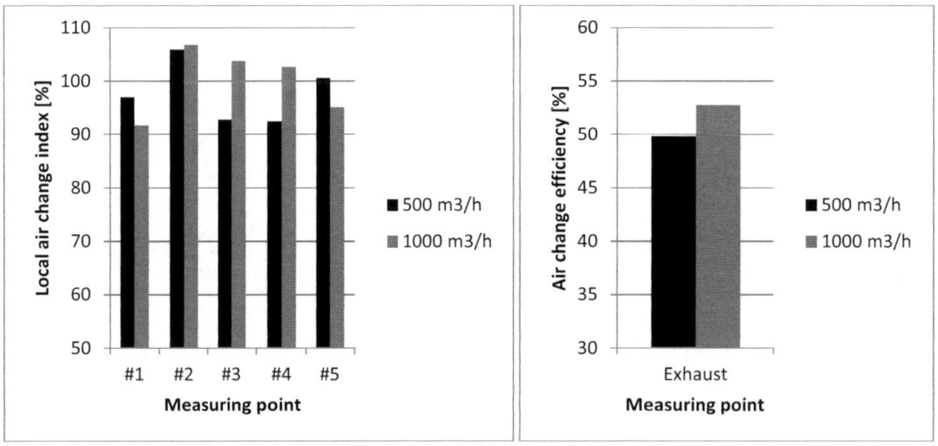

Figure 9. Left: Local air change index at sampling points. Right: Air change efficiency

Ceiling surface temperature

The infrared pictures in Figure 10 were taken with an inlet temperature of 10 °C to give a larger temperature difference and thereby clearer images.

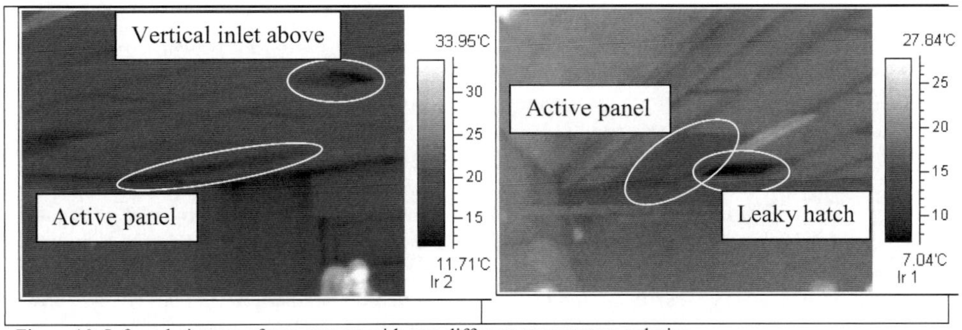

Figure 10. Infrared pictures of two corners with two different temperature coloring

The pictures clearly show that the active panels have a lower surface temperature than the passive panels. The temperature of the active panels is around 14 °C even though the supply air temperature is 10 °C. This shows preheating of the air when it passes the panels. This effect is due to radiation exchange of the ceiling with the rest of the room surfaces [6]. The panel joints are clearly visible due to the heat transmission of the metal suspension system. Most of ceiling has equal temperature with the room temperature and the radiant assymetry is deemed negligible.

CONCLUSION

The objective of the work in this paper was to validate the performance of diffuse ceiling ventilation in practise and document its applicability in school classrooms. Based on the investigation the following can be concluded on diffuse ceiling ventilation.

- The experimental results are in good agreement with previous results from test facilities and simulations.
- Virtually no risk of draught at low or high flow rates.
- Uniform airflow field with little difference between ankle and head height.
- Uniform temperature distribution with little difference between ankle and head height at both high and low flow rates.
- Negligible radiant asymmetry
- Perfect air mixing throughout the room independent of airflow rate.

Overall the results are positive and no negative aspects were detected. It is therefore concluded that diffuse ceiling ventilation in the studied form is applicable for use in school classrooms.

ACKNOWLEDGEMENTS

The trans-national R&D programme Eracobuild has financially supported this study as part of the SchoolVentCool project.

REFERENCES

[1] Jacobsen, L, Nielsen, P.V. and Morsing, S. 2004. *Prediction of indoor airflow patterns in livestock buildings ventilated through a diffuse ceiling*, in: Proceedings of the 9th International Conference on Air Distribution in Rooms – Roomvent, Coimbra, Portugal

[2] Chow, T.T. and Yang, X.Y. 2004. *Ventilation performance in operating theatres against airborne infection: Review of research activities and practical guidance*, J. Hospital Infection 56, 2, 85-92.

[3] Nielsen, P.V. and Jakubowska, E. 2008. *The performance of diffuse ceiling inlet and other room air distribution systems*, in: Proceedings of Cold Climate HVAC, Sisimiut, Greenland

[4] Hviid, C.A. and Svendsen, S. Accepted. *Experimental study of perforated suspended ceilings as diffuse ventilation air inlets*, Energy and Buildings.

[5] Jacobs, P., van Oeffelen, E.C., Knoll, B. 2008. *Diffuse ceiling ventilation, a new concept for healthy and productive classrooms*, in: Proceedings of Indoor Air, paperID#3, Copenhagen, Denmark.

[6] Hviid, C.A. 2010. *Building integrated passive ventilation systems*, Ph.D. thesis, Dept. of Civil Engineering, Tech. Univ. of Denmark, Lyngby, Denmark.

[7] CEN. 1998. CR 1752, *Ventilation for Buildings: Design Criteria for the Indoor Environment.* Brussels: European Committee for Standardization.

[8] Sandberg, M., Sjoberg, M. 1983. *Use of moments for assessing air quality in ventilated rooms*, Building and Environment 18 (4) 181-197.

UNCERTAINTIES IN AIRFLOW NETWORK MODELLING TO SUPPORT NATURAL VENTILATION EARLY STAGE DESIGN

Annamaria Belleri*[1, 2], Roberto Lollini[1]

1 EURAC research
Druso, 1
39100 Bolzano, Italy
**annamaria.belleri@eurac.edu*

2 Università degli Studi di Bergamo
Salvecchio, 19
24129 Bergamo, Italy

ABSTRACT

Despite a lot of Integrated Design Process guidelines and procedures have been developed in the last few years, more specific energy design procedures are needed to push the implementation of passive design techniques.
As natural ventilation design influences strongly the building shape and aspect, it has to be considered since the early design stages and its effect should be correctly predicted and proved by means of suitable tools and methods. In this respect, airflownetwork models seems a promising modelling techniques as they are already integrated in the existing dynamic building simulation tools and they have a quick solver. On the other hand, the simplicity of these models implies uncertainties in a lot of input parameters, first of all the wind conditions. The urban wind environment is a stochastic phenomenon, and as consequence the ventilation performance in a naturally ventilated building changes. An accurate wind analysis should be supported by weather data collected on site and by an external CFD simulation at urban scale. Discharge coefficients and external convection coefficients could be accurately estimated by laboratory tests. This would mean additional design costs (and time) and need of expertise in the field. A key issue of this work is to assess the thermal-airflow model reliability in airflow prediction when accurate estimation of input data is not feasible.
This paper presents an uncertatinty analysis performed on a dynamic simulation model of a new office building naturally ventilated during the night.
Full-factorial parametric analysis have been performed to assess the influence of the input parameters on the model reliability. Possible input ranges have been estimated and organized in a tree-structure to investigate the effect of dependent parameters like wind velocity profile, wind pressure coefficients, discharge coefficients, and external convection coefficients.
Significant variations in air change rates are shown that reflect an uncertainty of +/- 2% on total cooling loads in respect to the base case simulation result. No direct correlations between outdoor environmental conditions and air change rates have been found as the discharge coefficients affect significantly the results.
The design proposal supported by quantitative analysis and results prediction uncertainty assessment can be taken into consideration more seriously by the design team. The obtained results can be generally extended to airflownetworks with similar airflow paths.

KEYWORDS
Airflow network, uncertainty analysis

INTRODUCTION

Since 1993 a lot of Integrated Design Process guidelines and procedures have been developed [1]. They are mainly focused on team work methods, design evaluation and design strategy and, to the knowledge of the authors, no pragmatic energy design procedures have been yet developed to push the implementation of passive design techniques (among which natural ventilation strategies) and to model their effectiveness.
As natural ventilation design influences strongly the building shape and architectural impact, it has to be considered since the early design stages and its effect should be correctly

predicted and proved by means of suitable tools and methods. To face this issue, airflownetwork models seems a promising modelling techniques as they are already integrated in the existing dynamic building simulation tools and they have a quick solver.
However, the model implies assumptions on pressure distribution on the building and its interaction with the internal airflows, that cannot be accurately assessed unless additional design costs (and time), that need expertise and possibly laboratory facilities, are invested. A key issue of this work is to assess the thermal-airflow model reliability in airflow prediction when accurate estimation of input data is not feasible.
The paper presents an uncertatinty analysis performed on a model of a new office building naturally ventilated during the night, simulated with a dynamic Energyplus. The obtained results can be generalized for further airflownetwork models.

BUILDING MODEL

Building description

The case study of the present work is one of the new office buildings placed in the new Technology Park in Bolzano (Italy). The building is architecturally conceived as a black monolithic block with a nearly-square plant. It has five above ground floors and an underground floor. The main entrance is on the north side of the ground floor and on the south side there is the expo area. The upper floors will host offices, meeting and service rooms, whereas in the underground floor there will be several conference rooms. In the centre of the building and through the full height, a green patio is designed as a buffer zone to improve indoor comfort and daylighting.
The envelope is a metal-glass façade with a solar shading system in the south façade and a black surface with different strips of horizontal windows in the other facades. The horizontal windows on north, east and west façade are positioned on the inner side of the external wall. In this way, the deep reveal due to the wall thickness and the low height of the windows work as a sun shading system and the glazed part of the façade will not be visible from the outside perspective.

Natural ventilation strategies

A night stack driven cross ventilation was chosen as the most effective configuration that balances performances needs with constrains given by fire compartments, acoustic comfort and privacy needs in the offices during the working hours. [2]
To increase the height difference between inlet and outlet openings, connecting floor vents will be applied. This solution fulfils the architect's requests by reducing the movable part in the façade and by keeping free the internal layout of the spaces. Inlet and outlet openings are automatically controlled top hung windows. The floor vents are closed during the working hours to avoid acoustic discomfort and maintain privacy between offices.
The foyer is directly connected to a lightwell and to the hall of every floor and is ventilated through a stack driven cross ventilation to avoid overheating situations.
Due to safety reasons underground floor and expo areas are mechanically ventilated. A small office area in the center part of the building is single-sided ventilated and connected with the green patio.

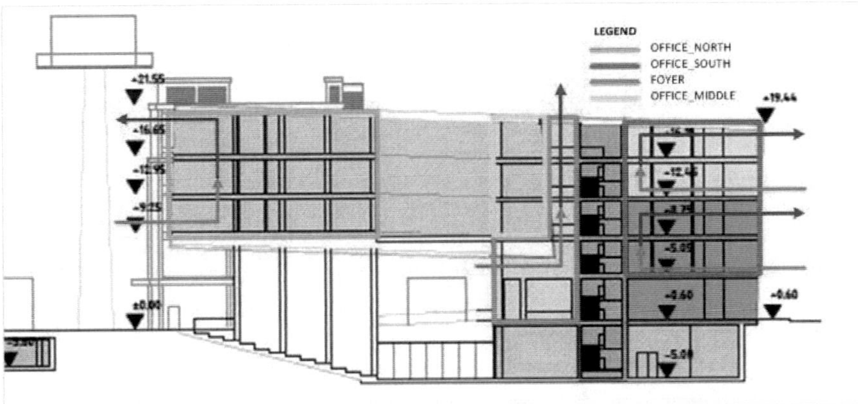

Figure 1. Cross section of the building fire regulation plan with fire compartments, model zones and a scheme of the selected stack-driven cross ventilation configurations for the considered zones.

Energyplus modeling to study airflows

The original Energyplus building model zoning has been re-thought to introduce an airflownetwork. Particular care has been taken to reduce computational time as the uncertainty analysis requires to perform parametric analysis running several simulation.

The building has been divided into thermal zones with the same temperature and pressures, occupation patterns, internal gains, major exposure, cooling setpoints. Thermal zones have been further on divided depending on the planned airflow paths and linkages. Building airflow zoning can be more detailed than building thermal zoning as Energyplus airflownetwork allows only one temperature node per thermal zone.

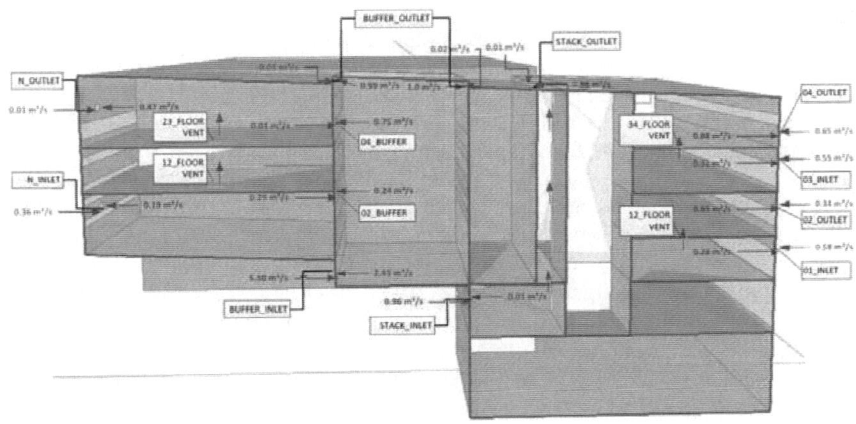

Figure 2. Section of the Energyplus geometry model in sketchup with average airflowrates during summer season. Red and blue arrows represent respectively positive and negative airflow directions.

According to fire compartments and natural ventilation strategies, three airflownetworks are set: a first network involves the office block on the south side of the building, the second network involves the atria zones and the stack and a third network involves the office block on the north side of the building.

It is important to detail the window area and the height from the ground level (Table 1). Horizontal window series and floor vents series have been modeled using multiplier as they have the same size and the same height from the ground level. In this way window shape effects are taken into account properly and less input objects are necessary to set window controls.

Zone group	Airflownetwork component name	Opening type	Height [m]	Area [m²]	Nr of windows	Tot. opening area [m²]
OFFICE_SOUTH	01_INLET	Top hung window	7.5	0.65	16	10.40
	02_OUTLET	Top hung window	10.9	0.65	18	11.70
	03_INLET	Top hung window	14.6	0.54	26	14.04
	04_OUTLET	Top hung window	16.8	0.80	26	20.80
	12_FLOOR VENT	Horizontal opening	8.2	1.60	9	14.40
	34_FLOOR VENT	Horizontal opening	15.6	1.60	10	16.00
OFFICE_NORTH	N_INLET	Top hung window	10.6	0.54	10	5.37
	N_OUTLET	Top hung window	19.0	0.74	10	7.39
	04_BUFFER	Top hung window	17.6	0.71	4	2.85
	02_BUFFER	Top hung window	11.7	0.71	4	2.85
	N23_FLOOR VENT	Horizontal opening	12.4	0.60	9	5.40
	N34_FLOOR VENT	Horizontal opening	16.1	0.60	9	5.40
BUFFER ZONE	BUFFER_INLET	Top hung window	21.5			8.00
	BUFFER_OUTLET	Top hung window	21.5			8.00
STACK	STACK_INLET	Top hung window	3.4	-	-	8.70
	STACK_OUTLET	Top hung window	22.0	-	-	8.70

Table 1. Airflownetwork surface components area and height from the ground.

An opening factor of 0.5 has been set for top hung window assuming a maximum opening angle of 45°.

This may cause inaccurate shadowing and daylight distribution inside the building, but the computation time decreases. However, daylight study is not one of the purposes of this work and opening area is only a small percentage of the whole glass area. For the same reason a full exterior solar distribution with no reflections has been set. Reflections would have required also no-convex zones, that would have meant setting more zones. [3]

Energy Managment System objects are used to introduce simple controls on windows and vents opening [4]. Indoor and outdoor temperature variables have been set as sensors and venting opening factors as actuators. A program was written to activate natural night ventilation between 8 pm and 8 am if the following conditions are fulfilled :

- indoor temperature is higher than 24°C;
- indoor temperature is higher than outdoor dry bulb temperature;
- outdoor dry bulb temperature is higher than 10°C.

External nodes at every floor height and with different orientations have been set to take into account more accurately wind pressure conditions and model stack effects.

Floor vent interzone surfaces references an horizontal opening component with a 90° sloping pane angle. Horizontally pivoted detailed openings are used to model inlet and outlet top hung windows.

An ideal loads air system template, with cooling setpoint temperature of 26°C during working hours, has been implemented to evaluate cooling loads without taking into account the plant system. Infiltration rates have been neglected, as the building tightness required by the local standard is restrictive and will be proven by blowerdoor test.

Simulations have been run from June to September. Average airflownetwork volume flow rates are shown in Figure 2. Simulation results of the base case model have showed that the airflow direction do not always follow the positive direction of the airflow path: it works in the 86% of activation hours on the upper floors and in the 46% of activation hours in the lower floors. This could be due to the lower opening area in the 1st and 2nd floor. The graph in Figure 3 show the airflow frequencies through the inlet-outlet components in the south office block airflownetwork. It could be noticed that low airflow rates are more frequent in the negative airflow direction.

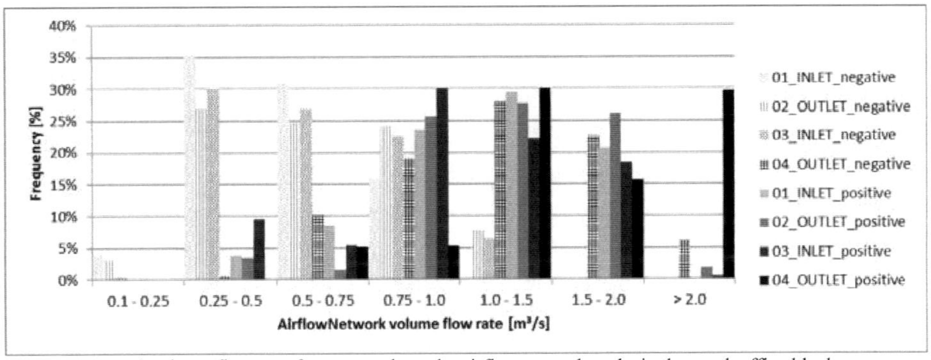

Figure 3. Natural volume flow rate frequency along the airflownetwork paths in the south office block.

MAIN UNCERTAINTY SOURCES ON AIRFLOW PREDICTION

Apart uncertainties derived from model assumptions and approximations, the main uncertainty sources in airflow prediction can be ascribed to the input data:

- Leakage's geometry, position and flow characteristics (discharge coefficient)
- External and internal temperatures
- Wind pressures

Opening position determines buoyancy pressures and external wind pressures. Discharge coefficients are considered in the airflow network models as a fixed property of an opening, which depends on its shape and Reynolds number. However, when an opening is installed in an envelope, the actual discharge coefficient may differ from the fixed one. Etheridge D. (2012) [3] provides some boundaries that can be placed on the uncertainties.

Whereas external temperatures are supposed to have low uncertainty, internal temperatures are affected by uncertainties due to the heat transfer - airflow model coupling. Hensen J.L.M. et al. (1995) [6] stated that coupling problems can be overridden by setting a proper time step. Therefore temperature uncertainties are affected mainly by convection coefficients. The external convection coefficients depend on inside-outside temperature difference, wind speed and direction and surface roughness. The internal convection coefficients depend on inside-outside temperature difference, zone airflow regime, surface orientation and heat flow direction. Furthermore vertical temperature profiles (stratification) cannot be reproduced by a multizone airflow model, as the temperature distribution is considered uniform in every zone.

The stochastic features of the urban wind environment affect the naturally ventilated building performances. An accurate wind analysis should be supported by weather data collected on site and by an external CFD simulation at urban scale, to assess wind pressure coefficients on the building envelope.

Parametric studies have been performed on Energyplus Airflow Network model by means of jEPlus program [7] to analyse the sensitivity of the model to different combination of dependent and independent parameters. Possible input ranges have been estimated and

organized in a tree-structure (Figure 4) to investigate the effect of these input parameters on airflow rate prediction.
The presented parametric analysis aims at assessing the model uncertainty by:

1) assigning a range and a discrete distribution for each input parameter;
2) executing the model in full-factorial design mode[1];
3) assessing the airflow rates prediction variation rates and its effect on cooling need calculation.

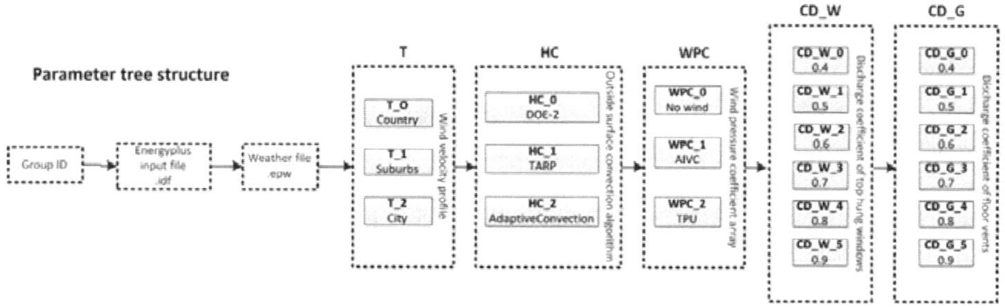

Figure 4. Parameter tree structure set in jEPlus. Each simulation job is a path from the root node of the tree to a leaf of the tree, with each node containing an optional value of the corresponding parameter. As a result, the total number of jobs encoded in the tree corrisponds to the total number of paths from the root to the leaves.

Wind velocity profiles

EnergyPlus converts wind velocity weather data through numerical method that considers the differences between the weather station location and the building site, according to:

$$U_{\infty} = V_{met}\left(\frac{\delta_{met}}{z_{met}}\right)^{a_{met}}\left(\frac{z}{\delta}\right)^{a} \qquad (2)$$

where z is the height from the ground, z_{met} is the height of the standard metereological wind speed measurement, and a and δ are terrain-dependent coefficients that determine the wind velocity profile.
Terrain type field in the building object associates different values of wind speed profile exponent and height. Three different wind velocity profiles (Table 2) have been taken into account in the parametric analysis.

Terrain type	Exponent, α	Boundary layer thickness, δ (m)
Country	0.14	270
Suburbs	0.22	370
City	0.33	460

Table 2. Wind speed profile coefficients. Source: energyplus engineering reference

Outside surface convection algorithm

Energyplus users can select which model equations or values to apply for the exterior convection coefficients calculation. There are five models based on ASHRAE correlations and different flat plate measurements: SimpleCombined, TARP, MoWitt, DOE-2 and AdaptiveConvectionAlgorithm.
Simple combined model returns higher values as also radiation to sky, ground and air is included in the exterior convection coefficient, whereas all other algorithms yield a purely

[1] A full factorial experiment is an experiment whose design consists of two or more factors, each with discrete possible values, and whose experimental units take on all possible combinations of these levels across all such factors.

convective heat transfer coefficient. MoWitt model can be applied only to very smooth vertical surfaces. DOE-2 model is a combination of MoWitt and BLAST. TARP model is very similar to BLAST and detailed convection models. The adaptive convection algorithm assigns default equations to surfaces depending on their outside face classification, heat flow direction and wind direction.

The graph in Figure 5 compares the exterior convection coefficients for a vertical surface calculated by TARP and DOE-2 model depending on wind speed, façade position respect to wind direction and different surface roughness. A fix temperature range has been assumed. DOE-2 model is more affected by wind speed than the TARP one.

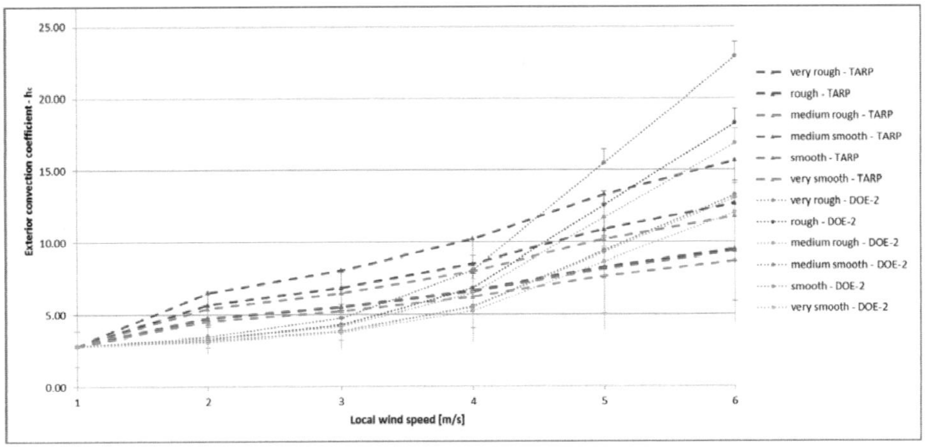

Figure 5. Vertical wall exterior convection model comparison depending on wind speed and surface roughness at a fixed temperature difference.

Wind pressure coefficients

Wind pressure coefficients are used to calculate wind-induced pressures on each façade surface during simulations and are defined as the ratio of static pressure to dynamic pressure at a given point on the façade:

$$c_p = \frac{P_x - P_0}{P_d} \tag{1}$$

where P_x is the static pressure at a given point on the building façade (Pa), P_0 is the static reference pressure (Pa), P_d is the dynamic pressure (Pa).

They can be evaluated in detail according to different approaches:

- full-scale measurements
- reduced-scale experiments in wind tunnels
- computational fluid dynamics (CFD) simulations

However these approaches are not suitable to early stage design simulations as they are time-consuming and requires additional costs. Therefore, analytical models and database are most often used in building energy simulation and airflownetwork tools [8].

For the present case study, surface-averaged c_p have been avoided as more openings per façade are present and the magnitude of their uncertainty is high [9].

C_p coefficient used in the parametric analysis are extracted from two main existing database.

The Air Ventilation and Infiltration centre (AIVC) database [10] is based on the interpolation of data from published material. Wind pressure coefficient datasets are available for low rise buildings with different exposion conditions and with different length to width ratio. The coefficient sets used for the analysis are extracted from the low rise building database with length to width ratio of 1:1 and surrounded by obstructions equal to the height of the building.

The Tokyo Polytechnic University (TPU) aerodynamic database [11] is a web based database on wind tunnel experiments on 12 test cases including low and high rise buildings with varied eaves. The coefficient sets used for the analysis are extracted from the low rise building database with flat roof, height to width ratio of 1:4 and length to width ratio of 2:2.
A wind pressure coefficient array of null values has been introduced in the parametric analysis to evaluate the influence of wind pressures on output results.

Discharge coefficients

The discharge coefficient is required as input in airflow models and is defined by Equation 1.

$$c_d = \frac{q}{A}\sqrt{\frac{\rho}{2\Delta p}} \qquad (2)$$

where q is the volume flow rate across the opening (m^3/s), A is a defined open area (m^2), ρ is the air density (kg/m^3) and Δp is the airflow difference across the opening (Pa).
Airflow predictions are linearly dependent on discharge coefficients and therefore they imply directly proportional results uncertainty.
The Equation 1 is applied to openings installed in a surface separating two much larger spaces in still-air conditions, with uniform and equal densities. In practice flows through openings are generated by wind and buoyancy forces. The wind modifies the external flow field, whereas the buoyancy forces cause different air densities. As Etheridge D. (2012) stated [4], the installation effects are negligible for air vents and small windows in case of low velocity ratio and for large stacks in case of inward flow only. In particular, this effect depends primarily on the magnitude of V/u_m (typical values range from 1.5 to 9), where V is the cross flow velocity and u_m is the spatial mean velocity through the opening.
As much as opening angles is smaller the uncertainty is higher, because leakages along all other sides of the opening account for a relative large part of the opening area.
It is to underline that these effects are dependent on wind velocity and direction. Therefore the parametric analysis has to be applied simultaneusly to discharge coefficients, wind velocity profile and wind pressure coefficient sets on the façade.
Johnson et al. (2012) compared airflow network predictions with measured airflow values and found that the discrepancy in the predicted value is due to inaccuracy in the discharge coefficient, which was an estimated value [12].
In the case of buoyancy driven cross ventilation, some measurements were performed on scale model tests [13] but no full-scale model have been yet analysed, especially in the case of horizontal openings.
Two different discrete variables have been set for the floor vents discharge coefficient and for the top hung windows discharge coefficients in the parametric analysis.
Stated Heiselberg's experiments [14], it has been assumed that discharge coefficents have a value bigger than 0.4.

RESULTS

The simulation model has been runned in full-factorial mode for the airflownetwork involving the office block on the south side of the building.
Analyzing the resulting air change rates at fixed outdoor wind and temperature conditions (Figure 6), significant variations with standard deviations up to 0.6 occur.
The lower values in the graph are refered to samples with null wind pressure coefficients. The upper values in the graph are refered to samples with wind pressure coefficients from AIVC database. The smaller variations are due to different discharge coefficient combinations. In presence of wind, discharge coefficient effect is higher than in absence of wind.

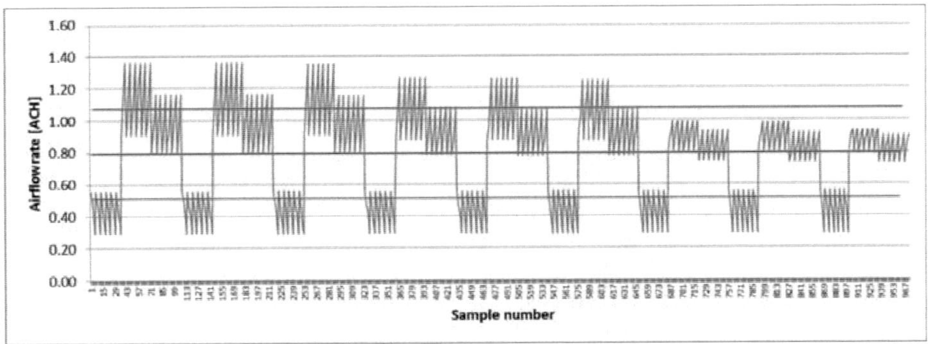

Figure 6. Resulting airflowrates at 22/06 h 21:00 from parametric analysis with 967 samples. Row lines represents average ACH +/- standard deviation.

Figure 7 show the airflow rates calculate for the base case with error bars that represent the variations calculated through the parametric analysis.

Total cooling loads of the office block in the south part of the building have been calculated with the result that the variations in air change rates reflect an uncertainty of +/- 2% on total cooling loads.

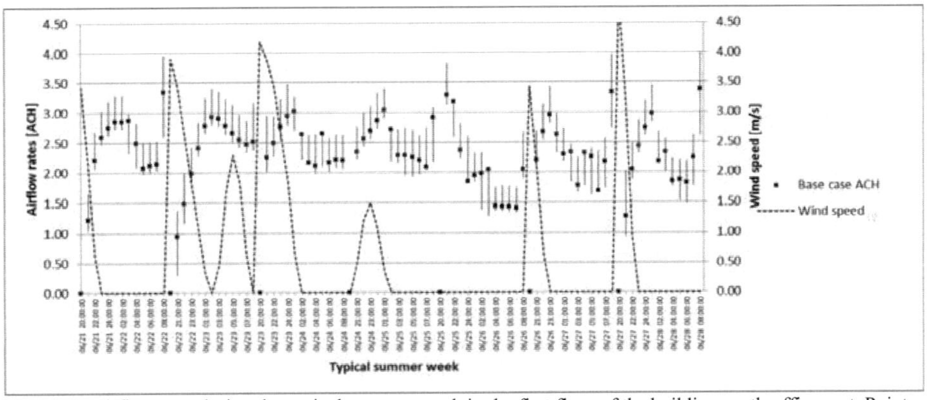

Figure 7. Airflowrates during the typical summer week in the first floor of the building south office part. Points represents the base case results and bars represent the airflow rate possible uncertainties.

Scatter plots in Figure 8 have been performed to find correlation between external environmental conditions and airflowrates. No direct correlation has been found, but it can be noticed that in absence of wind or in case of low outdoor temperatures, the range of standard deviations is larger. This means that, discharge coefficients may affect significantly the results.

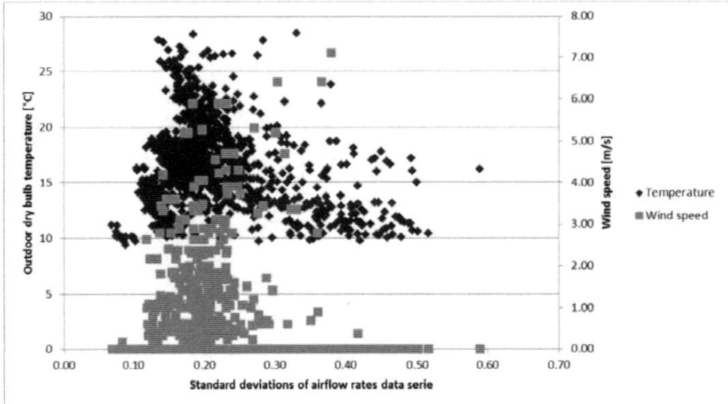

Figure 8. Scatter plot of standard deviations of airflow rate data serie related to wind speed and outdoor dry bulb temperature.

CONCLUSION

These results would be of particular interest for the natural ventilation design support during early design phases and can be generally extended to airflow networks with similar airflow paths.

Significant variations in air change rates are shown that reflect an uncertainty of +/- 2% on total cooling loads.

No direct correlations between outdoor environmental conditions and air change rates have been found as the discharge coefficients affect significantly the results. However, in absence of wind or at low temperatures the standard deviations have a wider range of variability.

Airflow network simulation results are also useful to control the efficacy of the natural ventilation strategy proposed. Simulation results of the base case model have showed that the airflow direction follow the positive direction of the airflow path in the 86% of activation hours on the upper floors and in the 46% of activation hours in the lower floors. Inlet and outlet opening area at 1st and 2nd floor should be increased.

Further developments of the process are planned to optimize opening area and opening controls to prevent natural ventilation strategy disfunctions.

Thanks to this quantitative analysis support, the mentioned natural ventilation strategies can be evaluated in a more rigorous way by the design team, for the building and architectural choices in the early design phases.

REFERENCES

[1] VV. AA. 2010. *Mapping of previous integrated energy approaches*, report EIE-06-021-INTEND.

[2] Belleri, A., Castagna, M., Lollini, R. 2012. *Natural night ventilation as passive design strategy for a Net ZEB office building*, Improving Energy Efficiency in Commercial Buildings, 18th-19th April, Frankfurt.

[3] VV. AA. 2012. *Input – Output reference.* Energyplus version 7.1 Documentation.

[4] VV. AA. 2012. *Energy Managment System – Application Guide.* Energyplus version 7.1 Documentation.

[5] Etheridge, D. 2012. *Natural ventilation of buildings: theory, measurements and design.* John Wiley & Sons, Ltd.

[6] Hensen, J.L.M. 1995. *Modelling coupled heat and airflow: Ping-pong vs onions*,16th AIVC Conference "Implementing the Results of Ventilation Research" p. 253-262, Palm Springs.

[7] Zhang, Y. 2011. *jEPlus – An EnergyPlus batch shell for parametric studies.* http://www.iesd.dmu.ac.uk/~yzhang/jeplus/docs_html/Users%20Manual%20ver1.1.html.
[8] Cóstola, D., Blocken, B., Hensen, J.L.M. 2009. *Overview of wind pressure coefficient data in building energy simulation programs*, Building & Environment 44, 2027-2036
[9] Cóstola, D., Blocken, B., Ohba, M., Hensen, J.L.M. 2010. *Uncertainty in airflow rate calculations due to the use of surface-averaged coefficients*, Energy & Buildings 42, 881-888
[10] Liddament, M.W. 1986. *Air infiltration calculation techniques*, AIVC.
[11] Tokyo Polytechnic University (TPU) aerodynamic database http://wind.arch.t-kougei.ac.jp/system/eng/contents/code/tpu
[12] Johnson, M.H., Thai, Z., Krarti, M. 2012. *Performance evaluation of network airflow models for natural ventilation.* HVAC & R research.
[13] Chiu Y- H., E. D. 2007. *External flow effects on the discharge coefficients of two types of ventilation opening.* Journal of Wind Engineering and Industrial Aerodynamics, 225-252.
[14] Heiselberg, P., Sandberg, M. 2006. *Evaluation of Discharge Coefficients for Window Openings in Wind Driven Natural Ventilation*, International Journal of Ventilation ISSN 1473-3315 Volume 5 No 1

IMPROVEMENT OF SUMMER COMFORT BY PASSIVE COOLING WITH INCREASED VENTILATION AND NIGHT COOLING

Tommaso Pellegrini*[1,3], Peter Foldbjerg[2], and Bjarne W. Olesen[3]

1 Department of Industrial Engineer, University of Padova,1 Venezia Street,35100 Padova, Italy
**Corresponding author: tommaso.pellegrini@gmail.com*

2 VELUX A/S, Daylight, Energy and Indoor Climate, 99 Ådalsvej Street, 2970 Hørsholm, Denmark

3 International Centre for Indoor Environment and Energy, Technical University of Denmark,b420 Nils Koppels Alle Street, 2800 Lyngby, Denmark

ABSTRACT
The present study describes the potential improvement of summer comfort and reduction of energy consumption that can be achieved by adopting passive cooling solutions, such as daytime comfort ventilation with increased air velocities and night cooling, in domestic buildings. By means of the IDA ICE based software EIC Visualizer, the performances of ten ventilation and cooling strategies have been tested in four different climatic zones across Europe (Athens, Rome, Berlin and Copenhagen). Thermal comfort and indoor air quality (IAQ) have been evaluated according to the standard EN15251 for the summer period of the year only. The study revealed that thermal comfort can be achieved by passive means in all four locations. It was also found that, with the exception of Athens, the initially investigated combination of ventilative and night cooling is too aggressive, causing overcooling and increasing the energy consumption. A moderate strategy performed well without overheating and overcooling in Rome, Berlin and Copenhagen. In general the natural ventilation turned out to be capable to achieve a very good IAQ and a reduction in energy consumption in all locations, when compared with mechanical ventilation or mechanical cooling.

KEYWORDS
Natural ventilation, ventilative cooling, night cooling, increased indoor air velocity, residential buildings.

INTRODUCTION
The 2007 report of the Intergovernmental Panel on Climate Change (IPCC) [1] stated that the warming of the climate global system is an ascertained problem. For this reason in 2008 the EU set the target of reducing by 20% the total energy consumption within 2020 [2]. According to the Promotion of the European Passive House (PEP) [3], buildings account for 40% of this total energy consumption, and through the application of the Passive House concept, a relevant reduction of the energy consumption, quantifiable in a CO_2 emission reduction between 50% and 65%, can be obtained. The aim is to lower the buildings' energy demand without affecting the thermal comfort or the IAQ. The comfort condition is function of different parameters and thermal comfort can be provided within a range of air temperatures. When the air temperature increases, the warm thermal sensation can be restored from warm to neutral by decreasing the mean radiant temperature or by increasing the convective heat exchange between the body and the surrounding ambient [4]. The reduction of the radiant temperature is achieved by mean of the *night ventilation*: cold air is circulated through the building during night, the building structure is then cooled, providing a thermal sink and a lower radiant temperature during the next day. The increase in the convective heat exchange is the basic idea of the *ventilative cooling*: thermal comfort is obtained by increasing the air velocity in the room through natural or mechanical ventilation.

The effectiveness of increased air velocity and night cooling in reducing the energy consumption has been proven by means of both field surveys and dynamic simulations. In particular E. Gratia et al. [5] showed it is possible to reduce the cooling needs by about 30% using the ventilative cooling strategy. S. Schiavon and A. K. Melikov [6] demonstrated that increased air velocities can improve the comfort and allows a cooling energy saving between 17% and 48% and a reduction of the maximum cooling power between 10% and 28%. Furthermore Yun et al. [7] stated that opening the windows at the ambient temperature higher than the indoor temperature does not help to cool down the office, but can still improve the thermal comfort providing direct cooling over the occupants.
According to many authors, night ventilation appears to be one of the most promising passive cooling techniques. The work of Böllinger and Roth [8] revealed that in Frankfurt a nighttime air flow rate of 3 ach can compensate for a specific load of about $35W/m^2$, while for a 6 ach the compensated specific load rises up to $41W/m^2$. Similarly Santamouris et al. [9] showed that night ventilation applied to residential buildings may decrease the cooling load up to 40 kWh/m^2. Santamouris [10] also found that, under free floating condition, the night ventilation decreases the next day peak indoor temperature up to 2.5°C and reduces the expected number of overheating hours between 64% and 84%. According to Shaviv et al. [11] depending on thermal mass, air flow rate and temperature swing, the night cooling can achieve a 3 – 6°C temperature reduction in the hot and humid climate of Israel.
The purpose of this project is to determine, by means of dynamic simulations with the EIC Visualizer software, under which climatic condition the passive cooling techniques are capable to reduce the energy need without compromising the occupants' thermal comfort and the IAQ, during summer. The tested scenarios include, beside a fully mechanical system regarded as a reference case, both natural and hybrid ventilation and cooling solutions.

THE METHODOLOGY

The building

A 1½-storey, single-family house with a 8x12 m footprint, corresponding to a 175 m^2 floor area, has been selected for the investigation (Figure 1). The dwelling has no internal partitions and has then been studied as a single-zone building. The building tightness allows an infiltration rate of 0.15 ach for an outdoor-indoor pressure difference of 50 Pa. The internal surface of roof, walls and floor is an exposed concrete layer whose thickness has been obtained by empirically optimizing the thermal mass.

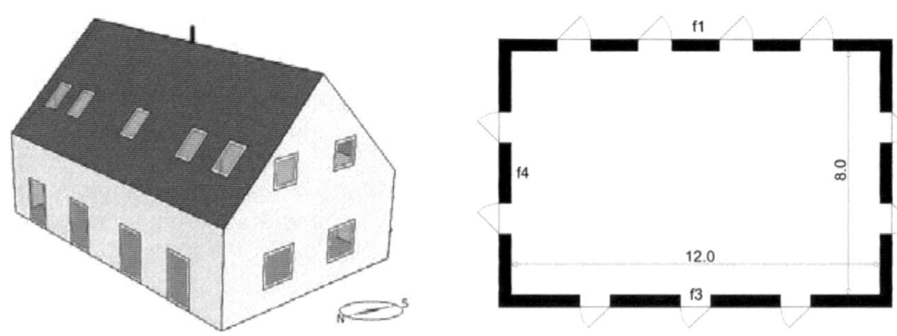

Figure 1. Visual representation (left) and footprint (right) of the building selected for the investigation.

To achieve a potential for sufficient natural ventilation and good daylight conditions requires a large windows area; the building has then a 30.4 m^2 glazed surface (corresponding to 17%

of the internal floor area): 23.1 m^2 are façade windows and 7.3 m^2 are roof windows. All windows are operable and equipped with an external sunscreen.
The thermal properties of opaque and glazed surfaces are summarized in Table 1.

	U-value [$W/(m^2K)$]	Solar heat gain coefficient (g)	Solar transmittance (T)	Solar transmittance for the visible spectrum (T_{vis})	Internal/external emissivity
Roof	0.23	-	-	-	-
Wall	0.34	-	-	-	-
Floor	0.32	-	-	-	-
Glass	1.47 (0.90)	0.6 (0.1)	0.54 (0.05)	0.77	0.84

Table 1. Thermal properties of opaque and glazed furface (in brackests the multiplier due to the sunscreen).

In the AHU a supply fan introduce fresh air from outside the building through a grid at the ground level and an extraction fan extracts the exhausted air through a grid located 2.5 m from the ground. Both the supply and the extraction fans give a pressure rise of 200 Pa with a 0.8 electricity-to-air efficiency and a 0.25 $kW/(m^3/s)$ specific power. A 5 Pa pressure loss has been assumed for the grid. No heat recovery has been used: the air is then supplied at the outdoor thermodynamic condition.
The heating system is an ideal heater with a 17500 W maximum power and a 0.9 generation efficiency, the cooling system is an ideal cooler with a 35000 W maximum power and a COP of 2.4. The distribution losses have been assumed to be equal to 1% of the heat delivered by the plant for the heating system and to 0.10 W /m^2 of the internal floor area for the cooling system.

Standard EN15251
Thermal comfort, IAQ and energy consumption are the three parameters used to quantify the performance of the tested ventilation and cooling strategies. The thermal comfort and the IAQ (the CO_2 level has been chosen as indicator of the IAQ) have been evaluated according to the standard EN15251 [12]. For the thermal comfort the standard prescribes two models. In buildings equipped with a mechanical cooling system the comfort categories are defined according to the non-adaptive model of the ISO7730. In naturally ventilated buildings the standard prescribes an adaptive model comparable to the one developed by de Dear and Brager [13]. In addition the EN15251 adopts the temperature offset that can be obtained by means of increased air velocities under summer comfort condition proposed in the ISO7730. Since only a graphical representation of the temperature offset is reported in the standard, an approximation had to be made. Four easy-to-identify points ((0.2;0), (0.3;1), (0.9;2.75) and (1.2;3.3)) have been isolated and interpolated with a logarithmic trend line. The equation of the trend line (1) has then been used to calculate the temperature offset.

$$\Delta T = 1.777 \cdot \ln(v) + 2.9782 \quad (1)$$

Where ΔT is the temperature offset and v is the indoor air velocity.
Thermal comfort and IAQ have been evaluated only during the *natural ventilation period*, which is defined as the period of the year that starts the day during which natural ventilation is used for the first time (i.e. the conditions for the window opening are met for the first time since the beginning of the year), and ends the last day of application of natural ventilation (i.e. the conditions for the window opening will never be met again for the rest of the year).

Assumptions and system setup

The simulations have been run with the software Energy and Indoor Climate Visualizer (EIC Visualizer), which is based on the commercial software IDA Indoor Climate and Energy 4 (IDA ICE).

The dwelling is occupied by four people who leave the house at 8:00 in the morning and return at 17:00 in the afternoon every weekday. During the weekend they spent 24 h/day inside the building. The occupants' clothing and activity levels have been set equal to 0.5 ± 0.2 clo (the clothing level is automatically adjusted between limits to obtain the best comfort) and 1.2 met respectively. Other contributions to the internal load are the constant 4 W/m^2 due to the equipment and the load from electrical lighting with an installed capacity of 4 W/m^2. The lights are switched on only when the average daylight level is below 50 lux and with a percentage of lighting on simultaneously equal to 75% (2.4 W/m^2). The sunscreens used to reduce the solar gain are operated by a PI controller and are automatically activated when the indoor air temperature rises above 23°C.

The windows controller is intended to simulate human behaviour [14]. During daytime (from 7:00 to 22:00) the window opening is modulated to maintain an air temperature set point by cooling when the outdoor temperature is lower than the indoor by 2°C. During night the windows are opened between 22:00 and 7:00, if at 22:00 the indoor air temperature is higher than the outdoor and it is above the given threshold.

When window opening is not required with regards to thermal comfort, the mechanical system supplies a constant air flow rate of 0.29 $l/s/m^2$ calculated, according to EN15251, as sum of the amount of fresh air needed to compensate for the pollution from the occupants (4 l/s/person for category III) and the amount of fresh air necessary to remove the building emission of pollutants (0.2 $l/s/m^2$ for a very low polluting building).

The set points for the heating and for the cooling systems are 21.0°C and 24.6°C respectively. An approximated method has then been used to calculate the mean indoor air velocity. The air flow rate through the windows has been divided into two contributions according to the direction, axial or transversal, with respect to the building footprint. For each direction two values of air velocity have been calculated: one on the windows threshold and one on the building cross section. Averaging the threshold air velocity and the cross-section air velocity the two components, namely axial indoor air velocity and transversal indoor air velocity, of the indoor air velocity have been obtained. Finally, averaging[1] those two components, an approximation of the indoor air velocity value has been obtained. The procedure just described has been adopted to determine the air velocity, and from it the temperature offset, hour by hour.

The temperature offset has then been subtracted from the operative temperature to calculate the *perceived operative temperature* (2), a parameter whose purpose is to express the temperature actually experienced by the body.

$$T_{op,\,perceived} = T_{op} - \Delta T \tag{2}$$

THE CLIMATIC ZONES

The analyses have been performed by testing the ventilation and cooling strategies in four different climatic conditions. The selected cities are Athens, Rome, Berlin and Copenhagen. According to the Köppen-Geiger climate classification system [15], both Athens and Rome have a Mediterranean climate, Berlin has a humid-continental climate and Copenhagen is in the oceanic climatic zone.

[1] The temperature offset has been calculated referring to the velocity-offset curve valid when the mean air temperature is equal to the mean radiant temperature. In our case the mean radiant temperature is lower than the mean air temperature for most of the time, then, for the same air velocity, the offset prescribed by the standard is lower. To compensate for it, the two contribution have been averaged and not summed as vectors.

In Athens the climate is hot and dry, with a maximum monthly average high temperature of 33.1°C in July. Rome is hot and humid; the highest monthly average high temperature (30.6°C) is reached in August. Berlin has a temperate climate; the monthly average high temperature rises up to 24°C in July. Lastly Copenhagen has a cold climate and the monthly average high temperature presents a maximum of 20.4°C in July.
Rome presents very good night cooling potentials because of a 12°C day-to-night temperature swing. Also in Athens the night cooling is expected to perform well. Berlin and especially Copenhagen present a lower day-to-night temperature difference. All the locations are windy enough to provide an adequate air flow rate. In particular Athens and Copenhagen preset a prevailing wind blowing from North and from West respectively.
The climatic data used in the simulations are obtained from the ASHRAE's International Weather for Energy Calculations (IWEC) database.

CASE STUDIES

The analysis can be divided in two steps. First a set of preliminary simulation has been run to empirically optimize the thermal threshold for the night ventilation, the orientation, defined referring to façade 3 (f3 in Figure 1) and the building thermal mass, with respect to the climatic condition of each location. After, ten different ventilation strategies have been tested. The examined cases are described inTable 2.

Case studies	Ventilative cooling	Night cooling	Mechanical ventilation	Mechanical cooling
N_02_H	Non-increased air velocities	Open all night from 22:00 to 7:00	When windows are closed	Not equipped with a mechanical cooling system
N_02_H_A	Non-increased air velocities	Modulated according to comfort requirements	When windows are closed	Not equipped with a mechanical cooling system
N_02_HC	Non-increased air velocities	Open all night from 22:00 to 7:00	When windows are closed	Daytime: when the temperature is above the set point. Nighttime: when the temperature is above the set point and the windows are closed
N_I_H	Increased air velocities	Open all night from 22:00 to 7:00	When windows are closed	Not equipped with a mechanical cooling system
N_I_H_A	Increased air velocities	Modulated according to comfort requirements	When windows are closed	Not equipped with a mechanical cooling system
N_I_HC	Increased air velocities	Open all night from 22:00 to 7:00	When windows are closed	Daytime: when the temperature is above the set point. Nighttime: when the temperature is above the set point and the windows are closed
M_HC	Never used	Never used	During the entire year	When the temperature is above the set point
M_HC_N	Never used	Open all night from 22:00 to 7:00	When windows are closed	Daytime: when the temperature is above the st point. Nighttime: when the temperature is above the set point and the windows are closed
M_HC_N_A	Never used	Modulated according to comfort requirements	When windows are closed	Daytime: when the temperature is above the set point. Nighttime: when the temperature is above the set point and the windows are closed

Table 2 – Case studies

RESULTS

For every location the performances of the analized cases are graphically represented by mean of the *individual signature*. In a 3D graph the thermal comfort, the IAQ and the energy consumption have been correlated. The data used to plot the individual signatures and to compare the performances of the different strategies are: for the thermal comfort the percentage of hours in category II (the static or adaptive model has been used depending on whether the mechanical cooling system had been used or not), for the IAQ the percentage of hours in category I and for the energy consumption the energy used on a year long period expressed as a percentage of the energy consumption of the M_HC scenario.

Athens

The tested night thresholds for Athens range from 23.0°C to 25.5°C with a 0.5°C incrase step. The sesnitivity analysis proved the 25.0°C threshold to be the best performing: when compared with the 23.0°C one, the number of hours of comfort increases by 16.7% and the energy consumption is reduced from 43.7 kWh/m^2 to 1.2 kWh/m^2 because of the overcooling prevention. Also it shows a 4.0% increase in the thermal comfort and the same energy consumption if compared with a scenario where the night ventilation is not used (the overheating is avoided). For the strategies which combine mechanical cooling and night ventilation the threshold has been lowered to 24.5°C.
Eight orientatios have been tested (N, NE, E, SE, S, SW, W, NW). The SW orientation, exposing to NE the façade with the largest glazed surface, allows to achieve a very good thermal comfort (for 98.5% of the time the building is in category II) by reducing the solar gain. The price to pay is that, with its 2.2 kWh/m^2, the SW presents one of the highest energy consumption among the tested orientations (the solar gain is reduced during winter as well, the heating system must then supply more heat to the dwelling). Also, with a 0.25 m/s average indoor air velocity, the SW orientation has the largest potential for ventilative cooling.
The thermal mass optimization evaluated the bulding performances for concrete layer thicknesses ranging from 0.08 m (415 kg/m^2 of floor area) to 0.24 m (1183 kg/m^2) with a 0.02 m increase step. Comparing the 0.24 m with the 0.08 m thickness, the thermal comfort increases by 1.5% (both overheating and overcooling are reduced) and the energy demand is decreased by 37%. Increasing the building mass is beneficial up to a 0.20 m concrete layer thickness (991 kg/m^2), a further increases gives only a negligible performance improvement. For the daytime ventilation a 24°C threshold has been chosen.
With the selected parameters the mean air velocity for the non-increased air velocity cases is 0.12 m/s, while for the increased air velocity ones is 0.25 m/s. The natural ventilation period goes from April 20th to October 30th, that is 194 days of natural ventilation (53% of the year). The individual signatures (Figure 2) show that the N_02_H_A scenario is the best performing. When compared with the fully mechanical system N_02_H_A grants only a slight improvement in the thermal comfort (1.2%), but an 83% decrease of the energy consumption. The IAQ is much higher as well: the mechanical system supplies 0.5 ach only, while the natural ventilation strategy provides 4.8 ach during nightime and 1.7 ach during daytime, which results in a 26% increase of time in category I. It is also true that the N_02_H_A requires an automatic controller for the windows opening and such systems are not very common in domestic buildings. If we limit the choice to the manually operated systems the best performing is the N_I_H. Infact during the transition seasons the increased velocities of the N_I_H scenario are capable to maintain the indoor temperature within an acceptable range, thus limiting the use of night ventilation and, with it, the overcooling (task that in the non-increased air velocities scenario required the installation of the automatic controller). The N_I_H, when compared with the M_HC case, presents a slight decrease in the thermal comfort (0.6%), an improvement in the IAQ (28%) thanks to the increased air flow rates (4.1 ach during night and 5.6 ach during day), and a reduction in the energy consumption (83%).

Among the mechanically cooled building the N_I_HC ensures the best comfort conditions: the thermal comfort is increased by 0.6%, the IAQ by 5.9% (the hybrid ventilation system supplies 2.6 ach during night and 1.5 ach during day) and the energy demand is reduced by 20% (the consumption is lowered by 8.3 kWh/m^2, of which 8.0 kWh/m^2 for cooling needs). This proves that if the mechanical system is assisted by passive ventilation and cooling strategies, even based on a very simple logic, the result is a relevant reduction in the energy consuption and a potential increase in the quality of the indoor environment.

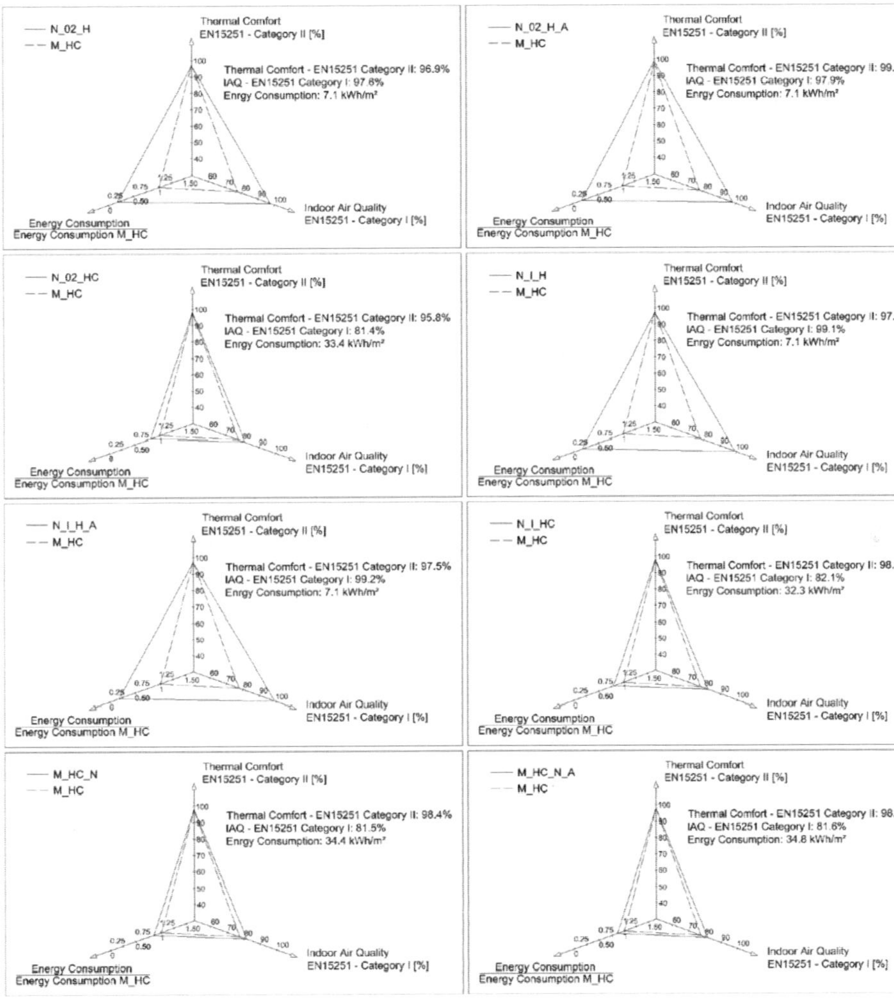

Figure 2. Individual signatures for the case studies in Athens, reference case: M_HC (Thermal Comfort – EN15251 Category II: 97.8%, IAQ – EN15251 Category I: 77.5%, Energy consumption: 40.6 kWh/m2).

Rome

The same night cooling thresholds used in Athens have been tested in Rome (i.e. 23.0°C to 25.5°C with a 0.5°C increase step), where the best thermal comfort has been obtained when the nigth cooling strategy is not applied. In all cases the discomfort is caused by the

overcooling of the building, due to the large day-to-night temperature swing mentioned before. Nonetheless a 24.0°C night ventilation threshold has been chosen since the prevention of the summer overcooling is a priority.
The E orientation has been considered the best performing.
For the thermal mass analysis the considerations made for Athens are valids for Rome as well, i.e. increasing the thermal mass is beneficial up to a 0.20 m concrete layer thickness.
For the daytime ventilation a 24°C threshold has been chosen.
With the selected parameters the mean air velocity for the non-increased air velocities cases is 0.16 m/s while the mean air velocity for the increased air velocities ones is 0.28 m/s. The natural ventilation period starts on April 7th and ends on October 25th, the natural ventilation strategies are then applied for 202 days over 365 (55% of the year).

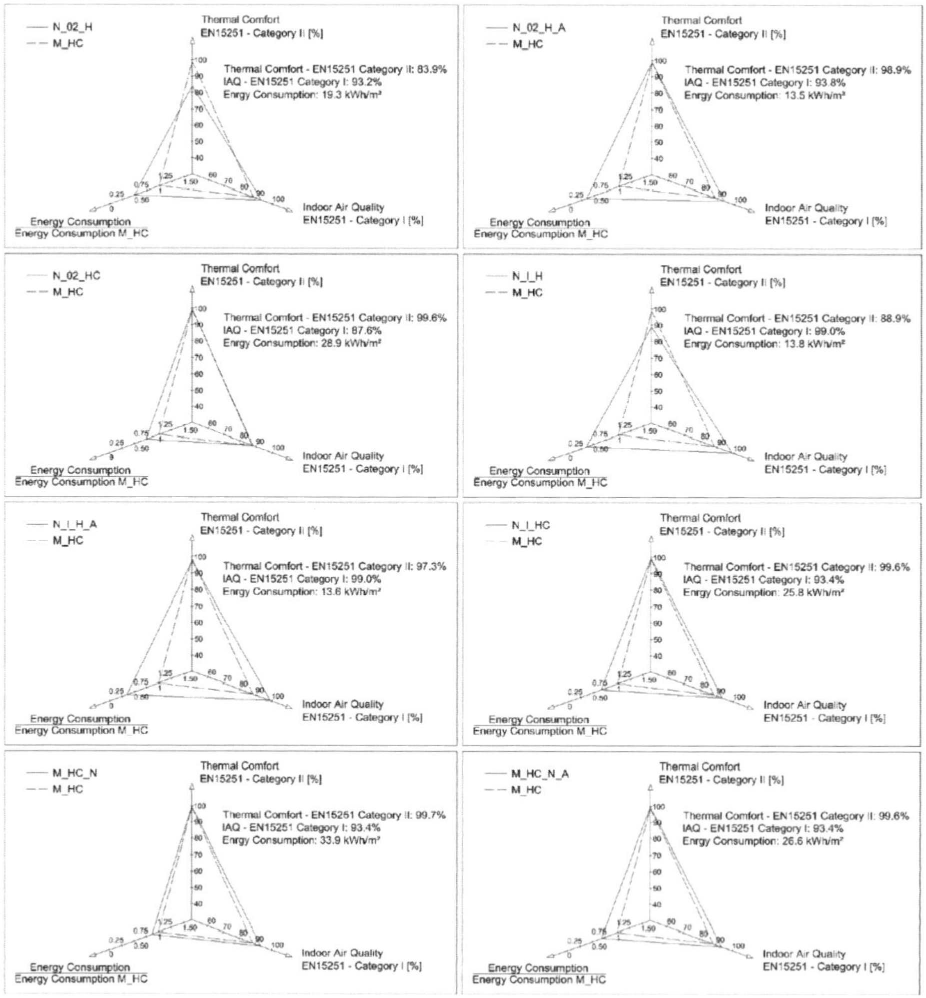

Figure 3. Individual signatures for the case studies in Rome, reference case: M_HC (Thermal Comfort – EN15251 Category II: 99.0%, IAQ – EN15251 Category I: 88.2%, Energy consumption: 39.0 kWh/m2).

For Rome in the mechanically cooled building the thermal comfort is generally higher than in the naturally ventilated and cooled ones. The cause is the overcooling of the building during nightime and indeed the introduction of the automatic controller, creating a constrain on the night air flow rate, improves the thermal comfort in both the N_02_H (17.9%) and the N_I_H (9.4%) solutions. Among the passive cooling systems the one that give the best results is the N_02_H_A. From a comfort point of view the solution performs as good as the M_HC system (there is a negligible 0.1% decrease), the IAQ is increased by 6.0% because, again, the natural ventilation provide much higher air flow rates (4.2 ach during night and 1.5 ach during day), and the energy consumption is reduced by 65%. In Rome when the mechanical cooling system is assisted by an automatically controlled night ventilation strategy, the improvements are relevant: the consumption is reduced by 31.8% and the indoor environment is more comfortable (+0.6% thethermal comfort and +5.9% the IAQ). The energy demand can be further decreased if the mechanical system is assisted by ventilative and night cooling (-33.8%), in which case the automatic control becomes unnecessary.

Berlin

In Berlin the upper limits for the comfort categories defined according to the adaptive model are quite lower than in Athens and Rome, then the tested night ventilation thresholds have been proportionally decreased. For the sensitivity analysis the potential thresholds go from 22.0°C to 24.5°C with a 0.5°C increase step. As in Rome, the solution that generates the best thermal comfort is the one without night cooling and, as in Rome, the overcooling prevention has been considered a priority. Then a 23.5°C threshold has been adopted.
In Berlin the influence of the orientation on the building performance is negligible, then the N orienation, being capable to provide a 0.26 m/s average indoor air velocity, has been chosen. A 0.20 m concrete layer has been considered sufficient (2.3% increase in the thermal comfort and 4.3% reduction of the energy consumption when compared with the 0.08 m thickness). For the daytime ventilation a 23°C threshold has been chosen.
With the selected parameters the mean air velocity for the non-increased air velocitites cases is 0.12 m/s while the mean air velocity for the increased air velocities ones is 0.26 m/s. The natural ventilation period starts on May 7th and ends on October 5th, the natural ventilation strategies are then applied for 152 days over 365 (42% of the year).
In Berlin the energy demand for cooling accounts for only 5.4% of the total consumption, then the passive cooling strategies are not paricularly beneficial from an energetic point of view: the highest reduction in the energy consumption is indeed equal to 5.6% (N_02_H_A). If we limit the choice to the solutions without mechanical coolig, N_02_H_A is the scenario with the highest performances: the 0.2% reduction in the thermal comfort is compensated by the improved IAQ (1.7%) and by the decreased energy consumption. Among the mechanically cooled buildings the optimum is reached by the N_I_HC system, which increases the IAQ by 3.6% and reduces the consumption by 2.6%, mantaining the building in category II for 100% of the occupancy time.

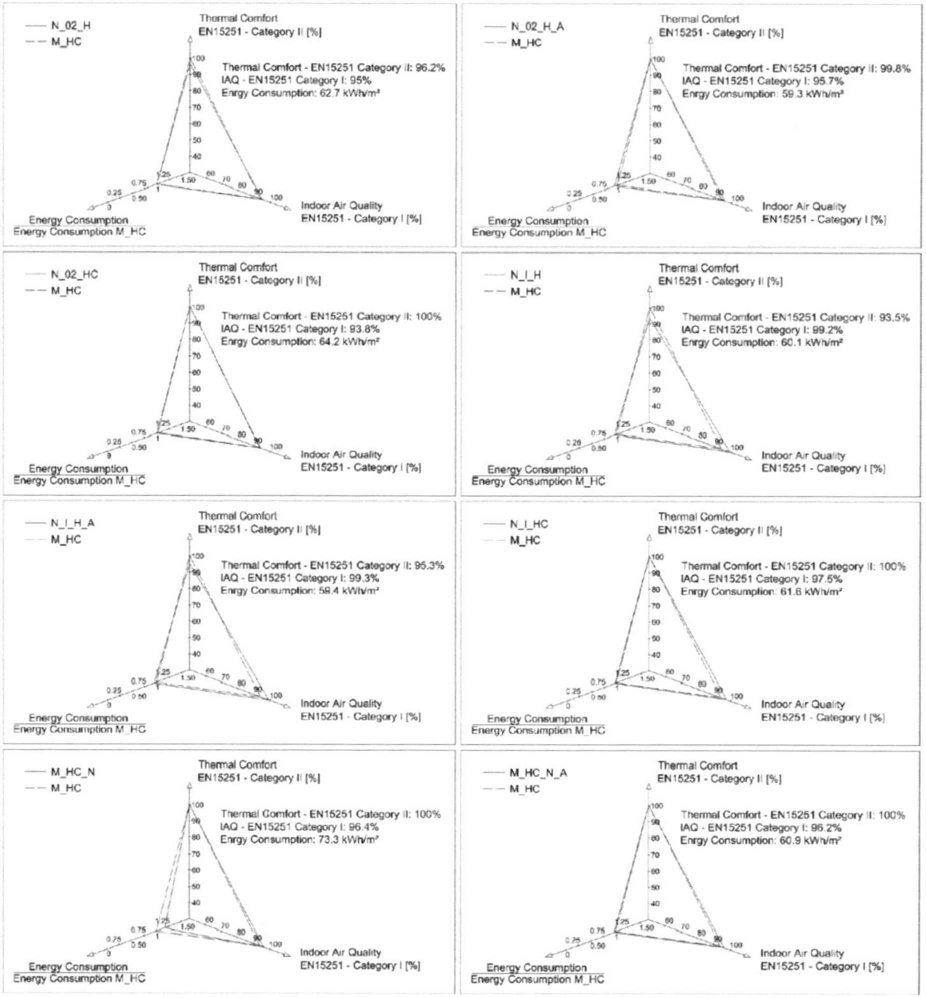

Figure 4. Individual signatures for the case studies in Berlin, reference case: M_HC (Thermal Comfort – EN15251 Category II: 100%, IAQ – EN15251 Category I: 94.1%, Energy consumption: 62.8 kWh/m^2).

Copenhagen

For the night thresholds optimization the temperatures taken into account range from 22.0°C to 24.5°C with a 0.5°C increase step. The analysis revealed that for thresholds higher than 23.5°C (included) the night cooling strategy is never applied during the year, while for thresholds lower than 22.5°C (included) the use on night ventilation increases the discomfort, causing the overcooling of the dwelling, and the energy consumption for heating. The 23.0°C threshold can be then considered the best option, allowing the application of the night ventilation only when strictly necessary.

In Copenhagen the risk of overheating during summer is limited to very few days, during most of the year the discomfort is caused by the overcooling of the building. Thus the NE orientation has been selected since it maximize the solar gain, improving the thermal comfort

and reducing the energy consumption. During summer, when needed, the solar gain can be decreased by activating the solar shading.
As the windows are almost always closed during night, no improvement at all in the thermal comfort have been obtained by increasing the thermal mass, then, in accordance with the other locations, a 0.20 m concrete layer thickness has been chosen.
For the daytime ventilation a 23°C threshold has been chosen.

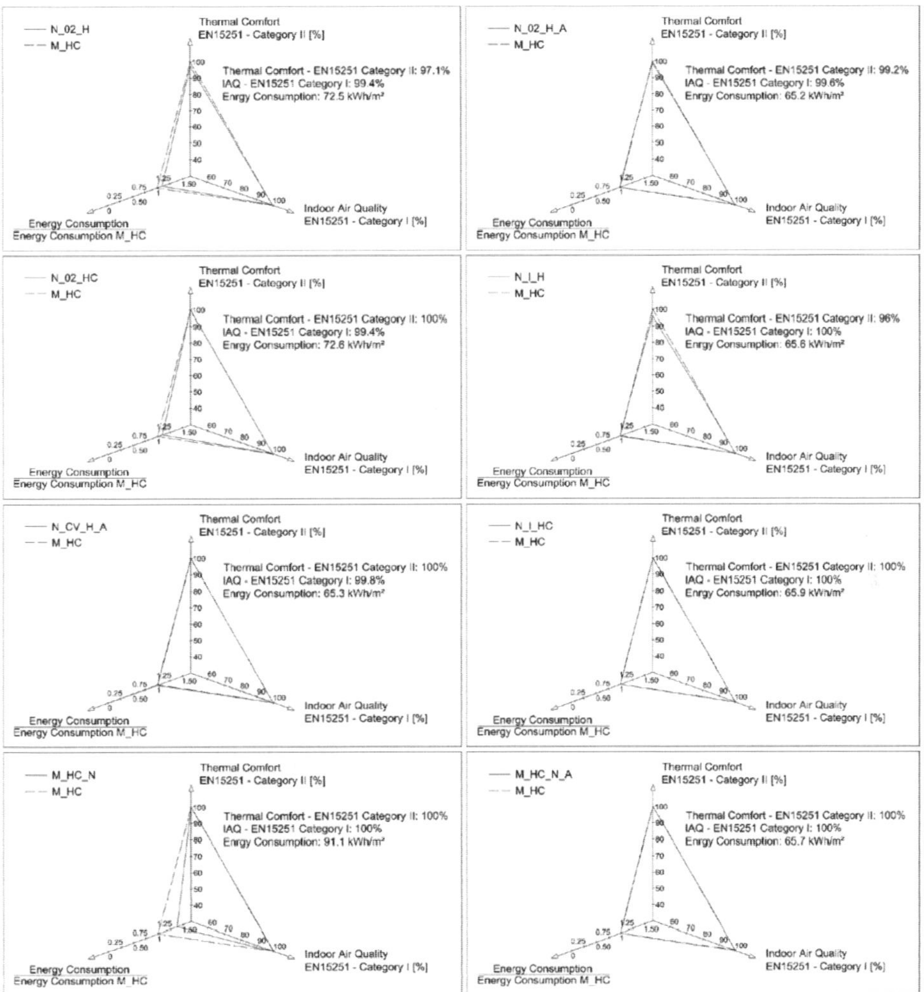

Figure 5. Individual signatures for the case studies in Copenhagen, reference case: M_HC (Thermal Comfort – EN15251 Category II: 100%, IAQ – EN15251 Category I: 99.7%, Energy consumption: 66.1 kWh/m2).

With the selected parameters the mean air velocity for the non-increased air velocitites cases is 0.16 m/s while the mean air velocity for the increased air velocities ones is 0.30 m/s. The natural ventilation period starts on April 30th and ends on September 17st, the natural ventilation strategies are then applied for 141 days over 365 (39% of the year).

As a matter of fact the night ventilation is almost never used: over the entire natural ventilation period the windows are opened for 13% of the nights in the N_02_H scenario and for 1.5% of the nights in the N_I_H one. Indeed in Copenhagen only 0.6% of the energy consumption is for cooling, it means that the cooling load is almost zero. This is the reason why all the passive cooling strategies, beeing too efficient for the local climate conditions, reduces the thermal comfort (from a minimum of 0.8% for N_02_H_A, to a maximum of 4.0% for N_I_H), sometimes increasing the thermal comfort (e.g. by 9.7% for N_02_H). The M_HC_N_A solution has been designed as a thermostatically controlled one, but the energy consumption reveals that the mechanical cooling system is never turned on and there is no extra heating demand: the night cooling seems then capable alone to reduce the cooling load to zero without causing overcooling. The observation suggested us to test one more ventilation strategy in Copenhagen, namely N_CV_H_A. The strategy is based on daytime mechanical ventilation and automatically controlled nigth cooling, but the windows can be opened, according to the occupants comfort, for a short period of time (15 min.) in the early morning (8:00 a.m.) and when the occupants go back home (17:00 in the afternoon) for airing the dwelling. The daytime natural ventilation achieved in this way does not provide ventilative cooling, but allows the occupants to better control the indoor environment. The solution performs excellently: the building is for 100% of time in category II (thermal comfort), for 99.8% of time in category I (IAQ) and requires 1.3% less energy than the thermostatically controlled one. Beig capable to prevent both the nighttime overcooling that affects the non-increased air velocities scenarios, and the draft sensation that affects the increased air velocities ones, it is the only passive cooling solution which does not decrease the thermal comfort.

DISCUSSION AND CONCLUSIONS

The project here reported investigated the potential energy saving and summer comfort improvement that can be achieved by mean of passive cooling strategies such as solar shading, ventilative cooling and night cooling. In general the passive approach seems capable to ensure a good indoor environment in terms of high IAQ and prevention of both overheating and overcooling, as well as a reduction in the energy consumption.
In Athens the increased air velocities, being capable to maintain the mean air temperature within an acceptable range, are efficient in limiting the use of night cooling without the need of an automatic controller. The combination of the two strategies seems then capable to achieve a very good thermal comfort.
For Rome, Berlin and Copenhagen the same combination of daytime increased air velocities and night cooling turned out to be too aggressive, causing some overcooling and an increase in the energy consumption for heating that ranged from the 1.6% of Rome to the 9.7% of Copenhagen. A moderate approach showed good results.
In Rome and Berlin a constraint to both the daytime air velocity and the nighttime air flow rate was capable to provide a very good indoor environment, even more comfortable than the one obtained by mechanically cooling the building. For colder climates, such as the one of Copenhagen, the best performance on thermal comfort (100% of the time in category II) was obtained with the use of the night cooling strategy only.
The hot Mediterranean climate of Athens and Rome presents a very high cooling load, thus the adoption of the passive technique leads to a consistent reduction in the energy consumption (83% for Athens and 65% for Rome). In Berlin and Copenhagen the reduction in the energy demand (5.6% and 1.3% respectively) corresponds entirely to the energy used by the chiller, showing that the strategy is capable to lower the cooling load, in particular in Copenhagen the selected strategy is capable to reduce the cooling load to zero.

When the IAQ is looked at, the natural ventilation performs much better than the mechanical one in all cases. The air flow rate ranges from 1.1 ach (Copenhagen) to 4.8 ach (Athens) during nighttime and from 0.7 ach (Copenhagen) to 1.7 ach (Athens) during daytime.
In general, the scenarios that performed best on all three parameters (thermal comfort, IAQ and energy), were the ones based on natural ventilation.

ACKNOWLEDGEMENT
This study was sponsored by the SINO-DANISH research project "Activating the Building Construction for Building Environmental Control" under the Programme Commission for Sustainable Energy and Environment, the Danish Council for Strategic Research.

REFERENCES
[1] IPCC. 2007. *Climate Change 2007: The physical science basis*. Working Group I of the IPCC.
[2] European Commission. 2008. *20 20 by 2020: Europe's climate change opportunity*. Commission of the European Communities.
[3] Joosten, L., Strom, I., and Boonstra, C. 2006. *Energy saving potential*. Promotion of European Passive houses (PEP).
[4] AIVC. 1996. *The role of natural ventilation in cooling non-domestic buildings*. Technical note AIVC 48.
[5] Gratia, E., Bruyère, I., and De Herde, A. 2004. *How to use natural ventilation to cool narrow office buildings*. Elsevier, Building and Environment 39.
[6] Schiavon, S., and Melikov, A.K. 2008. *Energy saving and improved comfort by increased air movement*. Elsevier, Energy and Buildings 40.
[7] Yun, G.Y., Steemers, K., and Baker, N. 2008. *Natural ventilation in practice: linking façade design, thermal performance, occupant perception and control*. Routledge, Building Research and Information 36(6).
[8] Böllinger, A., and Roth, H.W. 1993. *Benefits and limits of free cooling in non-residential buildings*. 14th AIVC Conference: Energy Impact of Ventilation and Air Infiltration.
[9] Santamouris, M., Sfakianaki, A., and Pavlou, K. 2010. *On the efficiency of night ventilation techniques applied to residential buildings*.Elsevier, Energy and Buildings 42.
[10] Geros, V., Santamouris, M., and Guarraccino, G. 1997. *Applying night venetilation techniques in office building*. 18th AIVC Conference: Ventilation and Cooling.
[11] Shaviv, E., Yezioro, A., and Capeluto, I.G. 2001. *Thermal mass and night ventilation as passive design strategy*. Pergamon Press, Renewable Energy 24.
[12] CEN. 2007. *CEN Standard EN15251:2007, Indoor environmental input parameters for design and assessment of energy performance of buildings addressing indoor air quality, thermal environment, lighting and acoustic*. European Committee for Standardisation.
[13] Brager, G.S., and de Dear, R. 2001. *Climate, Comfort & Natural Ventilation: A new adaptive comfort standard for ASHRAE Standard 55*. University of California, Berkeley.
[14] Sahlin, P., Bring, A., and Eriksson, L. 2009. *Real controllers in the context of full-building, whole-year simulation*. Eleventh International IBPSA Conference.
[15] Peel, M.C., Finlayson, L.B. and McMahon, T.A. 2007. *Updated world map of the Köppen-Geiger climate classification*. Hydrology and Earth System Sciences.

HYBRID VENTILATION – THE VENTILATION CONCEPT IN THE FUTURE SCHOOL BUILDINGS?

Simone Steiger [1], Jannick Karsten Roth[*2] and Lennart Østergaard[2]

1 Fraunhofer Institute for Building Physics
Fraunhoferstraße 10
83626 Valley, Germany

2 WindowMaster A/S
Skelstedet 13
Vedbæk, Denmark
**Corresponding author: jkr.dk@windowmaster.com*

ABSTRACT

Hybrid ventilation (HV), as a combination of automated natural ventilation (NV) and balanced mechanical ventilation (MV), provides opportunities to use the advantages of both ventilation systems during the seasons in order to reduce energy demand and at the same time obtain comfortable indoor climate.

For this study a comparison of NV, MV and HV systems applied to an existing school building has been made by means of detailed modelling in the widely adopted simulation program IESVE. The energy demand for heating and ventilating the building in the different ventilation modes was calculated for three key European cities; Munich, Copenhagen and London. Control strategies were set to obtain the same indoor climate for all ventilation systems, and the indoor climate classification was made according to EN15251 [4].

The overall ambition in the study has been to make a very realistic modelling of state-of-the-art MV and NV systems and apply these systems to an existing school building renovated to fulfill expected energy performance requirements for 2015 buildings.

It is noteworthy that the results show that the energy performance of the MV and NV systems are nearly the same in terms of primary energy, while it is showed that HV enables energy savings of 44-52% compared with MV and NV. This corresponds to a total primary energy consumption for the NV and MV of 18-21 kWh/m² per year in all three locations, while the HV consumption was only 9-10 kWh/m² per year.

Calculation of the total investment of the different systems including maintenance, operation (electricity and heating) and capital cost (products and installation) showed that in the first year MV was found to be 2.5 to 4 times more expensive than NV. By selecting HV 25% of the total investment could be saved compared to a MV system. The differences between the systems were found to be the same in a 20 year timeframe

KEYWORDS

Hybrid ventilation; natural ventilation; mechanical ventilation; ventilation in schools; comparison of different ventilation strategies; Energy; IAQ.

INTRODUCTION

From the literature it is known that the combination of automatic NV and MV is a promising opportunity to achieve significant energy savings in buildings while maintaining a comfortable indoor climate. HV might as such be a key technology for designers in need of means to fulfil the ever stricter energy requirements while at the same time provide the user with a healthy and comfortable indoor climate.

Many studies of HV are available in the literature. The Reshyvent project [11] investigated HV in residential buildings, while the Hybvent project [7] studied the applicability for non-residential buildings. A number of case studies investigated in the international project IEA ECBCS-Annex 35 [7] show that there can be achieved significant energy savings in hybrid ventilated buildings, especially through reduction in fan and cooling energy demand. The case studies for school buildings show that the HV system saves 17-55 % in a year compared to a solitary mechanical system. Examples also include Cron & Inard [3] who investigated classrooms and found the best results for HV in the warmer regions, compared to MV without heat recovery, and in the warmest regions, compared to NV with heat recovery. Emmerich [5] found that heating demand is reduced most in cold regions, and fan electricity is reduced to a maximum in warm regions. Sowa & Karas [9] found a reduction of about 60 % for heating demand and about 40 % for fan electricity in a HV simulation for a real school. Heikkinen et al. [6] who investigated ventilation concepts for a school in Finland, only found a very little possible reduction in heating demand, but also a reduction of 70 % in fan electricity.

Thus, the literature contains a number of publications and in general, HV is demonstrated to result in significant energy savings. The present study investigates whether this conclusion will also be valid in schools in the near future (2015), using state-of-the-art MV and NV systems and a school building fulfilling expected 2015 performance requirements. In order to give a true comparison between the energy performances of the systems, nearly identical indoor air quality and thermal climates in the buildings have been created. The simulations are carried out for three large European cities with different climates; Copenhagen, London and Munich. The study also calculates the expected CO_2 emissions and the economical costs from selecting the different systems.

SCHOOL BUILDING GEOMETRY, PROPERTIES AND LOCATION

Building layout

The school building has one storey with eight classrooms, four on each side of the corridor and is oriented North/South. The ceiling height varies from 2.8 to 4.5 m for all ventilation types and the area of one classroom is 76 m^2, the volume is 278 m^3. The geometry of the building can be seen in Figure 1.

Construction properties

Some of the construction properties attached to the models are listed in Table 1.

The glass ratio of the outer façade is 45 % and the g-value is 0.63 and light transmittance is 0.74. Only the lager windows have an external sun screening with a shading coefficient of 0.2.

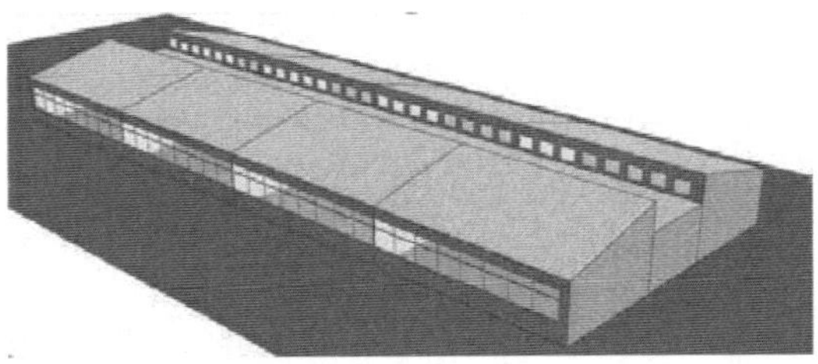

Figure 1. Layout of the school building.

Building Element	U-Value [W/m²K]
Ground Slab	0.08
Exterior Walls	0.12
Roof	0.08
Windows	0.9-1.1

Table 1. Constrution properties

Internal heat loads

There are 28 students and one teacher in each classroom, resulting in an occupancy density about 2.6 m^2/person, which is a typical occupant density for schools. There is assumed 95 % occupancy during lessons (Monday to Friday from 8 am to 2:50 pm) taking into account absent persons. Vacation time is 12 weeks per year in total (week 7, 14, 20, 26 - 31, 42 and 51 - 52). The occupancy during vacation is set to 10 % from 8:00 am to 2:50 pm from Monday to Friday, which takes into account summer courses or maintenance. There is no occupancy during weekends.

Each person has a heat load of 75 W sensible heat and 50 W latent heat corresponding to an adult with an activity level of 1.2 met. This assumes a heat emission of 70 W/m^2 skin surface and a skin surface of 1.8 m^2. Children with a lower body mass normally also have a higher level of activity of about 1.4 met (81 W/m^2skin surface). Assuming a skin surface of 1.5 m^2 per child, the heat emission for all persons is quite the same.

Each student and teacher is expected to have a computer, which is switched on 50 % of the time during occupancy. There are typically not so many computers in a classroom today, but it is expected that the use of computers will increase in the future.

The lighting (fluorescent lighting) shall provide a luminance intensity of 300 lux at the table and has a maximum heat load of 15 W/m^2, which corresponds to an effective lighting system.

Outdoor climatic conditions

The locations chosen for the comparison are Copenhagen, Munich and London. These three cities are typical European cities with different climates and therefore different possible opportunities for HV. Copenhagen has a cold winter and a cool summer, whereas Munich has a colder winter and a warm summer. London, located close to the sea, has a maritime climate with a mild winter and a cool summer.

REQUIREMENTS ON INDOOR AIR QUALITY AND THERMAL COMFORT

Many studies have shown that existing schools have a poor indoor climate with CO_2 levels sometimes exceeding 2-4,000 ppm [2]. These levels are clearly adversely affecting the learning ability of the school children [10][12][13] - and these levels must be improved. However, general adoption of the current Category II requirements in EN15251 with a maximum CO_2 concentration of 900ppm seems unrealistic in schools for two key reasons: First, the air exchange rate for a typical 60 m2 class room with a room height of 2.8m and 29 persons needs to be at least 6-7 ACH. This will create problems with air speeds in the comfort zone in the majority of existing schools. Second, it is also noted that the financial abilities of the public authorities in most countries does not support such strict requirements. In fact, they could prove to be a barrier against improving the indoor climate in existing schools simply because the systems become too expensive.

The classification of the thermal comfort and indoor air quality in the buildings are based on EN 15251. For the school building Category III with an acceptable level of expectation is applied for the assessment of indoor climate.

DIMENSIONING OF VENTILATION SYSTEMS

To maintain the air quality according to Category III of EN 15251 the necessary air flow rate was calculated by the air flow rates per m^2 given in the standard for persons in a classroom and low emissions from the building. This results in a flow rate of 2.4 l/sm^2 and a total air flow rate of 180 l/s, 648 m^3/h. The total flow rate for all 8 classrooms is 1,440 l/s or 5,148 m^3/h. For maintaining temperature different air change rates were tested in the simulation. Due to these results a maximum air exchange of 4.6 per hour was chosen for summer and night ventilation. This is a flow rate of 360 l/s or 1,296 m^3/h for one classroom and 2,880 l/s or 10,368 m^3/h for all classrooms.

Natural ventilation

For the NV every second high level window on both sides of the room can be opened with motors to realize cross ventilation. The resulting openable window area for the automated windows is 4.1 m^2 and 5.4 % of the room area.

A temperature difference of 1 K and 5 K between inside and outside is resulting in a nearly 4 and 9 fold air exchange, respectively. A wind speed of 0.5 m/s and 1 m/s is resulting in a 5 and 10 fold air exchange rate, respectively (calculated according to the British Standard Method [1]). Outdoor conditions with 0.5 m/s wind speed and a temperature difference above 1 K should be available during most time of the year for all three locations and are also adequate for temperature maintaining ventilation in summer.

Mechanical ventilation

For the MV in the school building four smaller decentralized units are utilized. The system was dimensioned for the maximum air flow rate according to air quality and indoor temperature (10,368 m^3/h). The specifications of the four units are selected amongst available market products, resulting in a total flow rate of 15,680 m^3/h.

The pressure loss for the supply and the exhaust pipe system is only 80 Pa for the supply system and 80 Pa for the exhaust system. The filter classes were F7 for supply air and F5 for exhaust air causing an additional pressure of 40 Pa due to dirt. The SFP value for the whole system is 993 J/m^3 - which probably is among the best available on the market currently. There were assumed no additional heating unit and no cooling unit. As the 'Demand

Controlled Ventilation' is working with a constant pressure loss in the main pipes the setting for the external pressure was hold constant.

The heat recovery system is a state-of-the-art counter-flow plate heat exchanger with low-energy deicing function and the sensible heat effectiveness in the simulation was set to 92 %, which is the temperature efficiency including the effects of the motor heat at 1800 m^3/h, 75 % of the design flow rate.

Hybrid ventilation

For the HV there are applied two decentralized units dimensioned for the air flow rate only according to air quality (5,148 m^3/h). To maintain indoor temperature in summer NV is utilized with a flow rate of 10,368 m^3/h.

The pressure loss in the MV system for the supply and the exhaust pipe system is then about 132 Pa for the supply system and 143 Pa for the exhaust system. Pressure loss from filters, sensible heat effectiveness and heating/cooling unit is the same as the MV. SFP value for the whole system is 1135 J/m^3.

CALCULATION METHOD

The energy demand and the indoor climate of the building were simulated in the widely adopted simulation the program VE-Pro (version 6.4.0.7, Integrated Environmental Solutions Limited, Glasgow, UK). This program has a special device for calculating more complex HVAC systems (ApacheHVAC) and also a very reliable calculation tool for NV (MacroFlo), which is able to calculate NV and effects from wind turbulence on air exchange, considering special features like the aspect ratio and sash type of the opening. The calculation was done in 1 min steps to achieve realistic results for natural and especially natural pulse ventilation. The results are derived from 6 min averages of the calculation.

For the assessment of indoor climate, CO_2 levels inside the building were used as indicator for indoor air quality and operative room temperature was used as indicator for thermal comfort. The values were obtained during occupancy in one representative room, and the requirements to the thermal comfort and indoor air quality were based on EN15251.

CONTROL STRATEGIES

The transfer of the control strategy to the simulation models was done as close as possible to WindowMaster's control strategy. Sometimes changes were necessary due to the restrictions of the simulation software or to obtain a similar thermal comfort and indoor quality.

Natural ventilation

NV in this investigation is defined as automated NV through high level windows on both sides of the rooms. The windows are opened and closed to a specific amount with small motors. The opening width is defined by a controller, which uses indoor and outdoor climatic parameters to calculate the appropriate opening width. This precise opening is necessary, because the resulting air flow rate is not only depending on the climatic parameters, but also very much on the opening width of the windows. A precise air flow is necessary to avoid too high ventilation rates, which cause additional heat losses or a bad thermal comfort due to low temperatures or a high draught ratio, and to provide a good air quality at the same time.

There is implemented three different opening strategies; continuous ventilation with a varying opening degree, pulse ventilation with the maximum opening degree calculated due to weather for a short time and night ventilation. The first strategy is utilized for control of air quality during the whole year and indoor temperature in summer. The second strategy is only for additional control of indoor air quality during winter and transient times, where the opening width for continuous ventilation is restricted due to comfort reasons. The third strategy is utilized for an additional cooling of the rooms in summer. In addition, the windows are opened to maximum after occupancy to ventilate the rooms completely with fresh air until outdoor air quality is reached.

Mechanical ventilation

The flow volume of the MV is defined due to improvement of indoor air quality and reduction of overheating. Therefore the maximum flow volume is utilized, when either the flow rate due to carbon dioxide level or the indoor air temperature rises above a certain point. Furthermore night ventilation is only active during the warmer periods

Hybrid ventilation

The HV control strategy is a combination of the natural and mechanical control strategy. The main strategy is to use the best aspects of both systems in order to reach the best possible values for energy consumption and indoor quality.

During winter season only MV is activated as the heat recovery of the system helps saving energy to heating. NV has best results during summer where good indoor air quality and temperatures easily can be reached. In addition, the motors of the windows need much less electricity than the fans for MV and the flow rate can be raised only by opening the windows a little more without using additional energy. This is also the benefit from NV during night time.

During the transient season it depends most of the times on internal situations, if MV or NV is the best solution. Therefore the system chooses automatically between the two systems depending on indoor temperature as indicator for heating or possible cooling demand.

Heating, shading and lighting

The heating is activated from October to May. The heating is set to avoid too low temperatures in accordance to Category III during occupancy (8 am 2:50 pm). During hours with no occupants in the building the set-point is 18°C.

The setting for the shading devices is controlled due to outdoor and indoor parameters. This is done to avoid overheating, which may affect indoor temperature and thermal comfort also a few days later. The device rises with a wind velocity above 12 m/s and/or an outdoor temperature below -6°C to avoid damages on the devices. The device lowers with a solar radiation above 100 W/m^2 and if the indoor temperature is above 23.5°C.

The dimming of the artificial light is controlled due to occupancy and to keep 300 lux in the rooms.

RESULTS

Temperature and CO_2

The results of thermal comfort and indoor air quality was evaluated for one south facing classroom as there only were found very small differences between a north- and south facing classrooms. Figure 2 shows the relative frequency of indoor temperatures during occupancy in accordance to the categories in EN15251. This is displayed for all three ventilation types in each location.

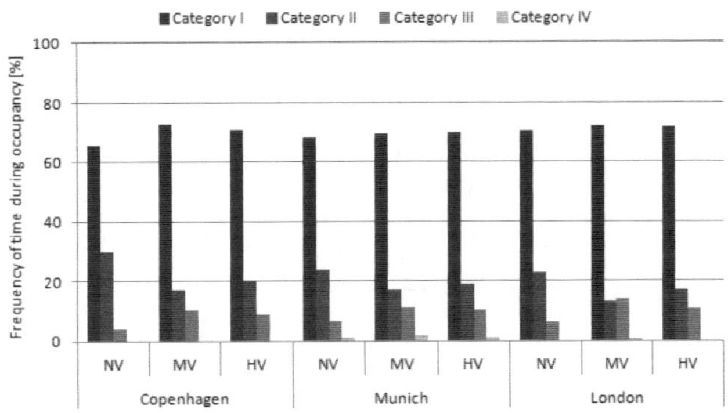

Figure 2. Relative frequency of indoor temperatures during occupancy in accordance to EN15251

Only very small temperature variations were found for the different locations. This implies that each of the three ventilation systems has the same thermal comfort level. A similar picture was found when comparing the CO_2 levels. The result showed that Category II could be reached in 45-55% of the time depending on the ventilation system, while the remaining time was fulfilling the requirements to Category III.

Primary energy

For the calculation of the total primary energy demand (sum of heating and fans electricity demand multiplied with primary energy factors) the nationally adopted primary energy factors have been used for the different locations; Munich (district heating: 0.7; electricity 2.6), Copenhagen (0.8; 2.5) and London (1.2; 2.92).

Figure 3 shows the primary energy consumption. Comparing the primary energy demand it can be seen, that heating demand can be reduced by nearly 70 % for HV compared to NV. Fans electricity can be reduced with 75 % for HV compared to MV. Total primary energy is almost the same for MV and NV, but can be reduced up to 50 % for HV.

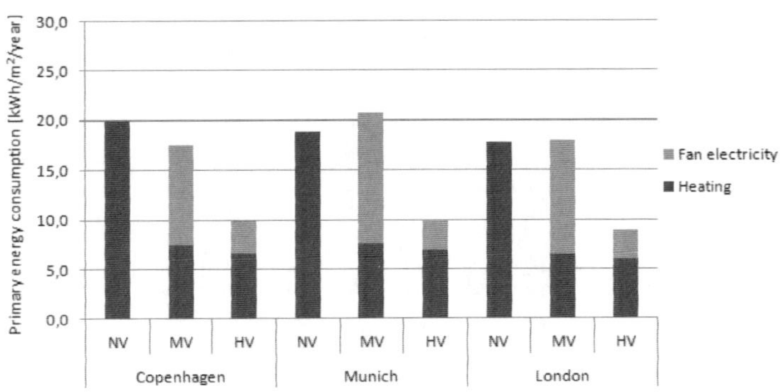

Figure 3. Primary energy consumption

CO_2 EMISSION

Calculation of the CO_2 emissions are based on the following figures; Munich (district heating 200 g/kWh and electricity 606 g/kWh), Copenhagen (104; 425) and London (206; 517). CO_2 emission due to electricity and heating is almost the same for NV and MV. Depending on the location the CO_2 emission ranges from 2.6-5.4 kg CO_2/m^2 per year. HV makes it possible to reduce this CO_2 emission up to 50%.

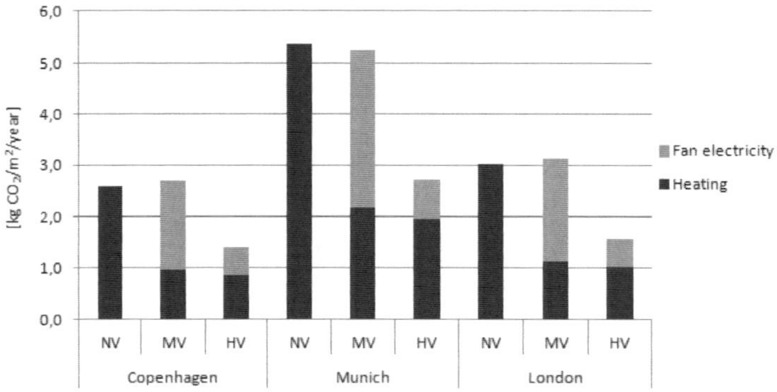

Figure 4. CO_2 emission

COST

The total investment of the different systems has been evaluated including costs for maintenance, operation (electricity and heating) and capital cost (products and installation) for the first year of operation (Figure 5) and during a period of 20 years (Figure 6). The prices are calculated by WindowMaster in close collaboration with a Danish ventilation contractor [8].

The maintenance cost for HV is almost the same as MV. Choosing NV this cost could be reduced with 70%. No significant difference was found between NV and MV for the operation cost during the first year. However, using HV this cost could be reduced with 50%. One of the major differences of the three systems is the capital cost. Here it was found that a MV system is more than four times as expensive as a NV system. For HV this was only a factor three. As a result of this big capital costs the total investment the first year is still in favour of NV with a factor four compared to MV and a factor three compared to HV. HV was found to be 25% cheaper than a full MV system.

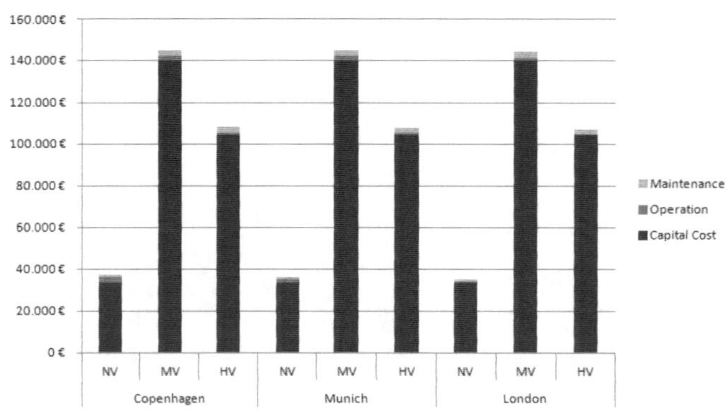

Figure 5. The total investment during the first year of operation

The total investment over a time period of 20 years showed almost the same pattern as the first year of operation. MV was found to be 2.5 to 3 times more expensive than NV on the total investment during a 20 year period. 25% could be saved choosing a HV system compared to a MV system.

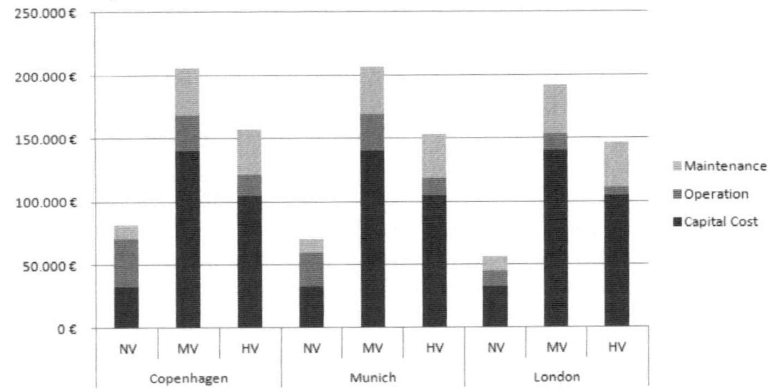

Figure 6. The total investment during a period of 20 years

DISCUSSION

The main ambition for the control strategies is to obtain the desired thermal comfort and indoor quality defined in EN15251 Category III and to obtain very similar indoor air quality for all three ventilation types. This is necessary for a useful comparison of energy demand resulting from the ventilation types. It should be noted that better indoor air quality and thermal comfort could have been obtained for all of the three ventilation systems if other requirements for CO_2 and temperature was chosen.

The transfer of the control strategy to the simulation models was done as close as possible to WindowMaster's control strategy. Sometimes changes were necessary due to the restrictions of the simulation software or to obtain a similar thermal comfort and indoor quality. It is believed that these simulations still are in accordance to the WindowMaster control system.

HV control strategy is a combination of the NV and MV control strategy. The main strategy is to use the best aspects of both systems, in order to optimise the balance between indoor climate and energy consumption. This is possibly the greatest challenge with HV and it is therefore necessary to have a control strategy that can take this into consideration.

During winter and summer the HV control strategy is almost straight forward. MV has the best results during cold periods, when there is a heating demand. The heat recovery of the system then helps saving energy of the building. During the summer period is it NV that has the best results. Good indoor air quality and thermal comfort can easily be reached without using any fan energy. The flow rate can be raised only by opening the windows a little more whiteout using additional energy. NV has also the ability to benefit from night ventilation without using any additional energy.

The transient season is, however much more complicated and the most of the time it depends on the internal situation, if MV or NV is the best solution. Therefore it is very important to have an automatic system that can choose between the two systems depending on indoor temperatures as an indicator for heating or possible cooling demand. This is perhaps not that complicated. The complicated part is to make for instance MV stop and then start up the NV system due to the fact that the internal environment has change throughout the period where MV has been used. This strategy has been developed in these calculations.

CONCLUSION

The total primary energy demand (sum of heating and fans electricity demand multiplied with primary energy factors) for the NV and MV systems ranges from 18-21 kWh/m^2 per year in all three locations. For HV the total primary energy demand was only 9-10 kWh/m^2 per year.

The result shows that HV enables energy savings of 44-52% compared with MV. Compared to NV an energy saving from 46-50% could be reached. The heating demand can be reduced by nearly 70% for HV compared to NV. Fans electricity can be reduced with 75% for HV compared to MV. Total primary energy is almost the same for MV and NV, but can be reduced up to 50% using HV.

One of the major differences was to be found in the total investment of the different systems including maintenance, operation (electricity and heating) and capital cost (products and installation). Looking at the first year of operation and during a period of 20 years MV was found to be 2.5 to 4 times more expensive than NV. By selecting HV 25% of the total investment could be saved compared to a MV system.

The results demonstrate clearly that HV should be considered for schools in addition to NV and MV. Overall the HV makes it possible to save money for heating and electricity during operation time and to save up to 50% of the CO_2 emissions.

ACKNOWLEDGEMENTS

First and foremost, we would like to thank WindowMaster for the valuable guidance, advice and knowledge sharing. Besides, we would like to thank the *Fraunhofer Institute for Building Physics* for providing a good environment and facilities to complete this project.

REFERENCES

[1] F. Allard. *Natural ventilation in buildings*. A design handbook. MPG Books Ltd, London, 1998.

[2] Byg DTU (2009) *Elever undersøger indeklima i klasselokaler - rapport om resultater fra Masseeksperiment*

[3] F. Cron, C. Inard. *Analysis of hybrid ventilation performance in france*. Proceedings of Eighth International IBPSA Conference 2003, Eindhoven, 243 - 250.

[4] EN 15251: 2007. *Indoor environmental input parameters for design and assessment of energy performance of buildings addressing indoor air quality, thermal environment, lighting and acoustics.*

[5] S. J. Emmerich. *Ventilation Systems in an Office Building*. HVAC & Research 12 (2006), 4, 975 - 1004.

[6] J. Heikkinen, J. Heinonen, T. Laine, K. Liljeström, M. Vuolle. *Performance simulation of hybrid ventilation concepts*. IEA - Technical report in Annex 35 Hybrid Ventilation, 2002.

[7] P. Heiselberg. *Design of Natural and Hybrid Ventilation*. DCE Lecture Notes, Aalborg University, 2006.

[8] Jannick Karsten Roth (2012) *Cost calculation for the school investigated in the Fraunhofer study.*

[9] J. Sowa and A. Karaś. *Whole year simulation of natural and hybrid ventilation performance and estimation indoor air quality for modernized school building*. Proceedings of Clima 2007 WellBeing Indoors, 1 - 8.

[10] M.J. Mendell and G.A. Heath (2005) *"Do indoor pollutants and thermal conditions in schools influence student performance? A critical review of the literature"*, Indoor Air, Vol. 15, 27-52.

[11] Reshyvent (2004) *Demand controlled hybrid ventilation in residential buildings with specific emphasis on the integration of renewables.*

[12] P. Wargocki and D.P. Wyon (2007) *"The effects of outdoor air supply rate and supply air filter condition in classrooms on the performance of schoolwork by children"*, HVAC&R Research, Vol. 13(2), 165-191.

[13] Wargocki P. , D.P. Wyon, K. Lynge-Jensen, K. and C-G. Bornehag (2007) *"The effects of electrostatic filtration and supply air filter condition in classrooms on the performance of schoolwork by children"*, HVAC&R Research, submitted.

ABSTRACT SUBMISSION FORM FOR THE 2012 AIVC-TIGHVENT CONFERENCE

Send your abstract using this form at info@inive.org

PAPER TITLE
Numerical predictions of the discharge coefficient of a centre-pivot roof window

MAIN AUTHOR
Ahsan Iqbal

OTHER AUTHORS
Peter Vilhelm Nielsen, Alireza Afshari, Per Heiselberg

ABSTRACT
The centre pivot roof window is the dominating roof window in Nordic region. This paper focuses on the discharge coefficient of the centre pivot roof window. Focus is given on unidirectional flow and bidirectional flow. CFD techniques are used to predict the airflow through the model window. The objective of the present research is to find out the accurate method to determine the discharge coefficient, hence the airflow rate using orifice flow equation, of centre-pivot roof window. The k-ε turbulent model can only predict the fully developed turbulent flow, that's why the results are only valid for higher pressure difference. It is concluded that the single value of the discharge coefficient leads to deceptive estimation of airflow rates.

EXHAUST VENTILATION UNDER 5 VENTILATION STANDARDS : A PERFORMANCE ASSESSMENT

Jelle Laverge*, Xavier Pattyn and Arnold Janssens

Ghent University
Jozef Plateaustraat 22
9000 Gent, Belgium
**Corresponding author: jelle.laverge@ugent.be*

ABSTRACT

Sizing rules in residential ventilation standards lack uniformity in both methodology and resulting design flow rates. Additionally, mere comparison of design flow rates is case sensitive and, due to effects of infiltration, adventitious ventilation and occupancy, ill-suited to assess performance of an exhaust ventilation system with regard to the achieved indoor air quality and energy cost in terms of heat loss. This paper presents a multi-zone simulation based performance assessment of residential mechanical exhaust ventilation systems, using five common dwelling typologies and the sizing rules put forward in the Belgian, British, Dutch, French and ASHRAE residential ventilation standards. The performance of the different cases proved to be substantially different, with an occurrence of poor perceived air quality in 5% or less of the occupation time for the Belgian, Dutch and French standard, and about 15% for the British and ASHRAE standard. When the trade-off between indoor air quality and heat loss is considered, the cases with the Dutch and ASHRAE standard did not achieve pareto-optimal performance.

KEYWORDS

Standard, Exhaust Ventilation, Residential, Simulation, IAQ

INTRODUCTION

The 1970's oil crisis caused the first wave of energy conservation campaigns in buildings. Improved airtightness of newly built dwellings and intensive weatherisation actions considerably reduced the amount of fresh air infiltration. As an unintended consequence of this, the incidence of indoor mould problems peaked and reports on high prevalence of occupants complaining of a wide variety of symptoms or physical discomfort, baptised 'sick building syndrome', emerged.

As a reaction to these problems with indoor air quality, ventilation standards were established in most western countries. Unfortunately, this did not happen on an internationally coordinated level, giving way to the introduction of a wide range of sizing rules. As there is no common methodology, like the one that was developed for non-residential buildings by CEN [1], that is used for the different standards, the flow rates proposed in them can't be compared easily. AIVC listed the requirements of 15 standards without attempting to analyse their performance [2]. the sizing rules to a reference dwelling and found that the design air change rate in the majority of standards is around 0.5 ACH.

In the moderate climate region of West-Europe, especially in Belgium, the Netherlands, France and the UK, simple exhaust mechanical ventilation systems dominate the residential ventilation market [3-5], while heat recovery ventilation and natural ventilation are the most common residential ventilation systems in northern and southern Europe respectively. Such simple exhaust systems are composed of a mechanical exhaust fan, ducted to a series of vent holes in the different 'wet' spaces in the dwelling such as kitchens, toilets, bathrooms and service rooms, combined with externally (trickle ventilators) and internally (transfer grilles) mounted air transfer devices [6]. Since the introduction of ASHRAE 62.2, this kind of ventilation system is also rapidly achieving a dominant position in the US residential ventilation market, although the use of trickle ventilators is usually omitted and not treated as such in the standard. The sizing rules for the trickle ventilators in the standards of the 4 European countries also demonstrate little uniformity, requiring the design flow rate, which itself is different for all standards, to be achieved at a different design pressure difference across the ventilator, ranging between 1 and 20 Pa.
The total air change rate achieved by simple exhaust ventilation systems can be considerably different from the flow rate of the fan due to adventitious ventilation and infiltration [7]. The importance of the extra flow rate is mainly related to the sizing of the trickle ventilators relative to the flow rate of the fan [8]. Therefore, the ventilation heat loss of exhaust ventilation systems can't be assessed comprehensively by simple comparison of the design flow rates. In addition, the air flow in the system is controlled by the mechanical flow rate only in the 'wet' spaces, whereas the flow rate in the rest of the dwelling, which comprises the main living spaces, is governed by much less stable driving forces such as wind and buoyancy. Since the occupants spent the vast majority of time in the livings spaces [9], the indoor air quality (IAQ) achieved in these spaces will be the dominant contributor to perceived air quality [10]. Again, the design flow rates will not be a good metric for assessing the performance of simple mechanical exhaust ventilation systems.
Presenting the results from a multi zone simulation based performance assessment of simple mechanical ventilation systems sized in accordance with the Belgian [11], British [12], Dutch [13], French [14, 15], and ASHRAE [16] residential ventilation standards, this paper aims to provide a comprehensive analysis of the performance of the ventilation systems proposed in the standards. The 4 european countries are chosen because of the dominance of exhaust ventilation in their ventilation market and their geographical clustering. Although exhaust ventilation historically also represents a large part of the residential ventilation market in the Nordic countries, their cold climate [7] and recent market evolutions favour heat recovery ventilation. Therefore they were not included. The ASHRAE standard was chosen for its large geographical applicability and it's authority in HVAC design. Additional motives include the fact that it's promotion of exhaust ventilation is novel in the US and it's recent publication. The sizing rules of each standard are applied to 5 common dwelling typologies and monte carlo analysis is used to consider the sensitivity of the results to the boundary conditions used.

SIZING IN THE STANDARDS

As was explained in the introduction, the sizing rules for simple exhaust residential ventilation systems put forward are different in the Belgian, Dutch, French, UK and ASHRAE standards. In this section, the specific rules found in each of the standards are summarized. If different sizing rules are provided for continuous and demand controlled systems, only those for continuous systems are considered.

Belgium

The Belgian standard requires a design flow rate of 1 $l/s*m^2$ for each occupied space. For the main living space, this design flow rate should be at least 21 l/s and can be limited to 42 l/s,

while for bedrooms, studies… the minimum value is 7 l/s and the design flow rate can be limited to 20 l/s. For kitchens, bathrooms and service rooms, a minimum design flow rate of 14 l/s should be taken into account, while it can be limited to 21 l/s. The design flow rate for a toilet is 7 l/s.
The occupied spaces and the wet spaces should be connected to each other or via circulation spaces by transfer grilles sized at 7 l/s at 2 Pa pressure difference, which corresponds to 70 cm², except for the kitchen, in which the transfer grille should be sized twice as large. Each living space, bedroom, study… should be connected to the outdoor environment by a trickle ventilator sized at the design flow rate for that space at 2 Pa pressure difference.

The Netherlands
With a design flow rate of 0.9 l/s*m²for each occupied space and minimum design flow rates of 7 l/s in bedrooms, studies and toilets and 14 l/s in bathrooms and service rooms , the Dutch standard's sizing rules are quite similar to those in the Belgian standard. The minimum design flow rates for the kitchen, however, is set at 21 l/s instead of 14 l/s, while in the main living space, only 7 l/s is required as opposed to 21 l/s in the Belgian standard. Furthermore, Trickle ventilators should be sized at the design flow rate at 1 Pa pressure difference and transfer grilles should have a free face area of 12 cm² multiplied by the design flow rate for that space. As a consequence, the size of the trickle ventilators and transfer grilles is larger compared to the Belgian standard's sizing rules.

France
The design flow rate for each of the 'wet' spaces in the French standard depends on the number of 'main' spaces in the dwelling, eg. living spaces, bedrooms, studies.... The design flow rates of the trickle ventilators in the remaining spaces are also defined as a function of the number of 'main' spaces. For dwellings with only 1 or 2 'main' spaces, the design flow rate is increased for higher total design flow rates in the 'wet' spaces.
The trickle ventilators should be sized to the design flow rate at 20 Pa pressure difference, while the transfer grilles should be sized to the design flow rate at 5 Pa and 2.5 Pa for 'wet' and 'main' space grilles respectively. As a consequence, the size of components is typically smaller compared to the Dutch and Belgian standard's sizing rules.

The UK
Simple exhaust ventilation is denominated 'extract ventilation' in the British standard. Design flow rates of 13, 8 and 6 l/s are required for kitchens, bathrooms and toilets respectively. Service rooms are treated as bathrooms. In addition to these design flow rates per space, the total extracted flow rate should not be less than 9 l/s, increased with 4 l/s for each bedroom.
The equivalent, referenced to a round sharp edged opening, free face area of transfer grilles is set at 76 cm², that of the trickle ventilators at 25 cm².

ashrae
The design flow rate for kitchens proposed in the ASHRAE standard is 5 ACH, while 10 l/s is required for bathrooms. The total design exhaust flow rate for a dwelling is at least 0.05 l/s*m², increased with 7 l/s for the first bedroom and 3.5 l/s for each additional bedroom. No requirements for trickle ventilators or transfer grilles are included.

SIMULATION MODEL

The results presented in this paper are based on airflow simulations. These were executed in the multi-zone airflow simulation package Contam [17], which takes effects of buoyancy, wind and fan pressure into account and is used in numerous ventilation studies [eg. 18, 19].

Building Geometries

To assess the quality of the sizing rules in the standards discussed above, simple exhaust residential ventilation systems have been designed is accordance with the different standards for five different dwelling typologies. Their size and layout is based on an extensive survey of 200 dwellings in Belgium built in the 1990's [20]. Their characteristics have been checked regularly with the evolution of newly built dwellings and still correspond well with current building practise. Four dwellings are single family houses, one is a flat. All dwellings have the same useful floor area corresponding to the mean from Belgian national statistical figures. All houses comprise a living room, 3 bedrooms, kitchen, bathroom, toilet, service room, and hall way, with a total net floor area of about 150 m². The detached, semi-detached and terraced house hold a separate study. The dwellings differ in building compactness, ranging from a detached bungalow to a flat in a 6-floor apartment building. The compactness is defined as the ratio of the volume to the heat loss area (Table 1).

	Bungalow	Detached	Semi-detached	Terraced	Flat
Compactness	0.9 m	1.3 m	1.6 m	2.1 m	3.8 m
Heated volume	557.3 m³	528.7 m³	521.0 m³	493.6 m³	450.0 m³
Heat loss area	611.3 m²	395.4 m²	330.1 m²	231.9 m²	118.4 m²
Number of floors	1	2	2	3	1 (of 6)

Table 1. Geometrical characteristics of reference dwellings.

Building envelope and ventilation system model

The airflow in the dwellings has been modelled taking into account both the ventilation system and leakage. Overall leakage, characterized by the v_{50} value, is modelled by means of cracks in the roof and wall surface. The v_{50} value is the ratio of the air leakage rate at 50 Pa pressure difference and the building envelope heat loss area. According to observations by Bossaer [21], the specific leakage rate through roof and walls has a 2/3 ratio, which has been implemented in the model. Each wall is fitted with two cracks, one at 1/4 of its height and the second one at 3/4. The internal doors are simulated with additional cracks in the walls. For the indoor walls, a fixed specific leakage value is assumed. This methodology is in agreement with guidelines given in EN 15242 [22]. In the results presented, a specific airleakage (v_{50}) of 3 m/h is used, representing the best quartile of measured airtightness values in a measurement campaign in Flanders in the late 90's [21]. A recent measurement campaign [23], shows a tendency towards this level of airtightness in newly built dwellings.

The design exhaust flow rate for the wet spaces in the dwellings according to the different standards are listed in Table 2. Since the design exhaust flow rates in most standards are size independent, the flow rates are mostly the same for all 5 dwellings. If this is not the case, the range of the design flow rates for that space in the 5 dwellings is mentioned. Since trickle ventilators are sized with respect to different reference pressure differences in the five standards, the flow coefficient at 1 Pa pressure difference, assuming a flow exponent of 0.5 and a simple power law flow profile, for the trickle ventilators in the various living spaces of the dwellings according to the different standards are listed in Table 3. Note that the ASHRAE standard does not require the installation of trickle ventilators.

Space	Belgium	France	Netherlands	UK	ASHRAE
Kitchen	14 l/s	38 l/s	21 l/s	13 l/s	33 - 48 l/s
Bathroom	14 l/s	8 l/s	14 l/s	8 l/s	10 l/s
Service room	14 l/s	4 l/s	14 l/s	8 l/s	10 l/s
Toilet	7 l/s	8 l/s	7 l/s	6 l/s	-
Total exhaust flow rate	49 l/s	58 l/s	56 l/s	35 l/s	53 - 68 l/s

Table 2. Exhaust flow rates in the different standard.

All mechanical exhaust vents were modelled as constant volume flow rate components in the respective zone node, while transfer grilles and trickle ventilators were modelled with single direction power law flow components with a flow exponent of 0.5 [24]. All systems were modelled with windows and internal doors closed, in order to simulate the performance of the systems as such, without user interaction.

Space	Belgium	France	Netherlands	UK	ASHRAE
Living room	20 - 29 l/s/Pa	3 l/s/Pa	26 - 40 l/s/Pa	2 l/s/Pa	-
Master bedroom	11 - 14 l/s/Pa	2 l/s/Pa	14 - 21 l/s/Pa	2 l/s/Pa	-

Table 3. Trickle ventilator flow coefficient in living room and master bedroom in the 5 standards.

Boundary conditions and Monte Carlo

As Contam is a ventilation model only, it cannot calculate transient room or duct temperatures. Therefore, for simplicity, the temperature inside the building and all ducts has been set to 18 °C, the inside temperature fixed by the Belgian EPBD calculation procedure , which corresponds to the average temperature measured in Belgian dwellings [21]. The test reference year for Ukkel, Belgium was used as the outdoor climate for all simulations, with hourly mean values for temperature, humidity, wind speed and direction.

The production of CO_2 within the model is only related to the occupants' metabolism and corresponds to their whereabouts. A constant outdoor background concentration of 350 ppm is assumed.

For most of the other boundary conditions, a Monte-Carlo (MC) approach, as proposed by Van Den Bossche et al. [25, 26], has been used in this study. In this approach, instead of fixing 1 value for each input data, a distribution is determined for the key parameters and multiple simulations are carried out with different values of these parameters. The following input variables are considered with a probabilistic approach (Normal distributions are mentioned as N(mean, standard deviation):

- Façade orientation - interval [0°; 359°]
- C_p coefficients - interval of the 6 AIVC tables [27]
- Terrain roughness α , partially correlated with the C_p coefficients – interval [0.149 – 0.377]
- Sunday is the ...th day of the year - interval [1;7]
- Moisture production from domestic activities - normal distribution (see below)
- Production of moisture and carbon dioxide by occupants - normal distribution (see below)
- Number of occupants - specific distribution
- Weekday / weekend occupancy schedules - specific distribution

The number of parameters can be considered to be small, so 100 datasets will be used to perform the simulations. Moisture production for domestic activities is based on data available in the EU technical report on design and dimensioning of residential ventilation systems [28]. The production in the bathroom is N(0.5, 0.05) l/s, in the service room cloth drying is N(1, 0.05)l/s and for cooking, a half hour cycle of N(0.6, 0.05) l/s, N(1, 0.1) l/s and N(1.5, 0.1) l/s for 10 minutes each is used. The production of moisture and carbon dioxide by occupants is modelled as a linear function of the metabolism, which varies for each activity (eg. N(0.8, 0.05) Met for sleeping, N(2, 0.1) Met for cooking). Based on EN 15251[29], the production rate is 11.875 l/h/Met for CO_2 and 34.375 g/h/Met for moisture. The number of occupants and the occupancy schedules are considered with a specific distribution based on the social demography and time use studies in Belgium. Based on the available data, 100 different data sets were compiled with different occupancy schedules. The number of occupants in the building varies from one to six (1: 3%, 2: 21%, 3: 31%, 4: 32%, 5: 10%, 6: 3%), with an average of 3.34 persons per building.

Assessment parameters

Through the correlation between excess CO_2 concentration and mean percentage of dissatisfied [30], excess CO_2 concentration is now widely accepted as a proxy for perceived indoor air quality [1], especially if the main pollution sources are related to the human metabolism. In contrast to the basic model, steady state conditions are rarely applicable to real ventilated environments. CO_2 concentrations are inherently transient, due to changes in environmental boundary conditions. Additionally, the relevant CO_2 sources tend to constantly move around in the multi-spaced dwelling, introducing discontinuous sources and further increasing the transient character of the indoor air quality. The amount of time an occupant spends in an environment within the different IDA classes [1] and the heating season averaged CO_2 concentration to which an occupant is exposed were selected as IAQ metric for this study. The best IDA class, IDA 1, corresponds to exposure lower than 400 ppm excess CO_2, while the lowest class, IDA 4, exposure to concentrations in excess of 1000 excess CO_2, is considered to correspond to poor perceived indoor air quality.

The total, heating season averaged, convective heat loss through the combination of intended ventilation, adventitious ventilation and infiltration is used to assess the energy performance of the different sizing rules. Fan power was not taken into account because it is very system specific.Since heat loss and exposure reduction are opposing interests, they have to be trade off against each other. This trade-off is addressed by means of the concept of pareto optimality. Pareto optimal cases are cases where none of the other standards achieve better results on both indoor air quality and heat loss.

RESULTS AND DISCUSSION

Exposure to carbon dioxide

Table 4. lists the time fractions spent in the different IDA classes considering all 334 occupants from the 100 simulations in the monte-carlo analysis for all 5 standards in all 5 geometries. The Belgian, Dutch and French standard consistently achieve similar indoor air quality, at a level that is considerably higher than that achieved by the British standard. The performance of the systems sized according to the ASHRAE standard, relative to the other standards, is much more prone to variation due to the fact that the flow rate is mainly concentrated in the kitchen and expressed as a function of its volume. Although the flow rates are similar in magnitude to those prescribed in the French standard, the lack of transfer grilles in the ASHRAE standard prevents a good distribution of this flow rate through the rest of the dwelling. Position and size of the kitchen relative to the other spaces therefore has a large influence on the achieved performance.

		Belgium	**France**	**Netherlands**	**UK**	**ASHRAE**
Flat	IDA 1	0.662	0.757	0.670	0.430	0.701
	IDA 2	0.184	0.185	0.162	0.221	0.114
	IDA 3	0.133	0.055	0.140	0.236	0.174
	IDA 4	0.021	0.002	0.028	0.114	0.010
Terraced	IDA 1	0.714	0.724	0.679	0.391	0.527
	IDA 2	0.164	0.130	0.151	0.191	0.197
	IDA 3	0.098	0.136	0.125	0.252	0.150
	IDA 4	0.024	0.011	0.046	0.166	0.127
Semi-Detached	IDA 1	0.695	0.702	0.719	0.452	0.634
	IDA 2	0.162	0.185	0.152	0.195	0.182
	IDA 3	0.116	0.106	0.118	0.227	0.161
	IDA 4	0.027	0.008	0.011	0.126	0.024
Detached	IDA 1	0.696	0.705	0.706	0.429	0.509
	IDA 2	0.163	0.142	0.148	0.183	0.185
	IDA 3	0.102	0.128	0.097	0.238	0.175
	IDA 4	0.039	0.025	0.048	0.149	0.131

Bungalow	IDA 1	0.719	0.666	0.693	0.436	0.482
	IDA 2	0.171	0.180	0.161	0.190	0.173
	IDA 3	0.103	0.144	0.137	0.215	0.205
	IDA 4	0.006	0.010	0.010	0.158	0.140

Table 4. Time fractions spent in the different IDA classes considering all 334 occupants from the 100 simulations in the monte-carlo analysis for all 5 standards in all 5 geometries.

Air change rates and ventilation heat loss

The median as well as first and third quartile values of the air change rate for all dwellings and all standards are listed in table 5. A clear distinction is seen between the Belgian and Dutch standard on the one hand and the British, French and ASHRAE standard on the other. The air change rate in the latter group is much less susceptible to variation due to changing boundary conditions due to the smaller sizing of trickle ventilators or the absence thereof compared to the Belgian and Dutch standard that require relatively large trickle ventilators. The Belgian, Dutch, French and ASHRAE standard all achieve median air change rates close to 0.5 ACH, that, as was mentioned in the introduction, can be considered a consensus value for residential buildings, while the system sized according to the British standard consistently renders about 40% lower values.

Figures 1 shows the cumulative distribution of the ventilation heat loss for both the detached house for all 5 standards, taking into account both intended and adventitious ventilation as well as infiltration. The same conclusions as with the air change rate apply.

		Belgium	**France**	**Netherlands**	**UK**	**ASHRAE**
flat	median	0.58	0.69	0.66	0.42	0.80
	Q1	0.57	0.68	0.65	0.41	0.79
	Q3	0.59	0.69	0.67	0.42	0.80
terraced	median	0.50	0.51	0.54	0.31	0.50
	Q1	0.46	0.51	0.50	0.31	0.50
	Q3	0.68	0.52	0.72	0.32	0.50
semi-detached	median	0.47	0.48	0.54	0.29	0.53
	Q1	0.42	0.47	0.48	0.29	0.53
	Q3	0.66	0.50	0.72	0.32	0.54
detached	median	0.64	0.59	0.72	0.39	0.55
	Q1	0.54	0.57	0.61	0.36	0.53
	Q3	0.85	0.67	0.96	0.45	0.61
bungalow	median	0.55	0.53	0.61	0.36	0.59
	Q1	0.47	0.51	0.54	0.33	0.57
	Q3	0.75	0.61	0.80	0.45	0.66

Table 5. Median, first quartile and third quartile air change rate from the 100 simulations in the monte-carlo analysis for all 5 standards in all 5 geometries.

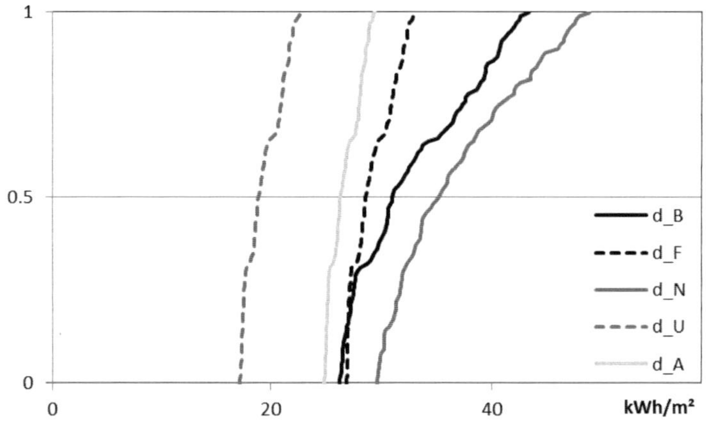

Figure 1. Cumulative distribution of ventilation heat loss in the detached dwelling for all 5 standards.

Trade-off ventilation heat loss and indoor air quality

If the trade-off between heat loss and indoor air quality is considered (Figure 2), using the average ventilation heat loss for the former and the average carbon dioxide to which the occupants are exposed as the criterion for the latter, the French and British standard provide pareto optimal solutions for each dwelling, although the fact that the indoor air quality achieved by the British standard is to be considered 'poor' 15% of the time is a cause of concern. Compared to the French standard, for example, the exposure to carbon dioxide of the cases using the ASHRAE standard is on average 40% higher, with higher or comparable heat losses (+16 to -8 %). Similarly, the ventilation heat loss in 4 cases using the Dutch standard is on average 20 % higher than that in the cases with the French standard for higher or comparable carbon dioxide exposure (+10 to -5%). In the flat, the heat losses using the French standard were comparable to those using the Dutch standard (+4%) with lower exposure to carbon dioxide (-26%).

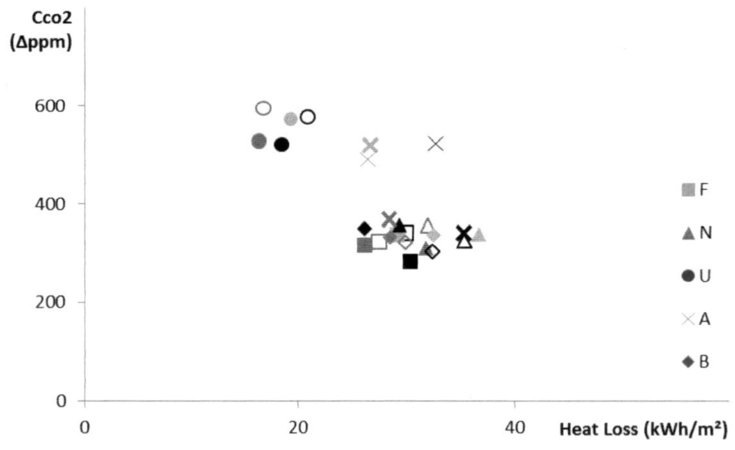

Figure 2. Heating season averaged ventilation heat loss traded-off against average carbon dioxide concentration to which occupants are exposed during the heating season for all 5 standards in all 5 dwellings (flat – black solid

fill, terraced – grey line, semi-detached – dark grey solid fill, detached – light grey solid fill, bungalow – black line, symbols correspond to the standards, as shown in the legend)

CONCLUSION

Sizing rules in residential ventilation standards lack uniformity in both methodology and resulting design flow rates. Mere comparison of design flow rates is case sensitive and, due to effects of infiltration, adventitious ventilation and occupancy, ill-suited to assess performance of an exhaust ventilation system with regard to the achieved indoor air quality and energy cost in terms of heat loss. A performance assessment of residential mechanical exhaust ventilation systems using five common dwelling typologies and the sizing rules put forward in the Belgian, British, Dutch, French and ASHRAE residential ventilation standards in multi-zone simulations with Monte-carlo based sensitivity analysis presented above showed that the performance of the different cases proved to be substantially different. An occurrence of poor perceived air quality in 5% or less of the occupation time for the Belgian, Dutch and French standard, and about 15% for the British and ASHRAE standard was found.

The spread observed in the performance of the cases using the ASHRAE standard can be attributed to the larger impact of geometrical parameters on the system design in this standard.

The total air change rate was close to or greater than the consensus value of 0.5 ACH in most cases, except in the cases using the British standard, where it was consistently about 40% lower. The cases using the Belgian and Dutch standards, with relatively large trickle ventilators, rendered the air change rates most sensitive to changes in boundary conditions. When the trade-off between indoor air quality and heat loss is considered, the cases with the Dutch and ASHRAE standard did not achieve pareto-optimal performance.

Considering the performance spread observed, harmonization of residential ventilation standards is to be recommended. The design philosophy of the French standard proves to be a good basis for exhaust ventilation design with high occurrence of good perceived air quality, minimized ventilation heat loss and robust performance. It's combination of moderately high exhaust flow rates, large transfer devices and small trickle ventilators should explored further when new, more uniform standards are developed.

REFERENCES

[1] CEN, Ventilation for non-residential builidings - performance requirements for ventilation and room-conditioning systems, in, Brussels, 2007.

[2] M.J. Limb, A Review of International Ventilation, Airtightness, Thermal Insulation and Indoor Air Quality Criteria, in: AIVC Technical Notes, AIVC, 2001, pp. 203.

[3] F. Durier, Trends in the French building ventilation market and drivers for changes, in: AIVC (Ed.) Ventilation Information Papers, 2008.

[4] W. De Gids, Ventilation in Dutch houses - a study in a representative sample of the dutch housing stock, in: 24th AIVC conference: ventilation, humidity control & energy, AIVC, Washington DC, 2003.

[5] K. Clarys, EPB: analyse van Vlaamse woningen en overheidsmaatregelen op energetisch vlak, Architecture and Urban Planning, Ghent University, 2012.

[6] CEN, Ventilation for buildings - Performance testing of components/products for residential ventilation - part. 1: externally and internally mounted air transfer devices, in, Brussels, 2004.

[7] J. Laverge, A. Janssens, Heat recovery ventilation operation traded off against natural and simple exhaust ventilation in Europe by primary energy factor, carbon dioxide emission, household consumer price and exergy, Energy and Buildings, 50 (0) (2012) 315-323.
[8] J. Laverge, Comparison of the use of trickle ventilators in european residential ventilation standards, in: Roomvent 2011, Trondheim, 2011.
[9] I.M. Glorieux, J., Belgisch tijdsbudgetonderzoek, in, Brussels, 2008.
[10] P.O. Fanger, Introduction of the olf and decipol unit to quantify air-pollution perceived by humans indoors and outdoors, Energy and Buildings, 12 (1) (1988) 1-6.
[11] BIN, Ventilatievoorzieningen in woongebouwen, in, Brussels, 1991.
[12] Building regulations 2000: approved document F - means of ventilation, in: O.f.t.D.P. Minister (Ed.), 2006.
[13] NNI, Ventilatie van gebouwen - Bepalingsmethoden voor nieuwbouw, in, Delft, 2006.
[14] Arrêté du 24 mars 1982 relatif à l' aération des logements, modifié par Arrêté du 28 ocotobre 1983, in, journal officiel de la république française du 15 novembre 1983, 1983.
[15] Installations de ventilation mécanique contrôlée – Règles de conception et de dimensionnement, in: XP P50-410 (DTU 68.1), 1995.
[16] Ashrae, Ventilation and Acceptable Indoor Quality in Low-Rise Residential Buildings, in, Atlanta, GA, 2010.
[17] W.S. Dols, A tool for modeling airflow & contaminant transport, Ashrae Journal, 43 (3) (2001) 35-+.
[18] G. Nirvan, F. Haghighat, L. Wang, H. Akbari, Contaminant transport through the garage – House interface leakage, Building and Environment, 56 (0) (2012) 176-183.
[19] Y.L. Chen, J. Wen, The selection of the most appropriate airflow model for designing indoor air sensor systems, Building and Environment, 50 (0) (2012) 34-43.
[20] WTCB, WENK, VLIET, SENVIVV: Studie van de Energieaspecten van Nieuwbouwwoningen in Vlaanderen: Isolatie, Ventilatie, Verwarming, in, (in dutch), Brussels, 1998.
[21] A. Bossaer, J. Demeester, P. Wouters, B. Vandermarke, W. Vangroenweghe, Airtightness performances in new Belgian dwellings, in: 19th AIVC conference: ventilation technologies in urban areas, AIVC, Oslo, 1998, pp. 77-84.
[22] CEN, Ventilation for buildings - Calculation methods for the determination of air flow rates in buildings including infiltration, in, Brussels, 2007.
[23] J. Laverge, S. Vandevelde, T. Debrauwere, M. Delghust, A. Janssens, Airtightness assessment of newly build single family houses in Belgium, in: Buildair 2010, Copenhagen, 2010.
[24] P. Karava, T. Stathopoulos, A.K. Athienitis, Investigation of the performance of trickle ventilators, Building and Environment, 38 (8) (2003) 981-993.
[25] N. Van Den Bossche, A. Janssens, N. Heijmans, P. Wouters, Performance evaluation of humidity controlled ventilation strategies in residential buildings, in: Thermal performance of the exterior envelopes of whole buildings X, Clearwater, 2007, pp. 7.
[26] J. Laverge, N. Van Den Bossche, N. Heijmans, A. Janssens, Energy saving potential and repercussions on indoor air quality of demand controlled residential ventilation strategies, Building and Environment, 46 (7) (2011) 1497-1503.
[27] M.W. Liddament, A guide to energy efficient ventilation, in: AIVC Guides, AIVC, Brussels, 1996.
[28] CEN, Ventilation for buildings - Design and dimensioning of residential ventilation systems, in, Brussels, 2005.
[29] CEN, Criteria for the indoor environment, inluding thermal, indor air quality, light and noise, in, Brussels, 2005.
[30] CEN, Ventilation for buildings - Design criteria for the indoor environment, in, Brussels, 1998.

DESIGN OF HVAC SYSTEMS FOR DEPRIVED COMMUNITY HOUSES IN YORKSHIRE AND THE HUMBER REGION IN THE UK

Hasim Altan[*1], Zakaria Emankaf[1], Young Ki Kim[1] and Mohamed Refaee[1]

1 Building Environments Analysis Unit (BEAU), The University of Sheffield
ICoSS, 219 Portobello, Sheffield S1 4DP, United Kingdom
**Corresponding author: h.altan@sheffield.ac.uk*

ABSTRACT

The stock housing of England (UK) constitutes the oldest housing stocks in the world. Indeed, 63 per cent of all dwellings were built before 1960s and thus most of the people in the UK live in an old house or at least a house that is more than 50 years old. The most common dwelling types in the UK are the semi-detached and terraced houses, and particularly within deprived communities. In deprived communities, houses suffer from poor indoor conditions and building standards of energy performance. They always have the issue of having to be well heated in winter and have to burn more fuel as a consequence. Heating and ventilation are the biggest part of energy consumption in a house. In this study, an ongoing investigation is whether the current houses in deprived communities, in a situation of pre-refurbishment, are within the standards of the Chartered Institution of Building Services Engineers (CIBSE) recommendations in terms of heat, ventilation and possibly cooling. The methodology used is to model different kind and the most common dwellings, and to conduct dynamic computer simulations, for each one, in terms of energy consumption and performance analysis. As a result, this would help to highlight the current energy consumption, and to find out the weaknesses in terms of energy and comfort parameters such as indoor conditions of temperatures and relative humidity levels. In addition, further studies to investigate issues related to indoor air quality and ventilation aspects have been carried out. Furthermore, several design scenarios of a 'Heating, Ventilation and Air Conditioning' (HVAC) system, less energy-consuming and in accordance with the CIBSE guidelines in order to improve the indoor comfort of deprived community houses while reducing the energy consumption and the carbon footprint, has been presented in the study.

KEYWORDS

HVAC systems, airtightness, old dwellings, energy efficiency, housing retrofit, deprived communities.

INTRODUCTION

In deprived communities, houses suffer from poor standards of energy performance and they always have the issue of having to be well heated in winter and have to burn more fuel. The carbon footprint is higher than an equivalent energy efficient dwelling. One of the main consequences is that occupants fall in fuel poverty and may also experience health problems [2, 7]. In this study, the approach consists of analysing the current dwellings' situation, which is in pre-refurbishment stages. The analysis has been carried out using dynamic simulations with IES VE software [10] for different kind of dwellings that can be found in deprived communities. These dwellings are semi-detached or terraced houses. The computer models of the houses have been assigned with all fabric and constructions properties, defining the current thermal condition and ventilation. Moreover, the simulation studies would enable to find out the weaknesses in the current heating and ventilation systems, and see whether these fulfil the UK compliance and ratings. As a result, if the system consumes too much energy, an improvement of the system of heating and ventilation with a design of HVAC system less energy-consuming would be suggested.

The BIG Energy Upgrade (BEU) programme is a consortium programme in the Yorkshire and the Humber region in the UK. It is also known as the Energy Innovation for Deprived Communities (EIDC). The total investment is about £14.9 million which is partially funded by the European Regional Development Fund (ERDF) as this is a project part-financed by the European Union. This is a support for the region's economic development through the Yorkshire and the Humber between 2007 and 2013 [13]. The aim of this innovative programme is to deliver energy efficiency and renewable energy measures in deprived houses. The measures are going to be applied to a minimum of ten thousand deprived communities across six local authorities within Yorkshire and the Humber region in the UK. In this study, case study houses have been selected from this programme.

HOUSING IN ENGLAND

The end of World War I social housing was built in mass scale. After the World War II, the need for mass housing was even greater especially after all damages due to the aircrafts bombarding. England has one of the oldest housing stocks in the world; 63% of all dwellings were built before 1965, 39% before the World War I and with 4% before 1851 [6] (see Figure 1).

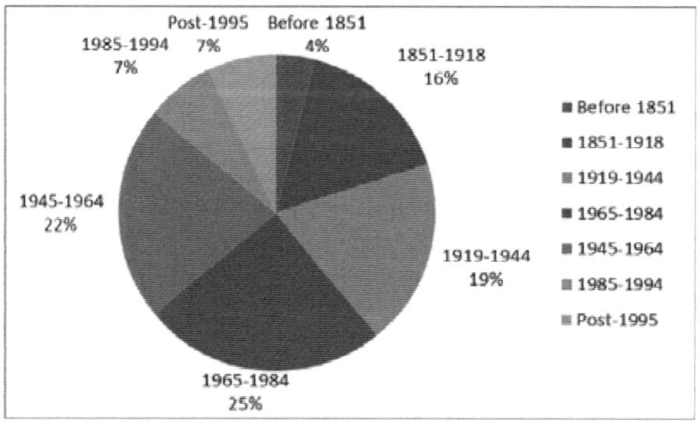

Figure 1. Age of the housing stock in England.

New houses that have been built after 1995 added only 7% to the existing housing stock. As a result, existing houses would require energy conservation and therefore most improvements can be and should be made by upgrading old houses. Accordingly, most of people living in the UK are in an old house or at least 50 years old house [6, 7, 8]. These houses are not energy-efficient even though improvements were made such as installing double glazing, new boilers or adding insulation, especially after the oil crisis in 1970s.

There are nearly 25 million of dwellings in England. One third of these dwellings are semi-detached houses and other third is terraced houses. In this study, a particular attention would be made to the latter type of dwellings and they represent 56% of the dwellings UK-wide (see Table 1). In Yorkshire and the Humber region, there are a total of 2,294,400 of dwellings [7].

Dwelling type	Number (000s)	Percentage
Semi detached	7,052	28
Terraced	6,876	28
Flats	4,716	19
Detached	4,021	16
Bungalow	2,086	8
Other	74	1
Total	**24,825**	**100**

Table 1. Roadmap to 60%: Eco-refurbishment of 1960s flats [7].

Terraced Houses

The terraced house is a very compact model for mixed use and very affordable house type. The population growth at the time of the Industrial Revolution was huge, and as a result, the migration of workers from the countryside to the cities had caused housing booms which enabled the creation of millions of houses. These houses were for middle-classes and poor communities. The terraced houses were small, especially in deprived areas. A lot terraced houses survived because of their flexibility and the popularity with the public (see Figure 2).

Figure 2. Front of a mid-terraced house in North East Lincolnshire (left) and an end-terraced house in Leeds (right).

The row of terraced houses is built, side by side with mid-terraced house in the middle row. The last house of the block is an end-terraced house and has normally three facades (see Figure 2).

Semi-detached Houses

A semi-detached house is a house built side-by-side with a party wall in between. It is a pair of houses with the layout of a mirror image. They have front, rear and any one side open spaces. This type of dwellings can be thought as being row housing and detached homes (see Figure 3).

Figure 3. Rear of a semi-detached house in Doncaster (left) and a single family detached house in Yorkshire (right).

The semi-detached house is the most common dwelling in England with 28% of the total housing stock which represents more than 7 million of houses [6, 7, 8] (see Table 1).

METHODOLOGY

In housing, it is very important to take an integrated design approach to make sure that insulation and HVAC would work optimally well together. Below are more details about the key aspects have been looked at in the study:

Insulation

One of the main action items, to be carried out, concerns insulation. A good insulation will slow down heat losses which will reduce the heat requirement to keep the internal temperature at an acceptable set point. The insulation of the loft space is a cheap and easy way to minimise the overall heat losses. Also, thermal bridges must be considered to avoid all condensation problems. Condensation occurs when warm air is in contact with a cold surface and this occurs often on single-glazed window. A lot of old houses had been constructed too draughty to avoid the moisture from rainfall and ventilated to the outside. The latter fact is one cause of heat losses [7, 8]. Condensation issues can build-up moisture which is damageable to the building fabric such as structural timbers used for the roof structure. An old house needs to breathe even if the insulation and the airtightness have been improved, to avoid all condensation issues. A controlled-ventilation is required in any house because moisture will always be created in bathroom, kitchen by occupants.

Heating

Historically houses were heated to a much more low temperature than today. People wore more layers of clothes. In 1970s, the average UK houses were heated to 12°C to reach 18°C in 2003 [4, 5, 7, 8]. In this study, the temperatures are going to be set in the simulation, according to the building recommendations by CIBSE.

Ventilation

The ventilation in old houses has always been a problem. The latter is due to uncontrolled air infiltration and heat escaping through openings such as windows, gaps around doors, or even due to building fabric that can be leaky such as in timber-framed [4, 5, 7, 8]. Usually, an airtightness test has to be performed to determine accurately where draughts are located. This test can target the improvements in the specific areas. An efficient draught-stripping is a good energy efficient measure especially in old houses. However, it should allow the evaporation of moisture and to dry out the traditional construction such as rain-soaked solid walls. There are two kinds of ventilation; controlled and uncontrolled. A good strategy of ventilation has to be thought in purpose of being energy efficient dwellings.

Energy Performances

Since the introduction of the home information packs to the UK's housing market, all dwelling transaction requires an Energy Performance Certificates (EPCs). The rating of energy efficiency is based on A to G scale (see Figure 4). This system is based on the legislation from the European Energy Performance of Building Directive (EPBD) [14]. EPCs gives information about the energy use and the typical energy costs as well as a recommendation with suggestions to reduce energy use and save money by making homes more energy efficient. All homes bought, sold or rented is required to have an EPC.

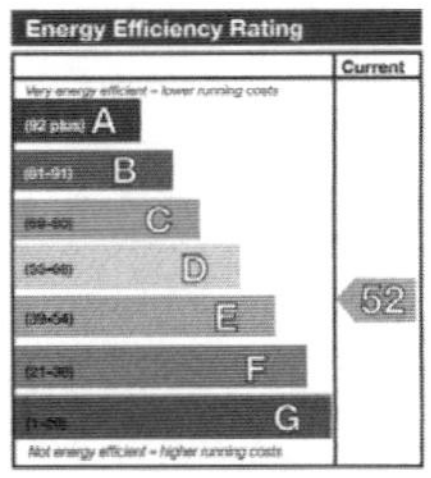

Figure 4. EPC rating for a house type [3]

BUILDING RECOMMENDATIONS

In the study, building recommendations would be mainly provided by the CIBSE guides and some by ASHRAE.

Thermal Comfort

The comfort is defined as the condition of mind that expresses satisfaction with environment [1, 4, 5]. The indoor environment should be designed and controlled to assure the comfort. There are differences in the perception of each one and the evaluation is subjective, what can bring dissatisfaction in buildings or dwellings. The main goal of the design is to minimise this dissatisfaction as much as possible. The environmental factors considered here in the study, include the thermal and indoor environment. There are a couple of parameters that affects the thermal comfort such as the air temperature, the mean radiant temperature, the humidity or still besides these factors there are personal factors as the metabolic heat production or the clothing [1, 4, 5]. The humidity has effect of warmth tough for sedentary people may become apparent if the operative temperature is above 26 to 28°C. The influence of humidity on warmth may be ignored if it is in the range between 40 and 70% [1]. In general, bathroom and kitchen may be prone to a high humidity due to evaporation from moisture and from poor ventilation.

Overheating

For some buildings, there are some risks of overheating, especially in summertime [4, 5, 11, 12]. There are limited recommendations to decide whether or not some cooling is required. The general summer indoor comfort temperature for non-air conditioned building is 25°C for the living areas and 23°C for the bedrooms [4, 5, 11, 12]. In some cases, sleep may be impaired if the temperature is above 24°C. The benchmark for the summer peak temperature is 28°C for living areas and 26°C for bedrooms. The overheating criterion is that if 1% of occupied hours have an operative temperature above 26°C for bedrooms or 28°C for living spaces, then some cooling system will be required. Even if there is no overheating in most of the dwellings selected for simulation studies, the effect of climate change can make domestic buildings prone to temperatures above 28°C.

Carbon Dioxide

Carbon dioxide is a constituent that people exalt while breathing. It is measured to evaluate whether volumes of fresh outdoor air are being introduced into indoor air adequately. The outdoor level of CO_2 is usually from 300 ppm to 400 ppm. Usually, inside buildings, the CO_2 levels are greater than outside. The indoor CO_2 has to be below the guideline of 1,000 ppm otherwise some complains can be prevalent such as headaches, fatigue and irritation of eyes or throat [1]. If the recommendation is not met, the space should have better ventilation and in some case opt for a mechanical one.

CASE STUDIES

In this study, two case studies have been selected for the simulation analysis as representative of the UK existing housing stock and the typical dwelling of social housing.

Case Study 1: End-terraced House

This case study is in Leeds. It is a house with the ground floor and the first floor. It is fully double glazed. The house is a 40 years old dwelling which means that it has been built in 1970s. Accordingly with building surveys undertaken as part of BEU programme, there are 2 occupants who are a married couple and they are both smokers. The ventilation is made by opening one window at least and one kept open during nights. They have one pet and one cat. All this data would be useful to define the internal gains or the ventilation rate in the computer model constructed (see Figure 5). Again, it is assumed that the house is made with cavity wall as external walls and more information was provided from other sources (see Table 2) that for a 1970s house the cavity width is 50mm.

Decade	Development of cavity walls	Cavity width
1920s	Solid walls still dominate, but cavity walls grow in popularity.	Typically between 50mm and 100mm
1930s	Cavity walls become main form of construction, but some solid walls still built.	Typically between 50mm and 100mm
1940s and 1950s	Cavity width becomes standardised.	50mm
1960s	Concrete blocks used to inner leaf.	50mm
1970s	Lightweight blocks are introduced.	50mm
1980s	Partial fill cavity wall insulation introduced, cavity widths increased.	60-70mm
1990s onwards	Full fill cavity wall insulation becomes dominant.	50-100mm

Table 2. Cavity width by decades [8].

Figure 5. Case study of the end-terraced house in Leeds and the model built in IES VE.

Case Study 2: Mid-terraced House

The house is located in Leeds. It is a house with the ground floor and first floor. It is fully double glazed. The house is a 40 years old dwelling which means that it has been also built in 1970s. Again accordingly with building surveys, there are 2 occupants with one of them a smoker. The ventilation is made by opening one window at least and one kept open during nights. They have also 2 cats. All this data would be useful to define the internal gains or the ventilation rate in the computer model constructed (see Figure 6). As previously, it is assumed that the house is made with cavity wall as external walls with a thickness of 50mm.

Figure 6. Case study of the mid-terraced house in Leeds and the model built in IES VE.

RESULTS AND DISCUSSION

After several simulations with each scenario considered, a summary listing of the different scenarios have been presented (see Table 3). The scenario A and B are the cases without external insulation and with classic heating system. The situation C and D are the cases with external insulation and with classic heating system. The two last scenarios, E and F, are with alternative heating systems.

Case Study 1 & 2		Scenarios A	B	C	D	E	F
External insulation		no	no	yes	yes	yes	yes
CIBSE recommendations		no	yes	no	yes	yes	yes
Heating system	Central heating (boiler band A/radiators)	yes	yes	yes	yes	no	no
	Central heating (boiler band A/underfloor heating)	no	no	no	no	yes	no
	Ground source heat pump (underfloor heating)	no	no	no	no	no	yes

Table 3. Summary of the different scenarios undertaken.

Case Study 1 - Scenario A vs. B

The scenario A is the house without any external insulation and with boiler band A. Accordingly with building surveys; the house was under heated and therefore not complying with the CIBSE recommendations. The house in the scenario A consumes yearly 10.6MWh just for the heating system. On the other hand, the scenario B is exactly the same with scenario A but with the only exception that here the house would comply with the CIBSE recommendations, which means more comfortable for the occupants. To have this thermal comfort, the energy consumption in this case is 11.4MWh, which also means more than 7.5% higher than the scenario A. Figure 7 shows this higher consumption of the scenario B except from May to August which corresponds to the switch off period of heating system. With this comparison, it can be said that, to have a comfortable indoor environment, it is needed to increase the energy consumption which means not always affordable, especially in deprived community housing.

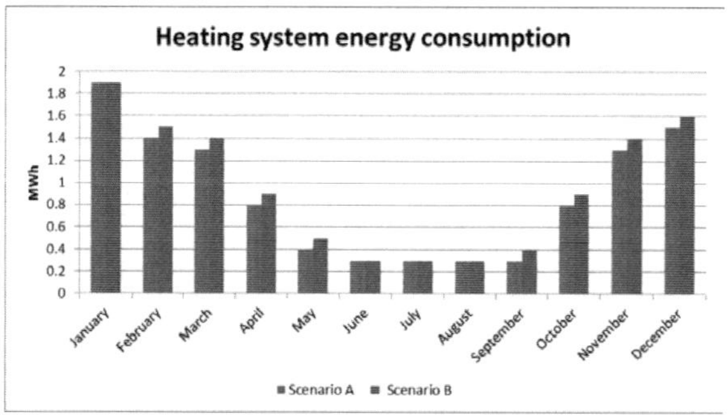

Figure 7. Energy consumption for the heating system, scenarios A and B.

Case Study 1 - Scenario A vs. C

The scenario C represents the same case that scenario A but with an energy saving action that consists in external insulation. Figure 8 shows that the energy consumption has decreased installing the insulation. The yearly consumption is about 9.7MWh, which represents a reduction of 8.5 % in comparison with the scenario A. As a result, it is not far off what was said about the insulation affect; it could reduce the heat losses by more than 10%.

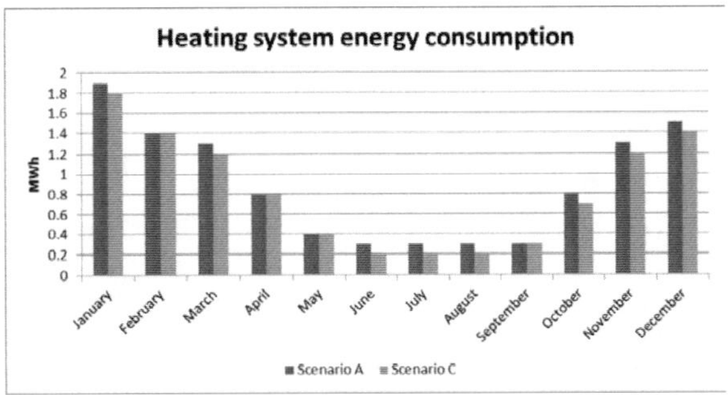

Figure 8. Energy consumption for the heating system, scenarios A and C.

Case Study 1 - Scenario D vs. F

The scenario D represents the case where the insulation is installed and that the CIBSE recommendations are complied with. The scenario F, is an alternative using the Ground Source Heat Pump (GSHP) with the underfloor heating. Figure 9 illustrated the big difference between the two scenarios, which is 67.6% on the yearly consumption.

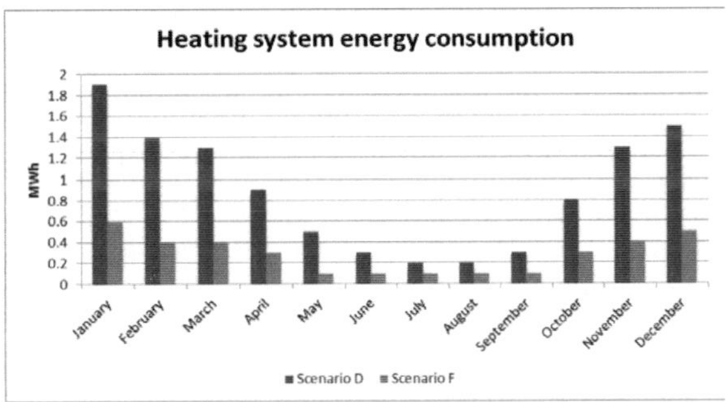

Figure 9. Energy consumption for the heating system, scenarios D and F.

Case Study 1 - Scenario C vs. F

This is the most important comparison. The last situation of the house is represented by the scenario C with fabric refurbishment and an efficient boiler. Figure 10 shows that the energy consumption can be reduced considerably. The scenario C has a consumption of 9.7MWh and the scenario F is about 3.3MWh. This is a reduction of 66%, which would be a very energy-efficient alternative. The light energy consumption for all scenarios is 1.5MWh and 1.3MWh for the appliances. The heating ratio is almost equal to the sum of electrical goods and appliances together.

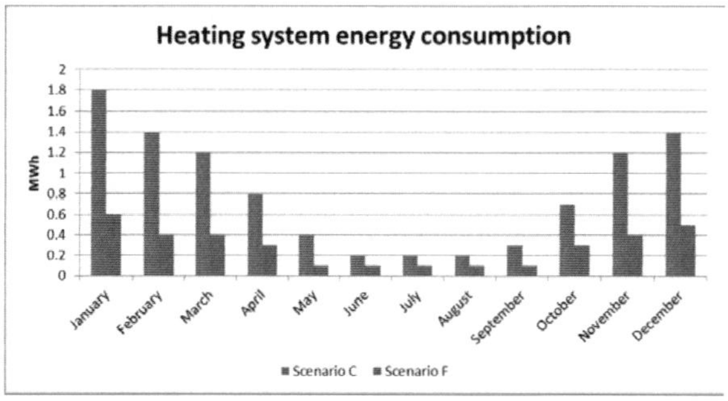

Figure 10. Energy consumption for the heating system, scenarios C and F.

The following figure illustrates the CO_2 emissions for the different scenarios. It can be noticed that for the scenario A, CO_2 emissions are about 2,165 kg CO_2 and that for the scenario C is about 2,324 kg CO_2, which represents an increase of 7.9%. There is a noticeable difference between the scenario A and F which is a reduction of 31% emitting 1,994 kg CO_2. The electrical goods and appliances emit 614.1 and 563.9 kg CO2. Although the reduction of energy between scenarios A and F were shown before which is about 66%, the reduction of CO_2 is reduced in proportion difference. The increase is that the GSHP uses electricity for the scenario F and gas for the boiler for the scenario A, and it is known that the production of electricity emits more CO_2 than gas (the emissions kg CO_2 per KWh is 0.198 for gas and 0.517 for electricity) [3].

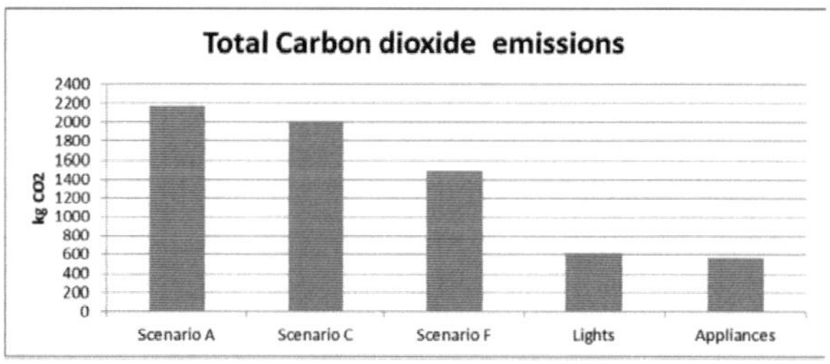

Figure 11. Total emissions of CO_2 for all scenarios.

Case Study 2 - Scenario A vs. B

The scenario A is the house without any external insulation and with the boiler band A. The house in the scenario A consumes yearly 9.6MWh. In the scenario B, the energy consumption in this case is 10.7MWh, which is more than 11.5% higher than the scenario A. Figure 12 shows this higher consumption of the scenario B expected. Also, the energy consumption is lower because it is a mid-terraced house whereas the case study 1 is an end-terraced house. Further analysis will show in terms of how the number of exposed walls affects the energy consumption.

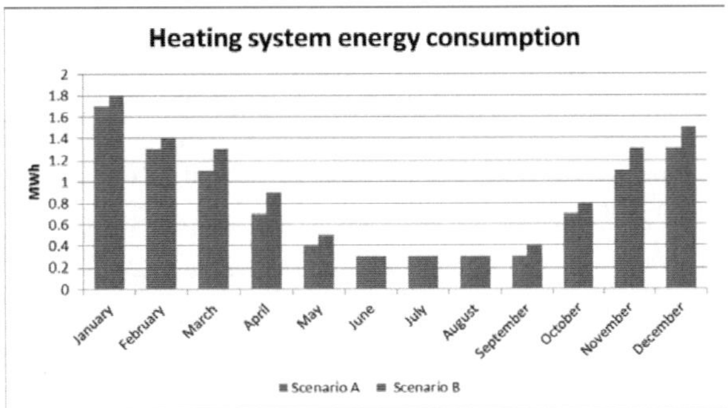

Figure 12. Energy consumption for the heating system, scenarios A and B.

Case Study 2 - Scenario A vs. C

The scenario C represents the same case that scenario A but with external insulation boards. Figure 13 shows that the energy consumption has decreased with the installation of the insulation and the yearly consumption is about 9MWh, which represents a reduction of 6.3% in comparison with the scenario A. In this case, the installation of external insulation limits the energy consumption to be reduced because only 2 facades would benefit from the insulation.

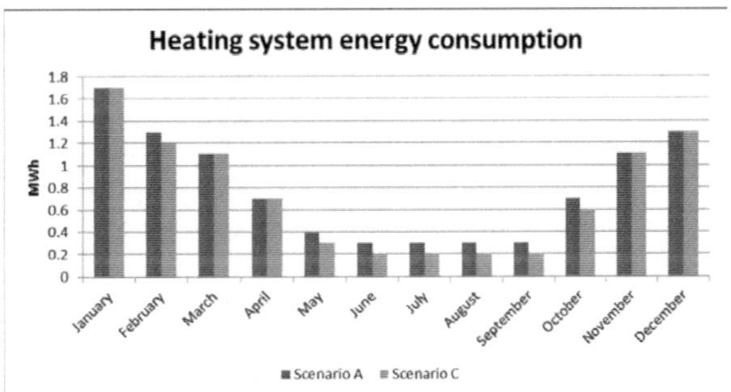

Figure 13. Energy consumption for the heating system, scenarios A and C.

Case Study 2 - Scenario C vs. F

With this comparison, the use of a GSHP enables the energy consumption to be reduced considerably as shown in Figure 14. The energy consumption for the scenario F is about 3.1MWh which is a reduction of 65.6%.

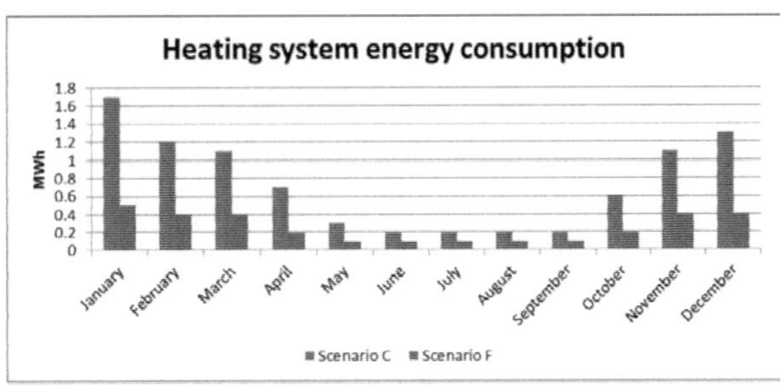

Figure 14. Energy consumption for the heating system, scenarios C and F.

The following figure illustrates the yearly consumption of energy dedicated to the heating, lighting and appliances. It can be seen that measures of retrofitting can be taken such as insulation or double glazing of windows but only the change in the system can bring a considerable reduction of energy (see Figure 15).

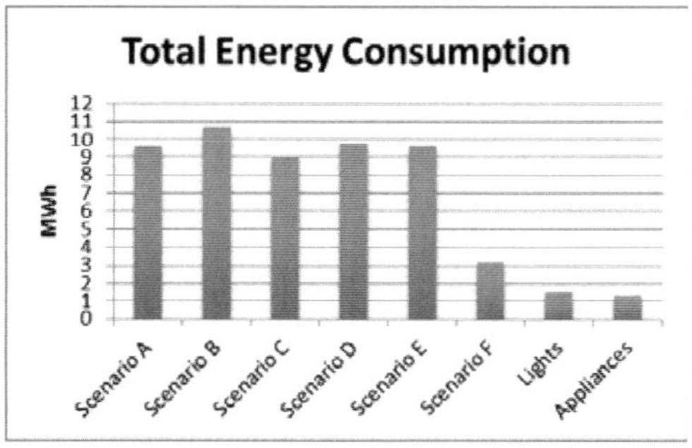

Figure 15. Total energy consumption for all scenarios.

It can be seen that for the scenario A, the CO_2 emissions are about 1,959.9 kg CO_2 and that for the scenario C about 1,849.1 kg CO_2, which represents an increase of 5.7%. There is also a noticeable difference between the scenario A and F which is a reduction of 28.8% emitting 1,395.9 kg CO_2. Although the reduction of energy between A and F as shown before is about 66%, the reduction of CO_2 is reduced in proportion different.

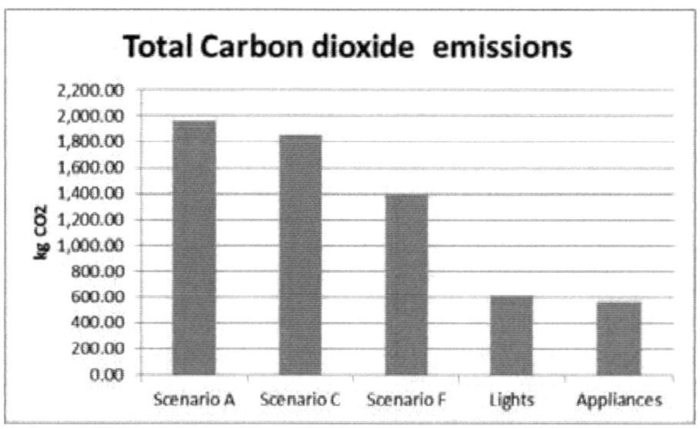

Figure 16. Total emissions of CO_2 for all scenarios.

In a perspective to reach the goals of the UK government about reducing by 80% the carbon footprint by 2050, some renewable sources should be considered to provide the electricity to run GSHPs. This is going to have a big impact on the CO_2 emissions and to evolve in the direction of producing low carbon houses.

CONCLUSION

After several simulation studies, it was possible to reduce the energy consumption dedicated to the heating. This reduction of energy can be performed complying with the building recommendations by CIBSE and ASHRAE. This way, the comfort of the indoor environment is improved which would bring more benefit to the occupants. In both case studies, it can be seen that to comply with the recommendations, more energy had to be used, about 7.5% in the case study 1 and 11% in the case study 2. This increase is not negligent and it would be a significant increase in the running costs as well which needs to be considered.

For the BEU (EIDC) project, there are some measures that can be taken and has been taken such as installing a new boiler or adding some insulation to the external walls, but the reduction still remains as low as around 10%. The significant reduction of energy consumption and CO_2 emission has to come from the system of heating. The best alternative found, for the houses in deprived communities, is to use a ground source heat pump (GSHP) as a source of heat in combination with the underfloor heating system. In both case studies, the energy reduction was more than 65% which is a very huge reduction. This percentage can even be higher, because in the simulation, the COP (Coefficient of Performance) implemented was the minimal value of 2.5. Usually, the GSHP is operating with a COP of 4 more or less. This was important to show that even in the worst case situation; a very big energy saving percentage can be achieved.

In the selected houses, natural ventilation was enough to avoid overheating during the summertime where the CIBSE criterion of overheating has been complied with. However, the prediction of future climate analyst asserting that houses in the UK and generally in Europe would more and more prone to be overheated; this should be taken in account for any further analysis including computer simulations. Therefore, with the choice of a GSHP, a reversible mode, it would be possible to have a cooling system for the houses. The GSHP can be applied at wide-scale as it is a realistic solution. On the other hand in this study, the running costs were not considered to see if this was economically beneficial, however this alternative of using a GSHP is feasible as the technology is already very well developed.

Finally, the energy reductions are still possible to be achieved by complying with all building recommendations. This is also synonym of a reduction of the energy bills to have affordable warmth. It would be in the same way, the solution to achieve the goal of UK government's that is aiming to reduce the CO_2 emission by 80% by 2050.

ACKNOWLEDGEMENTS

This paper is the result of the work undertaken as part of the BIG Energy Upgrade Programme (BEU). The project has attracted £7 million from the European Regional Development Fund through the Yorkshire and Humber ERDF Programme 2007-13.

REFERENCES

[1] American Society of Heating, Refrigerating and Air Conditioning Engineers (ASHRAE). (2012). http://www.ashrae.org/, Accessed on September, 2012.

[2] Building Research Establishment (BRE). (2005). *Estimates of Hot Water Consumption from the 1998 EFUS. Implications for the Modelling of Fuel Poverty in England.* June 2005. A Summary Report Presenting Data from the 1998 EFUS, BRE Housing Centre.

[3] Building Research Establishment (BRE). (2009). *The Government's Standard Assessment Procedure for Energy Rating of Dwellings*, 2009 Edition, BRE, Garston.

[4] Chartered Institution of Building Services Engineers (CIBSE). (2006). *CIBSE Guide A: Environmental Design*, London.

[5] Chartered Institution of Building Services Engineers (CIBSE). (2006). *CIBSE Guide F: Energy Efficiency in Buildings*, London.
[6] Cook, M.G. (2009). *Energy Efficiency in Old Houses*, The Crowood Press Ltd, Wiltshire.
[7] Energy Saving Trust (EST). (2008). Roadmap to 60%: Eco-refurbishment of 1960s Flats, CE294, Energy Saving Trust, London.
[8] Energy Saving Trust (EST). (2010). *Sustainable Refurbishment: Towards an 80% Reduction in CO_2 Emissions, Water Efficiency, Waste Reduction, and Climate Change Adaptation*, CE309, Energy Saving Trust, London.
[9] Energy Saving Trust (EST). (2010). *Domestic Low and Zero Carbon Technologies: Technical and Practical Integration in Housing*, CE317, Energy Saving Trust, London.
[10] Integrated Environmental Solutions (IES). (2012). IES VE Software, http://www.iesve.com/, Accessed on September, 2012.
[11] Porritt, S.M., Shao, L., Cropper, P.C. and Goodier, C.I. (2010). Ranking of Interventions to Reduce Dwelling Overheating during Heat Waves, Passive and Low Energy Cooling of Buildings (PALENC) Conference Proceedings, September 2010, Rhodes.
[12] Peacock, A.D., Jenkins, D.P. and Kane, D. (2010). Investigating the Potential of Overheating in UK Dwellings as a Consequence of Extant Climate, Energy Policy, Vol. 38, pp. 832-839.
[13] The BIG Energy Upgrade. (2012). http://big-energy-upgrade.com/, Accessed on September, 2012
[14] The European Parliament and the Council of the European Union. (2012). DIRECTIVE 2010/31/EU, http://www.epbd-ca.eu/, Accessed on September, 2012.

ASSESMENT OF PERFORMANCE OF INNOVATIVE VENTILATION SYSTEMS: USE AND LIMIT OF MULTICRITERIA ANALYSIS

Marc JARDINIER[1] (air.h), Laure MOURADIAN[2] (CETIAT), Pierre-Yves PAMART, Charles PELE[3] (CSTB), Anne-Marie BERNARD[4] (ALLIE'AIR)

1 air.h
C/O CETIAT
Domaine Scientifique de la Doua
25, avenue des Arts
BP 52042
69603 VILLEURBANNE Cedex, France

2 CETIAT
Domaine Scientifique de la Doua
25, avenue des Arts
BP 52042
69603 VILLEURBANNE Cedex, France

3 CSTB
84 avenue Jean Jaurès
Champs-sur-Marne
77447 Marne-la-Vallée Cedex 2, France

4 ALLIE'AIR
4 clos Ballet
01800 Meximieux, France
Annemarie.bernard@allieair.fr

ABSTRACT

Building sealing may affect the total air change by decreasing the leakages and question the ability for ventilation systems to reach their goal of providing an acceptable indoor air quality. Improving energy performance must not impair indoor air quality.

The QUAD-BBC study has led to define four groups of pollutants representative of similar behaviour, use or effect: CO_2 as a marker linked to human occupancy, NO_2, SO_2 (dwellings) and O_3 (offices) linked to occupants activities, CO and 7 VOC linked to materials, activities and behaviour and $PM_{2.5}$ and PM_{10}. The indexes related to these pollutants are calculated in occupancy periods only. Scenarios for pollutant emissions and occupancy have been determined and used as inputs for simulation through an improved version of SIMBAD (coupling airing and thermal effects). Calculation results have provided flows and patterns for each room, energy needs for auxiliaries and heating, pollutant concentrations for each species and in each room as well as more synthetics indexes.

The initial objective of the project was to develop a single criteria of indoor air quality (multi-pollutants) and to provide a reference system for each building based upon this index. Calculations have shown huge difficulties in reaching this objective. The lack of reliability of pollutant emissions' data in particular have led to propose a set of indexes and a method for comparison of systems rather than an absolute reference. Yet a graphical representation of these criteria helps a fast comprehension of the pros and cons of different systems and help comparison.

KEYWORDS

IAQ, criteria, pollutants, simulation

INTRODUCTION

Improving the airtightness of buildings modifies the distribution of air flow balanced between the air inlets and infiltrations. In this context, the issue of the ability of ventilation systems, as designed today, to ensure acceptable air quality is raised. Everyone agree that objective of energy efficiency shall not be to the detriment of the quality of indoor air; the matter is to determine how to evaluate it. Demand control ventilation is usually based on humidity ratio and CO_2 levels. Those criteria might not be enough to represent IAQ.

The QUAD-BBC project examines the building / user / system interactions using data sources of pollution related to the occupation, materials and user behaviour as well as several ventilation systems [1]. The initial goal of QUAD-BBC project was to propose an aggregated IAQ index from several pollutants and to provide a "reference system" for various buildings with an acceptable level of IAQ index. This project has been carried out on five different buildings: a single family house, a dwelling in a collective building, an office and a school. For each type of building, several ventilation systems have been designed, from very usual to very innovative ones in order to create various results: constant air flows, demand control ventilation (DCV) based on humidity, CO_2, and/or occupation, airing, ...

We used the simulation tool, SIMBAD, a Building and HVAC Toolbox developed by CSTB [2]. This tool implements multizone and nodal building models in MATLAB/Simulink environment by combining heat and mass transfer phenomena. On the one hand, the thermal model is composed of detailed wall models describing the material layers and their properties, window models, heating and cooling devices, lighting systems, etc. It so deals with conduction, convection and radiation phenomena for calculating surface temperatures, mean radiant and indoor air temperatures. Sources of pollutants were introduced in the model with their respective scenarios. Therefore, the level of pollution obtained in each room is due to the inner source itself, and also to the transfer of pollutants between parts of the building studied.

To achieve this goal several assumptions were set on the building (Nearly Zero Energy buildings with low level of infiltration), the use of it and the pollutants emissions and acceptable level. The use of the building is a major set of assumptions as the type of pollutants and their emissions levels are highly correlated to it.
Another major set of assumptions is the acceptable level for each pollutant and the way they are combined in a single index.

USE OF BUILDING – POLLUTANTS EMISSIONS

The uses of the building were described in terms of scenarios of presence and activities in the different rooms. For each type of building, scenarios describe week days, week-ends and holydays. Typical days have been cut into "slices" related to presence and type of activity in each room. The sum of these "slices" gives a yearly scenario which can be connected to pollutant emission for each species through the use of scientific literature and databases on pollutants (examples in Figure 1, Figure 2 and Figure 3).

For the IAQ evaluation of the QUAD-BBC project, five types of pollutants from indoor environments were identified:

- Moisture for its impact on the building, namely the risk of condensation on the walls, and occupant comfort;
- Carbon dioxide (CO_2 metabolic) considered as a tracer of the occupation. It is therefore linked to humans bio-effluents pollution;
- Volatile organic compounds (VOC) emissions due to materials and building equipment, whose production is proportional to the surface of the walls;
- Products of combustion activities (in kitchen and living room);
- Particulate pollutants from indoor air and incorporating sources deposition phenomena, coming from cooking and cleaning.

Pollutant source	Room	Week	Week-end
Cleaning	Living room, rooms	-	9:00-10:00 (Saturday)
Cigarette	Living room	- - 18:00-18:10 22:50-23:00	13:00-13:10 14:00-14:10 18:00-18:10 22:50-23:00
Incense	Living room Kitchen	- -	16:00-17:00 14:00-15:00

Figure 1 : Pollutant emission scenario relative to activity in dwellings

Room	Metabolism	Activities	Equipment	Construction materials	Cleaning products
Residential	H_2O, CO_2	$PM_{2,5}$, PM_{10}, H_2O, CO, NO_2, SO_2	HCHO, $PM_{2,5}$, PM_{10}	VOC	HCHO, VOC
Offices	H_2O, CO_2	O_3	HCHO, $PM_{2,5}$, PM_{10}		VOC
Classroom, nursery	H_2O, CO_2		HCHO, $PM_{2,5}$, PM_{10}		VOC

NB: pollutants from metabolism, coming from human occupancy, are found in all rooms.

Figure 2 : Examples of pollutants sources for each type of building

Room	H_2O	CO_2	VOC's	HCHO	$PM_{2,5}$	CO	NO_2	SO_2	O_3
Bedroom, living room	X	X	X	X	X				
Office	X	X		X	X				X
Classroom, nursery	X	X	X	X	X				
Hospital	X	X		X	X				
Kitchen	X	X	X	X	X	X	X	X	
Bathroom, shower	X	X							
WC	X	X	X						

Figure 3 : Examples of pollutants emitted in various rooms

The choice of emission levels is crucial but very difficult: the variability of data sources is a major complication. At the beginning of the study we chose one source almost arbitrarily for each pollutant. Our bibliographic references were principally the PANDORE [3] database, the IA-QUEST [4] tool and also Annex 27 European project [5].

IAQ LEVEL

The choice of maximum level was also difficult as the sources again are very different: they do not refer to the same duration (peak, average 1h, 2h, 8h, ...) and do not address the same level of concerns (WHO, OEHHA, Anses, ATSDR, USEPA, ...). The choice was made to use the most demanding level as a maximum for each pollutant (Table 1).

CAT.	POLLUTANT	Limit value for 1 h exposure
A	Carbon dioxide, CO_2	10 000 ppm
B	Nitrogen dioxide NO_2	0,20 mg/m^3
	Sulfur dioxide, SO_2	0,66 mg/m^3
	Ozone, O_3	0,12 mg/m^3
C	Carbon monoxide, CO	30 mg/m^3
	Formaldéhyde, HCHO	0,05 mg/m^3 for 2h
	Acétaldéhyde	0,47 mg/m^3
	Éthylbenzene	43,36 mg/m^3
	Styrene	21 mg/m^3
	Toluene	3,8 mg/m^3
	o-Xylene	8,7 mg/m^3
	Acetone	31 mg/ m^3
D	Particles 10 µm	50 µg/m^3
	Particles 2,5 µm	25 µg/m^3

Table 1: Reference values used for Indoor Air Quality index

The index for individual pollutant is $I = C_p/C_{\lim,p}$, where C_p is the average concentration of the pollutant on a specific period, and $C_{\lim,p}$ is the maximum value recommended for the same period. The index shall then be lower than 1 to be acceptable. The lower I is, the better is Air Quality.

The combination of indexes to provide a single index is another difficulty as several possibilities exist in the literature. We decided to use combination of all pollutants and additivity of effects. The initial global index is:

$$I_{QAI} = \sum_{p=1}^{Np} \left(\frac{C_p}{C_{\lim,p}} \right) \tag{1}$$

Each pollutant is taken in the index only during occupancy period and the index refers to room (one index per room). The main problem with this index is of course that it is very sensitive to the number of pollutants (20 at the beginning) for its absolute value.

All these arbitrary choices (emission levels, maximum concentration, index ...) were supposed to be neutralized in the context of comparison of systems to a reference system, as well as simulation tool assumptions.

The first shots of calculation and their analyses raised a number of questions leading to change the initial goal.

One of the first difficulties found was to match consistent assumptions for calculation model of physical phenomena (for example treatment of humidity in the wall) and design of Demand Controlled Ventilation (DCV). Physical models might have an impact on indoor air environment and therefore on the reaction of DCV systems, their regulations, and so the airflows and the final result.

The construction of a reference system (included it own regulation) is also linked to the model and the assumptions. Therefore we decided not to define a reference system.

Another outcome from the calculations was the presence of a "main pollutant" which level was so important that the index became almost constant in some rooms. The impact of other pollutants, although their levels were different from one system to the other, disappeared in the global index. So, we decided to split the original global index in 4 specifics indexes related to common activity or impact and add specific information on humidity:

- $I_A = CO_2$ as index of confinement linked to occupation,
- $I_B = NO_2$, SO_2 (dwellings) and O_3 (offices) linked to occupant activities,
- $I_C = CO$ and 7 VOC linked to materials, activities and occupants behaviour,
- $I_D = PM_{2.5}$ and PM_{10} linked to activities.
- Relative humidity (number of hours above 80%) linked to occupants and their activities

Specific indexes are built with the same equation than the original one but their extent is limited to the same category of pollutant. The additivity of effects is there more relevant. The information is also more specific and usable for the choice of the system.

For example, the index for occupant activities I_B is calculated as followed:

$$I_B = \frac{C_{NO_2}}{C_{\lim, NO_2}} + \frac{C_{SO_2}}{C_{\lim, SO_2}} \qquad (2)$$

Humidity is taken specifically into account as it is not really a pollutant but it is still a major problem in specific conditions. The impact is calculated for a one year period, during both occupancy and no-occupancy periods.

The number of VOC has also been reduced to a short list for which the data seemed to be more reliable.

INFLUENCE OF ASSUMPTIONS

Equipment – occupant behaviour

Several shots of calculations with specific assumptions (climate, infiltration level ...) have shown some recurrent difficulties on the equipment and user activities levels of emission.

In the kitchen the presence and use of oven leads to an excessive value of the index, whatever the system (Figure 4). For the collective dwelling simulations by example, the 5 mechanical ventilation systems named LC 0 to LC 4 present almost same pollutant concentration during cooking activity. Those 5 ventilation systems are either single extract single flow system or balanced supply and extract systems, with constant airflows or control on humidity, CO_2, presence. In the living room the tobacco smoke leads to excessive values of the index, whatever the system. Figure 5shows for the example for one ventilation system the influence of activity pollutant source scenario on formaldehyde concentration. The value of the index can also be very high, because people smoke of use incense, and not because airflow from ventilation system is not enough.

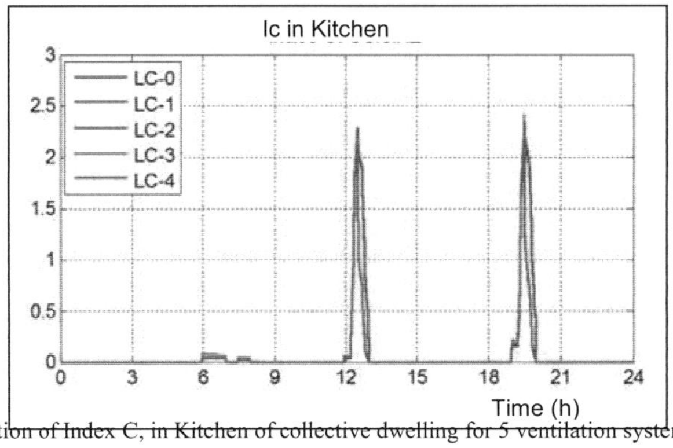

Figure 4 : Evolution of Index C, in Kitchen of collective dwelling for 5 ventilation systems (LC0 to LC4)

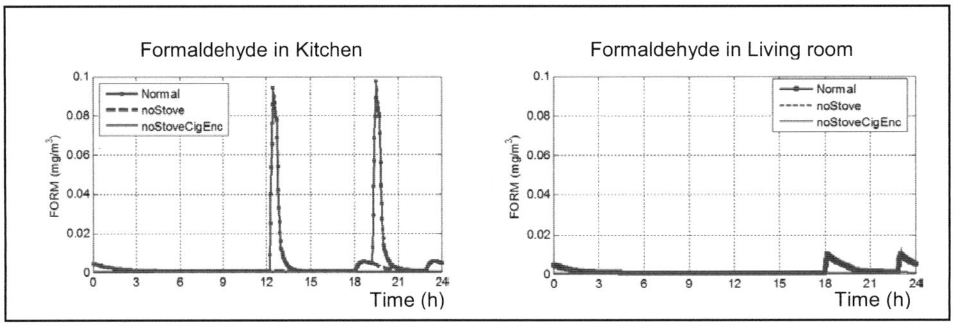

Figure 5: Formaldehyde concentration in the kitchen and in the living room, with constant airflow ventilation (LC0), comparison with emission from stove or not and with cigarette and incense or not

For particles from incense we found very different levels (10 mg/h to 240 mg/h) depending on the type or trademark, being very different from aromatic candles: the final concentration is more related to the user than to the ventilation system.

At this step it must be assert that the index will widely change with the choice of pollutants: for instance alpha-pinene or limonene are widely used in home perfumes: if they are included in the list, the index will be high if the occupants use this type of product. If they are not in the list there will be no effect on the index. Some products on the market detect "odours" with a VOC sensitive sensor and provide another VOC to mask the first one ...

The final index level may finally be more influenced by the occupant behaviour than by the system.

Material emissions

The use of database for material emission led to constant values, mainly for formaldehyde. The emission is constant whatever the indoor conditions are and not declining with time. The concentrations found with all systems are very low, far away from the acceptable limits (example in Figure 6 and Figure 7).

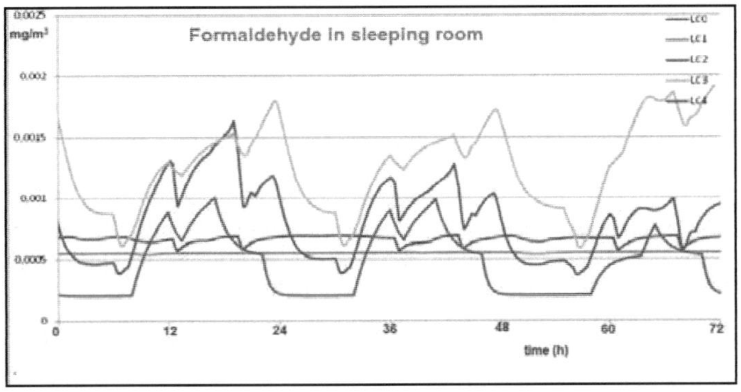

Figure 6: Formaldehyde concentration in a sleeping room for 5 ventilation systems

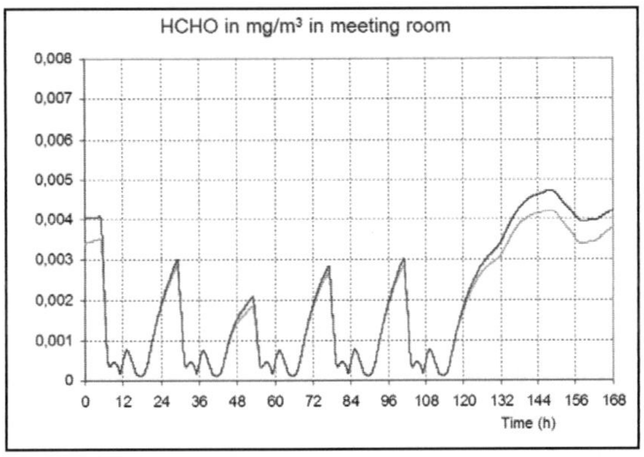

Figure 7: Formaldehyde concentration in a meeting room of an office building during one week, for 2 ventilation systems, with no airflow during night and week-end

This result is surprising as many measurement campaigns have shown relatively high values in real dwellings or offices, without a clear relationship with the system.
We can try to explain these differences in two ways:

- The measurements include all VOC, from the material (new or not) and from the occupant behaviour, and we know that the occupant behaviour can be of a major importance – this would be in line with the weak relationship between systems and levels.
- The data used for the calculation are wrong, either too low emission levels, either missing.

Both of these explanation lead to be very careful in giving any limit to a calculated result for a single system.

If we rely on the results, it could be possible to reduce the flows in bedrooms and living rooms by a factor of 10 without any problem on material related VOC levels. The temptation of reducing the flows from the study or from any calculation model including VOC must not be followed as some important VOC may be missing and the use of cleaning or masking products may not be correctly handled.

PROPOSAL FOR IAQ INDEX AND USE

The use of a single aggregated index is not pertinent, four groups of pollutant, plus humidity, are relevant for the analysis of the reaction of systems installed in a building. The proposal of one single reference system is also not relevant; it is linked to the type of pollutants and the building use.

A set of annual average indexes for the same building with the same conditions of use allows comparing systems between themselves in these conditions (the result cannot be extended as a general proposal). As energy efficient is now a major criterion to choose the system, IAQ indexes have to be associated with energy. The simulation, combined with climate files allows calculating the energy needs to heat air renewal. Assumptions shall be done on the heating system and its efficiency. This affects the weight of the heating consumption compared to fans consumptions.

The presentation of the indexes in the "radar" shape makes it relatively easy for systems comparison. The matter of this representation is the scale which should be the same for every branch. For the energy and the humidity criteria, we decide to divide by the greater one.

To facilitate the readability of the radar for comparison of systems, it is recommended to rescale it (0-100%) on each branch from the higher value of each pollutant category. In order to evaluate the importance of each branch between each over, it is then recommended to note the absolute value of the maximum. The absolute value is also useful to know if the differences are relevant (ie big scale) or if all systems are quite good (small scale). We need to remain that Index lower than 1 is consistent with acceptable air quality.
We chose an intermediate presentation, meaning the same scale for all the pollutant indexes. In this way, differences on indexes are not hidden by the value of energy, and the hierarchy between maximum levels of each category is still respected.

The interpretation of the radar whether the indexes are in absolute values or rescale for indexes will not be the same (Figure 8).

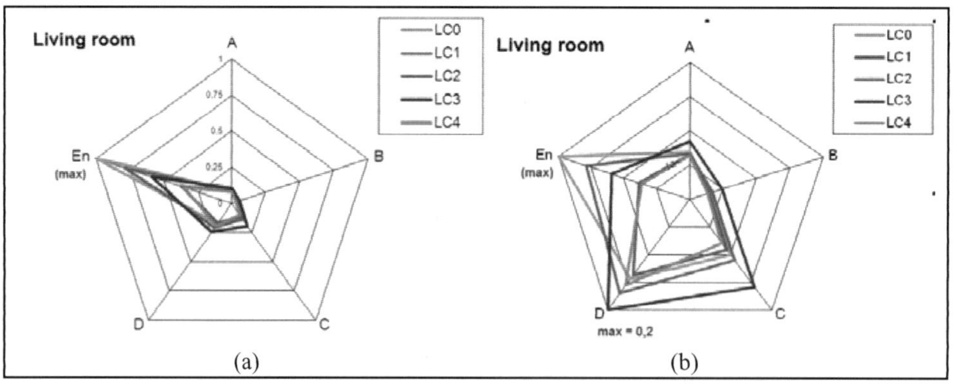

Figure 8: Multi criteria radars in living room; Energy in percentage of maximum and pollutant indexes A, B, C, D in absolute value (a) or rescale, same values for all indexes (b)

CONCLUSION

One of the conclusion of the study is to choose a set of pollutant which is not linked to the occupant behaviour regarding the emission of substances (tobacco, incense, candles, perfumes…), the ventilation systems would then be ranked with their ability to remove metabolism linked substances (H_2O, CO_2, human related VOC) and material emissions and to keep humidity in the safe range.

The following questions remain:
Which emission levels for material?
What about cleaning?
What about cooking?

If the occupancy and activities are documented enough, it is possible to use a set of indexes (per room) linked to the actual use of the building and to compare systems between themselves with these assumptions with a "radar" shape presentation that gives a quick global view.

Designers must be aware of the limitations of the method and the accuracy of the results, linked to the reliability of the emission data from sources. The radar presentation must also be taken with care as the rescaling of the branch may overestimate some impacts and differences between systems.

ACKNOWLEDGEMENTS
QUAD-BBC project partners are: air.h, CETIAT, CSTB, LaSIE, INERIS, ALLIE'AIR and we thank ANR, the French Agency for Research for its financial support.

REFERENCES

[1] Koffi, J., Allard, F. 2011. *Control and performance of innovative ventilation systems in low energy buildings.* Proceedings of 32nd AIVC Conference and 1st TightVent conference, October 12-13, 2011, Brussels, Belgium.

[2] Husaunndee, A., Lahrech, R., Vaezi-Nejad H., Visier J.-C. 1997, *SIMBAD a simulation toolbox for the design and test of HVAC control systems*, Proceedings of the 5th IBPSA Conference, Prague, Czech Republic, Vol. 2, pp 269–276.

[3] *PANDORE, comPilAtion des émissioNs Des pOlluants de l'aiR intériEur*, LEPTIAB, Université de La Rochelle, France. http://leptiab.univ-larochelle.fr/Presentation-PANDORE.html.

[4] *IA-QUEST version 1.1 Indoor Air QUality Emission Simulation Tool.* National Research Council Canada.

[5] *Simplified Tools Handbook: Evaluation and Demonstration of Domestic Ventilation Systems,* Edited by Lars-Goran Mansson, UK, Faber Maunsell Ltd on behalf of the IEA ECBCS Programme, 2002, ISBN 0-9542670-1-X, 136 pp. http://www.ecbcs.org/docs/annex_27_handbook.pdf

HEAT RECOVERY EFFICIENCY:
MEASUREMENT AND CALCULATION METHODS

Samuel Caillou[*1], Paul Van den Bossche[1]

1 Belgian Building Research Institute, Avenue P. Holoffe 21, B-1342 Limelette, ***Belgium***
www.bbri.be
**Corresponding author: samuel.caillou@bbri.be*

ABSTRACT

The efficiency of air-to-air heat recovery ventilation units is of great importance for EP calculations (energy performance of buildings) throughout Europe. Efficiencies compared on a reliable basis are also crucial for contractors and installers of such systems.
Different determination methods are in use across Europe (EN 308, EN 13141-7, NEN 5138 in The Netherland, Eurovent, Passive House Institute (PHI), Dibt in Germany, etc.). Heat recovery determination involves 2 main steps: (1) a measurement (test conditions, etc.) and (2) a further calculation of the result (definition of efficiency, possible corrections, etc.). The differences between the above mentioned methods concern the measurement conditions (air temperature and humidity, etc.) as well as the result calculation. In Belgium a determination method was recently developed in the context of the EP-regulation based upon measurement conditions of EN 308, but with some modifications of the measurement conditions as well as of the result calculation.

This paper gathers test data from more than 160 measurement points on real series products available on the European market. The aim was to compare and discuss the different ways of heat recovery measurement and calculation in order to identify the key points to improve the heat recovery determination methods, let's dream, towards a convergent and unique method across Europe.

Based on the temperatures measured in the 4 flows (outdoor, supply, extract and exhaust air), two different efficiencies (temperature ratios) can be calculated: on the supply side or on the exhaust side. The gap between both is directly related to the thermal balance, as defined in EN 308. This gap varies greatly from one product to another. Several hypotheses can explain this gap, such as: transmission heat fluxes through the casing, air leakages, unbalance of the flow rates during the test, heat from the fans (if not corrected), etc. Most of these effects lead to an overestimation of the supply efficiency and underestimation of the exhaust efficiency. To our knowledge, the Belgian calculation method is the only method which takes this effect into consideration (by using the average of supply and exhaust efficiencies).
The efficiency of a heat exchanger as a separate component is surely not enough to identify the performance of a whole AHU. The average efficiency determined on a whole AHU is always lower and depends on the quality of the AHU itself (internal thermal bridges, for example). The operation of the fan during the test seems playing also an important role. In real life, the efficiency of the whole AHU is the most relevant result.
Finally, the paper discusses also the impact of the different test conditions on the measured efficiency.

KEYWORDS
EN 308, EN 13141-7, NEN 5138, Dibt, balanced ventilation, supply efficiency, exhaust efficiency, thermal balance, flow balance, air leakages, fan heat, transmission heat losses.

INTRODUCTION

Ventilation systems with heat recovery are promising to reduce the energy use for building ventilation. Very high values of heat recovery efficiency up to 90% and higher are claimed by some manufacturers and popular information papers. However, these figures are often considered as overestimated. Reliable efficiency values are obviously necessary for

EP-calculations (energy performance of buildings) throughout Europe as well as for contractors, installers and users of such systems.
Different determination methods are in use across Europe (EN 308, EN 13141-7, NEN 5138 in The Netherland, Eurovent, Passive House Institute (PHI) [1], Dibt in Germany, etc.). Heat recovery determination involves 2 main steps: (1) a measurement (test conditions, etc.) and (2) a further calculation of the result (definition of efficiency, possible corrections, etc.). The differences between the above mentioned methods concern the measurement conditions (air temperature and humidity, etc.) as well as the result calculation. The EN 308 is a European standard with a large scope (heat exchanger (Hx) as a component or Air Handling Unit (AHU) as a whole; both residential and non-residential applications). But the EN 308 is not very clear and is often used out of its range of validity (leakages, thermal balance, etc.). In Belgium a determination method was recently developed in the context of the EP-regulation based upon measurement conditions of EN 308, but with some modifications of the measurement conditions as well as of the result calculation [2].

Moreover, there are different ways of calculating heat recovery efficiencies:

- "supply efficiency" based on the temperature measured on the supply side;
- "exhaust efficiency" based on the temperature measured on the exhaust side.

Previous studies [3] suggested that this "supply efficiency" can be largely overestimated due to uncontrolled heat fluxes. The standard EN 308 requires normally a thermal balance deviation lower than 5% but this requirement is very rarely fulfilled for most of tests carried out on the whole AHU. Most of the above mentioned methods use the "supply efficiency", which is probably overestimated. On the other hand, the exhaust efficiency is probably underestimated. In the new method in force in Belgium, the efficiency used in the EP-regulation is calculated with the average between the supply and the exhaust efficiency to take into account as much as possible these uncontrolled heat fluxes responsible for these large deviations of the thermal balance.

This paper gathers test data from more than 160 measurement points on real products available on the European market. The aim is to compare and discuss the different ways of heat recovery determination and calculation in order to identify the key points to improve heat recovery determination methods, let's dream, towards a convergent and unique method across Europe.

MATERIAL, METHODS AND DEFINITIONS

Test data of heat recovery units were obtained from different manufacturers present on the European market (38 different heat recovery units from 17 different manufacturers). The raw data were collected in the test reports from different well-known test laboratories across Europe (10 different test laboratories). For most of the heat recovery units, the test was carried out for different flow rates, giving more than 160 measurement points in total.
For most of the cases, the tests were carried out on the whole AHU with fans in operation. In a few cases only, the tests were carried on the whole AHU but with fans off, or on the heat exchanger as a separate component. All the tests were carried out according to one of the following test methods or standards: EN 308, EN 13141-7, Dibt, NEN 5138 or Passive House.

Based on the temperatures measured in the 4 flows (outdoor, supply, extract and exhaust air), two different efficiencies can be calculated: on the supply side, i.e. the heat which is added to the supply air, or on the exhaust side, i.e. the heat which is extracted from the extract air. The raw data from the test reports (temperatures in the 4 flows, flow rates, absorbed electrical power) were used to calculate the different parameters according to the definitions as follows.

Supply efficiency, not corrected: $\eta_{t,\sup} = \frac{t_{22} - t_{21}}{t_{11} - t_{21}}$	(1)

Exhaust efficiency, not corrected: $\eta_{t,eha} = \frac{t_{11} - t_{12}}{t_{11} - t_{21}}$	(2)

Supply efficiency, corrected for fan heat: $\eta_{t,\sup,corr} = \frac{t_{22} - \Delta t_{22} - t_{21} - \Delta t_{21}}{t_{11} + \Delta t_{11} - t_{21} - \Delta t_{21}}$	(3)

Exhaust efficiency, corrected for fan heat: $\eta_{t,eha,corr} = \frac{t_{11} + \Delta t_{11} - t_{12} + \Delta t_{12}}{t_{11} + \Delta t_{11} - t_{21} - \Delta t_{21}}$	(4)

Average efficiency used in EP-regulation in Belgium: $\eta_{t,average} = \frac{(\eta_{t,\sup,corr} + \eta_{t,eha,corr})}{2}$	(5)

where t are the temperatures of the air flows, as follows:

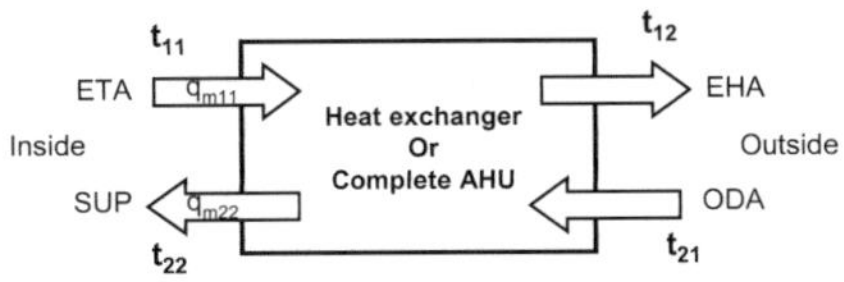

11: extract air (ETA);
12: exhaust air (EHA);
21: outdoor air (ODA);
22: supply air (SUP);

and Δt are calculated according to the position of the fan, by convention, as follows:

		Fan for extract air	
		In position 11 (ETA)	In position 12 (EHA)
Fan for supply air	In position 21 (ODA)	$\Delta t_{11} = \frac{0.5 \cdot P_{elec,ahu,test}}{0.34 \cdot q_{v11}}$ $\Delta t_{21} = \frac{0.5 \cdot P_{elec,ahu,test}}{0.34 \cdot q_{v22}}$ $\Delta t_{22} = \Delta t_{12} = 0$	$\Delta t_{12} = \frac{0.5 \cdot P_{elec,ahu,test}}{0.34 \cdot q_{v11}}$ $\Delta t_{21} = \frac{0.5 \cdot P_{elec,ahu,test}}{0.34 \cdot q_{v22}}$ $\Delta t_{22} = \Delta t_{11} = 0$
	In position 22 (SUP)	$\Delta t_{11} = \frac{0.5 \cdot P_{elec,ahu,test}}{0.34 \cdot q_{v11}}$ $\Delta t_{22} = \frac{0.5 \cdot P_{elec,ahu,test}}{0.34 \cdot q_{v22}}$ $\Delta t_{21} = \Delta t_{12} = 0$	$\Delta t_{12} = \frac{0.5 \cdot P_{elec,ahu,test}}{0.34 \cdot q_{v11}}$ $\Delta t_{22} = \frac{0.5 \cdot P_{elec,ahu,test}}{0.34 \cdot q_{v22}}$ $\Delta t_{21} = \Delta t_{11} = 0$

Each efficiency is given at one or more test flow rates. The test flow rate is defined as the lowest of the measured flow rates for supply and for exhaust.
Note that the term "efficiency" is used for simplicity rather than "temperature ratio". These temperature ratios are not exactly efficiencies from the physical viewpoint, because the physical heat recovery efficiency can only be simplified in the form of temperature ratio if (1) mass flow rates are in balance; (2) there is only exchange of sensible heat, no condensation occurs.

RESULTS

For all the tests carried out on the whole AHU with the fans in operation (Figure 1, Top), the calculated supply efficiency (not corrected) was always slightly (or even largely) higher than

the calculated exhaust efficiency (not corrected). The gap between the supply and the exhaust efficiencies decreased when these supply and exhaust efficiencies are corrected for fan heat as described above (Figure 1, Middle). However, even with correction for fan heat, this gap did not completely disappear at least for some tested products.
For the tests carried out on the heat exchanger alone or on the AHU with fans off (Figure 1, Bottom, no correction needed for fan heat), this gap was significantly lower than for tests carried out on the AHU with fans in operation. For 15 of these tests, the supply and exhaust efficiencies were very close from each other; the flow unbalance for these 15 tests was quite limited, lower than 1.5%. For the 5 other tests, this gap was slightly higher; but the flow unbalance for these 5 tests was higher than 5%.
This gap between the supply and the exhaust efficiency is directly related to the deviation to the thermal balance calculated as defined in EN 308. For the tests carried out on the AHU with the fans in operation, 84% of the tests showed a thermal balance deviation higher than 5%; the thermal balance deviation for all these tests was 13% on average. If the thermal balance is corrected for fan heat, 55% of the tests showed a thermal balance deviation higher than 5%, with a value of 7.5% on average. For the tests carried out according to the Dibt method or to the NEN 5138 standard, this average value of thermal balance deviation was significantly higher with more than 10% while it was lower for tests carried out according to EN 308 with less than 5% (only for tests carried out on the whole AHU with fans in operation).

Interesting observations can also be drawn from series of several tests carried out on the same individual product. Figure 2, top, middle and bottom, shows such comparisons for 3 different individual products.
For the first example (Figure 2, Top), comparing tests carried out on a same product, either on the heat exchanger alone or on the AHU with fans in operation, the calculated efficiencies (supply, exhaust and average) were higher for the heat exchanger alone than for the AHU with fans in operation. The gap between the supply and exhaust efficiency was also very low for the tests carried out on the heat exchanger alone and largely lower than for those carried out on the AHU with fans in operation.
For the second example (Figure 2, Middle), comparing tests carried out on a same product, either on the AHU with fans in operation or on the AHU with fans off, the gap between the supply and exhaust efficiency was largely lower for the AHU with fans off than for the AHU with fans in operation (with a thermal balance deviation of around 20%). The calculated average efficiency was also higher for the tests with fans off than for those with fans in operation. A similar effect was also observed for 3 other individual products for which the test data with fans off were available (data not shown).
The third example (Figure 2, Bottom) compared the results of several tests carried out on a AHU with fans in operation, obtained from different laboratories, according to different standards and on different product samples. The average efficiency (Figure 2, bottom left) from these different tests on the same product varied from 79% to 84% in the tested flow range, while the supply or the exhaust efficiencies (Figure 2, bottom right) vary to a larger extent. The average efficiency determined on the heat exchanger alone from the same product reached a higher value of 91%.

Finally, the influence of the flow rate on the heat recovery efficiency was also evaluated (data not shown). For all the tests with data available for several flow rates, the slope of the average efficiency as a function of the test flow rate ranged from a negative slope of -0.06%/(m^3/h) to a positive slope of 0.04%/(m^3/h) and was -0.02%/(m^3/h) on average. In more than 70% of the cases, this slope was negative and ranged between 0.00 and -0.04%/(m^3/h).

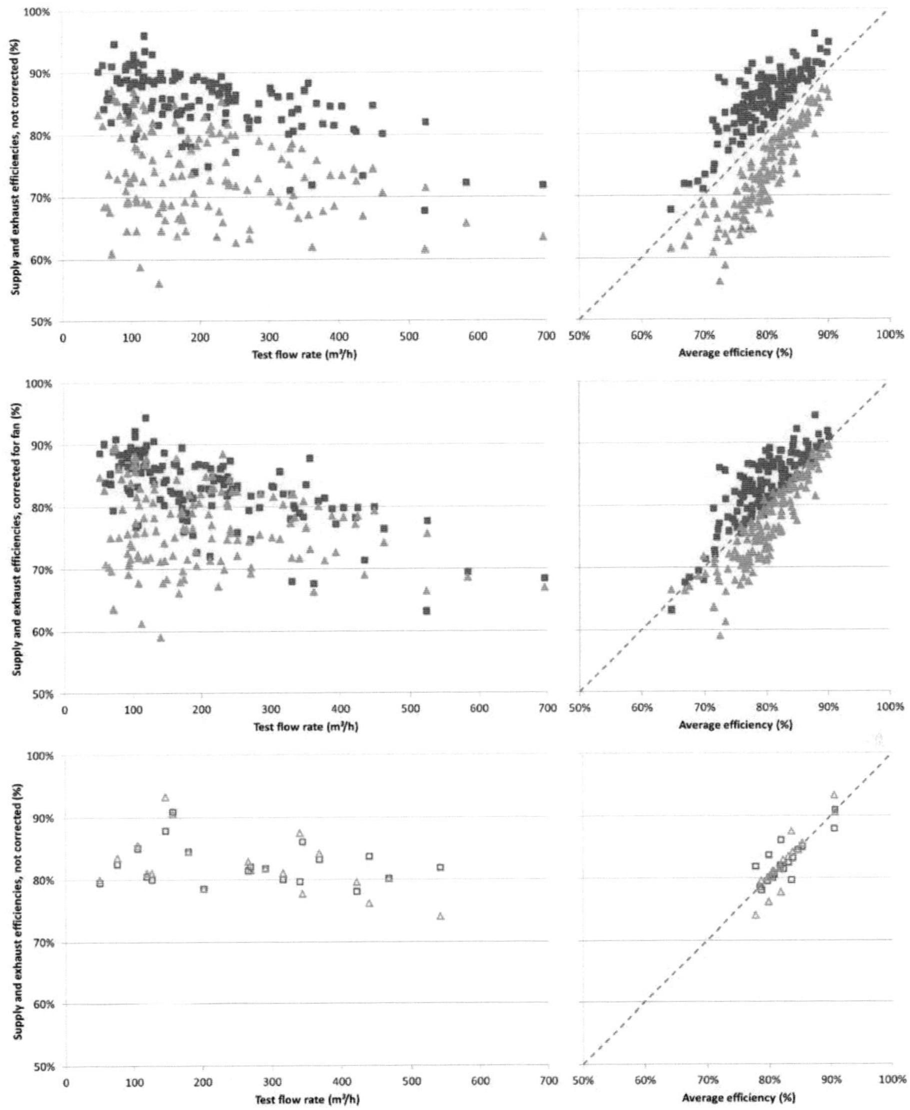

Figure 1. Supply efficiency (squares) and exhaust efficiency (triangles), as a function of the test flow rate (left) or as a function of the average efficiency calculated in the Belgian regulation (right): tests carried out on AHU with fans in operation (closed symbols), or testes carried out on the heat exchanger alone or the AHU with fans off (open symbols); not corrected values (Top), values corrected for fan heat (Middle), values without correction needed (no fan heat, Bottom).

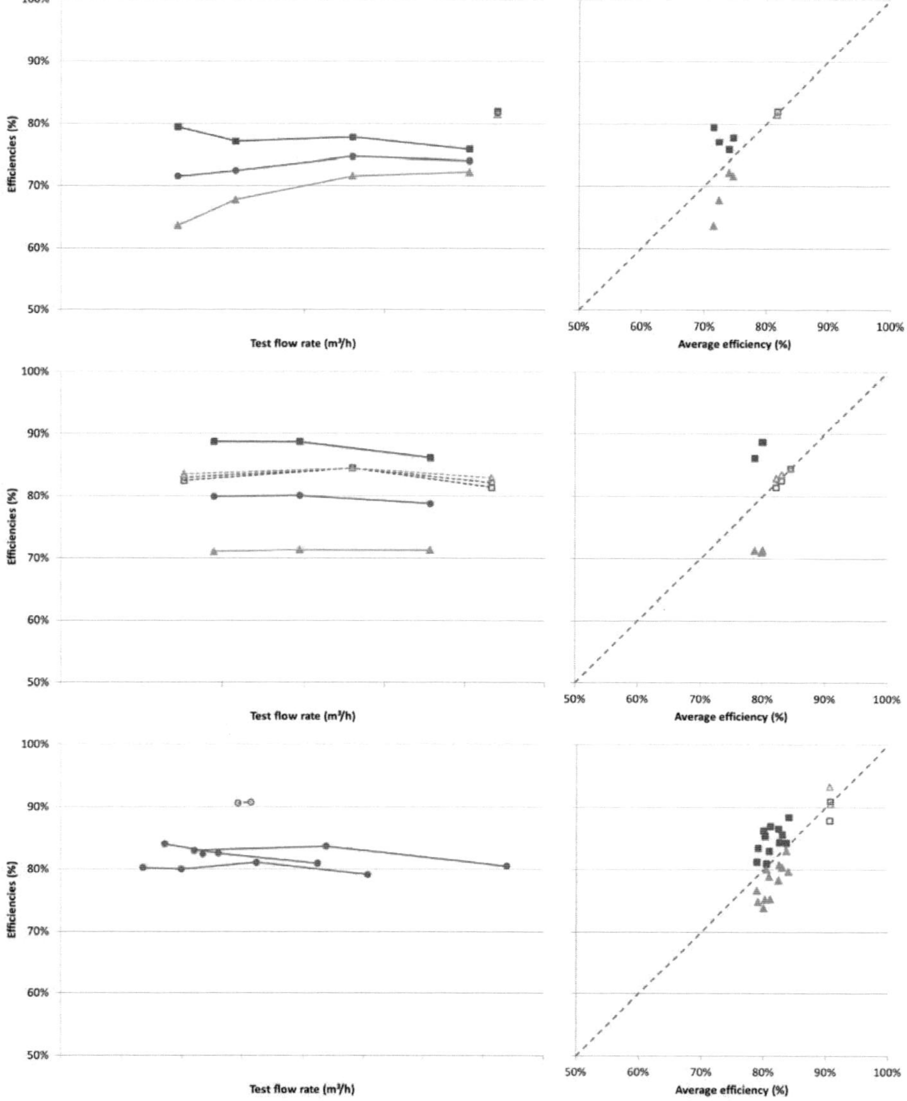

Figure 2. Supply efficiency (squares), exhaust efficiency (triangles) and average efficiency calculated according to the Belgian regulation (circles), as a function of the test flow rate (left) or as a function of the average efficiency (right), for 3 individual products (Top, Middle and Bottom): tests carried out on the AHU with fans in operation (closed symbols), or tests carried out on the heat exchanger alone or on the AHU with fans off (open symbols); supply and exhaust efficiencies were corrected for fan heat if applicable; for the product 3 (Bottom), several tests results were available from different laboratories (and according to different standards).

DISCUSSION

Thermal balance

In theory, the supply efficiency and the exhaust efficiency (if corrected for fan heat) should be equal (because no energy is created or lost in the system). However, the results demonstrated that these two efficiencies are nearly never equal! In most cases, the supply efficiency is higher than the exhaust efficiency. This means that apparently we win more energy in the supply air than we recovered from the exhaust air. This gap between the supply efficiency and the exhaust efficiency is directly related to the thermal balance, as defined in EN 308. The very high gap between the supply and exhaust efficiency observed for most of the tested AHU explains why the requirement of EN 308 for the deviation of the thermal balance (maximum 5%) is not satisfied for most of the products. Several uncontrolled heat fluxes in the system can be hypothetized to explain this gap, as follows.

Fan heat. First of all, the heat released from any fan will increase the temperature of the supply air, causing an apparent higher supply efficiency; and increase the temperature of the exhaust air, causing an apparent lower exhaust efficiency. A fan with a class SFP3 (according to EN 13779) will, for example, increase the temperature of the air of around 1 K (if all electricity is dissipated into the air flow). However, this effect can be easily corrected using the absorbed electrical power measured during the test. As shown in the results, such a correction decreases significantly this gap, leading to nearly equal supply and exhaust efficiencies for some products, while this correction is not enough to explain this gap for some other products.

Flow rate unbalance. Unbalance of the flow rates during the test is responsible for a deviation between the supply and the exhaust efficiencies. For example, if the extract flow rate is higher than the supply flow rate, the supply efficiency will apparently increase while the exhaust efficiency will apparently decrease: more heat is available to pre-heat the supply air. In theorie, the effect of flow unbalance could be corrected by calculation. However, it should be preferably to avoid such flow unbalance during the test itself.

Transmission heat fluxes from or to the surrounding (through the casing). With all the above mentioned methods, the efficiency test is carried out in a warm surrounding (between 18 and 25°C depending on the method used). In such conditions, the exhaust air can be warmed up due to heat transfer and can cause a lower exhaust efficiency. An increase of the supply efficiency can also occur to a lesser extent (smaller temperature difference). In real conditions, these transmission heat transfers correspond to real heat losses for the building if the AHU is placed inside the building for example.

Leakages. The effect of internal and external leakages is difficult to predict because not only the temperature can change, but also the mass flow rates in the 4 flows. For example in case of internal leakages from the extract flow to the supply flow, the supply efficiency will increase and the exhaust efficiency will decrease.

As a summary for the thermal balance deviation, there are several uncontrolled heat fluxes, possibly leading to an increase, apparent or not, of the supply efficiency and to a decrease, apparent or not, of the exhaust efficiency, as summarized in Table 1.

The supply efficiency, as used in most of the above mentioned methods such as EN 308, EN 13141-7, NEN 5138, etc., is thus highly probably overestimated for most of the tested heat recovery devices.

Note that the Passive House method is the only method using the exhaust efficiency (with additional calculation for the fan heat); this is specific to the conventions used in the EP-calculation method for Passive Houses (PHPP) and this will not be discussed further in this article [3].

Possible causes of deviation between the supply and the exhaust efficiencies		Estimated impact on the Supply efficiency	Estimated impact on the Exhaust efficiency
Fan heat	(depends on fan position)	↗	↘
Unbalanced flows:	Supply > extract	↘	↗
	Extract > supply	↗	↘
Leakages	Internal, extract to supply	↗	↘
	External	?	?
Transmission heat fluxes:	Test in warm surrounding	~ or ↗	↘↘

Table 1. Possible causes for deviation between the supply and the exhaust efficiencies and their qualitatively estimated impact on both the supply and the exhaust efficiencies.

To our knowledge, the new method in the Belgian regulation is the only method which takes both supply and exhaust efficiency into consideration, by using the average between the supply and the exhaust efficiency, in combination with the correction for fan heat. This average calculation assumes that the apparent overestimation of the supply efficiency and apparent underestimation of the exhaust efficiency are roughly symmetric. For the moment, this assumption is surely more reasonable than looking only at the supply efficiency, without any requirement on the deviation of the thermal balance. For some individual products on which several tests have been carried out, the variation of this average efficiency is also lower than the variation of the supply or exhaust efficiency between these different tests [3].

Test on the AHU or on the heat exchanger

It's clear from the results that the efficiency of a heat exchanger alone can be higher than the efficiency of a whole AHU equipped with the same heat exchanger.
The heat recovery efficiency of the whole AHU depends surely on the architecture and design of the heat exchanger itself (type of exchanger, e.g. counter flow vs. cross flow, size of exchange area, local exchange effectiveness, etc.) but depends also on the quality of the whole AHU, such as insulation, air leakages, thermal bridges, etc. The effect of the AHU can strongly degrade the whole efficiency even if the exchanger itself is very good. For example, certain thermal bridges will cool down the supply air because of contact with the colder exhaust air.
The bigger gap between the supply and the exhaust efficiency observed for tests carried out on the whole AHU than this for tests carried out on the heat exchanger alone could also be related to the possible uncontrolled heat fluxes described above. Air leakages as well as transmission heat fluxes from or to the surrounding can be largely higher for the whole casing of an AHU than for a heat exchanger as component.
It should therefor be advised to test the efficiency on a whole AHU only, as done in most of the above mentioned methods.

The effect observed for the tests carried out on the AHU with fans in operation or with fan switched off is however more surprising. While the gap between the supply and the exhaust efficiency was very low with fans off, it was largley higher with fans in operation. Among the hypothetic uncontrolled heat fluxes decribed above, the operation of the fan can affect the air leakages but not the transmission heat fluxes through the AHU casing. It should then be advised to test the whole AHU with fans in operation. Moreover, if the air leakages are well involved in this effect, it could be possible that not only the fan operation as such, but also the pressure difference across the AHU during the test could play a significant role. Pressure difference as close as possible to the real working pressure of the AHU should then be used during the test.

Influence of the flow rate

It is also usually known that the heat recovery efficiency of a given product decreases slightly as the test flow rate increases. As shown in the results, the slope of this decrease can vary largely from one test data to another, with even an increase of the efficiency with the flow rate in a few cases. E.g.: a negative slope of -0.02 %/(m^3/h), as average of the entire data, results in an efficiency drop of 3 % when doubling the flow from 150 to 300 m^3/h. For the determination of efficiency in the context of EP-regulation, it should be advised to carry out the test at least at a flow rate as close as possible to the maximum flow rate of the AHU, with eventually additional measurement points at lower flow rates.

Other divergences in test conditions

The other conditions of the test, such as the required temperature and the relative humidity for the outdoor air and for the extract air, play probably also a role in the result of the test. Fortunately, the divergences between the above mentioned methods are maybe not the most crucial point from the scientific point of view.
Among important differences in the test methods, the following can be underlined.
Temperature difference. For example in the new EN 13141-7 published in 2010, the temperature difference between outdoor and extract air has been lowered to 13 K instead of 20 K in the previous version of this standard (referring to EN 308). Nevertheless one can expect that more reliable results could be obtained using a higher temperature difference.
Relative humidity. In the Dibt method in Germany or the NEN 5138 standard in The Netherlands, the required relative humidity of the extract air is quite high leading to condensation in most of the cases. The higher thermal balance deviation observed with these methods compared EN 308 and EN 13141-7 might be possibly related to this condensation.

Proposal for future approach

Although European standards exist for quite a long time, a lot of member states still use their own test methods for heat recovery efficiency. This is maybe partly initiated by the fact that the EN 308 standard is not always very clear and partly due to different EP-calculation methods, using different efficiency figures. Having different test methods and different test conditions in each European country, for such very expensive tests (10 000 to 20 000 EUR), is not acceptable for the manufacturers in such an international and competitive market. Test methods should therefore converge towards a common approach as soon as possible, at least at European level. The test methods should result in the availability of relevant raw data that can be recalculated towards the required efficiency expression, adapted to the EP-calculation of each member state or to the requirements of the upcoming Ecodesign directive. We can hope that current development works in CEN standardisation (EN 308 revision) as well as in the context of the Ecodesign directive will help to facilitate this needed convergence.

Other challenges for the future

The last example product (Figure 2, Bottom) revealed also a certain variability of the results for different tests carried out on the same product. Besides the possible role of the test conditions (test method) and the possible variability between different test laboratories, the variability of the AHU production and the procedure of selection of the sample to be tested could be an important point of attention in the future.
Another important challenge concerns the custom products, used for example in large ventilation systems such as in commercial buildings. The high costs of the tests which can be

distributed over a high sales number for products in series are probably not acceptable for custom products. Another approach could be studied, such as the interpolation of calculated efficiencies for a range of custom products between 2 extreme products being effectively tested in the laboratory.

Finally, the heat recovery efficiency is not the only point of attention of ventilation AHU. More attention could also be drawn to other performances of AHU such as the electrical consumption of the fans, the automatic balancing of flow rates, the acoustical performances, IAQ related performances such as air leakages, type of materials, etc.

CONCLUSION

While ventilation with heat recovery is promising to decrease the energy use for building ventilation, more reliable and comparable values of heat recovery efficiency are needed in the context of EP-regulation as well as for contractors and users of these systems.

The presented results emphasized some attention points for the heat recovery efficiency measurement as well as calculation. The measurements should be carried out on the whole AHU with fan operating at a pressure condition as close as possible to the real conditions. Given the large divergence between the supply and the exhaust efficiency for most of products, alternative ways of calculation should be examined to avoid using overestimated values of supply efficiency. The recently developed method in the Belgian regulation, using the average between the supply efficiency and the exhaust efficiency, presents several advantages on this point.

Moreover, this paper pointed out also the need for convergence toward a unique and coherent test method across Europe. It is not acceptable for the manufacturers to have so many different test methods and test conditions in the different European countries for such international and competitive market. It can be expected that current works at the level of CEN standardisation and of the Ecodesign Directive will help to facilitate this need convergence.

ACKNOWLEDGEMENTS

The authors thank all the manufacturers which provided test data and constructive interactions in order to improve the test and calculation methods for heat recovery. Due to confidentially reasons, no names can be mentioned.

REFERENCES

[1] Feist, W. 2007. *Passive House Planning Package 2007. Specifications for Quality Approved Passive Houses. Technical Information.* Passive House Institute.

[2] www.epbd.be

[3] Caillou, S. Van den Bossche, P. 2009. *Towards Comparable and Relevant Heat Recovery Efficiencies.* Passive House Symposium 2009, Brussel, Belgium.

[4] Schild, P. 2004. *Air-to-Air Heat Recovery in Ventilation Systems.* Ventilation Information Paper n°6. Air Infiltration and Ventilation Center.

[5] Van den Bossche, P. 2007. *Air to air heat recovery – assessment of temperature efficiency.* Air Infiltration and Ventilation Center Conference 2007, Greece.

ABSTRACT SUBMISSION FORM FOR THE 2012 AIVC-TIGHVENT CONFERENCE

Send your abstract using this form at info@inive.org

PAPER TITLE
Integration of solar chimney with direct evaporative cooler as a passive ventilation technique under Egyptian climate

MAIN AUTHOR
Amr Sayed H.

OTHER AUTHORS
Yoshino Hiroshi , Napoleon Enteria, Magdy M. Radwan, M. Abdelsamei Eid

ABSTRACT

In Egypt's hot-dry climate, new housing is poorly adapted to the climate because architects lack of knowledge and adequate design tools. As a result, there has been a vast expansion in the use of air conditioning to cool buildings which expected to continue for sometime yet. The restriction on the use of CFC gases for cooling and the production of CO_2, which is a major gas contributing to the greenhouse effect, will encourage designers to apply more passive means for controlling the indoor environment. Also Egypt in general has rich sunny and clear sky. Therefore, these conditions encourage such a concept to enhance evaporating natural ventilation and save energy. This paper studies integration of direct evaporative cooler with a solar chimney to ventilate and provide thermal need for the occupants in the buildings. Simulation is done for the solar chimney with evaporating cooler in the hottest month in summer using commercial couple multi-zone airflow under COMIS-TRNSYS software to assess natural ventilation. The dependence on Egyptian outdoor air temperature has been studied to determine the thermal comfort criteria. This paper develops an integrated model incorporating these evaporative cooler components into couple multizone passive ventilation model. The temperature and airflow rates are predicted iteratively taking into account the zone pressure and the pressure drop in the evaporative cooler component. The simultaneous set of mass balance equations is typically solved using Newton-Raphson method to correct the zone reference pressures until the simultaneous mass balance of all flows is achieved. The result shows that room air temperature can be decreased by increasing the saturation efficiency and wet medium saturation of the evaporative cooler. As hot air passes over wet media, some of its energy goes to evaporate water from the media which decrease indoor temperature and increase relative humidity. According to the chimney effect, the buoyancy effect occurs due to the differences between inside and outside air that cause pressure gradient which sucks the cold heavy air through evaporative cooler. The findings show that natural ventilation with evaporative cooler can be a potential alternative to air conditioning systems in Egypt with hot and dry climate.

HYBRID VENTILATION AND COOLING TECHNICS FOR THE NEW NICOSIA TOWNHALL

Flourentzos Flourentzou[*1], Dickon Irwin [2], Margarita Kritioti [2], Tasos Stasis[3], and Nicholas Zachopoulos[3]

1 Estia SA
Par Scientifique EPFL
1015 Lausanne, Switzerland
Corresponding author:
flou@estia.ch

2 irwinkritioti.architecture
10 Gregoriou Xenopoulou
1061 Lefkosia Cyprus

3 Eliofotou, Zinieris & Stasis Partners
77 Strovolos Ave.,
Strovolos Center 4th Floor,
2018 Strovolos, Cyprus

ABSTRACT

The new Nicosia Town-hall is a very particular building. On the site where it is built, important antiquities were discovered during the first day of construction and the whole design was completely modified to fit to the new situation. The archaeologists continued to excavate 2/3 of the entire site and created an archaeological park in the centre of the town. The building area was constraint to the remaining land, and co-exists with the uncovered findings.

As a consequence, the building was split into 5 smaller units, 4 office and public service buildings and a municipal hall. Foundations were changed to a combination of piling between findings and large raft slabs sitting above the level of undisturbed ground. The design of office buildings followed the rules of bioclimatic architecture to meet the passive standards and the building is on process for Minergie® (Swiss) labelling. Massive buildings, naturally ventilated and cooled, offer a natural comfort with minimum energy consumption.

The hall follows completely different design principles. Above a large slab, sits a light structure and glazed façades, allowing maximum view and contact between the interior- where the municipal council meet- and the surrounding archaeological park.

According to good practice design rules, this building would be a bad building, especially in a hot climate. A completely glazed cube would certainly overheat and consume a lot of energy. To avoid this, the design proposes a hybrid ventilation system using a sophisticated natural air path to cool naturally the building. Several distinct air streams using smart stack effect path ventilate the building differently according to the time, to the use of the building and to the external climate, in order to reduce mechanical ventilation and air conditioning hours of use to the strict minimum. Ventilation system shifts automatically from natural to mechanical offering maximum comfort with minimum cooling, heating and fan energy consumption.

This original hybrid ventilation and cooling system made possible the particular architectural expression of the building with low energy consumption. The whole building complex makes a harmonious eco-neighbourhood with low-energy-consumption, confortable interiors and friendly shaded, wind-protected public spaces, open to the town, where urban life meets cultural heritage.

The article explains the ventilation concept, bioclimatic principles and the simulated comfort and energy performances.

KEYWORDS

Potential for ventilative cooling strategies; design approaches for ventilative cooling and case studies; summer comfort and ventilation; nnovative ventilation.

INTRODUCTION

The new Nicosia town hall (Cyprus) is not a simply green building showing several bioclimatic architecture principles. It is the first contemporary building in the island applying all the bioclimatic principles, which are necessary to meet the passive building standards (primary energy consumption for heating, ventilation, air conditioning and hot water production less than 30 kWh/m^2y).

The "town hall" is not a single building. Archaeological findings restricted the available land to the 1/3 of the initial available surface and the unique initial building is split to smaller units in order to fit in the remaining complicated site. 4 office buildings and a municipal hall, able to receive the council meetings in presence of 250 people, form a neighbourhood in Nicosia old town, just 100 m from the green line, where the war divided the city several decades ago.

Bioclimatic and sustainable architecture starts from the site use. The buildings respect the old town scales and they are integrated in the archaeological site not only preserving cultural heritage, but also making it available to the population, through walk paths, squares, and shaded patios. They create a public space with a social environment, where urban life meets culture and municipal services, in a marginalised district of the city, where social life is stopped for many years now. Orientation and disposition of the buildings group similar uses, separate polluting and noisy activities from office spaces, create natural shading to public space and neighbouring buildings.

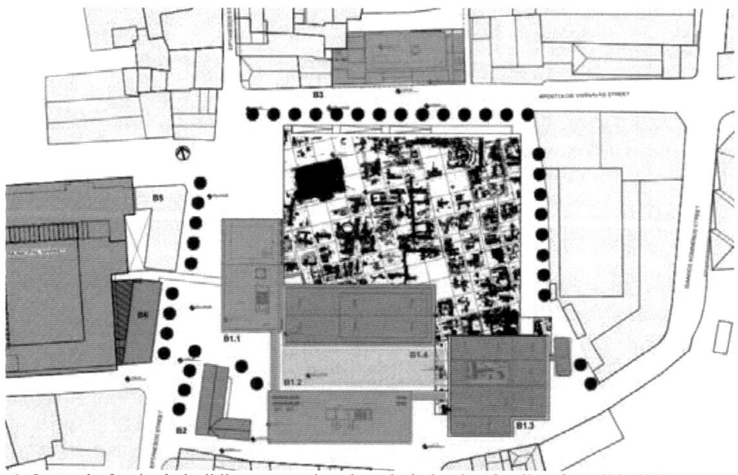

Figure 1. Instead of a single building occupying the whole land, a family of small buildings around the archaeological findings create interesting bioclimatic potential.

Figure 2. Panoramic virtual view of the building complex from the green roof of building 1.3; view of the municipal hall from the antiquities, view of the shaded patio between buildings B1.2 and B1.4. Only building B1.3 is finished. The other buildings are under construction.

Buildings B1.1, B1.2, B1.3 and B3 are massive and well-shaded office and public service buildings. Building B1.4 is the light structure, fully glazed 10 m height municipal hall. The square between buildings B1.2, B.3 and B 1.4 is shaded, providing solar protection to the three buildings and especially to the south glazed façade of the municipal hall. Building B1.3 is siting on 10 pillars over the archaeological site and it has an interior yard making available natural light and ventilation to the core of the building.

Regarding ventilation strategies, office buildings function with purely natural ventilation with specially designed vents. The municipal hall runs with a sophisticated hybrid system, with natural ventilation assuring air movement for free night cooling and mechanical ventilation distributing heat and mechanical cooling.

BIOCLIMATIC DESIGN, OF OFFICE BUILDINGS

The basic condition for a confortable thermal environment of offices is good insulation and solar protection. In south climatic conditions, with very hot summers and relatively cold winters, energy performance is necessary for both winter and summer seasons. A well-insulated building, with reasonable glazing orientation and solar protection, consumes 15-25% of the total thermal demand for heating and 75-85% for cooling. In the past, where buildings where not insulated, this ratio was inversed, with heating demand representing more than 75% of the total demand. This is illustrated on table 1 and figure 3.

Insulation	Heating demand	Cooling demand	Total demand
A. 0 cm, single glazing	149 (78%)	43 (22%)	192
B. 4 cm, double glazing - 3.5 W/m^{2}k	21 (25%)	64 (75%)	85
C. 10 cm, double glazing - 1.3 W/m^{2}k	8 (19%)	34 (81%)	42

Table 1. Heat and cooling demand of 3 building scenarios (building B3) simulated dynamically with DIA+ software: A-construction as it was usual before the entry into force of the energy Low (without insulation and with single glazing), B-with insulation and glazing respecting the minimum requirements of the local energy Law, C-with optimised insulation depth, high energy performance windows and with static solar shading, meeting passive standards. For all scenarios, there is no special free cooling strategy.

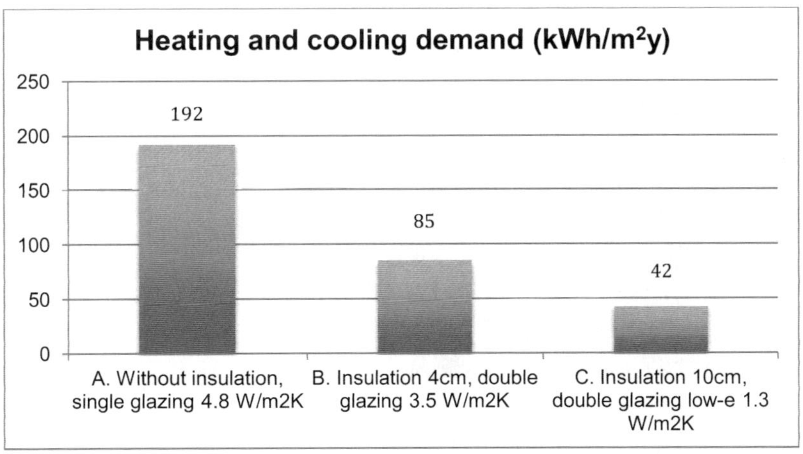

Figure 3. The graph shows heating and cooling demand, according to the 3 basic scenarios. Cyprus Energy Law reduces the energy needs to the 44% of those of a building, without any care for thermal insulation. Additional thermal insulation (10 cm instead of 4) and more insulated windows (U value 1.3 instead of 3.5), with 60 cm passive solar protection on the south façade, reduces the energy needs to the half of those of a building meeting the minimum legal insulation values. Passive buildings (C) have only 22% of the energy needs of non-insulated buildings (A).

Thermal insulation

After several optimisation dynamic simulations with DIAL+ software, we decided that the optimum insulation characteristics to meet the passive standards are 10 cm of rockwool for the roof and the facades, 5 cm for the periphery of the building and a U value of the windows of 1.3 W/m^{2}K. Thermal insulation on the ground does not change anything, as the mean ground temperature in Cyprus is high. A careful analysis and treatment of every joint between constructive elements minimises thermal bridges and heat losses in winter. External insulation gives the advantage of thermal mass inside the building.

Figure 4. Outside thermal insulation with minimised thermal bridges is composed mostly by 10 cm rockwool. Foundations are insulated with 5 cm xps. In some special cases it was necessary to use internal thermal insulation with 10 cm rockwool.

Thermal mass

An apparent cladded concrete ceiling and a floor composed with 4 cm anhydride screed over the concrete slab and rough concrete screed, offer a high thermal mass, absorbing excess heat during the day and restoring it during night. This optimises the use of internal heat gains during winter and reduces the pick temperature during summer.

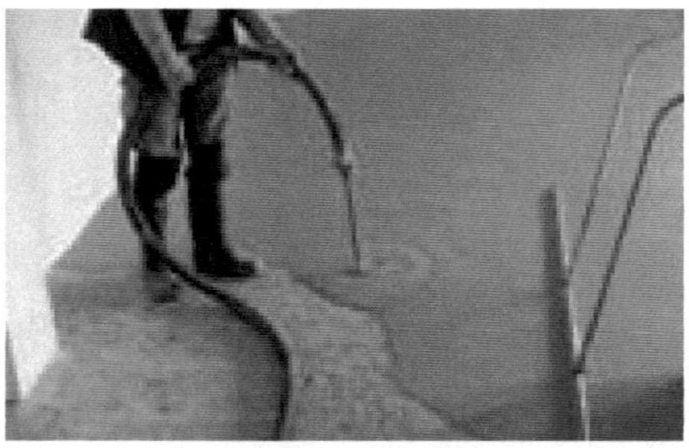

Figure 5. Floor and ceiling are composed by massive materials with high thermal mass.

Solar shading

As building B3 is north - south oriented, static solar protection of 60 cm is sufficient. As we see from table 1, comparing scenario B with scenario C, solar protection reduces cooling demand in summer by nearly 50% (additional wall and window insulation plays a small role for the cooling demand).

1st February	1st April	1st June	1st July
24 %	48 %	88 %	91 %

Figure 6. The south façade (15° South – West) at 14:00 is shaded with 60 cm static top and side solar protection. The choice of a glazing with g value of 0.4 is a good compromise between summer solar protection and winter useful solar gains. The façade white ceramics is a robust, inert, self-cleaning material with low sun absorption, offering low façade temperature rise to prevent overheating of incoming air during summer. The table shows the solar shading during the year. In winter solar shading is minimum and in summer maximum.

1st May	1st June	1st July	1st August
86 %	60 %	57 %	74 %

Figure 7. As the north façade is 15° oriented east, early in the morning during June and July, sun hits the façade during 2 hours – 05:00 – 06:00. Some solar shading is necessary to protect north offices early in the morning. The table shows the shading efficiency of a 60 cm vertical static solar protection at 6 h in the morning.

Natural and artificial lighting.

In a context where air conditioning represent ¾ of the total energy demand, internal gain control is capital. Lighting counts for around 50% of the installed electric power. The question is how many hours the users will need to turn it on. High natural lighting autonomy is easy to obtain in a climate with 300 days of sunshine per year. The designer should even pay attention so that there is not excessive light coming from windows. A delicate equilibrium should be found between daylight needs, winter solar gains for passive heating and summer solar gain control. A north office of 4 m large has 2 modules of glazing of 140 X 300 (70% glazing), while a south oriented one a single module of the same dimensions (35% glazing).

Figure 8. Natural light autonomy, simulated with DIAL+ software between 75 and 85% during working hours.

Several optimisation measures increase up to 30% the natural light autonomy: white colour walls and ceiling; reduction of glazing frames and rise of the window top, up to the ceiling; white colour external shading; glazing with high luminance transmission (g value 0.4, LT 0.7); light wash of the lateral white wall with the glazing moved to the side; These measures provide a natural light autonomy between 70 and 85% during working hours

In addition to natural light maximisation, artificial lighting uses high efficiency luminaires with installed power < 12 W/m^2. Light is automatically switched off when the office is empty.

Ecological materials and health.
The objective is not only a high-energy performance and confortable building but also a building using environmentally friendly materials, creating a healthy interior environment. Concrete for the structure, plain wood for the façade structure and the window framing (non treated larch), gypsum panels and rock wool for interior partitions, rock wool for the thermal insulation and ceramics for façade exterior facing, are the main materials used in the building. Joint synthetic substances are avoided. Instead of surface treatments and painting, the natural material colours create the chromatic synthesis and the aesthetical language. Wood is just oiled with linseed/turpentine/TiO mixture, avoiding varnishes and synthetic substances. The life cycle of the building elements is high, with low maintenance needs. Cyprus climate is very difficult for the exterior materials exposed to sun, dust and rain. The exterior façade facing with ceramics, white for façades with high sun exposure and coloured for other orientations, is a robust high life-span solution, adapted to the local environment.

There is no material in the building emitting VOC particles. The floor is done with anhydrite liquid screed, which is mineral and inert, offering high thermal mass and avoiding VOC emissions.

VENTILATION AND COOLING STRATEGIES OF OFFICE BUILDINGS

Before we adopt a ventilation strategy, we put on the balance 4 aspects: air flow necessary to assure air quality and occupant's health, energy consumption by fans or by thermal losses or gains because of excess ventilation, occupant's wishes / well-being. Some people, influenced by good practice in the North and Central Europe countries, concentrate on the possibility of heat recovery. They a priori consider that mechanical ventilation with heat recovery is a good practice for every climate and for every building use, extrapolating intuitively this conclusion from what happens in the cold climates.

Without excluding any solution, before adopting a ventilation strategy, we answered to 3 questions:

1. what are the wishes of the users and how do they feel in regards to the control of their environment;
2. what is the real impact of different ventilation strategies on heating, cooling and electricity demand;
3. what is the real risk of wrong use and bad ventilation control by the users ?

Question 2 and 3 may have a different answer according to climatic conditions, building function, physical characteristics of construction elements.

Users wishes and feelings in regard to ventilation systems.
The objective was not only a high-energy performance and a confortable building. A municipal building is a professional tool for public service. Well-being of the users is a key factor on productivity and service quality. Before the building design, the great majority of

the municipality personnel answered to a questionnaire about their current indoor environment quality and their expectations from their new place of work. The personnel showed a high degree of environmental consciousness with low CO2 emissions being their second concern. 63 people imagine an exemplary building of natural comfort and only 23 an exemplary fully air-conditioned building. 30% of the people consider mechanical ventilation as problematic. Less than 5% considered natural ventilation from the window as problematic. These results confirm the results of European research, showing higher acceptance and lower building sick syndrome index in naturally ventilated buildings.

Impact of ventilation strategy and heat recovery on energy demand.
In order to quantify the impact of different ventilation strategies, we have simulated a typical section of the building with two offices of 4 m large, and 10 m deep representing the total depth of the building from north to south. One office faces south and the other north with an interior buffer corridor zone in the middle. The thermal model considers the whole space as a single zone with dimensions, building thermal characteristics and shading as explained in the previous paragraphs. It considers also standard use conditions and occupation schedules according to the Swiss regulations SIA 2024. DIAL+ software simulates dynamically the solar gains, internal temperature, and natural ventilation airflow, cooling or heating load and energy consumption.

For the electricity consumption we used an optimistic hypothesis of a high efficiency fan, consuming 0.16 $W/m^3.h$ for simple extraction and 0.32 $W/m^3.h$ for a system with heat recovery. We used a high-COP energy system of 4 to calculate electricity consumption.

Column Title	Heating need kWh/m^2	Cooling need kWh/m^2	Total thermal demand kWh/m^2	Fan electricity demand kWh/m^2	Total electricity demand kWh/m^2
1. Mechanical, 36 m^3/h.pers	6.7	35.5	42.2	1.5	12.1
2. Mechanical + heat recovery 80%	3.4	33.1	36.5	3	12.1
3. Natural with 10 cm tilted window	6.0	31.8	37.8	0	9.5
4. Natural with 10 cm standard window	9.6	35.2	44.8	0	11.2
5. Excessive ventilation - 50 m3/h.pers	7.5	36.2	43.7	0	10.9
6. Natural with night ventilation	9.6	16.5	26.1	0	6.5

Table 2. Heating and cooling and ventilation thermal and electricity demand.

From this table we can learn several not very well known truths.

1. Heat recovery in south climates is not always energy effective.
As we can see from the results in the second line of table 2, heat recovery may reduce thermal demand by 13% (5.7 kWh/m^2y), but it consumes 3 kWh/m^2y of electricity instead of 1.5 of a simple extraction system and 0 of a natural ventilation system. If we measure electricity to produce 5.7 kWh of heat or coolness with a system of COP = 4, the heat recovery system does not recover even the energy that it needs to run. It could be energy effective for cases with direct heating or very low COP cooling, or for cases of non-insulated buildings with very high demand for heating as it was the case before the Energy Law.

2. Controlled natural ventilation is the most energy effective strategy for office buildings.
We can see this, if we analyse the results of the 3rd line. Controlled natural ventilation with a well-designed window, allowing small openings during hot or cold hours, does not spend more energy than mechanical ventilation. If we take into account saved electricity to run fans, the total balance is positive for natural ventilation, with 9 kWh/m^2y of energy consumption instead of 12.1 of a mechanical one. Dynamic simulations of the airflow showed that a vent of

40 cm large by 140 cm high, in tilted position, provides almost always the necessary airflow for pollutant evacuation without excessive ventilation.

3. Excess ventilation, if it is reasonable, does not destroy the energy balance of the building.
Minergie® standard counts for natural ventilation an overestimated 50 m^3/h airflow rate (simulation of line 5), but if we take a pessimistic hypothesis considering that the window is kept open all the time at 10 X 300 cm, independently of how cold or hot is the outside temperature, we can see that heat demand is not excessive. It creates extra 6% losses (44.8 kWh/m^2y of thermal demand instead of 42.2). This is because extreme temperatures (38 to 42°C in summer end -2 to 5°C in winter) take place very few hours during the office working hours. Most of the time, temperature difference is moderate, not creating excessive thermal losses.

4. The most interesting energy saving potential lies in night ventilation free cooling.
As we can see from the results of line 6, a night cooling strategy reduces cooling demand by 56% (16.5 kWh/m^2y instead of 35.5) and the total energy demand by 38%. **This is the key energy potential for south climates.** It is equivalent of the passive heating for cold climates.

As a result of these findings, we concentrated the efforts to design a smart, simple and user-convenient window. This window should provide easy and intuitive control for limited ventilation during office hours and high airflow rate ventilation during night in summer. In addition, users should not thing how to ventilate; they have just to open a window. The opening should be protected, in order to control the risk of intrusion by undesirable people, animals, insects rain and dust. People should feel safe to leave the vents open during night without any concern.

Natural ventilation design and ventilation strategies.
Vents are vertical, opening on the whole room height, in order to maximise stack effect. With 5°C temperature difference between inside and outside, a 40 by 300 cm vertical vent creates a stack effect of 611 m^3/h, while the same vent in horizontal position 300 by 40 cm creates only 223 m^3/h. The right disposition of the vent opening may boost ventilation airflow by 275%! High airflow rates are necessary only during night. During the day only 36 m^3/h per person are necessary. An opening of 40 by 140 cm height may provide 75m^3/h at a $\Delta T = 5°C$ and 47m^3/h at $\Delta T = 2°C$. These dimensioning calculations leaded us to divide the high vent in two parts and to make it open right or tilted. The user instructions become simple and easy to understand: "tilt the top vent during working hours winter or summer. During winter, you close it when leave the office and during summer, you open completely one or both vents, according to your cooling needs; you put it back to the tilted position in the morning."

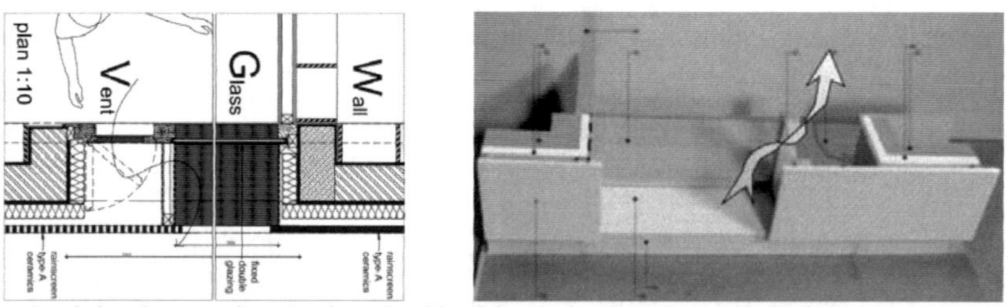

Figure 9. South façade vent. Air enters from the side of the glazing through a perforated protection sheet metal protection.

As we can see from the photos of figure 10, lighting openings are dissociated from air vents. This makes it possible to treat correctly the glazed part, hiding frames or any obstacles and divide, protect or hide the vent part. In the south façade, air comes from the side after passing through a perforated sheet metal. On the north light and façade air comes directly after the protection.

Figure 10. Tilted north façade vent, north façade open right, south window from outside, south vent from inside.

As we can see on the second picture of figure 10, all offices are equipped with a ceiling fan. This offers the possibility to the users not to use air conditioning up to 28-29°C of internal temperature and use the roof fan instead, reducing drastically the hours of use of air conditioning. It avoids also a wrong use of the window completely open when external temperature is around 30°C and users like wind breeze. If the user wishes an air movement, he may use the ceiling fan, avoiding excessive heat and dust entering in the office.

VENTILATION AND COOLING STRATEGIES FOR THE MINICIPAL HALL

The municipal hall is a light structure fully glazed building. According to what preceded in this article, bioclimatic architecture should exclude this design. However, the social needs constrained the design team to find special solutions for this building.

The constraints: light structure without thermal mass; high glazed-façades, difficult to equip with movable solar protection.

The advantages: rare and limited use during the day; a country yard in the south, possible to be shaded, creating a social place to meet and shading the exposed south façade; complete shading from the west building, a free massive technical space, under the bleachers, a high building offering high stack effect for ventilation and allowing a hot buffer zone outside the living zone.

The strategy: create a double skin façade using the acoustic element at the top part of the interior space; use a selective glazing with g value 0.4 to reduce sun thermal load; complete solar shading with an interior movable awing at the lower part of the façade and shade the south external yard; allow openings on to top and on the bottom of the façade canal, able to evacuate solar gains; allow a big opening behind the building for air inlet under the bleachers. We ventilate the building during night to cool the concrete slabs of the technical spaces under the building. Air may also come in the canal on the bottom of the double skin façade, when outside air is cooler than the interior air, in order to evacuate accumulated heat from solar gains locally very early in the morning in the north façade and during the morning in the east façade. When outside air becomes hotter than the interior, it comes in only from the back of the building. It is cooled by the coolness accumulated during night in the thermal mass. An automatic control system opens and closes the bottom and top openings of the space according to the desired strategy.

Simulations of interior temperature and stratification showed that with this strategy the interior climate is never worse than the exterior. The occupied space has the best climatic conditions, profiting from the precooled air under the building. By controlling stratification, we localise hot air in non-occupied spaces (within the double skin canal and on the top of the building behind the false ceiling).

Under these conditions, mechanical ventilation with cooled or heated air follows the same path with a part of natural ventilation. It comes to complete heating and cooling, when passive technics are not able to meet the demand. Mechanical ventilation recirculates air and recovers heat when this is beneficial (when stratification is low) but it may function as a hybrid system blowing only cooled air in cases where thermal gains are extreme and returning air is too hot.

By limiting mechanical cooling during some hours of the year, when the building is occupied and when bioclimatic technics cannot meat the demand, we limit energy consumption drastically and allow passive comfort without air conditioning for a large period of the year.

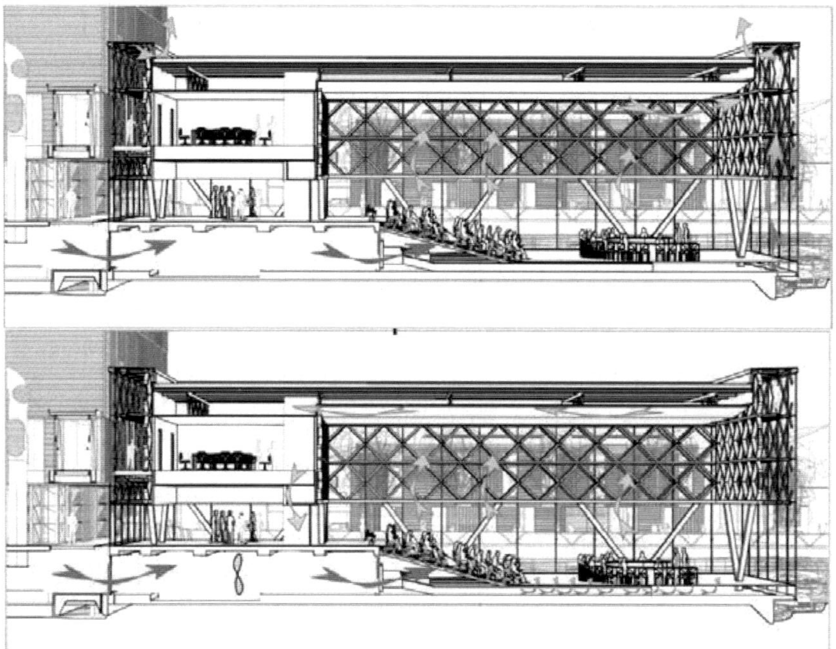

Figure 11. top natural ventilation patterns, bottom, mechanical ventilation patterns.

CONCLUSION

After thermal insulation and solar shading, free cooling is the key issue for low energy passive buildings in Mediterranean climate. Issues like high air-tightness and heat-recovery, which are important for North and Central Europe, have minor importance for southern Europe, simply because in the mild climates, most of the time the building should be open to benefit from the outside air, which is within or near the comfort zone.

Free cooling by natural night-ventilation is the simplest strategy, but it needs special design attention. Standard windows are not always the best way for natural ventilation. When ventilation strategy depends from the occupant's behaviour, simple smart windows with many opening possibilities, equipped with protections from insects, dust and vandalism, ensure the users and encourage a correct use. For common spaces and large halls, smart automation with the minimum number of openings and sensors is necessary, to achieve a sure result. Nicosia town hall is a good example illustrating both natural ventilation design principles on a very low energy consumption building.

THE INFLUENCE OF THE SELECTIVE VENTILATION IN THE THERMAL PERFORMANCE OF MODERN NATURALLY-VENTILATED HOUSES IN GOIÂNIA – BRAZIL

Leônidas Albano da Silva Júnior*[1], Marta Adriana Bustos Romero[2], Alberto Hernandez Neto[3]

1 Universitat Politècnica de Catalunya
Barcelona, Spain
**Corresponding author: leonidas.albano@gmail.com*

2 Universidade de Brasília
Brasília, Brazil

3 Universidade de São Paulo
São Paulo, Brazil

ABSTRACT

The aim of this paper is to investigate the influence of the selective ventilation in the thermal performance of modern naturally-ventilated houses built in the 1950's and 1960's in Goiânia, located in middle-west of Brazil. The selective ventilation is one of the passive thermal conditioning strategies recommended for buildings located in this city, in the summer. This study allows the analysis how much important is the selective ventilation as a bioclimatic strategy in the studied cases, considering the application of this strategy before the establishment of contemporary concepts as "zero energy building", helping us to reflect on the future from a historical perspective. Seven rooms in two houses were selected for in situ measurements and simulations. Indoor and outdoor air temperatures and relative air humidity were measured with data loggers HT-500, during 91 days in three different months in 2011: June (low air temperatures and low relative humidity); September (high air temperatures and low relative humidity); and December (medium air temperatures and medium relative humidity). The measured dates are used to calibrate the virtual model of the dynamic simulations. Based on this calibrated model, a reference model for computer simulations was defined in the EnergyPlus building energy simulation program, used also to solve the energy balances and to evaluate the thermal performance of the cases. Variations in the ventilation were carried out and hourly air temperature and relative air humidity were the output data for the same thermal zones studied in each measured case. By imputing such data in the Analysis Bio Computer Programme, the percentage of discomfort hours in the models was obtained. Correlations among the results were investigated considering different variables in the time and the space, allowing the comparison among different buildings and comfort zones, periods of the day and seasons. The main conclusion is that it is possible to verify the sensibility of the technical solutions applied in the cases in relation to the selective ventilation, demonstrating its importance in the building thermal performance and identifying important variables to improve the thermal performance of modern naturally-ventilated houses.

KEYWORDS
Selective ventilation, thermal performance, naturally-ventilated houses, modern houses.

INTRODUCTION

In Brazil, building codes about ventilation of dwellings require a very small area for natural ventilation and most of the dwellings around the country would benefit from larger areas. The minimum window areas that can be opened to be 1/6 (long occupancy spaces) or 1/8 (short occupancy spaces) of the floor area and openings are normally only half of this area. [1]

This paper presents the importance and the influence of the selective ventilation in the thermal performance of the buildings. Specifically, this paper focus the modern naturally-ventilated houses built in the late 1950's and early 1960's in the city of Goiânia, capital of Goiás state, located in the Midwest of Brazil. (Lat. 16.41S, Long. 49.25W, Alt. 741m/29,173.23in). Standard NBR15220 recommends for Goiânia (Bioclimatic Zone 6) medium ventilation openings (15%< A < 25%) but as it is only a recommendation, only a few buildings follow them. [1, 2] The selective ventilation is one of the passive thermal conditioning strategies recommended for buildings located in this city, in the summer. This study allows the analysis how much important is the selective ventilation as a bioclimatic strategy in the studied cases, considering the application of this strategy before the establishment of contemporary concepts as "zero energy building", helping us to reflect on the future from a historical perspective.

Due to be naturally-ventilated buildings, the study focuses on calculating variations of internal hygrothermal conditions and estimating the periods of lack of comfort, comparing the results with the external city conditions and also the immediate surroundings, and determining the thermal loads in reference to individual contribution of individual architectural elements of the two houses. Were considered the actual Brazilian normalization and the singularities indicated on the ResHB method about the internal thermal improvements, use of spaces and thermal zones differentiation [3].

The thermal performance of the modern houses located in Goiânia is a theme that transcends the technical study as it is related to a very important historical moment for Brazil in the 20th century. It is necessary to consider a historical, socio political and economic contextualization of a moment and place, so the theme can be addressed in its real complexity. This paper intends to study the object without losing this contextualized view.

ANALYSIS METHOD

Below are presented the basis for the study object definition, the methods and research procedures, the tools and instruments, also the general considerations regarding the analyses of the results.

The ABNT NBR 15575-1 (2008) [4] defines two levels of approach in assessing the thermal performance of houses: simplified normative and global informative, recommended on more detailed evaluations. In this paper we adopted the global informative evaluation that defines two procedures of requirements and criteria verification: by measurements on constructed buildings and through computer simulations.

Two houses were selected (1 and 2), starting from studies and surveys of the modern architectural heritage from the city of Goiânia [5]. To select the rooms to be measured we considered the requirements of the mentioned regulation that: relates the thermal conditions inside the houses with external conditions at shade; and focuses the study on prolonged stay rooms, living rooms and bedrooms, preferably those with the biggest surface exposed to direct sunlight and less favorable solar orientation. To define the measurement period we considered the typical conditions of the reference year in accordance to the meteorological data from 1961-1990 [6] as the summer and winter recommendations of the used regulations. This way, for each house were selected: 01 external environment shaded, 01 living room and 01 bedroom with easy access during the experiment. At the house 1, the central covered patio was also selected, for its peculiarity and strategic location. All in all, 07 environments were chosen in both houses for the measurements to be taken. Indoor and outdoor air temperatures and relative humidity were measured with thermal hygrometric Data Loggers HT-500, during 91 days in three different months in 2011: June (low air temperatures and low relative humidity); September (high air temperatures and low relative humidity); and December (medium air temperatures and medium relative humidity). The measured dates are used to calibrate the virtual model of the dynamic simulations. The environmental dates about the city of Goiânia were obtained by the measured dates of INMET (2011) [7].

All rooms of the residential units were simulated, considering the thermal exchanges between them, and the social and intimate functional sectors were also evaluated, with emphasis on the measured rooms. EnergyPlus building energy simulation program 4.0.0.024 was used to solve the energy balances and to evaluate the different contributions of each significant envelope component in the thermal performance of the houses, further allowing the comparison with the data obtained through measurements. These simulations were performed for the same period of the in situ measurements. The reference model for thermal performance simulations was based on the calibrated model, maintaining its volume and solar orientation. It was defined within the same thermal zones. The thermal properties of the materials and components used in the evaluation were determined as prescribed in the ABNT NBR 15220-3 (2005) [2], as shown in **Table 1.**

Variations in the ventilation were carried out and hourly air temperature and relative air humidity were the output data for the same thermal zones studied in each measured case. By imputing such data in the Analysis Bio computer programme, used to determine the periods of comfort, to obtain the percentage of discomfort hours and to compare the different variables allowing the analysis based on Givoni (1998) [8].

Data from the measurements and simulations have been treated to allow comparison such as: year period, territorial scales (urban/ immediate surroundings), type of spaces (internal or external), different houses (1 and 2), functional sectors (social/intimate) and internal rooms (living room/ bedroom). In this paper the data was analyzed from the thermo-hygrometric amplitude in accordance with the evaluation parameters of thermal performance existing on ABNT NBR 15575 (2008) [4], in a way to enable the analysis for surroundings correction and the thermal performance of the buildings in this scale. At the analysis of the results the particularities of residential typology in relation to other ones listed on ResHB Method were considered. The ResHB Method is an implementation of RHB Method, developed by the research project ASHRAE RP-1199, in order to adapt the HB Method to the particularities of residential typology, characterized by: less internal heat gain, mainly from the surrounding components, although it is generally more exposed to the other typologies [3]. The internal spaces were understood from the diversity and flexibility of related uses in these rooms, therefore higher tolerance of users to internal thermal fluctuations.

Correlations among the results were investigated considering different variables in the time and the space, allowing the comparison among different buildings and comfort zones, periods of the day and seasons.

Envelope components	U	φ
External Walls (thickness: 270mm) of solid brick (100x60x220mm)	2,25 W/m2K (0,39 Btu/ft2 h °F)	6,8h
Roof of fibro-cement roofing tile (thickness: 7mm) with slab of concrete of 200mm	1,99 W/m2K (0,35 Btu/ft2 h °F)	7,9h

Table 1. Description of the walls and roofs of the references models and its thermal properties: thermal transmittance (U) and thermal delay (φ)

CASE STUDIES

From a brief historic contextualization of the studied cases, below are presented the climatic characteristics from the city of Goiânia, followed by the case studies presentation.

Goiânia was founded in 1937 to be the new State Capital of Goiás, located in the Brazilian Midwest. In a context of national order characterized by political and economical changes, Goiânia eventually stood out as an expression of progress in a quest for modernization of the countryside. The early 50s were characterized by the countryside development, with the construction of a new capital for the country, Brasília, and the expansion of national infrastructure. It is in that decade that the first examples of modern architecture begin to appear in Goiânia [9].

The arrival of this new architecture transformed the architectural procedures that guaranteed the continuity of local tradition, changing the way these buildings adapted to climate, the factor which most interfere in the Brazilian architecture. Among the different typologies built, the residential ones were highlighted for their relevant role in the expression of the new architectural language. Among the strategies for climate adaptation used the highlights are: expansion of the openings in the building surroundings; constant use of cross ventilation in internal rooms; replacement of external porches and corridors for stilts, terraces and balconies; use of "*brise soleil*" as main solar protection element, extensively studied by architects and engineers of the time. The architectural changes mentioned affected the transparency, porosity and solar protection of the residential surroundings and directly influenced their thermal performance, mainly for being naturally ventilated.

The region where Goiânia is located is characterized by the continental and regular cyclic process of air masses displacements, implying a clear rainfall, causing the city climate to be formed by the composition of two main seasons: wet and dry (Aw according to Köppen). An important factor in relation to the dichotomy between the dry and wet seasons arise from the combined effects of nebulosity and insolation. Because of the nebulosity of 80% in December and 43% in June, although the difference between the number of daily hours of the summer and winter solstice is approximately 2h, the insolation in December is lower than in June, 161h/month and 275h/month respectively. Even thought the solar irradiation in December is higher than in June, 3361 W/m2 per day (1,066.13 Btu/ft2 h) and 2708 W/m2 per day (859 Btu/ft2 h) respectively, causes the south facade, struck by summer insolation, can be more transparent and free of solar protections. In opposite to the North facade, exposed to winter insolation, which should be well protected and more opaque [10].

According to ABNT NBR 15220-3 (2005) [2], the buildings located in the city of Goiânia (Bioclimatic Zone 6) must have: shaded openings with ventilation surface between 15% and 25% of the pave surface; heavy external walls, light and isolated roof. The passive thermal conditioning strategies recommended are: in the summer, evaporating cooling, thermal mass for selective cooling and ventilation; in the winter, internal thermal inertia. This way the characteristics of the surroundings have a greater influence in the environmental conditions during summer time, while the internal conditions of the buildings have bigger influence during wintertime.

Among the 78 modern houses currently identified in Goiânia, 2 houses in good state of conservation were selected, as shown in **Fig. 1**: House Abdala Abrão, projected by David Libeskind and built in 1961 (House 1), and House Eurípedes Ferreira, projected by Eurico Godoy and built in the late 50s (House 2), both located at Setor Sul, central area of the city, far from each other 280m (306 yards) approximately. Both houses have 2 floors and a functional shed on the back, focusing the residential areas (social, intimate and service) at the upper floor and leisure activities at the ground floor. Due to the topography, House 1 has the upper floor partially resting on the ground, allowing the main building access through this pave. Its intimate sector is located at the North side, oriented Northwest (Azimuth 343°), while the social areas are located at the South side. On House 2 is the opposite, the social areas are located at Northeast (Azimuth 44°) and the intimate at Southeast. As to shape, the design of House 1 is a square, with central covered patio while House 2 has an "L" shape. Externally both have a swimming pool, partially paved areas and permeable gardens with trees and grass.

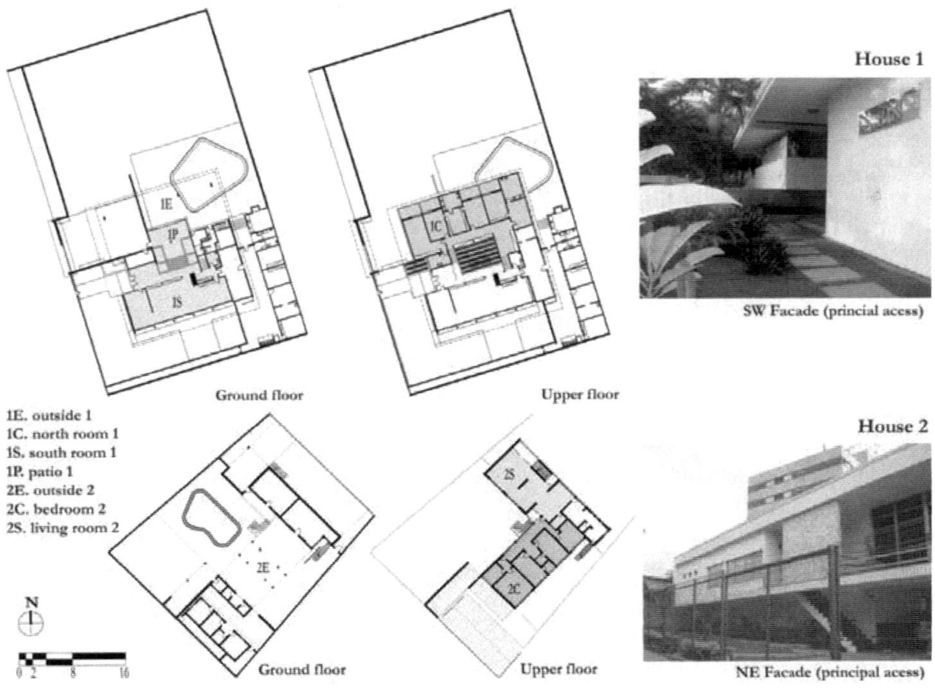

Figure 1 . Description of houses with plans and principal facade photo.

RESULTS AND DISCUSSION

The main results of the study are presented below comparing different results with Energyplus and AnalysisBio simulations.

The simulations done with AnalisisBio Programme pointed that passive solutions result in a zero energy demand in June. December presents more uncomfortable hours because the surround corrections are thought for dry seasons. Because this, the problem in December is the air humidity and it is not the air temperature. **Fig. 2** shows the Givoni graphic for the three sites and the three studied months. In September, the conditioning air was used in the bedroom of the House 1 (1C in Fig.1). It was the unique month and room that used this active system.

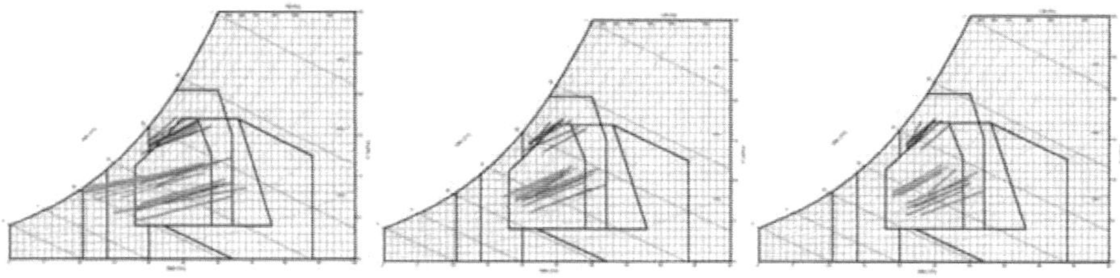

Figure 2 . Comfort graphics about the surround corrections in the different houses: on the left) Goiânia site measure dates; on the middle) House 1 outside; on the right) House 2 outside. Blue is June, red is September and lilac is December.

The simulations done with EnergyPlus Programme pointed the elements of thermal inertia as the factor of higher intervention rate on the thermal performance of the studied cases. The principal architectural components about its influence in the thermal performance of the building were: the ceiling and the walls. The internal ceiling was pointed as the main responsible for heat accumulation during sun exposure and for the thermal delay on the buildings, resulting it to be the main re-emitter of heat during the last hours of insolation, increasing the temperatures in this period of the day. The walls were the responsible for heat emission during the first daily hours, reducing the peaks of minimum air temperature during this time of the day. This combination between horizontal and vertical elements in the building confirms the importance of the integration of shaded and solar exposure architectural components, as the internal walls and the internal ceilings.

Variations in the ventilation were carried out and hourly air temperature and air relative humidity waere the output data for the same thermal zones studied in each measured case. Table 2 presents different ventilation schedules.

The schedule D represents a closed house and the schedule E represents an oppened house. Comparing the schedules D and E, variation in the ventilation can to oscilate the interior air temperature in almost 3°C (37.4 °F). The graphics presents that the oscilation is more evident in the last hours of the day. This evidence the needs of the night ventilation in this climate to help to reduce the effect of the thermal mass in the interior environment. Fig 3 shows the comparison between schedules D and E presenting the differences of the variation of the air temperature on September 21 and 22.

Schedule A	Schedule B	Schedule C	Schedule D	Schedule E
Schedule:Compact,	Schedule:Compact,	Schedule:Compact,	Schedule:Compact,	Schedule:Compact,
Dwell_DomBed_Cool,	Dwell_DomBed_Cool,	Office_OpenOff_Occ,	Office_all_closed,	Office_all_oppened,
Temperature,	Temperature,	Fraction,	Fraction,	Fraction,
Through: 31 Dec,	Through: 31 Dec,	Through: 31 Dec,	Through: 31 Dec,	Through: 31 Dec,
For: Weekdays	For: Weekdays	For: Weekdays	For: Weekdays	For: Weekdays
SummerDesignDay,	SummerDesignDay,	SummerDesignDay,	SummerDesignDay,	SummerDesignDay,
Until: 05:00, 0,	Until: 09:00, 1,	Until: 07:00, 0,	Until: 24:00, 0,	Until: 24:00, 1,
Until: 09:00, 1,	Until: 17:00, 0,	Until: 08:00, 0.25,	For: Weekends,	For: Weekends,
Until: 17:00, 0,	Until: 18:00, 1,	Until: 09:00, 0.5,	Until: 24:00, 0,	Until: 24:00, 1,
Until: 24:00, 1,	Until: 24:00, 1,	Until: 12:00, 1,	For: Holidays,	For: Holidays,
For: Weekends,	For: Weekends,	Until: 14:00, 0.75,	Until: 24:00, 0,	Until: 24:00, 1,
Until: 05:00, 0,	Until: 09:00, 1,	Until: 17:00, 1,	For:	For:
Until: 24:00, 1,	Until: 17:00, 0,	Until: 18:00, 0.5,	WinterDesignDay	WinterDesignDay
For: Holidays,	Until: 18:00, 1,	Until: 19:00, 0.25,	AllOtherDays,	AllOtherDays,
Until: 05:00, 0,	Until: 24:00, 1,	Until: 24:00, 0,	Until: 24:00, 0;	Until: 24:00, 1;
Until: 09:00, 1,	For: Holidays,	For: Weekends,		
Until: 17:00, 0,	Until: 09:00, 1,	Until: 24:00, 0,		
Until: 24:00, 1,	Until: 17:00, 0,	For: Holidays,		
For: WinterDesignDay	Until: 18:00, 1,			
AllOtherDays,	Until: 24:00, 1,			
Until: 24:00, 0;	For: WinterDesignDay			
	AllOtherDays,			
	Until: 24:00, 0;			

Table 2 . Different ventilation schedules simulated

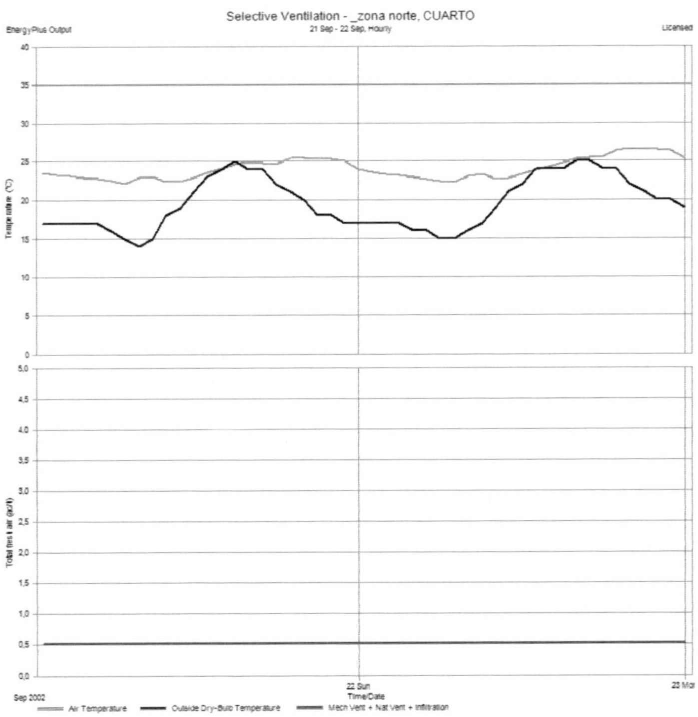

a)

b)

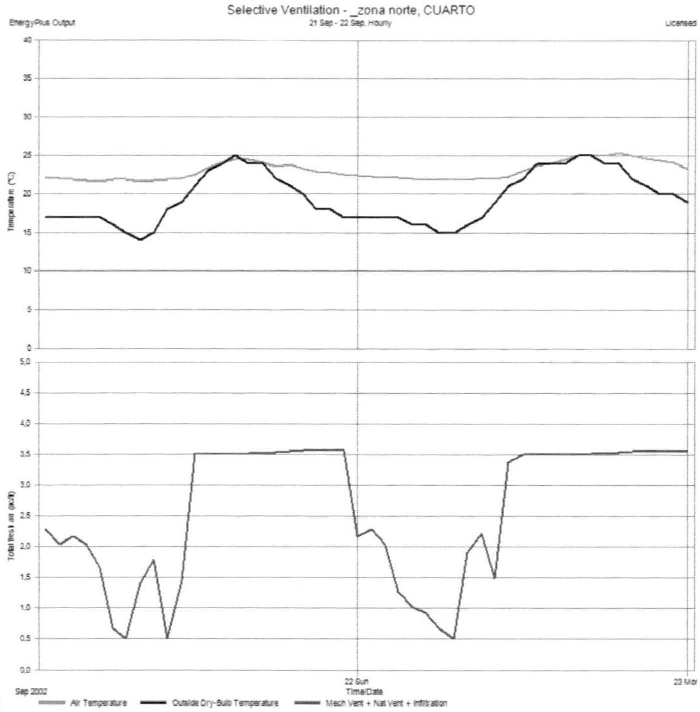

Figure 3 . Variations interior air temperature ºC (orange), outside dry-bulb temperature ºC (black), total air fresh ac/h (mechanic ventilation, natural ventilation and infiltration - lilac) of House 1: a) schedule D; b) schedule E.

CONCLUSIONS

The use of two methods of analysis, merging various computational tools and measuring instruments proven particularly adequate for the type of study, allowing: analysis of the cases from different scales, from global to specific, besides allowing the isolation, differentiation and comparison of various elements and intervening variables to the thermal performance of the studied cases. The results obtained with the different methods are complementary, indicating in both houses, that while some building components help to improve the thermal performance of the modern naturally-ventilated houses, others worsen this, resulting in a thermal performance below his potential.

June is a month with the best index with a good surround correction and good inside conditions. The most critic period of the year is September with higher air temperatures and low air relative humidity. The comparison between measure and simulated dates evidences the use of air conditioning in the bedroom of house 1 (1C in the **Fig. 1**) in some days of this month. Although, the air temperature peak does not represent an uncomfortable situation in the others studied rooms for people that live in this climate, as Givoni (1998) [8] indicates. Still thus, the air temperature peak occurs with 3 hours of delay, sufficient time: to reduce the inconvenient effects with a mix of external and internal air, and to supply the critical moment applying cooling systems with low energy demand. Despite September presents the worst situation, a simple strategy demostrates to be efficient in this month too. The control of the dwellings in the house is very important to guarantee the interior comfort during all the day.

It is possible to verify the sensibility of the technical solutions applied in the cases in relation to the selective ventilation, demonstrating its importance in the building thermal performance and identifying important variables to improve the thermal performance of modern naturally-ventilated houses. The reduction of the interior air temperature oppening the house in the correct hours is a easy strategy that can be used without complex active systems in the house but with the simple conscious of the inhabitant. Many times easy strategies are the best solution.

REFERENCES

[1] Lamberts, R.. 2008. *Trends in Brazilian building ventilation market and drivers for change*. Air Infiltration and Ventilation Centre AIVC. Ventilation Paper nº 23.

[2] ABNT. ABNT NBR 15220-3:2005. 2005. *Thermal performance in buildings. Part 3: Brazilian bioclimatic zones and building guidelines for low-cost houses*. Rio de Janeiro: Associação Brasileira de Normas Técnicas.

[3] ASHRAE. 2005. *ASHRAE Handbook-Fundamentals*. Atlanta: American Society of Heating Refrigeration and Air Conditioning Engineers, Inc.

[4] ABNT. ABNT NBR 15575-1:2008. 2008. *Residential Buildings up to 5 storied – Performance. Part 1: General requirements*. Rio de Janeiro: Associação Brasileira de Normas Técnicas.

[5] Moura, A. A. P.. 2009. *Arquitetura residencial moderna em Goiânia: delineando um cenário*. Programa de Especialização do Patrimônio – PEP. Goiânia: Instituto do Patrimônio Histórico e Artístico Nacional. Estado de Goiás – IPHAN-GO.

[6] INMET. 2009. *Normais climatológicas do Brasil 1961-1990*. Brasília: Instituto Nacional de Meteorologia.

[7] INMET. 2011. *Rede de Estações. Superfície Automática*. [online] Available at: <http://www.inmet.gov.br/sonabra/maps/pg_automaticas.php> [Accessed January-December 2011]

[8] Givoni, B. 1998. *Climate considerations in buildings and urban design*. New York: John Wiley & Sons Inc.

[9] Vaz, M. D. A. C. and Zárate, M. H. V.. 2005. *A experiência moderna no cerrado goiano*. Vitruvius. [e-journal] Arquitextos 067, año 06, December, [Accessed December 2010].

[10] Fernandes, A. M. C. P.. 2006. *Clima, homem e arquitetura*. Goiânia: Trilhas Urbanas.

AIR LEAKAGE CHARACTERISTICS OF DWELLINGS IN HIGH-RISE RESIDENTIAL BUILDINGS IN KOREA

Yun Jeong Choe[1], Hyun Kook Shin[2], and Jan Hun Jo[3*]

1 Inha University
Yonghyun-Dong, Nam-Gu
Incheon, Korea

2 Inha University
Yonghyun-Dong, Nam-Gu
Incheon, Korea

3 Inha University
Yonghyun-Dong, Nam-Gu
Incheon, Korea
**Corresponding author: jhjo@inha.ac.kr*

Note : the contact addresses may be re-arranged

ABSTRACT

Reliable airtightness data is needed to calculate the estimate of air infiltration and the thermal loads for building energy efficiency and indoor comfort. While useful information on air leakage in low-rise dwellings does exist, there is little data available on dwellings in increasing high-rise residential buildings (particularly ones with central core plan). In this paper, we conducted airtightness measurement using fan pressurization method for about 350 dwellings in 4 high-rise residential buildings in Korea. The results were compared to airtightness requirements for high performance buildings or several airtightness ratings. The measured results show that average ACH50 was 2.3, and the ACH50 value was within the range of 2~5 which are level of 'quite tight' on the basis of ASHRAE airtightness ratings. The results of the building component test show that the most leak parts of dwelling are the internal walls between residential units.

KEYWORDS

Airtightness, Air leakage distributions, High-rise residential building, Fan pressurization method

INTRODUCTION

Most residential buildings located in the United States, Canada, and Europe are low-rise dwellings. Many studies have been conducted on the airtightness level of low-rise dwellings. Sherman and Chan(2004) reviewed various airtightness research and practices. This report reviews the most important publications relating to the building airtightness. Many studies in this report mainly presented the data on airtightness of single-family house. Measurement methods of the airtightness of single-family house and the airtightness standard for each nation were also suggested. On the other hand, most residential dwellings in Asia are constructed as a multi-family house and high-rise buildings where many units are adjacent to each other. Most Asian countries including South Korea lack airtightness standards and data.

In addition, the measurement data or standards for low-rise dwellings of the United States, Canada, and Europe are not applicable. Therefore, airtightness data of high-rise residential buildings should be investigated to provide an airtightness standard.

This study presented the airtightness data and air leakage distributions of dwellings in 4 high-rise residential buildings in South Korea. Fan pressurization method with blower door was used to measure the airtightness value of about 350 dwellings, and the measurement results were compared with the standards of each nation. The airtightness of building components that form units of residential buildings—such as envelope, internal walls between dwellings units, and floor—were investigated.

TEST BUILDING DESCRIPTIONS

The core was located at the center of all the tested buildings, and hallways and each unite were on the surrounding plane. In addition, the floor structure was a flat slab, and walls between units and hallways were dry walls. Thus, compared to concrete walls, it is difficult for joints to be airtightly constructed in dry walls, which degrades airtightness. The exterior wall was a curtain wall type for buildings A and B and a punched window type for buildings C and D. The construction outline of the targeted building are summarizes in Table 1.

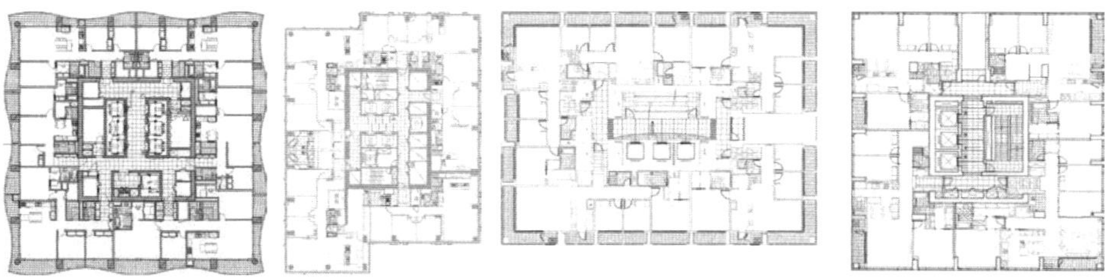

Building A Building B Building C Building D

Figure 1. Building Plan

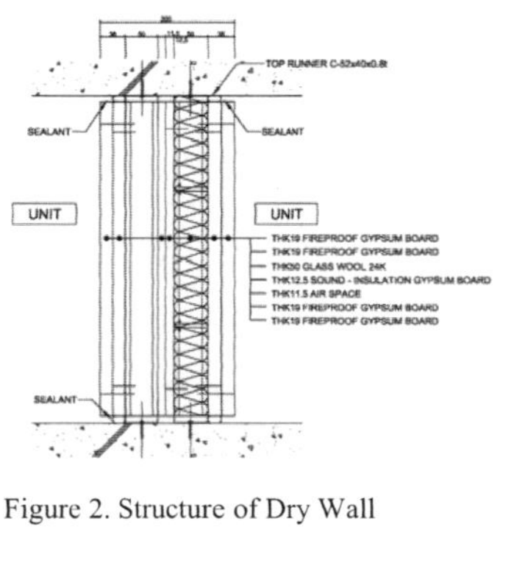

Figure 2. Structure of Dry Wall

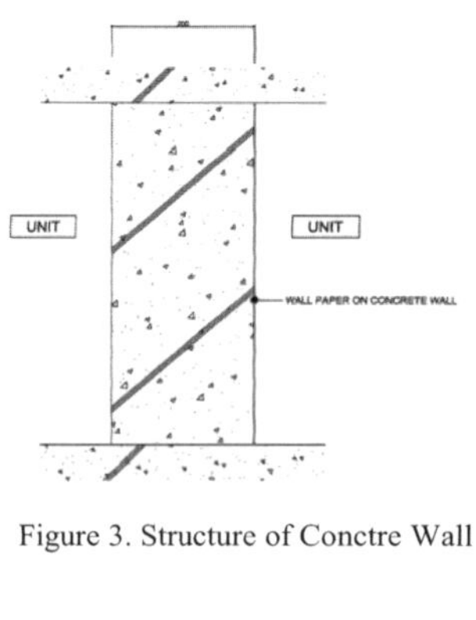

Figure 3. Structure of Conctre Wall

Classification	A	B	C	D
Building site	Incheon, SOUTH KOREA			
Construction	Flat Slab			
Number of stories	2 basements, 47 stories	2 basements, 42~49 stories	2 basements, 12~28 stories	2 basements, 11~33 stories
High	161.1	174.6	-	108.8
Building use	Multi-unit dwelling, commercial facility, neighbourhood living facility	Multi-unit dwelling, commercial facility, neighbourhood living facility	Multi-unit dwelling	Multi-unit dwelling
Exterior wall	Curtain wall type	Curtain wall type	Punched window type	Punched window type

Table 1. Test building summaries.

MEASUREMENT METHOD

The test conditions presented by the ATSM standard were followed to measure the airtightness of the unit and building components such as envelope, internal walls between dwelling units, floors, and ventilating equipment.

In order to measure the airtightness of a dwelling unit, a blower door was installed in the entrance door at each dwelling unit, and the pressure difference was controlled at 5–10-Pa intervals for measurement. In order to prevent the influence of the airtightness of adjacent units on the building, large openings (windows and doors) of units located above and below were left open so that the situation can be set based on ambient conditions. Building components were distinguished as envelope, internal walls between dwelling units, floors, and ventilating equipment for their air leakage distributions. In order to measure the airtightness of envelope, the airtightness data of envelope was measured two times (built condition and no air leakage condition). When the difference in the amount of air leakages is calculated before and after airtight processing, the airtightness of envelope could be identified. Airtightness of the ventilation equipment was measured with same.

To measure the internal walls between dwelling units and floors, two blower door sets were placed on either side of the measurement target. Then, pressurization and de-pressurization methods were conducted on the measured unit and adjacent units. If the pressure difference between the two dwelling units stayed at ±0 Pa, then air leakage did not occur from the measurement target. The amount of air leakage without air leakage through internal walls was measured. Finally, airtightness data of the internal walls were determined by the differences between in whole airtightness value of the dwelling unit and measured value without air leakage through internal walls.

MEASUREMENT RESULT

Analysis on airtightness of unit dwelling of each building

Measured ACH50 results for 350 dwellings of high-rise residential buildings are displayed in Figure 4, and the mean values for each building are marked. ACH50 range was about 1.9–3.8 for building A, 2.6–5.2 for building B, 1.4–3.8 for building C, and 1.4–3.7 for building D. Each the mean value was calculated as 3.1, 3.9, 2.5, and 2.3 ACH50 respectively. Thus, building D was the most airtight followed by buildings C, A, and B in order. Since the ground structure, internal walls between dwelling units, and ventilation types of the buildings were similar, the exterior wall was considered to be the factor with the greatest influence.

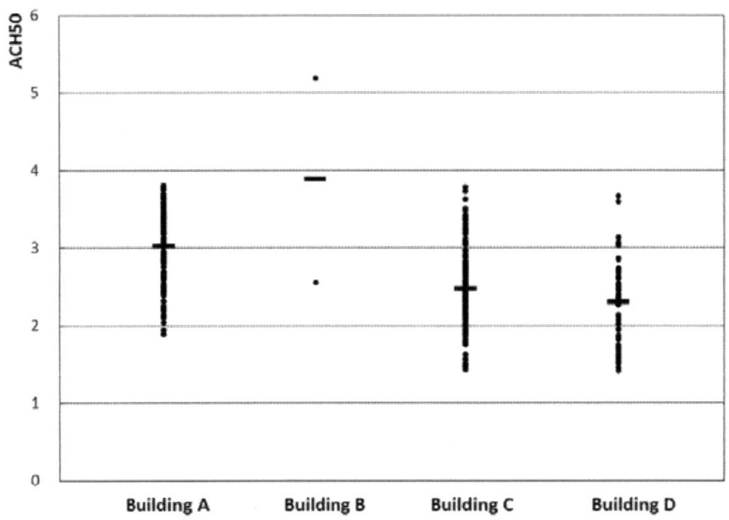

Figure 4. Results of airtightness for each building (ACH50)

When the four test buildings were evaluated according to the ASHRAE ventilation standard for residential buildings, they showed "airtight" or "quite airtight" level, which was sufficient airtight according to the standards of European nations such as Norway, the Netherlands, and Switzerland. However, the airtightness results were evaluated as insufficient airtight that is based on the standards for energy-conservative buildings such as the passive house of Germany and R-2000 of Canada. Figure 5 shows a comparison of airtightness of each nation's standard and the test buildings in this measurement study.

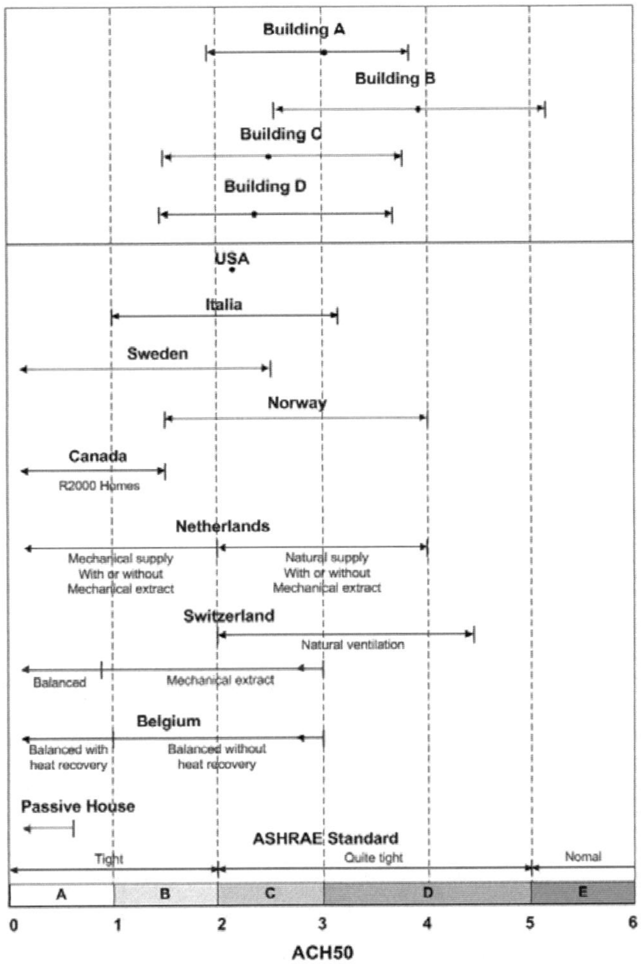

Figure 5. Comparison between measurement data and standard of each nation

Airtightness distributions of building components

The airtightness distribution of building components was measured for building A. Nine dwellings were targets for measurement in building A. Two dwelling types with general plans and one side facing the ambient were selected. Table 2 presents detailed information on the measured dwellings.

	Floor area (㎡)	Floor height (m)	Unite envelope area (㎡)	Volume (㎥)
Type 1	103.85	3.1	342.71	321.94
Type 2	132.53	3.1	418.17	410.84

Table 2.Test dwelling information

The airtightness measurement results were calculated as the air leakage distribution ratio for each building component and are summarized in Figure 6. The test residential units took up the highest amount of air leakage with 30%–58% of the overall amount of air leakage. Because of the trend of lighting structures of high-rise residential buildings, they were mainly constructed with dry walls. Thus, a great deal of air leakage occurred between wall joints or joints where columns and slabs came in contact with walls. The envelope took up 5%–30% of the overall air leakage of residential units. In addition, 3%–32% of air leakage occurred because of floors. Smoke inspection results showed that air leakage occurring from continuing curtain wall frames to adjacent dwellings. As a heat exchanger ventilation system was installed in the dwellings, the amount of air leakage coming through the ventilation equipment was 3%–19% of the overall air leakage of dwellings. The remaining 7%–30% of air leakage was considered to come from the entrance door of the each dwelling unit, equipment penetration, and electrical pipes. Further studies are needed to identify of specific air leakage areas as well as the air leakage distribution ratio.

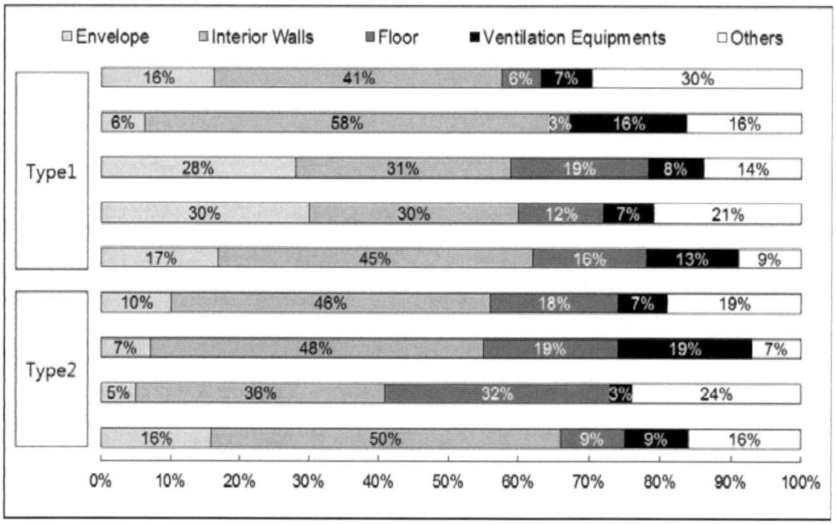

Figure 6. Infiltration (air leakage) distribution ratio for each building component

CONCLUSION

The airtightness data of 350 dwellings in 4 high-rise buildings in Korea was about 1.4–5.2 ACH50. Building A had an average ACH50 of 3.1, building B had an average ACH50 of 3.9, building C had an average ACH50 of 2.5, and building D had an average ACH50 of 2.3. The airtightness of the four test buildings was at the level of "quite airtight" and satisfied the standards of European nations such as Norway, the Netherlands, and Switzerland. By measuring the airtightness of building components of each dwelling and by calculating the air leakage distribution ratio to identify leaking parts where airtight-constructions are needed, the air leakage distribution ratio was determined to be the highest (31%–58%) for the internal walls between dwelling units. This is considered to be properties of high-rise residential buildings constructed with dry walls, and airtight construction is needed at internal walls between dwelling units.

ACKNOWLEDGEMENTS

This research is supported by a grant from High-Tech Urban Development Program (12CHUD-C059710-02-000000) funded by the Ministry of land, transport and maritime affairs.

REFERENCES

[1] Sherman, M.H. and Rengie Chan, *Building Airtightness: Research and Practice*, LBL Report No. 53356, 2004.

[2] Sherman, M.H. and D.J. Dickerhoff. *Airtightness of U.S. dwellings*, ASHRAE Transactions 104(2), 1998.

[3] AIVC, *An analysis and data summary of the AIVC's numerical database*, AIVC Technical Note 44, International Energy Agency Air Infiltration and Ventilation Centre, Coventry, U.K, 1994.

[4] Hyun-Hook Shin, Jae-Hun Jo, Su-Min Kim, Kwang-Woo Kim, 2010. *Field Measurement of the Pressure Distribution and Air Pressure Problems due to Stack Effect in a High-rise Building,* Clima

[5] Hyun-Hook Shin, Hyun Cho, Kwan Woo Kim, Hoi Soo Seo, Jae Hun Jo, 2010. Air Leakage Tests for the requirements of the LEED in two high-rise residential buildings, AIVC

[6] Hyun-Hook Shin, Hyun Cho, Kwan Woo Kim, Hoi Soo Seo, Jae Hun Jo, 2010. *A Study on Airtightness of Building Components in The High-rise Building,*

[7] Hyun-Hook Shin, Hyun Cho, Kwan Woo Kim, Jae Hun Jo, 2012. *Measurement of the Dwelling Unit Airtightness and Evaluation of Energy Standards*

[8] Hyun-Hook Shin, 2011. *Comparison of the Conditions of Fan Pressurization Measurements and Evaluation of Airtightness in Apartment,*

[9] ASHRAE, 2005 ASHRAE Handbook-Fundamentals. Atlanta: *American Society of Heating, Refrigerating and Air-Conditioning Engineers*, Inc., 2005.

[10] ASTM, ASTM E779-03 *"Standard Test Method for Determining Air Leakage Rate by Fan Pressurization" American Society for Testing and Materials*, U.S.A., 2004.

[11] ATTMA, *ATTMA TS1-Measuring Air Permeability of Building Envelopes*, 2007.

[12] ISO, ISO 9972:2006. *Thermal performance of buildings -Determination of air permeability of buildings-Fan pressurization method.*

[13] CMHC, 1996.*Energy Audits of High-rise Residential Buildings*, Technical Series 97-100

[14] Sherman, M.H., The Use of Blower Door Data, LBL Report No. 35173, 1998.

A SURVEY OF AIRTIGHTNESS AND VENTILATION RATES IN POST 1994 NZ HOMES

S. McNeil[1], L. Quaglia[1], M. Bassett[1], G. Overton[1], M. Plagmann[1]

1 BRANZ
1222 Moonshine Road, Porirua, New Zealand
Corresponding author: steve.mcneil@branz.co.nz

ABSTRACT

The airtightness of 36 houses built since 1995 and across four cities in New Zealand (NZ) was measured. In a subset of 31 of these homes, the average ventilation rate was measured over several weeks in the winter using a perfluorocarbon tracer technique (PFT). These results can be added to earlier airtightness data to provide a platform for improving the air quality and energy efficiency of residential ventilation in NZ.

Earlier airtightness data from the mid 1990's showed a trend for newer houses to be more airtight than older houses, largely as a result of sheet lining materials replacing strip flooring and the development of more airtight joinery. This trend continued in this study, even though there are no airtightness requirements for houses in New Zealand. The average N_{50} result for houses built since 1995 was 6.7 air changes/hour (ACH), down from 8.5 ACH for houses built in the previous decade.

Most new houses meet NZ building code requirements for ventilation (window and door openings that exceed 5% of the floor area), but the PFT measurements showed that most houses struggled to reach recommended levels of ventilation (0.35-0.5 ACH during the winter) because the windows weren't open often enough.

In recent years it has become common for occupants to add additional supply-only ventilation to control moisture. Houses with operational supply-only ventilation systems generally had more than enough ventilation but there were several cases where the ventilation system had been turned off.

There was clear evidence of moisture problems in several of the more poorly ventilated houses.

KEYWORDS
Airtightness, Ventilation, Residential, Tracer

1 INTRODUCTION

The aim of this study was to provide airtightness data for newer (post 1994) houses and to measure the achieved ventilation rate during occupancy. The measurements should provide some insight as to whether a ventilation scheme that relies on occupants to open windows can be relied on, or whether factors such as wintertime, security and noise intrusion prevent this from being so.

Earlier work at BRANZ [1, 2, 3] compiled an airtightness database of 137 homes built from the 1930's to the mid 1990's. A summary of this is shown in Figure 1.

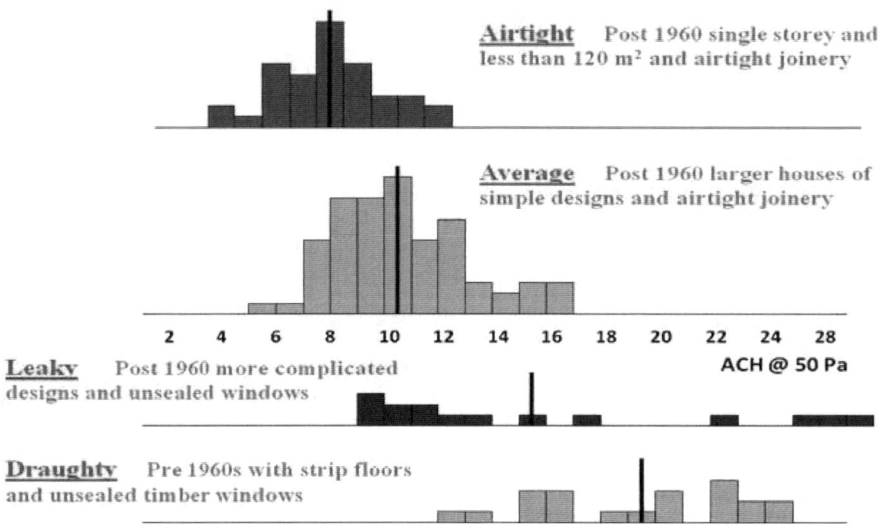

Figure 1 – Summary of earlier house airtightness in New Zealand

The airtightness of New Zealand homes has increased over time, even though there is no requirement for airtightness in the NZ building code. The average N_{50} result from houses built before WWII was around 19 ACH but this reduced dramatically to 8.5 ACH for houses built between 1960 and 1980. A significant contributor to envelope air-tightening around 1960 was the shift from suspended tongue and groove flooring to sheet floor construction and slab-on-ground floors. Another change at a similar time was the shift from timber joinery to aluminium framed doors and windows, as well as a reduction of open fireplaces. This brought the opportunity to fit better air seals around opening windows and doors, at the same time improving the weathertight performance of domestic joinery. By the mid 1990's the mean airtightness result was 6.7 ACH.

In the USA, Sherman [4] reported a similar two-fold reduction in average N_{50} measurements over a similar time period. In this case the changes were largely voluntary but were also influenced by the 'weatherization programme' to improve the energy efficiency of low income homes. In contrast, Stephen [5] reports little change in the N_{50} measurements in UK houses over a similar period. Much more dramatic changes in N_{50} are reported in Canada and Sweden where mandatory airtightness targets were adopted to reduce the energy loss consequences of uncontrolled ventilation. A second driver of airtight construction in these cold climates is the need to control exfiltration to prevent interstitial condensation.

In NZ, materials and construction practices are likely to have continued to influence the airtightness of houses. Recent examples of changes are the widespread use of bonded plaster cornice or a square stopped interior plaster finish, and the adoption of air seals around window and door assemblies. In 2005, the NZ Building Code - Clause E2/AS1[6], changed to require air seals to be fitted around door and window assemblies. The aim was to improve the degree of pressure moderation across the joints between window frame and cladding and improve the weathertightness of what was seen as a weak point in window installation.

In terms of ventilation, building code requirements for residential buildings are often quite unsophisticated in countries with temperate climates. In New Zealand, occupants are expected to open windows for ventilation and the NZ Building Code offers an acceptable solution 'G4 Ventilation'[7] requiring window and door openings to be at least 5% of the floor area. It is clear from the airtightness measurements of older houses that window opening may have been unnecessary to meet ventilation needs because of the background infiltration. In newer houses, the changes in construction discussed above have closed down natural ventilation paths and it may be necessary to actually open the windows to provide adequate fresh air.

Ventilation also has a large part to play in the control of indoor moisture. Indoor moisture has always been the most pressing indoor air quality issue in New Zealand houses. A 1971 survey [8] reported moisture problems in half of the surveyed houses and later surveys [9,10] have shown that little has changed.

In the past it has been difficult to measure ventilation rates in homes because conventional tracer methods were too intrusive and expensive to use in large numbers of houses. This study used passive samplers and emitters that are more easily deployed [11,12] and to provide the first survey of ventilation achieved in NZ homes.

The study described here will help to provide a picture of the current housing stock and how New Zealander's currently ventilate their houses. This information will then be used in the wider WAVE (Weathertightness, Air Quality and Ventilation Engineering) programme at BRANZ. One of the aims of WAVE is to provide guidance on suitable ventilation options that are optimised for moisture control, energy efficiency and the airtightness of the house.

2 EXPERIMENTAL METHODS

2.1 SELECTION OF HOUSES

The airtightness and ventilation survey was split across 4 different cities in New Zealand: Wellington; Palmerston North; Dunedin; and Auckland. A database of building consents was used to obtain a random sample of consents for houses built after 1994 and these homeowners were contacted via a letter, resulting in a final total of 36 houses.

Of these 36 houses, 8 had supply-only positive pressure ventilation systems installed in the roof space. These systems distribute filtered roofspace air throughout the home depending on temperature measurements in the living space and roofspace.

2.2 AIRTIGHTNESS MEASUREMENTS

A blower door test to EN13829[13] was completed on each of the 36 houses and then the opportunity was taken to measure the contribution of a range of different leakage paths. This was carried out by progressively sealing up openings in the envelope and repeating the blower door test.

In general, the following 3 tests were completed on each dwelling:

- A standard N_{50} test with no openings sealed
- A test with specific ventilation openings sealed. In most cases these ventilation openings consisted of extract fans in bathrooms and kitchens, some of which were simply ducted to the roof space (not outside).
- A final test with all obvious leakage openings sealed. The most obvious leakage openings to be sealed were around internal garage doors, and defective seals around attic access hatches.

The airtightness measurements were also used to give an estimate of the infiltration through the envelope using Equation 1.

$$Estimated\ Infiltration\ Rate = \frac{N_{50}\ result}{20} \qquad \text{Equation 1}$$

2.3 VENTILATION MEASUREMENTS

Ventilation measurements were performed in 31 of the 36 houses during winter. Winter was chosen because it was perceived that ventilation would be at its lowest i.e. windows are open less often.

A Perfluorocarbon Tracer (PFT) technique[12] was used and the equipment and analysis was supplied by the UK's Building Research Establishment (BRE). The technique involves deploying passive tracer gas sources and activated carbon sampling tubes in a building for a period of time. The resultant concentration of tracer in the sampling tubes can then be used to calculate an average ventilation rate.

Plans for each house were obtained to allow the room volumes to be pre-calculated. Key dimensions were measured upon arrival to ensure the plans matched the building and any differences were marked on the plans and the locations of sources and samplers modified accordingly.

The tracer sources were distributed around the home in a volume weighted manner, with the bathroom being chosen as a reference volume in all cases. Figure 2 shows a typical floor plan with tracer sources marked in red, and sampling tubes in blue.

Sampling tubes were placed in 4 rooms in each house, typically the lounge, bathroom, kitchen, and master bedroom. These were left in place for at least three weeks, but sometimes this was as long as 4 weeks because of occupants' unavailability. There were several important considerations when it came to the location of the source and sampling tubes:

- Source and sampling tubes need a good degree of separation to ensure the sampler collects tracer that has been well mixed in the zone.

- Both sources and sampling tubes needed to be located as far as practicable from windows/doors to allow incoming air to mix within the zone.

- Temperature has a direct influence on the emission rate; the sources should not be in direct sunlight or within 1.5 metres of heat sources. The temperature was also measured at each source location using Dallas DS1923 iButtons.

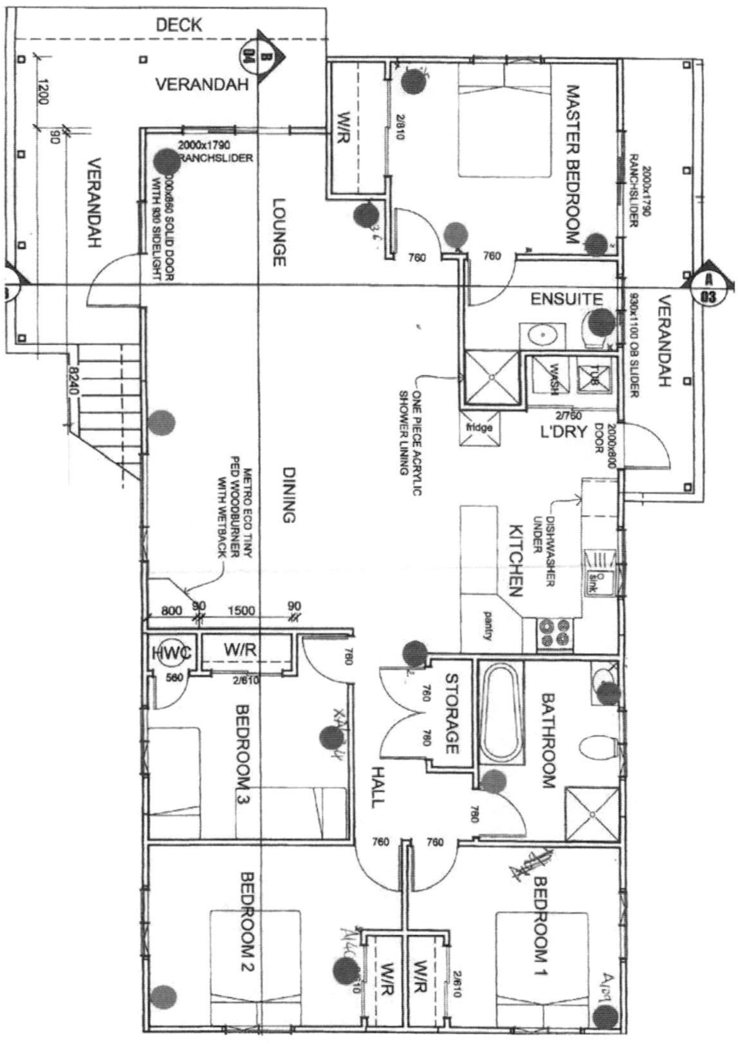

Figure 2 – Typical distribution of tracer sources and sampling tubes

3 RESULTS

3.1 AIRTIGHTNESS

The N_{50} results for the 36 houses are shown in Figure 3. Figure 4 shows the effect from sealing ventilation and obvious leakage openings (final test).

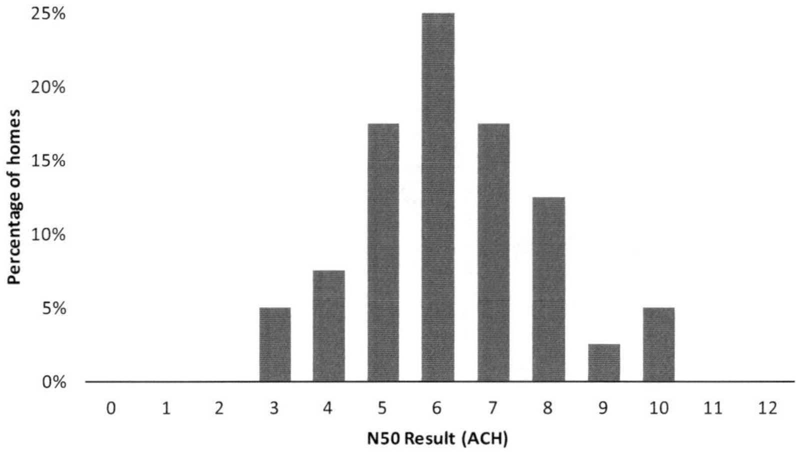

Figure 3 - Distribution of airtightness measurements for post 1994 homes

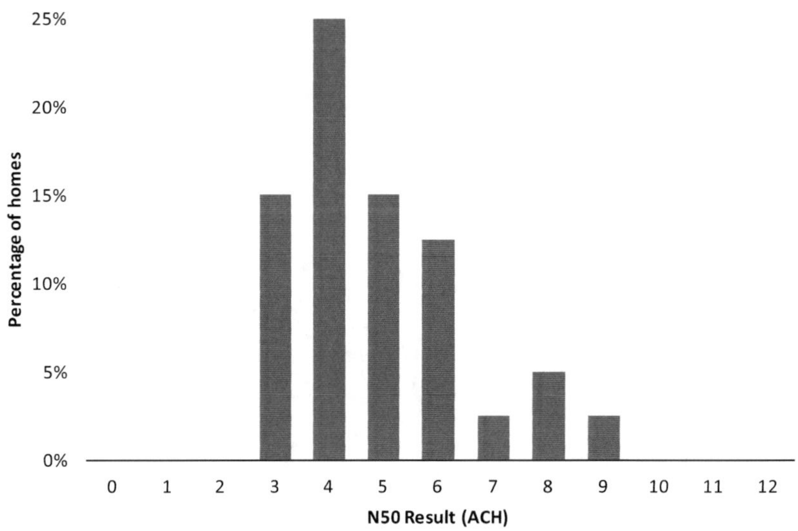

Figure 4 - Distribution of airtightness results, with specific ventilation and obvious leakage openings sealed

3.2 VENTILATION SURVEY

The results of the ventilation survey are plotted in Figure 5, against the estimated rate from the airtightness measurements (using Equation 1). Circled on the right are several outliers, three of which had a supply only roofspace sourced ventilation system. A line of slope 1 is also plotted.

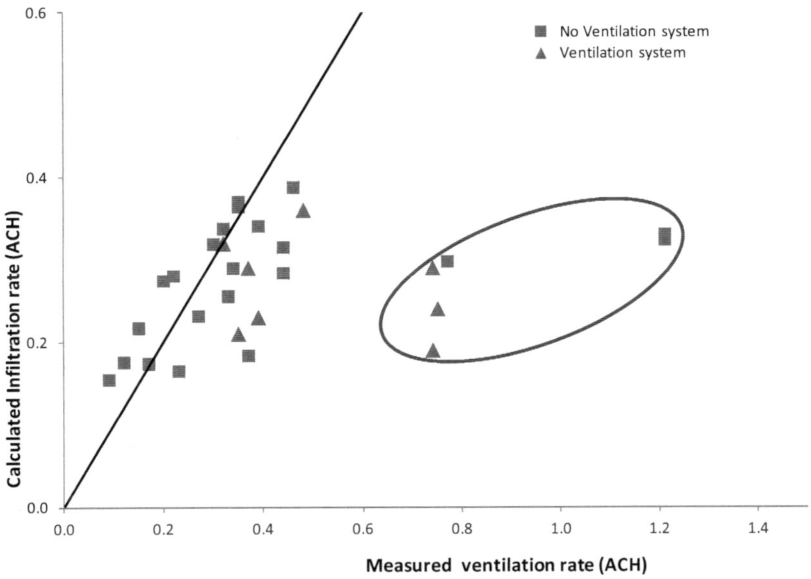

Figure 5 - Infiltration vs. measured ventilation (ACH), with outliers circled

4 DISCUSSION

4.1 AIRTIGHTNESS

The most important change from the earlier surveys was the significant reduction in the mean N_{50} result from 8.5 ACH to 6.7 ACH. The floor area of the newer houses was also bigger than those in the last survey, increasing from $115m^2$ to $155m^2$ (and not including internal access garages). The recent N_{50} results also fell in a much tighter range (7.8 to 3.1 ACH), suggesting more consistency in construction.

Much of the difference between Figures 3 and 4 was due to the leakage under internal access garage doors. On average, a drop in the N_{50} result of 1.4 ACH was noted when the internal access for the garage spaces was sealed from the rest of the house.

Internal garage doors therefore present an opportunity for increasing airtightness, and reducing infiltration from an unheated (and potentially polluting) part of the building.

4.2 VENTILATION SURVEY

There are clearly two groups of houses in Figure 5: those where the estimated infiltration rate and the measured ventilation rate are similar (25 cases) and those (6 cases) where additional ventilation (either from opening windows or supply-only ventilation systems) has been provided.

In the larger group, the small difference between the estimated infiltration rate (0.28 ACH) and the measured ventilation rate (0.32 ACH) indicates limited window opening by the occupants over the period. The measured ventilation rate of 0.32 ACH sits at the lower end of guidelines for acceptable indoor air quality [14]. In addition, observations of the presence of mould and mildew were made at several of the homes studied, evidence of excess indoor moisture.

The PFT technique is a longer-term, time-averaged measurement method and thus does not lend itself well to resolving small, short-term changes in ventilation performance. However, it is clear that window opening and the operation of extract systems in bathrooms and kitchens has added less than 0.2 ACH on top of the background air infiltration in most of these houses.

Overall, there is limited evidence of window opening providing the ventilation needed to control moisture and provide good indoor air quality in the more airtight homes constructed in the last 15 years.

Eight homes in Figure 5 were fitted with supply only ventilation systems. Three of these systems were shown to substantially increase ventilation above background infiltration to around 0.7 ACH. In the other 5 cases, little additional ventilation was provided by systems, several of which were apparently turned off to save energy during the period of PFT measurements.

5 CONCLUSIONS

This paper reports the results of a survey of house airtightness in New Zealand along with average ventilation rates measured using the PFT method. The conclusions of this study are as follows:

- **The airtightness of New Zealand homes continues to increase**. The average airtightness of houses built between 1994 and 2011 in this survey was an N_{50} of 6.7 ACH. Between 1960 and 1994 the average N_{50} was 9.7 ACH, dropping to 8.5 ACH for houses built in the early 90's. For even earlier houses it was 19 ACH at 50 Pa. This change has occurred without any intervention by the New Zealand Building Code.
- **Internal access garages have a large impact on airtightness**. The door into the garage was found to be a weak point in the envelope, contributing an average 1.4 ACH to the N_{50} result. A more effective door in this location would reduce the infiltration of potentially contaminated air from this unheated zone in the building.
- **Measured ventilation rates are similar to estimated infiltration in many houses**. Comparing the average infiltration (approximated as the N_{50} result divided by 20) and measured ventilation rates, resolves the ventilation added by small kitchen and bathroom ventilators and by opening windows. In 24 of the 30 cases the average infiltration rate of houses without mechanical ventilators is 0.28 ACH and the measured ventilation rate 0.32 ACH and in several of these houses indoor dampness was evident. In the six cases where infiltration had clearly been supplemented with additional

ventilation, three of these were fitted with a supply ventilation system and in the other three this was achieved by opening windows.

- **Reliance on open windows for ventilation may not be adequate** This passive ventilation solution appears to have worked adequately at times when New Zealand homes were not particularly airtight. Since 1960, a wider choice of large sheet lining materials and changes in the standard of interior finish have increased the airtightness to the point where window opening is now questioned as a ventilation source for moisture control and indoor air quality. Particularly with modern lifestyles, and when there are security concerns.
- **The control algorithms for supply only ventilation systems may have scope for improvement**. Eight homes were fitted with roof-space sourced supply only ventilation systems that distribute filtered roof-space air under a simple temperature controlled regime. Three of these boosted the average ventilation well above the background infiltration but in five cases there was little evident change. The data indicates significant differences in the operation and control of these supply only systems.

This work has provided a platform on which to discuss ventilation options for New Zealand housing. The survey has shown that the trend to more airtight houses has continued in the last decade and that that occupant controlled ventilation by opening windows is limited and too unreliable for indoor moisture control. The next steps in the WAVE programme will investigate alternative ventilation solutions that adapt to window opening in a temperate climate and are optimised for indoor moisture control [15].

6 ACKNOWLEDGEMENTS

This work was funded by the New Zealand Building Research Levy and the New Zealand Ministry of Business, Innovation and Employment.

The contribution and assistance of the following people is greatly appreciated:

Stuart Upton, Building Research Establishment, Garston, Watford, United Kingdom

Mikael Boulic, School of Engineering and Advanced Technologies, Massey University, Palmerston North, New Zealand

Inga Smith, Tim Bishop, Department of Physics, University of Otago, Dunedin, New Zealand

Paola Leardini, and Kara Rosemeier, School of Architecture and Planning, University of Auckland, New Zealand

The assistance of Kevin Smith, on secondment from Queens University, Belfast, working at BRANZ under the IAESTE scheme, is gratefully acknowledged

7 REFERENCES

[1] Bassett, M. R. with Synergy Applied Research Ltd. January, 1986. Survey of air tightness levels in Auckland and Christchurch homes, *New Zealand Energy Research and Development Committee Publication P87*

[2] Bassett, M. R. September 1985. The infiltration component of ventilation in New Zealand houses, *Proc. 6th AIC Conference. Ventilation Strategies and Measurement Techniques. Southern Netherlands*

[3] Bassett, M.R. 2001 Naturally Ventilated Houses In New Zealand Simplified Air Infiltration, *CIB World Building Congress, Wellington, New Zealand*

[4] Sherman, M.H. and Chan, R. 2003 Building Airtightness: Research and Practice, Lawrence Berkeley Laboratory report LBNL-53356

[5] Stephen R.K., 1998 Airtightness in UK Dwellings: BRE's Test Results and Their Significance, *Report 359, Building Research Establishment, United Kingdom*

[6] Department of Building and Housing, Compliance Document for New Zealand Building Code Clause E2, External Moisture

[7] Department of Building and Housing. 2005. New Zealand Building Code, *Approved Document G4, Ventilation. Wellington.*

[8] Trethowen, H. A. 1972 Dampness in housing. Report ; CR2 Building Research Association of New Zealand

[9] Clark, S J; Jones, M. Page, I 2005 House condition survey, *Study report no. SR 142 BRANZ Ltd*

[10] Buckett, N. R. Marston, N J Saville-Smith, K. Jowett, J. H. Jones, M.S. 2011 Preliminary house condition survey report *Study Report ; no. SR 240 BRANZ Ltd*

[11] Dietz, R.N., Cote E.A. 1982 Air infiltration measurements in a home using a convenient Perfluorocarbon tracer technique *Environment International, Vol. 8, pp. 419-433*

[12] Upton, S. and Kukadia, V. 2011. Ventilation Rate Measurement. BRE Information Paper IP 13/1. ISBN 978-1-84806-219-1Pla

[13] EN13829:2001 Thermal performance of buildings. Determination of air permeability of buildings. Fan pressurization method

[14] Limb, M.J. 2001. A review of international Ventilation, Airtightness, Thermal Insulation and Indoor Air Quality Criteria, AIVC Document TN55

[15] Plagmann M., McNeil, S., Bassett M., Overton G., Quaglia, L., Cunningham M., Construction and Set-up of a Full-Scale Experimental House for Ventilation Studies, AIVC 33rd Conference: Optimising Ventilative Cooling and Airtightness for [Nearly] Zero-Energy Buildings, IAQ and Comfort, 10-11 October 2012, Copenhagen, Denmark

NUMERICAL EVALUATION OF AIRTIGHTNESS MEASUREMENT PROTOCOLS

Adeline Bailly[1], Valérie Leprince[1], Gaëlle Guyot[1], François Rémi Carrié[2], Mohamed El Mankibi[3]

1 Centre d'Etudes Techniques de l'Equipement de Lyon
46 rue Saint-Théobald
38081 L'Isle d'Abeau, France
**Corresponding author: adeline2.bailly@developpement-durable.gouv.fr*

2 ICEE
93 rue Molière
69003 Lyon, France

3 Ecole Nationale des Travaux Publics de l'Etat
3 rue Maurice Audin
69518 Vaulx-en-Velin,

ABSTRACT

In France, starting January 1st, 2013, the energy performance regulation will impose an airtightness treatment for every new residential building. This translates into several tens if not hundreds of thousands of envelope airtightness measurements a year that will have to be performed. They will have to be performed by a certified operator and according to the NF EN 13829 standard. This ISO standard is being revised under the Vienna agreement to become an EN ISO standard. This revision should include changes in the measurement protocol to reduce the uncertainty for two indicators commonly used: the air change rate at 50 Pa and the air permeability at 4 Pa.

As far as it is quite impossible to determine the real airtighness of a building, the measurement error cannot be estimated only by a numeric protocol. Our approach relies on the simulation of the measurement protocol with the software CONTAM, varying wind conditions and airtightness levels.

This article addresses three issues that impact the uncertainty on these derived quantities: the wind speed, the distribution of the leaks, and the pressure correction with the zero-flow pressure difference. This implicitly entails the investigation of influencing factors such as airtightness level of the building. Based on the analyses of those simulations results, this paper proposes protocols for extracting the air permeability at 4 Pa with better accuracy.

KEYWORDS

Airtightness, Simulations, measurement protocols, uncertainty

INTRODUCTION

In France, the new energy performance regulation will start applying for every new residential building starting January 1st, 2013. It will impose an airtightness treatment and each building will have to justify a level for Q_{4Pa_Surf} (the air permeability at 4 Pa divided by the loss surfaces area excluding basement floor), which will have to be lower than 0.6 $m^3\ h^{-1}\ m^{-2}$ for houses. In most cases, this justification will involve an envelope airtightness measurement. It will have to be performed by a certified operator and according to the NF EN 13829 [1] standard and its implementation guide GA P50-784 [2]. The ISO 9972 [3] is the international standard associated to this European standard.

This ISO sets out the airtightness measurement protocol. Since the procedure for the review of this ISO is under way, various publications (e. g. [4] and [5]) look at the uncertainty of this protocol. The uncertainty is a crucial issue as soon as the measured value becomes a requirement. Various research works have shown that the uncertainty could be really important depending on the measurement conditions.

The objectives of this study are to determine the wind impact on the uncertainty of the measurement, and to find out a way to reduce it. So our approach relies on simulations of airtightness tests done with CONTAM[1].

This paper presents the impact of a constant wind during an airtightness measurement, depending on the airtightness level and the leaks distribution. Then, it explains how it is possible to drastically reduce the uncertainty due to the wind. The issue of the pressure correction with zero-flow pressure difference is also discussed.

METHOD

A numerical study has been carried out using CONTAM as multizone airflow calculation. In order to simulate airtightness measurement, a 1-zone model building has been designed. The simulated building envelope has 9 leaks: 2 leaks on the up-wind facade, 2 leaks on each of the 3 others facades and 1 leak on the roof. The following diagram describes the geometrical properties of the model.

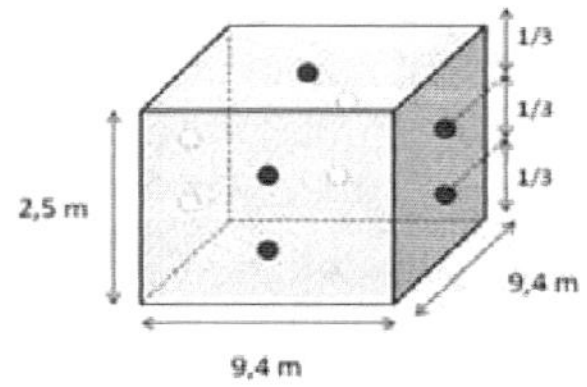

Local wind pressure coefficients:
C_p=-0.6 for the roof
C_p=0.5 for upwind façade
C_p=-0.7 for the three others facades

Figure 1 : Geometrical properties of the model

CONTAM calculates the flow through each leak using the principle of the mass conservation in each zone. For this model, each leak flow's derives from the following equation:

$$q_{mi} = \rho_{air} * C_i * \Delta P^{0.67} \quad (1)$$

with q_{mi}= flow through a leak *[$m^3\ h^{-1}$]*, ρ_{air}= air density *[$kg\ m^{-3}$]* and C_i = air leakage coefficient *[$m^3.s^{-1}.Pa^{-0.67}$]*

The pressure difference depends on the imposed pressure between the outdoor and the indoor, and the pressure due to the wind:

$$\Delta P = P_{out} - P_{int} - P_{wind} \quad (2)$$

[1] Multizone Airflow and Contaminant Transport Analysis Software

The wind pressure depends on:

$$P_{wind} = \frac{1}{2} * \rho_{out} * C_p * V_{wall}^2 \qquad (3)$$

with C_p = local wind pressure coefficient [-], and V_{wall}= Wind speed at the height of the wall [m s^{-1}]

The wind speed at the height of the wall depends on the wind speed at 10 m:

$$V_{wall}^2 = A_0^2 * \frac{H_{house}}{10}^{2\alpha} * V_{wind}^2 \qquad (4)$$

with $A_0 = 0.60$ *and* $\alpha = 0.28$ (coefficients for houses in a suburban area, [6]).

Three different geometric models have been tested. The nine leaks of the first model are all the same, i.e. the size of the "hole" is the same for all of them. For the second model, the size of the two leaks on the upwind side represents 75% of the leakage area. And for the third model, the size of the 2 leaks on the upwind side represents 5% of the leakage area.

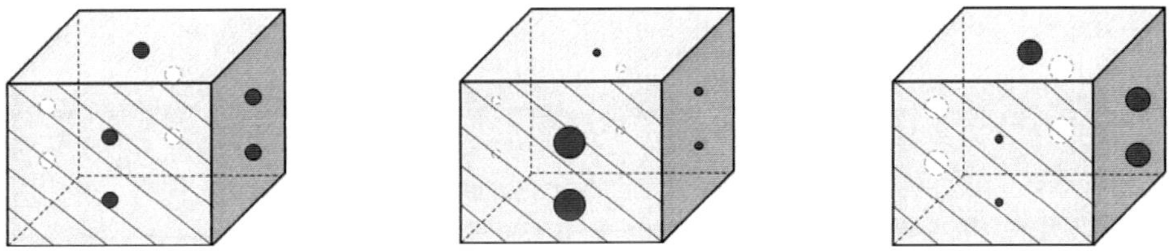

Model 1: equal leaks — Model 2: 75% on the up-wind side — Model 3: 5% on the upwind side

Figure 2 : Three different distributions of leaks

Each simulated airtightness measurement consists of 7 measurement points from 10 to 70 Pa for a pressurization test (or from -70 to -10 Pa for a depressurization test). With these 7 points, we made a linear regression according to the ISO9972. The impact of the stack effect is not studied, al the simulated tests are applied under isothermal conditions.

The major objective was to estimate the wind impact depending on the airtightness level. For three airtight levels (0.1 then 0.6 and 3 m^3 h^{-1} m^{-2}), the wind speed varies from 0 m s^{-1} to 8 m s^{-1} or 9 m s^{-1} (depending on the leaks' distribution).

RESULTS

Wind impact

For each leaks' distribution, the simulated measurement of Q_{4Pa_Surf} evolution depending on the wind speed is represented. For some wind speeds, the relative error between the simulated measurement Q_{4Pa_Surf} and the Q_{4Pa_Surf} assumed to represent the real leak, is estimated. The two following graphs show results for the Model 1: "equal leaks". The figure 3 represents the depressurization tests results, and the figure 4 represents the pressurization tests results.

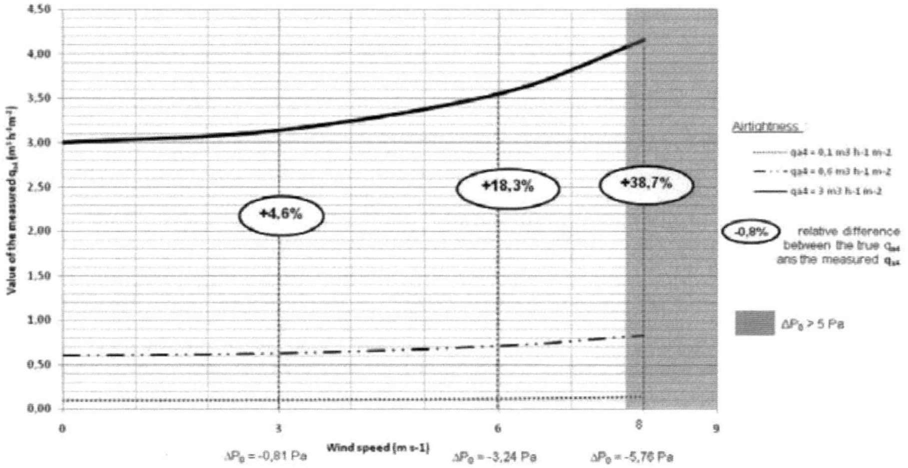

Figure 3: Wind impact on the value of the measured Q_{4Pa_Surf} for depressurization tests, with 9 indentical leaks

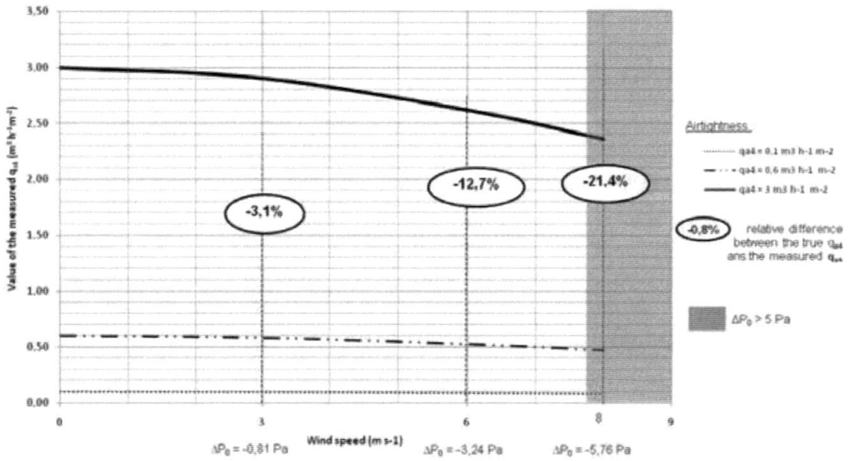

Figure 4: Wind impact on the value of the measured Q_{4Pa_Surf} for pressurization tests, with 9 indentical

According to the ISO 9972, if the zero-flow pressure difference is greater than 5 Pa, the test is not conform. For the distribution of the model 1, the zero-flow pressure difference exceeds 5 Pa when the wind speed is between 7 and 8 m s^{-1}. Thus, for a wind speed lower than 8 m s^{-1}, the test is conform (it has been checked that n is in range 0.5 to 1 and that r^2 is not less than 0.98 for each linear regression).
For the model 1, the relative error for a test declared valid could be more than 20% in pressurization and more than 35% in depressurization. For the three models, the relative error due to the wind is independent of the airtightness level. The following table gives the same key figures for the models 2 and 3.

	Model 2		**Model 3**	
Wind speed [m s^{-1}]	**Relative error in pressurization**	**Relative error in depressurization**	**Relative error in pressurization**	**Relative error in pressurization**
3	+1,3%	-0,2%	-0,2%	+1,4%
6	+3,5%	-1,3%	-1,8%	+3,5%
9	+6,1%	-3,5%	-4,5%	+8,9%

Table 1: Relative error due to the wind for models 2 and 3

The zero-flow pressure difference exceeds 5 Pa for wind speed higher than 12 m s^{-1} for the model 2, and for wind speed in range 8 and 9 m s^{-1} for the model 3. For all linear regressions, n and r^2 respect the ISO 9972 requirements. Table 1 shows that the impact of the wind depends greatly on the leakage distribution on the envelope. It highlights that the error drastically decreases when there is as much leakage on upwind façade (C_p>0) as in all others not upwind (C_p<0).
For one test, the ISO9972 recommends to make two sets of measurements: for pressurization and depressurization. With those figures, the average of a depressurization set result and a pressurization set results was estimated. The following table gives the relative error in this case for the three models.

Wind speed [m s^{-1}]	**Relative error for Model 1**	**Relative error for Model 2**	**Relative error for Model 3**
3	+0,8%	+0,6%	+0,6%
6	+2,8%	+0,3%	+0,8%
8	+8,7%	--	--
9	--	-0,9%	+2,2%

Table 2: Relative error due to the wind for two sets of measurements: for pressurization and depressurization

The simulated measurement of pressure differences correction with the zero-flow pressure difference

According to the ISO 9972, in order to obtain the induced pressure differences, the average zero-flow pressure difference is subtracted from each of the measured pressure differences. The measured Q_{4Pa_Surf} calculated without this correction and the measured Q_{4Pa_Surf} calculated with this correction have been compared. These figures are valid for each tested airtightness level.

	Wind speed: 3 m s^{-1}		Wind speed: 6 m s^{-1}		Wind speed: 9 m s^{-1}	
	Without correction	**With correction**	**Without correction**	**With correction**	**Without correction**	**With correction**
Depressurization	-2,6%	4,6%	-12,9%	18,3%	-24,1%	+38,7%
Pressurization	3,8%	-3,1%	13,1%	-12,7%	22,0%	-21,4%

Table 3: Impact of the pressure correction on the relative error du to the wind for the Model 1

According to these figures, the pressure correction with the average zero-flow pressure difference has a big influence on the result. Nevertheless, in these geometric models this correction does not decrease the error between the simulated measurement Q_{4Pa_Surf} and the Q_{4Pa_Surf} assumed to represent the real leak.

DISCUSSION

The ISO9972 explains that if the meteorological wind speed exceeds 6 m s^{-1}, it is unlikely that the zero-flow pressure difference can be lower than 5 Pa. Nevertheless, there is no wind speed limit. Considering a wind speed of 6 m s^{-1}, with the leaks distribution of the model1, the uncertainty of the measured Q_{4Pa_Surf} could be more than 18%. Moreover, the relative error on Q_{4Pa_Surf} could be more than 35% if a wind speed for which the zero-flow pressure difference is just under 5 Pa is considered. And none of the validation criteria of the ISO9972 could reject those tests.

These simulated tests have shown that making two sets of measurements in pressurization and depressurization is definitely a way to avoid the wind impact. For each leak distribution, if depressurization tests overestimate the Q_{4Pa_Surf}, then pressurization tests underestimate it (and vice versa), in the same order of magnitude. So, the average of the two results is far closer to the true Q_{4Pa_Surf}. This solution reduces significantly the uncertainty, which is not more than 9% in the worst scenario.

Another significant point of those results is that the uncertainty and the zero-flow pressure difference are independent of the airtightness level. This is true here because the *n* is the same for each leak. However, they depend on the leaks distributions. But, because the *n* of each leak and their distribution are unknown during a test, it is not possible to estimate for each measurement the mistake done because of the wind.

The final important issue raised by this article is the impact of the pressure differences correction. In this model, the impact is important, but does not reduce the relative error. Ideally, the correction should be done with the pressure difference at each leak, but it is not feasible. However, the difference between a result without and with correction shows that a better way to correct the measured pressure differences has to be found.

Those results have been obtained with a numerical study that does not exactly reflect what could happen during a true test. Firstly, the stack effect is not taken into account in those models. Secondly, the model is based on three hypothetical leaks and pressure coefficient distributions, and a flow exponent *n* constant for each leak. And thirdly, the wind speed is supposed constant during a test. Nevertheless, this simple model reveals some interesting results regarding the order of magnitude of the uncertainty due to the wind.

CONCLUSION

The objectives of this study were to determine the wind impact on the uncertainty of the measurement, and to find out a way to reduce it. Even if the model used to simulate a measurement of the airtightness of a building has some limits, it showed the wind could be responsible of significant errors (in some cases, more than 35%). Doing two sets of measurement in pressurization and depressurization could reduce this deviation in a very important way. This study also showed that the pressure differences correction imposed by the protocol might not be the better one to reduce the measurement error.

Finally, imposing the two sets of measurement and determining another way to correct the pressure differences should lead to reduce the errors during an airtightness measurement.

ACKNOWLEDGEMENTS

The contribution of CETE de Lyon is funded by the ministry for ecology, sustainable development and energy (MEDDE).

The sole responsibility for the content of this publication lies with the authors. It does not necessarily reflect the opinion of the Ministry.

REFERENCES

[1] EN 13829:2000. Therma l performance of buildings – Determination of air permability of buildings – Fan pressurization method (ISO 9972:1996, modified). European Committee for Standardization.

[2] GA P50-784. Performance Thermique des bâtiments - Guide d'application de la norme NF EN 13829:2001. European Committee for Standardization.

[3] ISO 9972:2006/Amd 1:2009. Thermal performance of buildings – Determiantion of air permability of buildings – Fan pressurization method. International Organization for Standardization.

[4] Christophe Delmotte, JelleLaverge. 2011. Interlaboratory tests for the determination of repeatability and reproducibility of buildings airtightness measurements. AIVC-TIGHTVENT Conference.

[5] Colin Genge. 2011. Modernizing ISO, EN and ASTM air leakage standards... more accuravy in less time. AIVC-TIGHTVENT Conference.

[6] ISO 9972:2006/Amd 1:2009. Thermal performance of buildings – Determination of air permeability of buildings – Fan pressurization method. International Organization for Standardization

BLOWER DOOR TESTS OF A GROUP OF IDENTICAL FLATS IN A NEW STUDENT ACCOMMODATION IN THE ARCTIC

Martin Kotol[*1], Carsten Rode[2], Jan Vahala[3]

1 Technical University of Denmark
Brovej 118
2800 Kgs.Lyngby, Denmark

2 Technical University of Denmark
Brovej 118
2800 Kgs.Lyngby, Denmark
**Corresponding author: mrko@byg.dtu.dk*

3 Brno University of Technology
Veveří 95
60200 Brno, Czech Republic

ABSTRACT

A new student accommodation for engineering students "Apisseq" was built in the town of Sisimiut, Greenland in 2010. Its purpose is not only to provide accommodation for students. Thanks to its complex monitoring system it enables researchers to evaluate the building's energy performance and indoor air quality (IAQ) as well as performance of some single components. In summer 2012 a blower door test was performed on all 37 living units out of which 33 are identical single room flats and 4 are larger double room flats. The purpose was to evaluate the air tightness of the envelope and to find out how much the flats differ from each other in terms of air tightness. The overall average specific leakage measured was $w_{50} = 2.05$ l/(s·m^2) of heated floor area corresponding to an air change n_{50} of 2.96 h^{-1}. Furthermore, the results showed that the difference between the most and the least tight flat is as high as 400%. This result is without consideration of one particular flat which had the extreme result of being 940% as leaky as the unit with the highest air tightness. The reasons for such poor air tightness are lack of the installation gap between the vapour barrier and the inner wall, and insufficient connections of the vapour barrier to the interior walls as explained in the paper. The large variation in results can be attributed to insufficient consideration of the importance of airtighness during construction of some parts of the building – despite of an intent to make a rather air tight building.

KEYWORDS

Blower Door Test, Air Tightness, Cold Climates, Residential Buildings

INTRODUCTION

In summer 2010 the new student accommodation for engineering students 'Apisseq' was finished in the town of Sisimiut, Greenland. The intention was to build an energy efficient building in which modern technologies, not yet commonly used in the Arctic, would be installed and which would provide its occupants with a healthy and comfortable indoor environment. Since balanced mechanical ventilation with heat recovery was installed, natural ventilation due to infiltration was no longer needed. In order to minimize infiltration heat losses, special attention was paid to the air tightness of the envelope.

There are no standard requirements on air tightness in the current Greenlandic building code, however the intention was to meet the current Danish requirement [1] which is that air changes through leakage in the building envelope must not exceed 1.5 $l/(s \cdot m^2)$ of the heated floor area when tested at the pressure of 50 Pa.
The aim of this study was not only to test the actual air tightness of the student accommodation, but also to study the distribution of the air tightness over a large number of identical flats by using statistical analysis.

Building key data

The floor plan of the building has the shape of an open circle, and has a partially heated ground floor and two upper floors. A main technical room and janitor´s office are in the heated part of the ground floor and small storage compartments for each flat are in the unheated part together with small technical rooms with ventilation units. The 1st and 2nd floor consist of 33 identical single room flats, and four dooble room flats at the gables of the building. In addition, there is a common room with a kitchen and a laundry room on the first floor (Figure 1 shows the floor plans). In the second floor, the common room and laundry is replaced with single room flats. There is also a glazed atrium with a staircase in the centre of the building. Each single room flat has a total floor area of 23 m^2 and consists of an entrance (3.3 m^2), a bathroom (2.8 m^2) and a living room with a kitchenette (16.8 m^2). The double room flats have a floor area of 50.2 m^2. All living units have a small balcony.

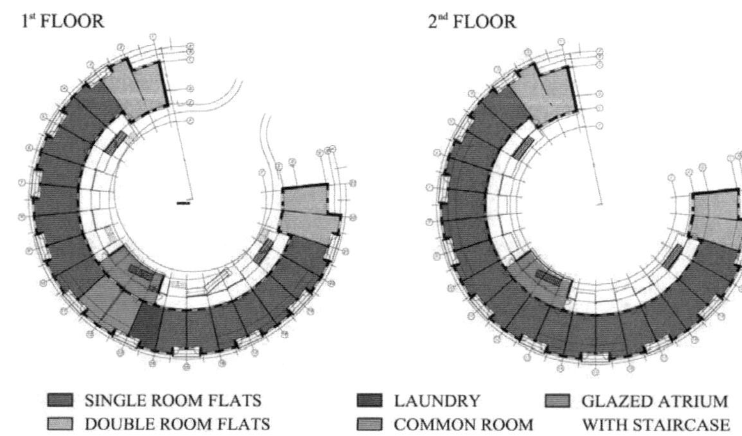

Figure 1. Floor plans of Apisseq

The aim to build an energy efficient building resulted in a well-insulated, air and vapour tight envelope supplemented by modern technology for space heating and mechanical ventilation of occupied spaces. The source for heating and domestic hot water (DHW) is district heating supported by evacuated tubular solar collectors connected to two accumulation tanks (2000 l each). The building is heated with radiators, floor heating is used in bathrooms and entrances.

Ventilation is provided by two identical ventilation units. Fresh air is delivered into the living rooms, and the poluted air is extracted through the kitchen hoods and exhausts in the bathrooms.

METHODS

Methodology of measurement

Standard procedure for measurements of air permeability of buildings and their parts in field specified in the standard [2] was followed. This standard offers two methods of air tightness measurement - method A where the air tightness of the object in use is measured and method B, when the air tightness of the building envelope is measured. Each of these methods requires a specific procedure of the object preparation before the measurement starts. Since the air exchange in all flats is ensured by means of mechanical ventilation there are not any ventilation elements or connections to the ambient, there is no difference between methods A and B in this case. All windows and doors to the ambient were closed, all air terminal devices were taped and internal doors were kept open to ensure equal pressure within the measured enclosure. The ventilation system was switched off.

Measuring equipment

The Retrotec Blower Door Test assembly was used to perform the tests. It consists of calibrated fan Retrotec 2200 Series, pressure gauge DM-2 and a cloth door panel. As the measuring and evaluation software was used the Retrotec FanTestic.

Measurement procedure

The fan was placed into the entrance door of an flat by using the cloth door panel. The measurement was automatically controlled by the software. The zero-flow pressure difference based on 10 baseline pressures taken for 10 sec each was taken at the beginning and at the end of every test. Subsequently the pressurization sequence was performed in 12 pressure steps by 5 Pa taken for 20 sec each from the initial level of 10 Pa to the final level of 65 Pa. After the pressurization sequence, the depressurization sequence was done. The results are the averages of these two measurements. The consistency of the measurements is given by correlation factor. The data are considered consistent when the correlation factor is 95% or higher.
In accordance with the standard, outdoor and indoor temperatures and the wind speed were monitored at the beginning and end of each test.
There have been changes of indoor and outdoor temperatures throughout the measurements. Calculation of airflow into the room through the fan is calculated can be affected by temperature fluctuations as they have effect on air density. However since the maximum difference between the temperatures had not been higher than 5K, the impact of these fluctuations is negligible.
In the case of flat 2.05 the blower door test was carried out on the balcony door after the first set of measurements. The second measurement was done through the balcony door. The intention was to compare the air tightness of the front door with the balcony door.
Some results were considered too far from normal. To enhance the preciseness of the measurements and to eliminate errors, the blower door test was repeated in flats 2.05, 2.12, and 2.20.

The characteristics of measured flats

The drawings of typical single and double room flats are shown in figure Figure 2 and the values used for the calculations are summarized in Table 1.

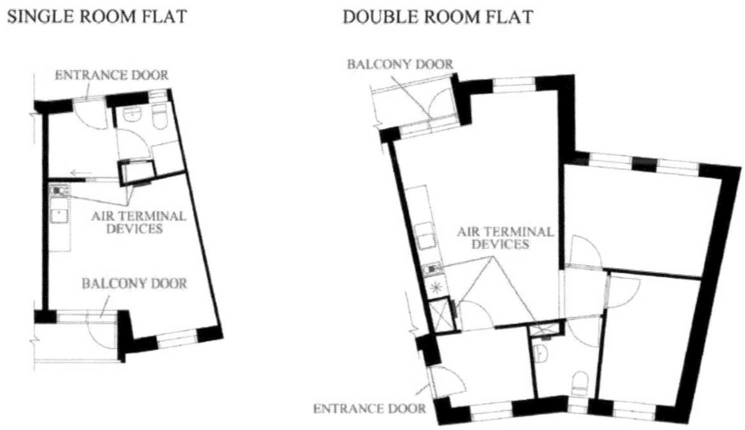

Figure 2. Drawings of the flats

	Single room flat	Double room flat
Volume [m^3]	57,5	131,8
Total Envelope area [m^2]	96	183,4
Floor area [m^2]	23	52,7

Table 1. Specification of flats

Evaluation of the measured data

Descriptive statistical analysis was performed on the results of specific air leakage. The possible relations in spesific leakage between nabouringh flats in certain part of the building were tested by means of the t-test and Pearson's correlation test. P-values of 0,05 were used to determine statistical significance. Statistical software R and MS Excel were used for the statistical analyses.

RESULTS

Overall results

The correlation factor is, except for three measurements, always higher than 95%. Only depresurization of flats 1.07 and 1.12 and pressurization of flat 1.10 is between 92% and 95%. The differences between pressurization and depressurization tests (see Figure 3) are on average 9.1%. When comparing the positive and negative differences, a two sample t-test yields a P-value of 0.95 which indicates that there is no prevalent trend of one of the tests (pressurisation or depressurisation) giving constantly higher or lower result.

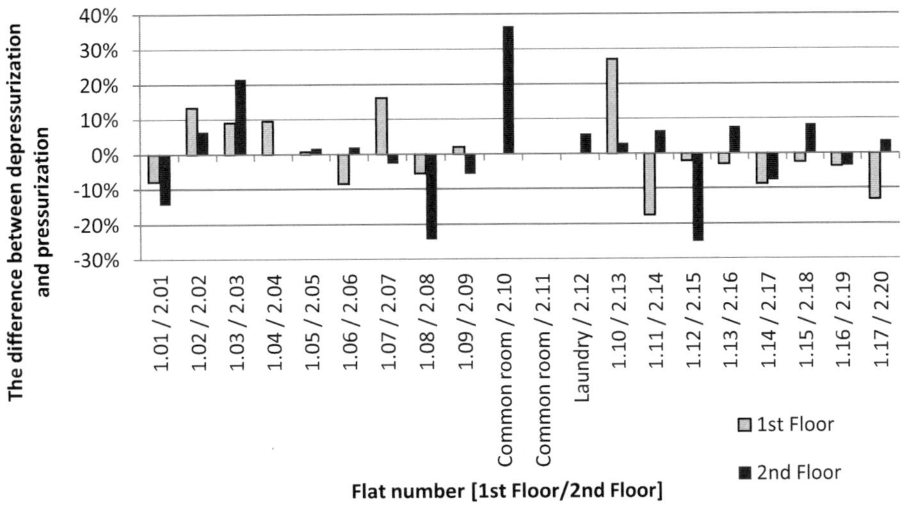

Figure 3. Negative values mean that the result from pressurization was larger than from depressurization

The mean value of specific leakages obtained from Apisseq is 2.05 l/(s·m^2) with standard deviation of 0.96 l/(s·m^2) corresponding to an air change n_{50} of 2.96 h^{-1} with standard deviation of 1.38 h^{-1}. The distribution can be seen from the box plot in Figure 4. It can be observed that the maximum value, which is the test result of flat 2.20, lies significantly above the 3rd quartile. To eliminate the measurement error we repeated the test next day. The result was only 3% different from the first test. This may indicate an abnormality due to construction problems in this flat. More discussion follows in the Discussion section.

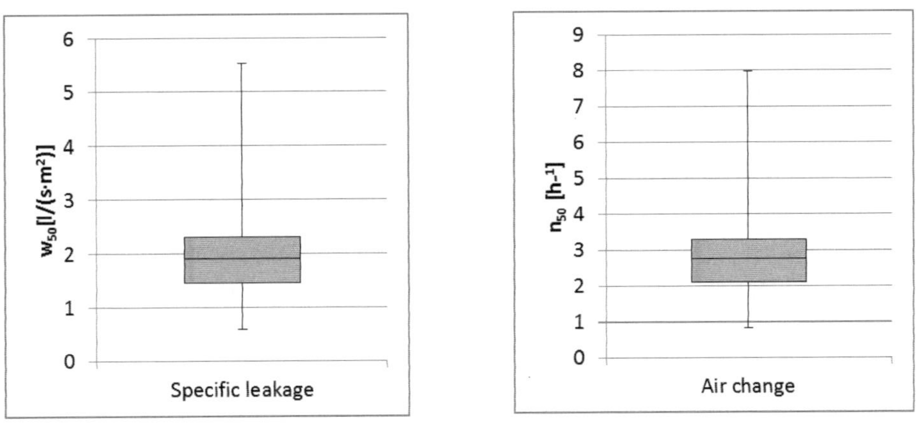

Figure 4. Distibution of overall desults of blower door test

The combined specific leakage in all the tested units is presented in Figure 5. When testing the correlation between the first and second floor by means of Pearson's correlation test, we found a positive correlation of 0.53 at 5% level of significance between the single room flats which are above each other.

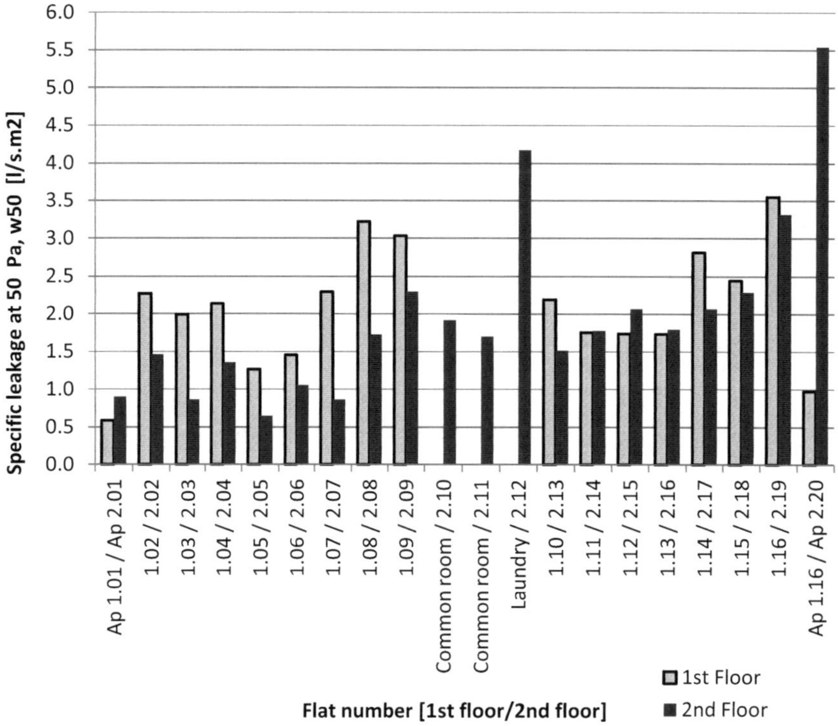

Figure 5. Combined results of testing all the units within the student accommodation

Comparison between flats inside and outside the atrium

The two sample t-test yields a P-value of 0.17 based on what the null hypothesis that there is no difference in air tightness between flats inside and outside the atrium cannot be rejected.

Flats outside of the glazed atrium		Flats behind the glazed atrium	
Mean	1.98	Mean	2.51
Median	1.89	Median	2.29
Standard Deviation	0.68	Standard Deviation	0.81
Variance	0.47	Variance	0.65

Table 2. The statistics of w_{50} [l/(s·m^2)] measured in flats inside and outside the glazed atrium

Single room vs. double room flats

The mean specific leakage of the four double room flats is 2.00 l/(s·m^2), which is not different from the mean specific leakage of the single room flats: 2.06 l/(s·m^2) (Table 3). However, excluding the abnormally high specific leakage of the double room flat no. 2.20, gives a mean leakage of 0.82 l/(s·m^2), which is significantly smaller than the mean specific leakage of the singe room flats (P-value of one tailed t-test < 0.01).

	Single room flats	Double room flats	Double room flats without no. 2.20
Mean	2.06	2.00	0.82
Median	1.99	0.94	0.90
Standard Deviation	0.72	2.37	0.21
Variance	0.51	5.59	0.04

Table 3. The statistics of w_{50} [l/(s·m^2)] measured in single and double room flats

1st vs 2nd floor

There is no significant difference in air tightness between the units in the first and second floor (two sample t-test P-value = 0.82) even when the worst flat (2.20) is excluded (P-value = 0.33).

	1st floor	2nd floor	2nd floor without 2.20
Mean	2.09	2.02	1.84
Median	2.14	1.78	1.77
Standard Deviation	0.79	1.10	0.75
Variance	0.63	1.22	0.56

Table 4. The statistics of w_{50} [l/(s·m^2)] measured in all units in 1st and 2nd floor

Flats that were tested twice

	Flat 2.05	Flat 2.12	Flat 2.20
1st measurement	0.65	4.18	5.36
2nd measurement	1.99	3.63	5.54
Difference	206%	13%	3%

Table 5. The results of w_{50} [l/(s·m^2)] in units which were measured twice

Test on balcony door

	Flat 2.05		
Blower door sitting in:	**Front door**	**Balcony door**	**Difference**
Specific Leakage	1.99	0.85	57%
Uncertainty	0.03	0.09	

Table 6. Comparison of main entrance door and balcony door

DISCUSSION

Overall results

The tests have shown, that the average specific leakage of the building is 2.05 $l/(s \cdot m^2)$ which would not fulfill the Danish requirement of 1.50 $l/(s \cdot m^2)$. Nevertheless 27% of all flats in the building had specific leakage lower than the requirement. This enhances the importance of large portion of flats in one building (even when they are identical) being tested when relevant results are sought.

The positive correlation between the single room flats above each other could be explained by the horizontal direction of the construction. The degree of dependence is however very low.

The reasons for poor air tightness are several. The lack of the installation gap between vapour barrier and inner surface plays a large role since all the installations have to penetrate the vapour barrier when entering the flats. Another reason is lack of overlapping flaps in corners where the vapour barrier connects to the concrete walls and floors/ceilings (see Figure 6).

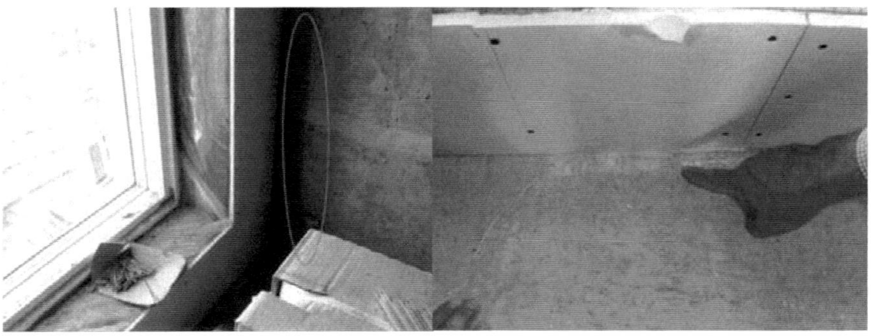

Figure 6. Left: Correct connection with overlap; Right: Missing overlap

Additionally an extra focus on air tightness had not been a part of the building tradition in Greenland until very recent years. Which can explain the insufficient consideration of its importance during construction and design phase.
We assume that if the blower door test was done during the construction phase, many errors would be explored and fixed which would have positive effect on the final air tightness.

Comparison between flats inside and outside the atrium

We have not found any evidence that the air tightness of flats inside the glazed atrium is significantly different from the rest of the building.

Single vs. double room flats

The reason why the double room flats have better air tightness than the single room flats (with one notable exeception) is the vapor barrier area/total area ratio which in single room flats is is 2x higher than in double room flats which gives higher risk of leakages.
There is probably some larger penetration of the vapour barrier in the flat number 2.20 which causes that high specific leakage. It is suggested to repeat the test together with smoke generating device in order to detect the leakage.

Flats that were tested twice

The 206% difference between first and second test of the flat number 2.05 can only be explained by a procedural mistake whereas the other two differences (13% and 3% in flats 2.12 and 2.20 respectively) are probably caused by combination of systematic and random errors.

Test on balcony door

The results show that the specific leakage when tested with the blower door equipmet in the balcony door is smaller than the leakage obtained from the test in the front door by 57%. It may imply that there is significantly higher air leakage through the balcony door than through the front door. To justify this hypothesis, repeated measurements and also measurements in other flats need to be done.

CONCLUSION

The air tightness of all 37 flats in the building was measured with the result which does not meet the current Danish requirements. There is however no such requirement in Greenland.
Bringing awarness of the necessity of air tightness to all parties involved in construction process is of very large importance.
Performing the blower door test during the construction phase is a way to avoid errors as well as shoddy work.
When the actual air tightness of buildings is to be determined, large portion of the whole building rather than just small sample needs to be tested.

In order to test the validity of measurement procedure multiple measurements of specific leakage of randomly selected flat should be carried out.
During the experiment period (03 - 13 Aug.2012), the weather varied from day to day (sunny, cloudy, rainy). For further studies, these factors should be considered.
More tests should be carried out to compare the specific leakage when tested both on the balcony door and on the front door.

REFERENCES

[1] Danmark. Erhvervs- og Byggestyrelsen 2010, *Bygningsreglement 2010 (BR10),* Erhvervs- og Byggestyrelsen, Kbh.
[2] Dansk Standard 2001, *DS/EN 13829 - Thermal performance of buildings – Determination of air permeability of buildings – Fan pressurization method,* Dansk Standard, Charlottenlund.

ASSESSMENT OF THE AIRTIGHTNESS AND AIR EXCHANGE IN POLISH DWELLINGS – MEASUREMENT EXPERIENCES AND PROBLEMS MET

Joanna Ferdyn-Grygierek*, Andrzej Baranowski

Department of Heating, Ventilation and Dust removal Technology, Silesian University of Technology
Konarskiego 20
44-100 Gliwice, Poland
**Corresponding author: joanna.ferdyn-grygierek@polsl.pl*

ABSTRACT
Indoor environment quality in buildings strongly depends on the proper ventilation. Still a large amount of single- and multifamily buildings are equipped with the natural ventilation system.
When the air exchange in the building is estimated, the main uncertainty concerns the air tightness of the given object. This parameter is used as the input data when the ventilation air flows in building are simulated, and therefore a reliable determination of the air tightness is essential.
The method of determining of the air tightness consists of the measurement of the air flows through the investigated flat or house for the given pressure difference. The procedure and techniques are well- known and the measurement is realized by so-called blower door.
The paper presents the course and results of experiments performed in several flats in different multifamily buildings as well as in the single-family house.

KEYWORDS
airtightness, air exchange, ventilation, blower door

INTRODUCTION

Airtightness of the building envelope is an essential parameter for assessment of the energy demand, for it determines the amount of air infiltration to the building, and therefore the amount of heat required for heating. Modern technologies and materials used in construction, allow the walls of the buildings for significant reduction of heat loss through transmition, however at the same time, due to tight windows, they limit the possibility of proper ventilation.

Large share of the housing stock in Poland are multi-family dwelling houses, ventilated by the natural duct ventilation based on the stack effect. This method of ventilation, cheap to run, results from existing regulations, which allow the system to be used in buildings up to 11 storeys. The system of natural duct ventilation is common in domestic single-family housing. More and more often passive single-family houses, as well as houses using renewable energy sources and mechanical ventilation systems are built.

Measurement of building airtightness can be used to verify the design assumptions, to allow comparisons with other buildings and to estimate the air infiltration.

The paper presents the results of the measurements of the airtightness in two types of dwelling houses: multifamily (two buildings) and single-family (one building). In addition the results of the measurement of exhausted air flow of the analysed multi-family houses and the air exchange simulation calculations on a model of one of the buildings, in which the results of previous measurements were used as an input, are presented.

POLISH REQUIREMENTS REGARDING BUILDING VENTILATION

Polish regulations are based largely on European standards, both for ventilation air flows and airtightness of buildings and flats. The selection of the ventilation method is based on Polish standard PN-83/B-03430/Az3 [1]: in buildings up to 11 storeys, both residential and public, gravitational ventilation can be used, whereas in higher buildings mechanical ventilation is required. In the case of dwelling houses, Polish standards set removed air flows on the following levels: 70 m^3/h for the kitchen, 50 m^3/h for the bathroom and 30 m^3/h for the separate lavatory. The minimum air change rate in dwellings should be 0.5 h^{-1} [2].

The airtightness of the building envelope (or the airtightness of the dwelling), measured by the n_{50} index, should be in accordance with PN-EN 12831 standard [2] (Tab. 1).

Building	**Airtightness requirement at 50 Pa pressure n_{50}, h^{-1}**		
	high airtightness	medium airtightness	low airtightness
Single family buildings	<4	4-10	>10
Multifamily dwelling houses	<2	2-5	>5

Table 1. Air change rate according to PN-EN 12831 standard

AIRTIGHTNESS AND AIR EXCHANGE MEASUREMENTS

The study aimed at: the measurement of the airtightness of the building envelope or its part, the evaluation of the ventilation air flow as well as the identification of uncontrolled air flow paths.

The measurements of the building airtightness were conducted in accordance with PN-EN 13829 standard [3], based on the fan pressurization method with the use of blower door device. The standard recommends measurements according to the method A or B. The main difference lays in the fact that heating and ventilation systems in method A are in the same condition as during regular utilization, while all intentionally made openings in the building envelope in method B should be closed. All the airtightness measurements in the research presented in this paper were carried out in accordance with method B.

Before performing the measurements of airtightness, areas were thoroughly examined in order to eliminate possible leakages, other than window leaks. All of the intentionally made openings in investigated areas were closed: windows, exterior doors, inlets of ventilation and exhaust ducts. All the interior doors were open.

Auxiliary measurements were performed prior to conducting the experiment: floor area, height and volume of the room were calculated; air temperature inside and outside the building, air pressure and wind speed were measured.

Subsequently, the device producing, depending on the flow direction, underpressure or overpressure in examined room, was installed in the door opening (Fig. 1). Airtightness of the joints in door opening was inspected, then the device measuring air pressure and the computer with an appropriate software were installed.

The measurements of the air flow and the air pressure difference between the zone and the environment at intervals of 10 Pa and in the range of 30 ÷ 70 Pa, were performed after the device initiating movement of the air was turned on.

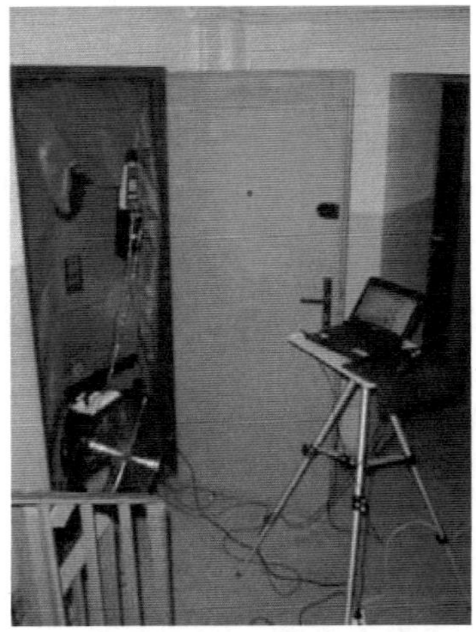

Figure. 1. The test stand for the measurement of the apartment airtightness in multifamily house

The main result of the study was a flat or building leakage curve (separately for pressurization and depressurization) in the form of the formula:

$$\dot{V} = C \cdot \Delta p^{n} \quad (1)$$

where:

$\dot{V}$ – leakage air flow rate, m^3/h

C – flow coefficient, $m^3/h \cdot Pa^n$

Δp – pressure difference induced by ventilator, Pa

n – exponent.

Additionally, the air flow V_{50} and the air change rate n_{50} for the air pressure difference of 50 Pa, were obtained.

Prior to airtightness tests in every analysed apartment in multifamily houses, the flows of the air blown out through ventilation ducts were measured with the use of barometer. Moreover, meteorological data, according to the local weather station, for the day and hour were recorded.

Multifamily dwelling houses

The measurements were performed in two multifamily dwelling houses: 5 and 11 storeys, equipped with gravitational natural ventilation. The buildings were built several decades ago. Gravity pipes are located in kitchens, bathrooms and separate lavatories in the case of 2-storey building (in each case one exhaust grille). There are several types of windows in buildings, both: relatively new, quite tight PVC windows with seals, as well as wooden windows made several decades ago. None of the windows is equipped with air inlets.

There are leaks in the form of cracks in building construction and poorly sealed verticals of central heating in apartments (Fig. 2). Serious leakages into the corridor in the place where gas pipes pierce to counters were noticed in some apartments of 11-storey building.

Figure. 2. The examples of leakages in the walls and ceiling of the apartments

In the 5-storey building, the measurements concerned three apartments: two on the 2nd storey (M1 and M3) and one on the 5th storey (M15). The selection was random and depended on the consent of the tenants to carry out measurements. Similar situation occurred in the case of 11-storey building, where the measurements were performed in apartments on the groundfloor (M4) and one on the 10th storey (M55).

It was necessary to seal some components of the electrical installation, as well as culverts of central heating verticals and culverts of gas installation. Grilles' outlets were also plugged. Large and hard-to-reach openings, caused by the culverts for exhaust pipes to the individual gas water heaters, were detected in bathrooms in 11-storey building. Due to the fact that bathrooms are interior rooms, without windows and external walls, they were isolated by sealing the door for the period of measurements (Fig. 3). Hereby, problems with all leakiness in these rooms were avoided.

The results of pressurization tests are presented in Table 2. Knowing the length of the window leakages and a flow coefficient C generated by the program, window airtightness factor a was calculated.

Figure. 3. Method of sealing of bathroom's door in the flat of the 11-storey building

The value of an index generated by the program should be about 0.67. If it is different, much lower than 0.67, it may indicate the presence of uncontrolled air flows through the envelope of a zone.

Unfortunately, during the study, it was not possible to seal the zones enough to obtain a desired value of n, therefore, to calculate airtightness factor a, the index of a flow characteristics (n) was corrected to a value of 0.67. With the use of spreadsheet, corrected values of flow coefficient C were obtained. Dividing the flow coefficient C by the length of the window cracks, the values of airtightness coefficients a were obtained and summarized in Table 2.

Building	Flat	Type of window	V_{50}, m³/h	n_{50}, h⁻¹	Airtightness factor *a*, m³/m·h·Pa^0,67	Air flow measured in air outlets (required air flow), m³/h
5-kond.	M1 (2st storey)	old	**415**	**3.3**	1.16	32 (120)
	M2 (2st storey)	new	**232**	**1.5**	0.54	92 (150)
	M15 (5th storey)	old/new	**715**	**3.8**	1.37	105 (150)
11-kond.	M4 (1st storey)	old/new	**320**	**2.7**	0.57	15 (120)
	M55 (10th storey)	new	**132**	**1.5**	0.32	10 (120)

Table 2. The results of the measurements of the airtightness and ventilation air flow

Single family building

Figure 4. The example of leakiness in the building

The measurements were performed in 2-storey building with a cubage of 570 m3. The building is equipped with a mechanical ventilation system providing required ventilation air flow. All windows are in very good condition, relatively new and tight.

Before the measurements, the inspection of rooms was conducted to search and seal potential leakages. Air intake and exhaust air device, as well as fireplace doors were detached (Fig. 4). The results of pressure tests are: V_{50}=983 m³/h, n_{50}=1.73. Analysed building is tight, obtained factor n_{50} describes the building with high degree of envelope tightness in terms of PN-EN 12831 [2] standard. In the case of the building with mechanical ventilation, this coefficient can be considered as satisfactory.

ASSESSMENT OF THE AIR INFILTRATION

One of the goals of the study was to estimate of the air change rate in the analysed buildings. Knowledge of the characteristics of airtightness and the total length of the leakages in the windows made it possible to calculate the rate of air infiltration.

In addition to measurements related to building ventilation, numerical simulations of ventilation air flow on the model of the analysed 5-storey building were carried out. Each flat and corridor was modelled as a separate zone, resulting in a total of 30 zones. Calculations were performed using the CONTAM [4] software for the meteorological data recorded by the local weather station. The air infiltration coefficients, calculated on the base of airtightness measurements, were used with the assumption of their repeatability in the same type of windows in other flats in the building. Air flow simulations were carried out for the period of the heating season. Fig. 5 presents the course of air infiltration variability on the day of measurements in three analysed apartments. Fair compatibility of the results of measurements of the air flows exhausted from two apartments with an air infiltration can be noticed. Differences do not exceed 12%. In the case the 3rd apartment, large (3-fold) difference may result from substantial contamination of the ventilation grille, so that the air flows largely through apartment's door to the staircase, which is not included in the measurement.

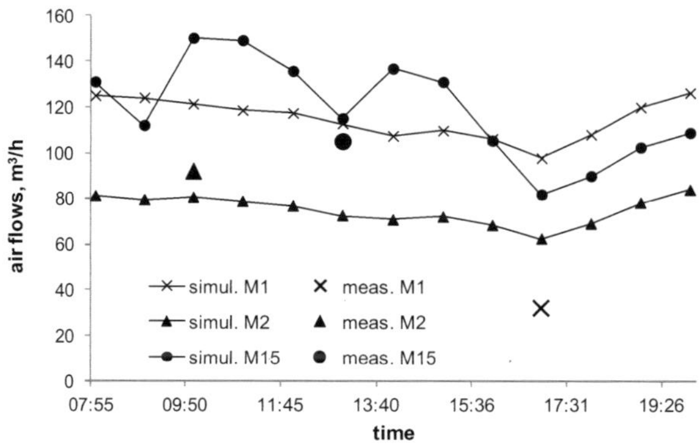

Figure. 5. Air infiltration in analysed flats – simulation versus measurement

CONCLUSIONS

The obtained results of measurements of airtightness are within the ranges specified in the standards. The measurements indicate high airtightness of apartments with new PVC windows. The value of n_{50} does not exceed 2 h^{-1}.

The study confirmed that many uncontrolled leakages, which impede measurements and increase their uncertainty, exist in old multifamily houses. Uncertainty as to the results may also result from the fact that the measurements were performed in completely random houses, not including possible connections to the neighbouring houses.

Ventilation air flow, measured directly in the exhaust grille, is small and substantially deviates from the Polish Standard (with one kitchen and one bathroom required air flow is 120 m^3/h). The maximum measured air flow for this type of housing was 32 m^3/h. The ventilation air flow decreases with increasing storey. Unfortunately, the amount of gravitational ducts on the highest storeys of the buildings has not been increased. However, it should be noted, that the given air flow was measured only in the exhaust grilles, without taking into account the air flow through the apartments doors to the building staircase.

ACKNOWLEDGEMENTS

The work is part of the project financed by the National Centre for Research and Development (NCBiR) agreement SP/B/4/77113/10 and was also supported by the Polish Ministry of Science and Higher Education – statutory research BK-364/RIE-1/2011.

REFERENCES

[1] Polish Standard PN-83/B-03430/Az3: 2000 *Ventilation in dwellings and public utility buildings* (in Polish)

[2] Polish Standard PN-EN 12831: 2006 *Heating systems in buildings. Method for calculation of the design heat load* (in Polish)

[3] Polish Standard PN-EN 13829: 2002 *Thermal performance of buildings. Determination of air permeability of buildings. Fan pressurization method* (in Polish)

[4] Walton W.G.N., Dols S. (2005) CONTAM 2.4 *User Guide and Program Documentation* National Institute of Standards and Technology, Gaithersburg

A numerical study on the role of leakage distribution and internal leakages under unsteady wind conditions

Dimitrios Kraniotis*, Tormod Aurlien, Thomas Kringlebotn Thiis

Department of Mathematical Sciences and Technology, Norwegian University of Life Sciences
Drøbakveien 31, P.O. box: 5003, IMT, N-1423, Ås, Norway
**Corresponding author: dimitrios.kraniotis@umb.no*

ABSTRACT

The existence of air leakages in a building has been very clearly stated as an important reason for energy loss. The decrease in the efficiency of the mechanical ventilation has also been clarified. The global demand for achieving nearly zero-energy buildings makes the uncontrolled leakage paths even more undesired. Despite the fact that steady state measurements of in- and exfiltration rates offer a simple and easy way of estimating the airtightness level of an eclosure, a supplement to those methods might be imposed.

While a significant amount of studies points out the key role of the 'artificial' unsteady conditions to the actual leakage rates of a building, there are only few that discuss the influence of natural unsteady phenomena. In this context, the correlation between the dynamic characteristics of the wind and the leakage numbers of a building, should be more studied. Computational Fluid Dynamics (CFD) could be employed in order to investigate the role of the air flow mechanisms.

In the current numerical study, unsteady wind conditions are performed around an one-storey building-model of size 5m x 10m x 3m. Variable leakage areas A_{leak} around windows are simulated and solved in a transient mode aiming to investigate the role of the distribution of the leakages under natural conditions. A ratio ($0 \leq \alpha \leq 1$) that represents the portion of leakages (distribution) per surface is employed and the infiltration rates respect to this ratio are shown. Different situations of the enclosure volume (from the perspective of internal wall airtightness) are assumed in order to investigate the influence of the latter to the infiltration rate of the building's envelope. The impact of the internal leakages is proven and the importance of controlling them is discussed.

KEYWORDS

Leakage rate, air flow, air infiltration, leakage distribution, internal leakages, unsteady wind conditions, wind gust, computational fluid dynamics, shear-stress-transport

INTRODUCTION

Air infiltration has been recognized as one of the major reasons for energy loss [1]. The decrease in the efficiency of the mechanical ventilation has also been clarified [2]. Uncontrolled leakage paths have very clearly stated as pervasive, resulting in severe consequences [3]. The nature and extent of uncontrolled air flow have also been studied through testing, measurement and monitoring. Many researchers have also stated the uncertain phenomena that are connected to the airflow through leakages located on a building envelope. The dynamic characteristics of air infiltration have been pointed [4] and therefore challenges arise upon that field. The role of the climate parameters and location characteristics on average infiltration rates has also been studied [5]. Turbulence causing wind gustiness is recognized as one major factor that affects infiltration [6]. In addition, building aerodynamics contributes to air infiltration too. In that context, modelling approaches have been presented [7], [8]. Although, the air leakage of a building envelope can be determined from fan pressurization measurements with a blower door, estimating in a simple and easy way an enclosure's airtightness level [9], further research that takes the latter phenomena into account should be done.

Furthermore, leakage distribution has been mentioned as important factor towards the annual infiltration rate calculation [10]. Models have been developed towards the estimation of leakage distribution [11]. In addition, the later affects even the air pressure conditions in building and the wind-induced internal pressure fluctuations [12], [13]. In the same manner, the role of internal volume has been mentioned [14] as well as the influence of internal air leakages [15].

Computational fluid dynamics (CFD) could be employed to investigate the role of the air flow mechanisms from the perspective of the phenomena presented above, especially under unsteady conditions. Numerical studies could contribute to an estimation of the impact of potential leakages areas in the building envelope as well as in internal elements. Facing the global demand for achieving nearly zero-energy buildings, a more holistic and detailed approach of the phenomena linked to air infiltration should be given through both measurements and numerical simulations.

CASE STUDY

The current numerical study deals with the influence of unsteady wind to the instanteneous infiltration (exfiltration) rates of an one-storey building-model (of size 5m x 10m x 3m) on which variable leakage areas around windows are simulated. Two 'windows' are used on each side. The size of each 'window' is 0,8*m* x 0,8*m*. The leakages are supposed to be 'cracks' along their frame. The total leakage area (whole building-model) is assumed to be $64cm^2$.

The leakages are located on the windward and on the leeward side of the model. Seven different cases of distribution (windward vs leeward) are solved. For the representation of the latter, a ratio α is defined as follows:

$$a = \frac{A_{leak,front}}{A_{leak,total}} * 100\ [\%] \tag{1}$$

where

$A_{leak,front}$: the leakages located on the windward side (front side) for the building-model, expressed in *cm2* and

$A_{leak,total}$: the total leakage area of the model in cm^2 (both on windward and leeward sides), which as mentioned above equals $64cm^2$.

In fact, the ratio α expresses the leakages located on the windward side as fraction to the total leakage areas of the building. The α takes the values: α = 5, α = 15, α = 30, α = 50, α = 70, α = 85 and α = 95 [%]. To give a magnitude of order of the amount of the leakages:

$$\frac{A_{leak,total}}{total\ model\ surface} = 4{,}57 * 10^{-5} \tag{2}$$

Furthermore, since the influence of wind gust frequency ω has been discussed (especially for single-side airflow) [14], studying of its connection to the leakage distribution would be useful. Thus, two different gust frequencies are assumed, ω_{high} and ωl_{ow} and they are implemented in the wind profile forluma as a sinusoidal factor (explained in the 'methodology').

Finally, three different 'situations' (S_1, S_2, S_3) regarding the internal volume are simulated in order to research the influence of the internal leakages and their connection the external ones. The first case S_1 refers to an internal volume without internal walls ('uniform', single space)

(fig. 1). The second and the third cases (S_2 and S_3 respectively) both assume the existence of internal wall that divide the whole space in two 'rooms'. The difference is that in S_2 a leakage area of $4cm^2$ is assumed to be located on the low level of the wall, allowing the inter-flow between the two rooms (fig. 3a and 3b), while in S_3 there are not internal leakages at all (assumption of completely tight internal wall) (fig. 2).
Summarizing, 42 cases are studied in total:
Σ = 7*(leakage distibution cases)* * 2*(wind gust frequency cases)* * 3*(internal space cases)*.
A notation described by the following rule is employed and used hereinafter:
(Internal volume case S_i) - (leakage distrubion α_j) - (frequency of the wind gust ω_k),
where:

i = 1, 2 or 3,
j = 5, 15, 30, 50, 70, 85 or 95 and
k = 'high' or 'low'.

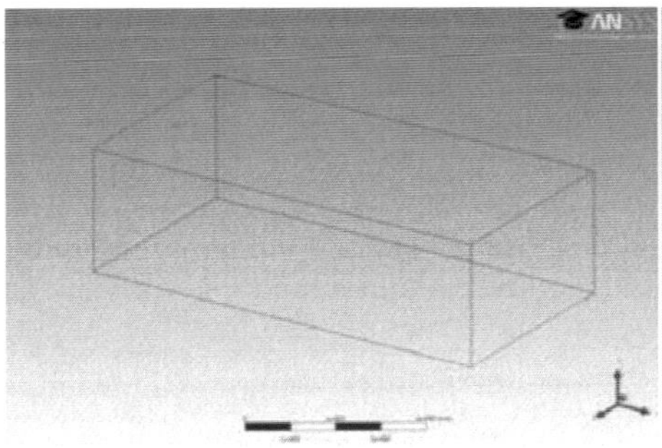

Figure 1. The case S_1: 'uniform' single space (no internal wall).

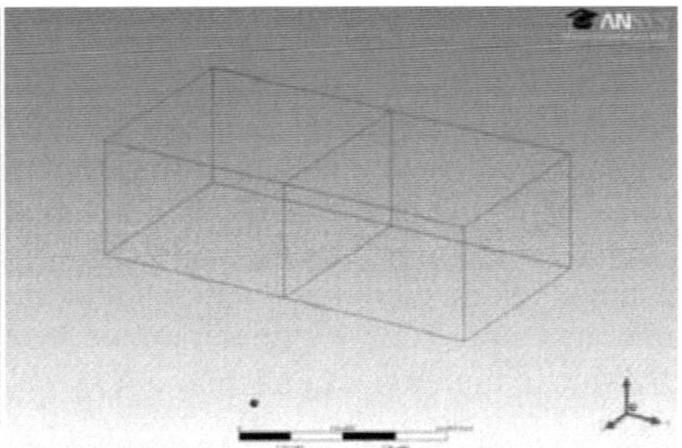

Figure 2. The case S_3: two spaces separated by an internal wall. No internal leakages.

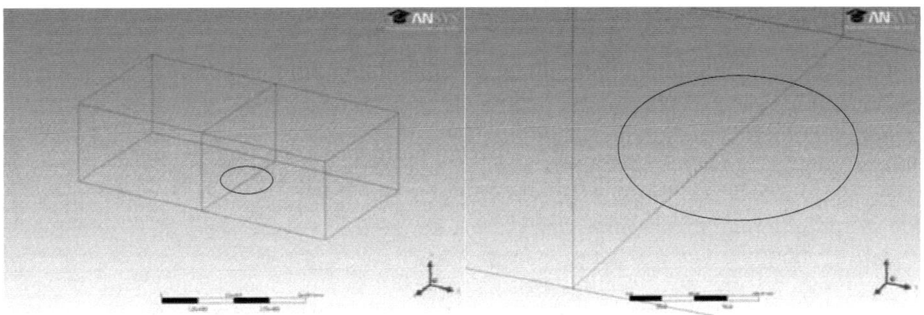

(a) (b)
Figure 3. The case S_2: two spaces separated by an internal wall. Internal leakages are located on the low level of the wall (circle area).

The table 1 shows, as example, the notation for leakage distribution of α = 5% and α = 70% for all the subcases of the internal volumes (and in the high gust frequency).

Case	Internal volume	Leakage distribution	Wind gust frequency
S_1-α_5-ω_{high}	'uniform' space	$\alpha = 5\%$	high
S_2-α_5-ω_{high}	two spaces – with internal leakages	$\alpha = 5\%$	high
S_3-α_5-ω_{high}	two spaces – no internal leakages	$\alpha = 5\%$	high
S_1-α_{70}-ω_{high}	'uniform' space	$\alpha = 70\%$	high
S_2-α_{70}-ω_{high}	two spaces – with internal leakages	$\alpha = 70\%$	high
S_3-α_{70}-ω_{high}	two spaces – no internal leakages	$\alpha = 70\%$	high

Table 1. Example of the notation followed. Here, the notations for the leakage distribution $\alpha = 5\%$ and $\alpha = 70\%$ in the high gust frequency ω_{high}.

METHODOLOGY

The CAD model was developed in ANSYS Design Modeler™ 12.1. The CFX-mesh method of the ANSYS Mesh program (involved in ANSYS Workbench) was employed for commiting the meshes (fig. 4). The fluid dynamic package ANSYS CFX 14.0 was used as solver for the numerical simulations. Pressure distribution around a building is in general important to get correct prediction of the pressure gradients and consequently of the air infiltration through the envelope. Among the available turbulence models, the Shear-Stress-Transport (SST) model, a two equation k-ω based model [16], was imposed. The reason for that is the inclusion of transport effects into the formulation of the eddy-viscosity. This results in a major improvement in terms of flow separation predictions [17]. In addition, other relevant studies have shown a good agreement between SST model and full scale data, better rather than compared with standard k-ε and RNG k-ε models [18].
A period of 30*sec* was assumed to be the total time per run, while a fine timestep of 0,25*sec* was selected. At the inlet of the domain, a logarithmic wind profile was assumed based on the equation (fig. 5):

$$\frac{u}{u_*} = \frac{1}{\kappa}\ln[\frac{z}{z_0} - \Psi_m(\frac{z}{L})] + 2\sin(2\pi\omega t) \quad (3)$$

where u is the wind velocity at height z, u_* is the shear velocity, κ von Karman's constant, z_0 the roughness length and Ψ_m a stability function. The stability function can be evaluated directly from the Monin and Obukhov length L, knowing the flux of sensible heat, or indirectly through simultaneous measurements of air temperature profiles [19]. Under neutral stability conditions Ψ_m and $\frac{z}{L}$ vanish.The second term in the right side of the wind profile represents the wind gustiness frequency ω.
As mentioned above, two different gust frequencies ω have been employed, $\omega_{high} = 0{,}5Hz$ corresponds to the high frequency and $\omega_{low} = 0{,}1\ Hz$ to the low one (fig. 6). Thus, the period of the wind velocity on a certain height is $T_{high} = 2sec$ and $T_{low} = 10sec$ respectively.
The leakage area along each window side represents a 'crack' of 0,8*m* long, simulated by a row of 5 circural 'holes'. The latter are equally ditributed along the window side and their total 'opening area' equals the the leakage area of the relevant crack.
The instantaneous mass flow rate Q_m is solved numerically and extracted. Thus, the instantaneous volumentric flow rates Q_v across the leakage areas are calculated (based on the transient, local density field) for the interval run time (30*sec*) for every case. Assuming that

the dynamic mathematical and physical behavior of the model does not change within an hour, the equivalent air change rate ΣACH_i extrapolated over time $t_{tot} = 1h$ is calculated:

$$\Sigma ACH_i = \frac{3600}{t_{run}} * \frac{\left(\int_0^{t_{run}} Q_v\,dt\right)}{V} \tag{4}$$

where

t_{run} is the total run time per case, means $t_{run} = 30sec$ and
V the volume of the enclosure.

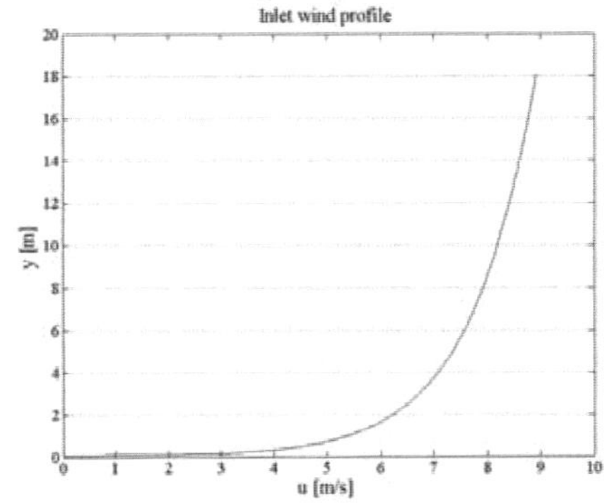

Figure 5. The selected inlet wind profile.

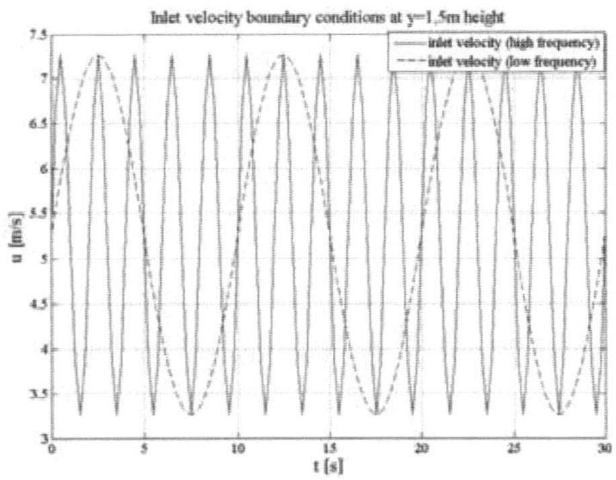

Figure 6. Inlet velocity boundary conditions at the height of $y = 1{,}5m$ as they have been defined for the high frequncy ω_{high} and the low ω_{low}.

RESULTS

The equivalent air change rates ΣACH_i for all the cases S_1, S_2 and S_3 are plotted against the leakage ratio α and shown in figures 7, 9 and 12 respectively. In each graph, two lines appear representing the rates under high and low wind gust frequency.

In the figure 7, it is clear that the strong 'cross ventilation' that takes place in the case of the single space results to relatively very high infiltration rates. Especially under the conditions of the high freguency gustiness the air exchange becomes even more severe. Having employed the assumption of 'one room' with no internal wall, there are no serious resistance against the flow. The ΣACH_i increases respect to the ratio α, and appears the maximum value when the leakages are equally distributed on the windward and leeward façade of the model ($\alpha = 50\%$). The air change rates seem to get lower again when α increases more. The role of the inertia forces of the enclosure appear to be in general weak in the case S_1. However, it would be reasonable to claim that when the leakages are mostly located either on the windward or on the leeward side (the ΣACH_i seems to have fairly symmetric picture), the compressibility of the volume tends to reduce the actual leakage rates, as the model behavior is getting more similar to single-side infiltration (fig. 8).

In the case of S_2 the existence of a relatively tight internal wall (the internal leakages are only 6,25% of the leakages of the envelope) seems to have a dramatic impact (drop) to the air change rates (fig. 9). The ΣACH_i appear (for both the gust frequencies ω_{high} and ω_{low}) to be much lower. Although a 'cross ventilation' takes place even in this case, the quite high level of tightness of the interior element 'activates' the inertia forces of the enclosure, resulting to lower infiltration rates. Especially for the high wind gustiness, it seems that the most unbalanced the leakage distribution, the lowest the infiltration (exfiltration) rates that are caused. When the gustiness of wind is getting more mild (ω_{low}), the influence of the relatively

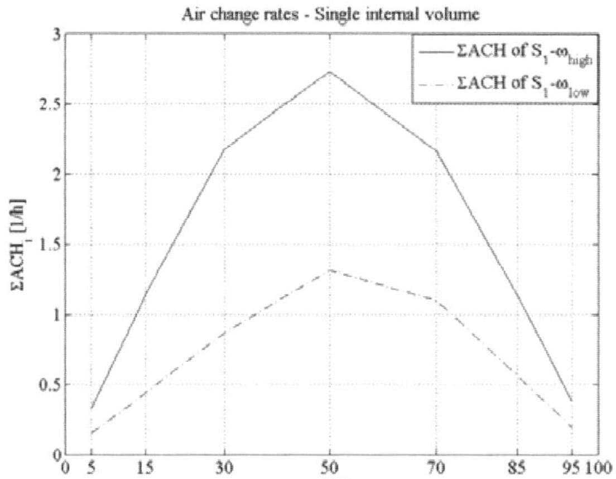

Figure 7. The air change rate for the case of the 'single space'.

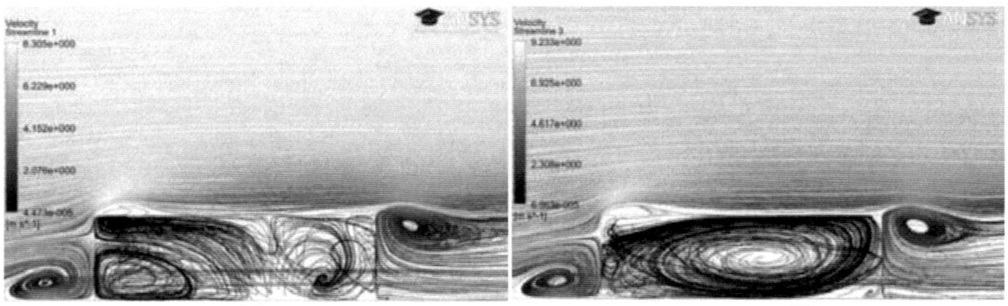
(a) (b)

Figure 8. The velocity steamlines using a symmetry plane (timestep : $t = 2s$). (a) case of S_1-α_{95}-ω_{high}, (b) case of S_1-α_{30}-ω_{high}.

tight wall is becoming even more significant, resulting in air change rates within the acceptable range, as described in buidling regulations [20].

However, a difference can be pointed out when comparing the cases of high and low frequency. Under ω_{high}, the infiltration rate of S_2-α_5-ω_{high} is higher than the 'inverse' case of S_2-α_{95}-ω_{high}, as well as S_2-α_{15}-ω_{high} > S_2-α_{85}-ω_{high}, as well as S_2-α_{30}-ω_{high} > S_2-α_{70}-ω_{high}. In contrast, in the low wind frequency ω_{low}, higher air change rates appear as result of leakage concentration mostly on the windward side. The flow patterns in figure 10(a) show a mild air circulation during ω_{high} even when the size of the leakages on windward is relatively big. The reason could be the size of the internal leakages compared to the external ones on the leeward façade; when they have the same magnitude of order, the inertia forces of the 'second room' seem to increase, preventing the air to flow from the 'first room', even though there is a significant amount of air that enters from the environment to the latter. But when the leeward leakage areas are getting larger, the pressure field in the 'second room' changes, forcing the air to flow out through them.

In case the unsteady wind is more mild (low frequency), the size of the 'inlet' on windward dominates the airflow (fig. 11). The reason could be that the wind gusts are not anymore so strong enough in this case that they would force great amount of air to flow to the 'secnd room', which remains more 'neutral' compared to the 'first room'.

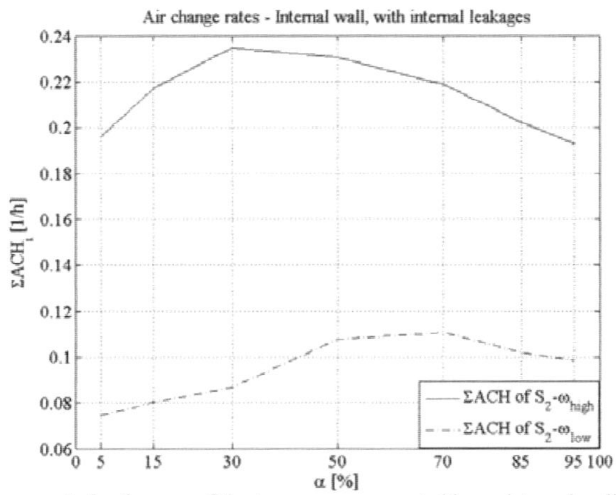

Figure 9. The air change rate for the case of the two rooms, separated by an internal wall where leakages are located on.

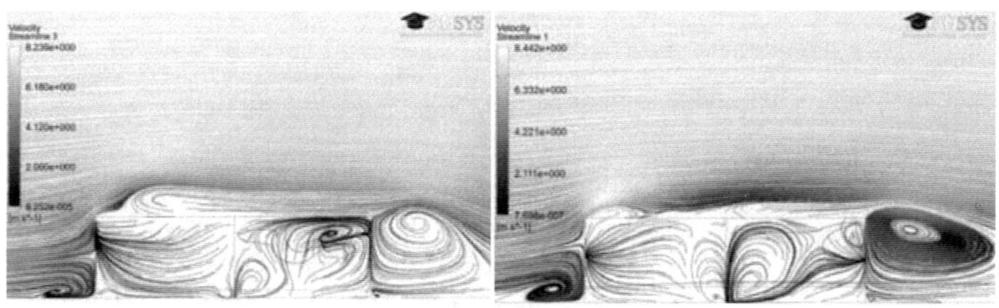

(b) (b)

Figure 10. The velocity steamlines using a symmetry plane (timestep : $t = 2s$). (a) case of S_2-α_{95}-ω_{high}, (b) case of S_2-α_{30}-ω_{high}.

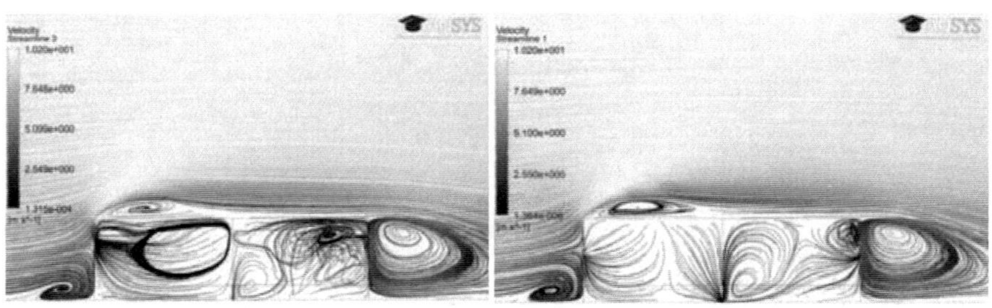

(a) (b)

Figure 11. The velocity steamlines using a symmetry plane (timestep : $t = 3s$). (a) case of S_2-α_{95}-ω_{low}, (b) case of S_2-α_{05}-ω_{low}.

The role of the internal leakages is even more clearly shown in the figure 12 that represents the situation S_3. Assuming that the internal wall is completely tight, the air change rates seem to become even lower compared to S_2 highlighting the importance of controlling the internal

leakage paths. Furthermore, reading the infiltration rates from the perspective of the blower door 'rule-of-thumb' ($ACH_{50} = \Sigma ACH_i * 20$), they fullfil requirements of a 'passive house airtighness level' [21]. The inertia forces of the first 'room' are in this case higher because of the single-side infiltration and the compressibility of the volume decreases. The second 'room' has gotten 'isolated' in this case, so there is not significant air exchange through the leeward leakages. Thus, the most favorable case seems to be when the leakages are mostly concentrated on this façade (α_{05} - leeward).
The ΣACH_i increases with the ratio α in both the wind frequencies studied. The maximum value appears when the leakage area on windward is getting big enough (α_{95}).

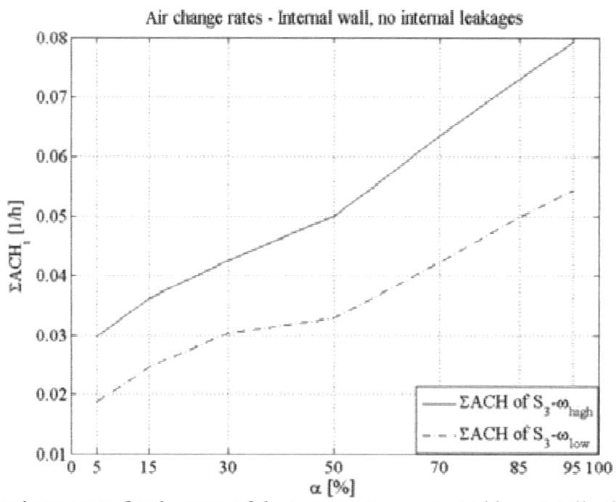

Figure 12. The air change rate for the case of the two rooms, separated by a totally tight internal wall.

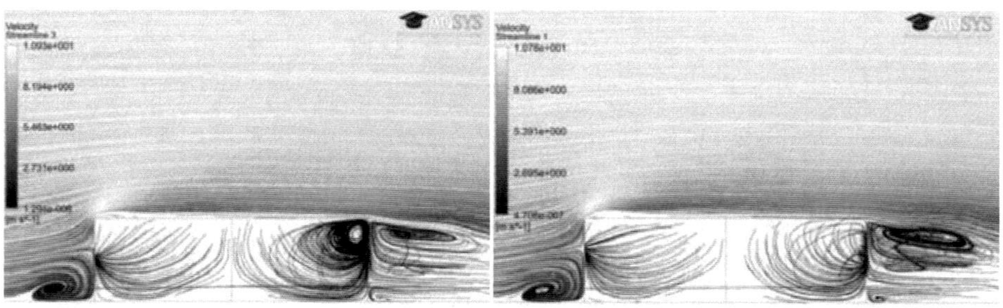

(a) (b)

Figure 13. The velocity steamlines using a symmetry plane (timestep : $t = 3s$). (a) case of S_3-α_{05}-ω_{high}, (b) case of S_3-α_{95}-ω_{high}.

In all the cases (S_1, S_2 and S_3), the impact of the gust frequency ω_i seems to be very important resulting to increased infiltration rates. The normalized difference δ between the air exchanges during the high frequency ω_{high} and those during the low one ω_{low} is drawn against the ratio α [fig. 14].

$$\delta = \frac{\Delta(\Sigma ACH_{high} - \Sigma ACH_{low})}{\Sigma ACH_{low}} \tag{5}$$

In the figure 14, it is clear that the influence of wind gustiness, in both the cases S_1 (single space) and S_2 (internal wall with leakages), is more significant compared to the case S_3

(completely tight internal wall). Again, increasing the tightness of the internal elements, the impact of the wind unsteadiness becomes less important. In addition, in the first cases the normalized difference δ has similar behavior, showing that higher wind gustiness results to even higher infiltration rates when most leakages are concentrated on the windward façade. In contrast, in case S_3 (internal wall with no leakages), the increase of wind frequency has similar results to the ΣACH_i in the whole range of the leakage distribution that is studied. The dynamic characteristics of the wind and the inertia of the enclosure mass seem to influence in an analoge manner the actual air change rates.

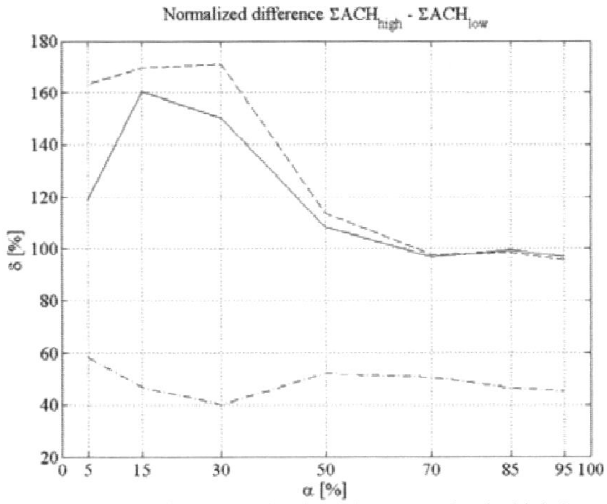

Figure 14. The normalized difference δ between the air exchanges under the high frequency ω_{high} and those under the low frequency ω_{low} is drawn against the ratio α.

CONCLUSION

An one-storey building-model with variable leakage areas on the windward and the leeward side was simulated and studied numerically under unsteady wind conditions. Two wind gust frequencies (ω_{high} = $0{,}5Hz$ and ω_{low} = $0{,}1Hz$) were used to describe the inlet boundary conditions. A ratio α [%] (5% < α < 95%) was employed to describe the leakages located on the windward side as fraction to the total leakage areas of the building. Three different situations of the internal volume were assumed; a single space (S_1), an enclosure with an internal wall with leakages (S_2) and an enclosure similar to the latter but without internal leakages (S_3).

In total 42 cases were solved using the shear-stress turbulent model (SST). The equivalent air change rate ΣACH_i, extrapolated over time $t_{tot} = 1h$, was calculated and was shown against the leakage distribution α. The leakage distribution seems to govern the infiltration rates in case of a strong cross 'ventilation' (S_1). The most severe situation appears to be when the leakages areas on the windward and the leeward façade are of the same magnitude of order. Again, the most 'unbalanced' the way that the leakages are distributed the least air exchanges that take place.

Existence of relatively tight internal walls (S_2) decrease dramatically the leakage numbers. Even though a 'cross ventilation' takes place even in this case, the quite high level of tightness of the interior element 'activates' the inertia forces of the enclosure (of the 'front' room). Fullfilling high tightness of the internal elements (S_3), the air change rates decrease even more, reaching almost passive house airtightness standards (even under more severe

wind gustiness). In addition, in the latter case, it seems to be of relatively high importance to eliminate as possible the leakages on the windward façade (according to the main wind direction of a location).
It would be reasonable to claim that internal leakages seems to be a major parameter towards the demand of decreasing the infiltration rates. Gustiness of wind is also a critical factor that results to higher leakage numbers. However, increasing the tightness of the internal elements, the impact of the wind unsteadiness becomes less severe. To determine even further the influence of wind frequency, a graph that represents the normalized difference between ΣACH_{high} and ΣACH_{high} is defined and is shown against the ratio α.
The study sets up issues regarding the uncontrolled leakages on the building envelope. The detection of leakages and their distribution should migh be considered as critical factor. Furthermore, internal leakages seem to play an important role towards the nearly-energy-zero building target. Further research needs to be done, in order to investigate the connection between internal and external leakages in a detail way.

REFERENCES

[1] Shaw C.-Y. 1981. *A correlation between air infiltration and air tightness for houses in a developed residential area*, National Research Council Canada, Ottawa.

[2] Månsson L.G. 2002. *IEA ECBCS Annex 27 – Evaluation and demonstration of domestic ventilation systems – Simplified tools Handbook,* FaberMaunsell Ltd, United Kingdom.

[3] Cummings J.B., Withers C.R., Moyer N., Fairey P., McKendry B. 1996. *Uncontrolled air flow in non-residential buildings – Final report*, Florida Energy Office - Department of Community Affairs, Tallahassee, Florida.

[4] Hill J.E., Kusuda T. 1975. *Dynamic characteristics of air infiltration*, ASHRAE Trans. 81 (1) 168–185, ASHRAE, New York.

[5] Sherman M. H. 1986. *Estimation of infiltration from leakages and climate indicators*, Energy and buildings Vol. 10, 81–86.

[6] Haghighat F., Brohus H., Rao J. 1999. *Modelling air infiltration due to wind fluctuations – a review,* Building and environment Vol. 35, 377–385.

[7] Haghighat F., Rao J., Fazio P. 1991. *The influence of turbulent wind on air change rates – a modelling approach*, Building and environment Vol. 26, 95–109.

[8] Haghighat F., Rao J., Riberon J. 1992. *Modelling fluctuating airflow through large openings*, 13th AIVC conference, Nice, France.

[9] Sherman M.H. 1995, *The use of blower-door data*, Indoor Air Vol. 5, 215–224.

[10] Dubrul C. 1988. *Inhabitants behavior with respect to ventilation. Technical note 23*, Air infiltration and Ventilation Centre AIVC, pp. 63.

[11] Chan W.R., Price P.N., Sohn M.D., Gadgil A.J. 2003, *Analysis of U.S. Residential Air Leakage Database*, LBNL Report Number 53367, Lawrence Berkeley National Laboratory, Berkeley, U.S.A.

[12] Guha T.K., Sharma R.N., Richards P.J. 2011. *Internal pressure dynamics of a leaky building with a dominant opening*, Journal of Wind Engineering and Industrial Aerodynamics, Vol. 99, 1151-1161.

[13] Kalamees, T., Korpi, M., Eskola, L., Kurnitski, J., Vinha, J. 2008. *The distribution of the air leakage places and thermal bridges in Finnish detached houses and apartment buildings*, Proceedings of the 8th Symposium on Building Physics in the Nordic Countries NSB2008, Copenhagen, 1095-1102.

[14] Kraniotis D. Thiis T.K., Aurlien T., 2011. *Behavior of leakages exposed to dynamic wind loads. A numerical study using CFD on a single zone model*, Proceedings of the 32nd AIVC and 1st TightVent Conference, Brussels, Belgium, 80-82.

[15] Aaltonen A., Lähdesmäki K., Vinha J. 2011, *Air tightness of structural elements and internal air leakages in a multi-apartment building*, Proceedings of the 9th Nordic Symposium on Building Physics NSB2011, Tampere, Finland, 79-86.

[16] Menter F.R. 1994. *Two-equation eddy-viscosity turbulence models for engineering applications*, AIAA Journal, Vol. 32, no 8. 1598–1605.

[17] ANSYS Inc. 2011. *ANSYS CFX-Solver Modeling Guide,* ANSYS CFX 14.0 help manual.

[18] Guha, T.K., Sharma, R.N., Richards, P.J. 2009. *CFD modeling of wind induced mean and fluctuating external pressure coefficients on the Texas Technical University building*, EACWE 5.

[19] Grimenes A.A., Thue-Hansen V. 2004. *Annual variation of surface roughness obtained from wind profile measurements*, Theoretical and Applied Climatology V. 79, 93-102.

[20] Direktoratet for byggkvalitet 2010. *Byggteknisk forskrift – TEK 10, FOR 2010-03-26 nr 489: Forskrift om tekniske krav til byggverk (Byggteknisk forskrift)*, Oslo, Norway.

[21] Feist W. 2010. *Active for more comfort – The Passive House*, International Passive House Association, Darmstandt, Germany.

THE DISCHARGE COEFFICIENT OF A CENTRE-PIVOT ROOF WINDOW

Ahsan Iqbal[1], Alireza Afshari[1], P.V.Neilsen[2] and P. Heiselberg[4]

1 Danish Building Research Institute
Aalborg University
Dr Neergaards Vej 15
2970 Hørsholm, Denmark
**Corresponding author: ahi@sbi.aau.dk*

2 Indoor Environmental Engineering
Aalborg University
Sohngaardsholmsvej 57,
9000 Aalborg, Denmark

ABSTRACT

Windows are normally used in residential buildings for natural ventilation. There are several possibilities of selecting a window for a residential building. The centre pivot roof window is often used in residential buildings of Nordic countries. This study is about the performance of this type of window. Accuracy in estimation of airflow is the key parameter for modelling and designing of naturally ventilated buildings. The flow through the window is usually described by the orifice flow plate equation. This equation involves the discharge coefficient. The discharge coefficient is one of the main parameters used to calculate the airflow rates. The building and energy simulation tools use (often) a constant value of discharge coefficients. The constant value of discharge coefficient leads to deceptive airflow estimation in the cases of windows with movable flaps. The literature for the discharge coefficient of façade window with movable flap is available. However, the discharge coefficient of roof windows, especially centre pivot roof window has not been studied much. The object of this paper is to study and evaluate the discharge coefficient of the centre pivot roof window. Focus is given on unidirectional flows i.e. inflow and outflow. CFD techniques are used to predict the airflow through the modelled window. Analytical orifice flow equation is used to calculate the discharge coefficient. The k-ε turbulent model can only predict the fully developed turbulent flow, that's why the results are only valid for higher pressure difference. Results are compared with experimental results. It is concluded that the single value of the discharge coefficient leads to ambiguous estimation of airflow rates. The discharge coefficient decreases with increase in flap opening area. The discharge coefficient also depends upon the flow direction

KEYWORDS

Centre pivot roof window, discharge coefficient, CFD

INTRODUCTION

Building simulation and calculation programs are not generally capable of accurate calculation of airflow rates through either intentional or unintentional openings. The existing programs either assume to be fixed air-change rates like Be06 or calculate air-change rates by applying empirical equations like in the programs, Contam and IDA. Especially, the simple methods applying empirical techniques can lead to unreliable results [7]. The amount of air entering through natural ventilation (or in hybrid systems) is extremely difficult to predict accurately as airflow depends on unknown wind and buoyant effects. In practice, the orifice flow equation is used to compute the airflow through the intentional openings and windows. The discharge coefficient (C_D) in this equation is usually taken as constant. The constant value of the discharge coefficient is valid only for constant opening areas [12], [7]. Hence, the use of constant value of discharge coefficient for operable windows leads to deceptive results. The exactness of the coefficient can have a significant impact on the ability of a mathematical model to predict the airflow rates [7], [9].

There is a need of evaluation of discharge coefficient of operable (i.e. with flap) windows. Operable windows are broadly used in residential buildings for ventilation. Scientific literature on façade windows is somehow available. However, the literature on roof windows (especially center-pivot roof window) is not much discussed. This paper focused on the discharge coefficient of a center-pivot roof window.

The computational fluid dynamics (CFD) techniques are used to evaluate the C_D of the centre pivot roof window. A commercial CFD package i.e. Star CCM, is used for this purpose. The use of CFD allows direct evaluation of various parameters [7], like ventilation rates in buildings, wind pressure coefficients etc. Moreover, separate evaluation of wind and buoyancy effect on ventilation is possible. The predictions of airflow rates in natural ventilation systems are very sensitive to geometry of the openings. If the area and the C_D of openings are known then airflow rates can be well modelled [7],[9]. The CFD results are also compared with the onsite experimental results.

BACKGROUND

Airflow rate through the opening is the integral of velocity over the opening area i.e. $q = \int vdA$ [14]. In practice it is difficult to do this integration. Therefore an alternate way has to be adopted. The airflow passet through the opening acquires the shape of a jet [5]. Therefore the volume flowrate at the vena contracta of the jet is the actual volume flow rate through the opening. Area (A_c) of vena contracta is defined as:

$A_c = C_c A$; Where C_c is the contraction coefficient and A is the area of opening.

Velocity (v_c) in the vena contracta is defined in terms of a theoratical (frictionless flow) velocity as:

$v_c = C_v v_{th}$; Where C_v is the contraction coefficient and v_{th} is the theoratical velocity.

The velocity in the vena contracta is constant (chocked flow) therefore the flow rate (q_c) in vena contracta is:

$q_c = A_c v_c = C_c C_v A v_{th}$

The theoratical velocity (v_{th}) is mainly due to pressure difference (ΔP) whereas; the product of velocity coefficient and contraction coefficient is called the discharge coefficeint (C_D).

$v_{th} = \sqrt{\frac{2\Delta P}{\rho}}$ and $C_D = C_c C_v$

The cotractraction coefficient and velocity coefficient are mostly discussed in literature but in practice, especially with operable windows, they are exteremly difficult to estimate. Therefore, the discharge coefficient is usually used. From above mentioned corelations, the airflow through an opening is defined as:

$$Q \equiv C_D A \sqrt{\frac{2\Delta P}{\rho}} \quad (1)$$

This is generally referred as orifice flow plate equation.The discharge coefficeint can also be defined as a ratio between the actual airflow rates and the ideal airflow rates. The discharge coefficient is very important coefficient for estimation of correct airflow rates through the window. The colossale literature is available for estimation of C_D. However, predominantly a constant value of discharge coefficient is used in practice. These constant values are derived from the data used to estimate the flowrate in pipes [12].

LITRATURE SURVEY

Bot performed [11] a full scale measurements of flowrate through the one side mounted casement windows (façade window). The authour defines the resistance coefficient/friction factor in terms of aspect ratio of the window and opening angle. *The resistance coefficient(ζ) is a coefficient that defines the pressure drop due to friction in the opening and flow.*

Theoratically, it is $\Delta P_{fr} = \frac{1}{2}\zeta\rho v_c^2$ [i]. The authour uses the corss sectional opening area, and from the results of the research it can be concluded that the overall discharge coefficient of the top hinged window increases with the increase in opening angle.

P. Heiselberg [12] uses the minimum opening area to estimate the discharge coefficient of a façade window with movable flap. Experimental results showed that the discharge coefficient is not constant for different flap opening angles. The authour concludes that the value of discharge coefficient is approaching to 1 with the decrease in flap opening angle (consequently with the minimum opening area). Whereas, only for large opening angles the value of 0.6 can be used as a discharge coefficient of a window with movable flap.

Andersen [3],[4] discussed theoretically, friction and contraction coefficients of openings with movable flaps. The authour used the artificial as well as pure resistance coefficients, along with artificial and real opening angle, to calculate the contraction coefficient. The authour concluded that (for centrally hinged flap) the contraction coefficient decreases with increase in the flap opening angle. Whereas, resistance coefficient (both artificial and theoratical) increases with the increase in flap opening angle. The real opening angle is dependent on aspect ratio of the window therefore, the contraction coefficient, and consequently the discharge coefficient, is also dependent on the aspect ratio of the window. For sharp edged openings, the discharge coefficient is 0.61 [3], [4], [5].

Hult [7] determines the discharge coefficient of the façade window using CFD. The authour concluded that the discharge coefficent of a façade window is reliant upon aspect ratio and window opening angle. However, for larger opening angle it approaches to the commonly used value of 0.6. According to the research, the CFD results suggest that the actual flow through a top-pivoted window may be as much as twice the flow predicted by EnergyPlus.

METHODS

CFD techniques was used to test the dependence of discharge coefficient (C_D) on the flap opening angle(α). The CFD domain was defined in such a way that on right side of the domain, the outflow through the window could be examined. Whereas, the inflow through the window could be examined from the left side of the domain. For reducing the processing time, only half part of the window was examined by using symmetric boundary condition. Height and width of the domain was selected in such a way that the size of the domain had no influence on the local velocities and the pressure distribution around the window. The model room was defined as shown in Figure 2. InVent was the opening in the model room with the window and OutVent was the opening without any window. Both InVent and OutVent were on the roof with slope/pitch of 45°. The window and flap geometry were kept simple because the details, i.e. minor bends on flap, of window has insubstantial affect on overall discharge coefficient [2].

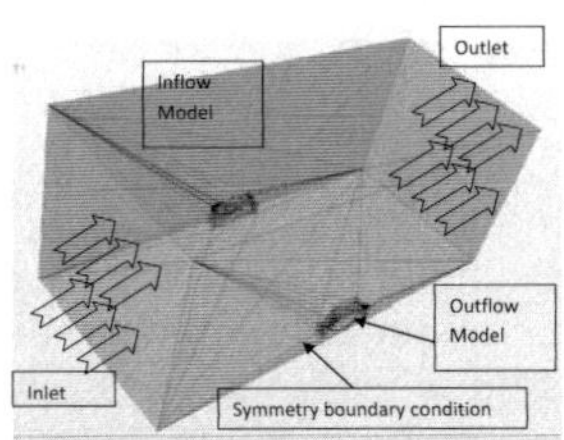

Figure 1 CFD Domain

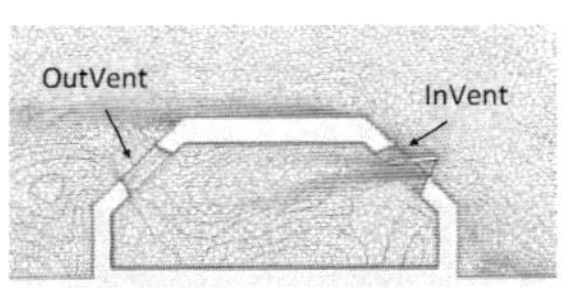

Figure 2 Model room with InVent and OutVent

The polyhedral meshing scheme was used with 8 prism layers mesher at the boundaries. The base size was 1 m. The prism layers were 8% of the base cell size. The surface growth rate was 1.3. Allowable skewness for cells was 85°. Several parts of the domain had customised surface mesh size to ensure proper mesh quality. The Inlet was the velocity-inlet, the Outlet was the pressure-outlet. The domain top, left and the

[i] *The definition is from theoratical books and from (ANDERSEN 1996)*

right boundaries were symmetric boundaries and all other boundaries were walls with no slip conditions.

Physics:

A body interacts with the surrounding fluid through pressure and shear stress, and the resultant force in the direction of stream is the drag force. The drag coefficient D_f is used to define the drag force when the detailed information about pressure and shear stress is not known i.e. it is a ratio between the drag force and the wind pressure force [14]. The minimum number of cells was selected in such a way that by increase in the number of cells, there is no effect on the D_f of the model room. This means that the further decrease in cell sizes had no effect on the local velocity and the pressure distribution around the building.

The "Two layer realizable k-ε turbulent" model was used to compute the airflow and its physical behaviour [9]. The working fluid was incompressible ideal gas. The flow was 3D steady state. Flow and energy were both modelled using the segregated approach. The second-order upwind discretisation scheme was used for both flow and energy. Under relaxation factors for velocity, pressure and energy were 0.7, 0.3 and 0.9 respectively. Inlet condition (Inlet -Figure 1) for the turbulent kinetic energy ($k_o = 1.5(T_iU_o)^2$) and the dissipation rate ($\varepsilon_o = k_o 1.5/l_o$) were according to Nielsen [13] recommendations. Where, T_i is the turbulence intensity and was taken as 4% with inlet temperature of 293K. U_o is inlet velocity in {m/s}. l_o{m} is length scale and was taken as one-tenth of the height of the inlet. The inlet velocity was selected in such a way that for each simulation the airflow through the window was fully developed turbulent flow. The C_D value for fully developed turbulent flow does not vary with Reynolds number [9]. The equation (1) was used to find out the C_D for the window (Figure 4).

The flowrate at InVent (Figure 2) was estimated by the flowrate at the OutVent (Figure 2). The flow Q{m³/s} at OutVent was calculated by integration of velocity over the area of OutVent i.e. the face area of the cells at the interface (OutVent and external region) times the perpendicular component of the velocity i.e.

$$Q \equiv \sum_{i=1}^{n} A_i v_i = \frac{1}{\rho}\sum m_i \qquad (2)$$

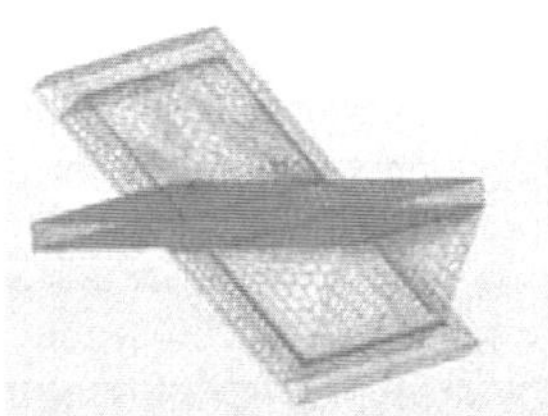

Figure 3 Outside pressure (average)

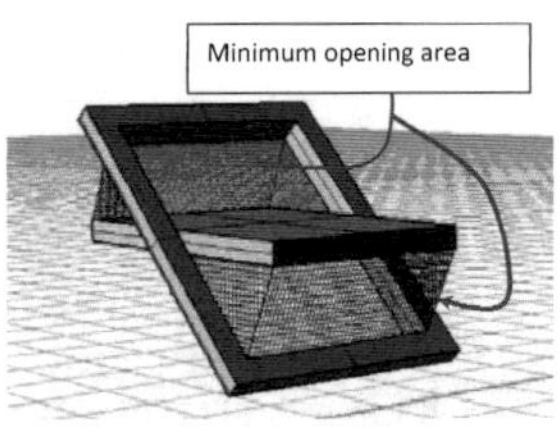

Figure 4 Centre-pivot roof window and minimum opening areas

Where, A_i{m²} is the face area of cell at the interface and v_i{m/s} is the perpendicular (to A_i) component of the velocity. The airflow rate can also be calculated by the mass flow rate divided by the density as shown in Equation (2). The density ρ is constant i.e. 1.2 kg/m³, m_i is mass flow rate in kg/s.

Pressure difference:

To measure the pressure difference (ΔP) across the InVent one probe was measuring the pressure inside the room (in the centre of the room). The outside pressure was an area weighted average of outside pressure at the InVent opening as shown in Figure 3.

Opening area:

The C_D was also dependent on the opening area. Therefore it was evaluated for two different opening areas. One was the minimum opening area (A_{min}). The minimum opening area is shown in Figure 5. The sum of two minimum opening areas is the total minimum opening area (see Figure 5). For flap opening angles of 50° and greater, the minimum opening area is the sum of two face cross sections of the window.

Another way to define the opening area is the gross face area (A_{face}) of the opening i.e. 1.14 x 1.4.

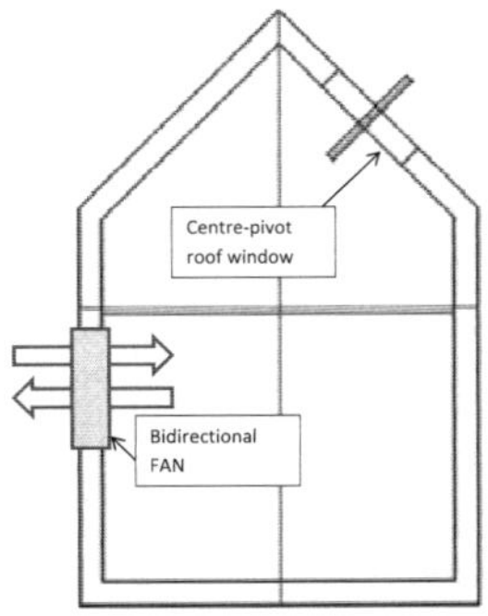

Figure 5 Schematic of on-site measurements

On-site measurements:

On-site measurements were performed in the Energy Flex House (EFH) of the Technological Institute of Denmark (Copenhagen). The house was equiped with the VELUX centre-pivot roof window. Figure 5 shows the schematic diagram of the measurements setup. Only outflow measurements were compared in this study, because this study is mainly focused on CFD. The measurements were performed only for validation purpose. The flow through the window was the sum of fan flow and infiltration/exfiltration. The discharge coefficient was calculated by using equation (1) and minimum opening area. The outside pressure was the area weighted average pressure around the window. Inside pressure was the average inside pressure of the house.

The centre-pivot roof window in the EFH was fully automatic and it was not possible to open the window more than 17°. Therefore, discharge coefficient only for very small ($\alpha<17°$) opening angles was measured.

RESULTS

Figure 6 illustrate the discharge coefficients ($C_{D,min}$) of a centre-pivot roof window when minimum opening area is used in equation (1). The $C_{D,min}$ of inflow through the window (flow from outside to inside) is represented by the blue line. The $C_{D,min}$ of outflow (flow from inside to the outside) is represented by the red line. The green line is $C_{D,min}$ that is obtained by the on-site measurements. As mentioned earlier, it was not possible to measure (onsite) the discharge coefficient for angles greater than 17°. Therefore, measurement data is available only for 17°, 14°, 9.3° and 4.6°. It should be noted that the minimum opening area is taken from the manufacturer catalogue.

On the contrary, with available computing power, it was very time consuming to go below 15° of opening angle. That's why the CFD simulation was performed for 15°, 25°, 35°, 45°, 50° and 90°. It should be noted that the minimum opening areas were calculated through the CAD drawings. Figure 7 illustrate the discharge coefficients ($C_{D,face}$) when the gross face opening area is used in equation (1). The blue line shows the $C_{D,face}$ when flow is directed inward. Whereas, the redline is the $C_{D,face}$ when flow is directed outward.

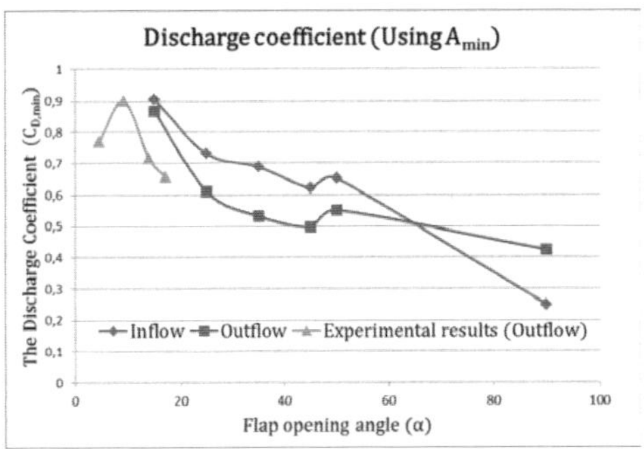

Figure 6 Discharge coefficient of the window using minimum opening area

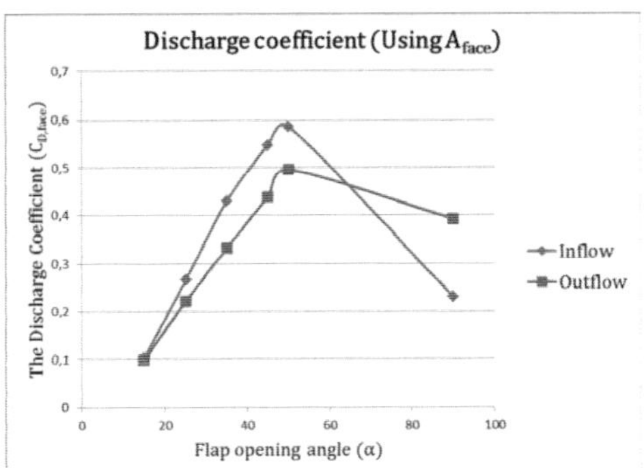

Figure 7 Discharge coefficient of the window using face area

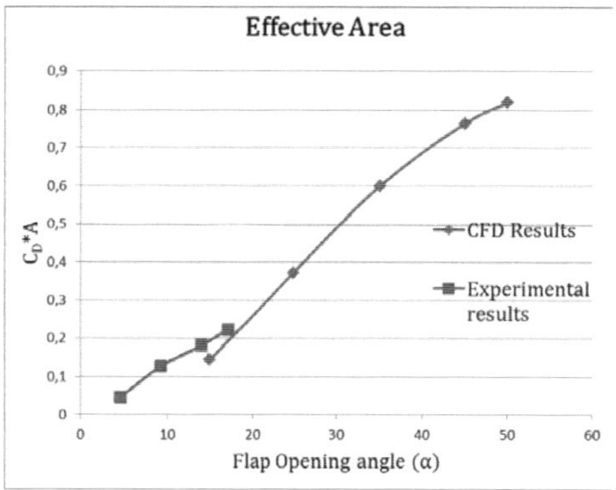

Figure 8 Comparison of CFD results with experimental results

DISCUSSION

From Figure 6, apperantly there is a difference between the $C_{D,min}$ values evaluated by the measurements and by the CFD. One of the reason is ideal versus real condition. However, the trend of decrease in $C_{D,min}$ is almost the same. Ergo, the problem might also be in calculation of the minimum opening area. Therefore, another quantity, C_D x A (so called effective area) of both, measurements and the CFD are compared. This comparision is shown in Figure 8. The difference in measurements and the CFD results are now minimum. The CFD results are in sound concurrence with the measurements. Hence, the realizable k-ε turbulent model predicts the flow (across the window) in a good agreement with the measurements.

The $C_{D,min}$ curve represents the change in the discharge coefficient due to (mainly) flow effect. Whereas, the $C_{D,face}$ curve represents the change in the discharge coefficient due to, both, flow and area effects.

The $C_{D,min}$ (as shown in Figure 6) decreases with the certain pattern until it reaches to 50° of the flap opening. At this point the A_{min} shifted from the position shown in Figure 4 to the face cross sectional area. Then the pattern of decrease in $C_{D,min}$ also changes. This change in $C_{D,min}$ is due to the change in A_{min} and due to the fact that pressure field on the roof surface (closed to the window) is disturbed.

In the caseof inflow through the window, the flow in each section of the window is not evenly distributed, except for very small opening angles. Flow in the lower section is much higher then in the upper section. At the angle of 90° there is no flow in the upper section of the window. In the case of outflow through the window, the flow in each section of the window is somewhat evenly distributed. Therefore, the $C_{D,min}$ inflow is higher than the $C_{D,min}$ values of outflow. This phenomenon negates the assumption of stagination condition at the inlet in determination of C_D. However, the slope/trend of the outflow curve is same as of inflow curve. After 50° of flap opening, the slope of outflow become lower than the inflow. Therefore, the $C_{D,min}$ at 90° of flap opening of outflow direction was higher value than that of inflow direction. The reason for difference in $C_{D,face}$ curves (for inflow and outflow direction) is also the same.

For more concrete conclusion, the results have to be evaluated for several openings with different aspect ratios. However, these results are only subjected to the particular window type and for natural ventilation caused by wind effect.

CONCLUSION

It is concluded that the k-ε turbulent model predicts the flow in sound agreement with the measurements. It is also concluded that in the orifice flow plate equation, the discharge coefficient is not a constant quantity. The $C_{D,min}$ decreases with flap opening angles. The trend of decrease in $C_{D,min}$ changes after 50° of flap opening angle. Likewise, the discharge coefficient also depends upon the flow direction. The $C_{D,min}$ is higher for inflow and the $C_{D,min}$ is lower for outflow. For higher opening angles (e.g. higher than 65°) the criteria is swapped.

ACKNOWLEDGEMENTS

This paper is based on research conducted in a PhD project, which is a part of the Strategic Research Center for Zero energy Buildings at Aalborg University and financed by Velux A/S, Aalborg University and The Danish Council for Strategic Research (DSF), the Programme Commission for Sustainable Energy and Environment. Furthermore, the authors gratefully acknowledge the assistance of Technological Institute of Denmark during measurements in Energy Flex House.

REFERENCES

[1] Fundamentals, inch-pound edition. 2. printing ed. Atlanta, Ga.: American Society of Heating, Refrigerating and Air-Conditioning Engineers; 2004. Rygtit.: 2001 Fundamentals; ASHRAE handbook ; 2001.

[2] Computational fluid dynamics in ventilation design REHVA guidebook no 10. - International Journal of Ventilation(- 3):- 291.

[3] ANDERSEN, K. (2002). Friction and contraction by ventilation openings with movable flaps. Room Vent 2002 (8th International Conference on Air Distribution in Rooms), Copenhagen. .

[4] ANDERSEN, K. (1996). Inlet and outlet coefficients - A theoretical analysis. ROOMVENT, 5th International Conference on Air Distribution in Rooms, Yokohama.

[5] ANDERSEN K. Theoretical considerations on natural ventilation by thermal buoyancy. 1995;101(2):1103-17.

[6] Baturin VV. Fundamentals of industrial ventilation. 8th ed. Pergamon press; 1972. .

[7] E. Hult, G. Iaccarion and M. Frischer, "Using CFD simulation to improve the modeling of window discharge coefficients," in SimBuild, Madison, 2012.

[8] Chiu Y, Etheridge DW. External flow effects on the discharge coefficients of two types of ventilation opening. J Wind Eng Ind Aerodyn 2007 4;95(4):225-52.

[9] A. Iqbal, A. Afshari, P. V. Nielsen and P. K. Heiselberg, "Numerical predictions of the discharge coefficient of a window with," in Healthy Buildings, Brisbane, 2012.

[10] Heiselberg P. Modelling of natural and hybrid ventilation. Aalborg University; 2006. .

[11] de Jong, T., & Bot, G. P. A. (1992). Flow characteristics of one-side-mounted windows. Energy and Buildings, 19(2), 105-112.

[12] Heiselberg P, Svidt K, Nielsen PV. Characteristics of airflow from open windows. Build Environ 2001 8;36(7):859-69.

[13] P. V. Nielsen, "Specification of a two-dimentional test case," International energy agency, energy conservation in buildings and comunity systems, p. Annes 20: Airflow pattern withing buildings, 1990.

[14] Munson, B. R., Young, D., & Okiishi, T. (1999). Fundamentals of fluid mechanics John Wiley & Sons.

EFFECT OF MEASUREMENT LOCATION OF AIR AIR-TIGHTNESS PERFORMANCE ON APARTMENT UNITS IN KOREA

Cheol-Woong Shin[1], Yun-Gyu Lee[*1]

1 Korea institute of Construction Technology
283, Goyangdae-Ro, Ilsanseo-Gu, Goyang-Si
Gyeonggi-Do, Korea
**yglee@kict.re.kr*

ABSTRACT

The purpose of this study is to evaluate the effects of measuring position of air-tightness performance in Flat-type and Tower-type apartments. Air-tightness performance was measured on entrance doors using the Blower Door System in accordance with CAN/CGSB 149 and on the windows using air-tightness Measuring (KNS-serise) in accordance with JIS A 2201. The air-tightness test was performed with newly builted apartments in 2011. The air-tightness test results on location were converted into ACH50 for comparison. The result on the windows was higher compared to the result of the entrance door. According to the results, the entrance door has greater air-tightness performance than the window which has less air-tightness. In conclusion, it is appropriate to take measurements of air-tightness performance on the windows of apartments in Korea

KEYWORDS

Airtightness, Field studies, Measuring Position, Apartments

INTRODUCTION

Recently, there is an increase in air-tight and high insulation construction material being used in order to save energy. Accordingly, the airtightness of materials applied to the exterior of buildings is becoming increasingly important. Korea's apartment housing are generally constructed using a wet-construction method into flat-type and tower-type buildings. Compared to the probabilty of leakage or infiltration from gaps in the building structure, the probability of leakage or infiltration from windows, front doors, construction material itself or copula is much higher. This study has used the airtightness performance measurement method that is most used worldwide, the fan pressurization / depressurization method, on Korean apartment buildings to measure the airtightness performance of houses from various locations within the house and analyzed the airtightness performance based on the position of the measurement equipment.

MEASUREMENT EQUIPMENT AND METHOD

Air-tightness Measurement Equipment

This study applied "ISO 9972 (Thermal performance of buildings - Determination of air permeability of building-Fan Pressurization method)" as the air-tightness measurement standard. Table 1 show on site pictures of the equipment used when measuring air-tightness. The two methods shown here are similar to the measurement method where the air volume under fixed differential pressure levels is measured to measure air-tightness performance. The Blower Door System is a measurement equipment that is generally installed on the entrance door and measures the air volume depending on the rotation count of the fan and shows air-tightness performance results in accordance to ISO 9972[2], ASTM E779 and CGSB-149. KNS-serise is a measurement equipment that is installed on small windows in the kitchen or entrance door which measures the air volume by the differential pressure sensor installed on the whole blower and shows air-tightness performance results in accordance to Japan's JIS A 2201[3]. Blower Door System was installed on the entrance door while, the KNS series were installed on the small windows in kitchens. The depressurization measurement method was in accordance to ISO 9972.

Content	Blower Door System	KNS-serise
Equipment		
Measurement Flow Rate	34 ~ 10,700 m^3/h	1 ~ 9,999 m^3/h
Measurement Pressure	± 1,250 Pa	± 147 Pa
CODE	ISO 9972, ASTM E779, CGSB-149	ISO 9972, JIS A 2201
Equipment position	entrance door	Window

Table 1. Airtightness Measurement Equipment

Description of the apartment

Air-tightness performance was measured in 36 apartment houses among 6 complexes that were to be lived in starting 2011. Complexes A, B, C and D had mechanical ventillation equipment while complexes E and F had natural ventilation equipment. The table below shows an outline of the air-tightness measurement subject complexes.

Content	A (Daejeon)	B (Pusan)	C (Bucheon)	D (Seoul)	E (Suwon)	F (Daejeon)
Measured units	6	6	6	6	6	6
Floor space	$132m^2$	$84m^2$	$84m^2$	$115m^2$	$51m^2$	$84m^2$
Number of floors	21	23	17	26	15	22
Ventilation systems	mechanical ventilation	mechanical ventilation	mechanical ventilation	mechanical ventilation	natural ventilation	natural ventilation

Table 2. Outline of the apartment buildings subject to measurement.

Comparative analysis of test result

Generally, air-tightness is expressed as CFM50, ACH50, EqLA, ELA and etc. in literature in and out of Korea. There are small differences in measurement equipment, standard pressure differences and detailed application methods within different literature. However, in general they have similar concepts or rules based on the differential pressure measurement method.
The Blower Door System can find ACH50, EqLA (10Pa), ELA (4Pa) and more which show the rate of ventilation at differential pressure 50Pa. On the other hand, KNS serise shows the amount of ventilation (Q) at the standard differential pressure 9.8Pa measured at JIS A 2201's standard differential pressure levels. This study used Equation (1) to convert the amount of ventilation measured by the equipment stated above into the amount of ventilation at 50Pa to find ACH50 (h^{-1}) per house.

$$Q = C(\Delta P)^n \quad (1)$$

RESULTS

The comparison of average air-tightness performance of Korean apartment buildings with air-tightness performance of major developed countries[5],[6] can be shown as in the figure below. The average ACH50 of the 6 complexes subject to the air-tightness test was 2.89 h^{-1} when the Blower Door System (BD) was used and 3.12 h^{-1} when the KNS series (KNS) was used. Compared to the air-tightness performance of major developed countries such as the US, Japan, France, Czech Republic and etc. the results were relatively satisfactory. However in comparison to countries such as Norway, Germany, Finland and Canada, the results were somewhat low.

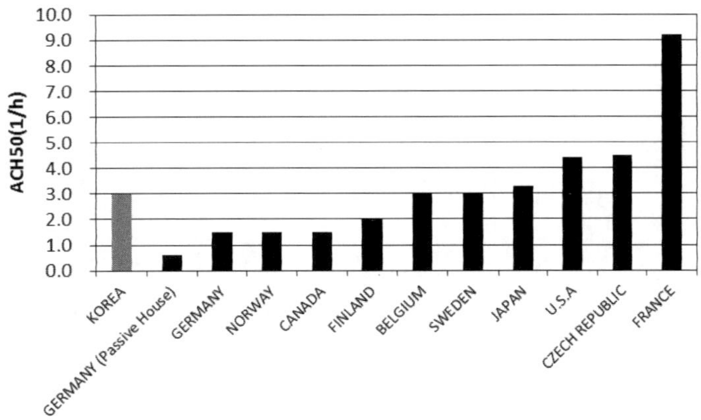

Figure 1. ACH50 Measurement results (Three-sided exposure appear on the outside)

One representative flat surface type house (2 or 3 sides exposed to the outside) from each of the subject apartment complexes was chosen and one house from each of the low, middle and top floors in the same vertical line were chosen. A total of 33 houses were measured and after comparing and analyzing these houses, the toal air-tightness measurement value was converted into ACH50 and is shown in Fig. 1. Because the areas of the representative houses of each complex are different, the air-tightness performance measurement values were different. Comparing the air-tightness performance measurement values of the low, middle and top floors, it could be seen that air-tightness was generally higher or similar in the top floors compared to the low floors and a few houses showed different results.

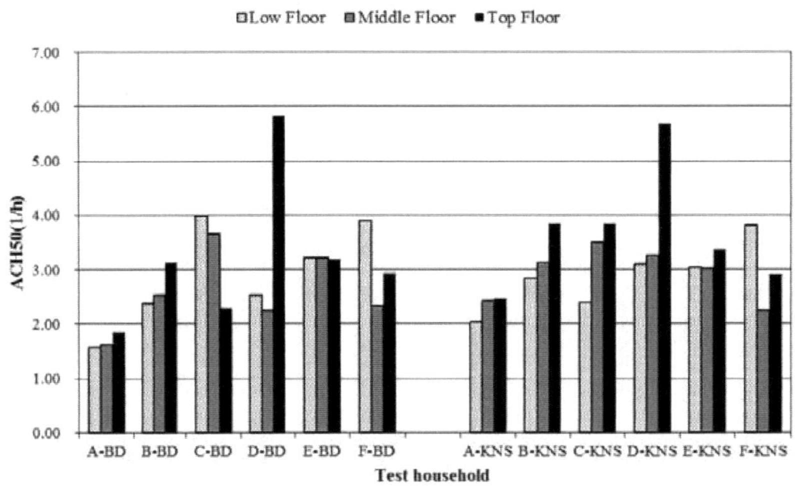

Figure 2. ACH50 Measurement results (Two-sided exposure appear on the outside).

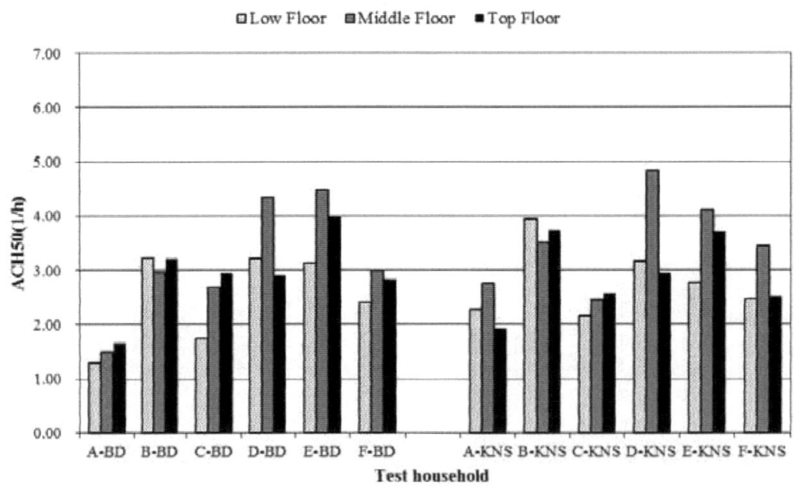

Figure 3. ACH50 Measurement results (Three-sided exposure appear on the outside)

In addition, in apartment D, whrere houses had two sides exposed to the outside, the top floor measured using the BD was extremely high compared to the other measurement values. This is most likely because the top floor has the highest altitude and is exposed to the outside through the ceiling. The average ACH50 values of the different houses based on the installation position of the measurement equipment is shown in the table below. Comparing the ACH50 values of the BD installed in the entrance door and the KNS series installed in the kitchen windows, it can be seen that the KNS series values were higher in four complexes compared to the BD values. The BD ACH50 for apartments E and F had a tendency of being about 0.06 ACH50 h^{-1} higher than the KNS series values but this difference is considered to be insignificant.

Content	A	B	C	D	E	F
Measuring at entrance door (BD)	1.68	2.69	3.31	3.54	3.21	3.06
Measuring at Window (KNS)	2.31	3.28	3.25	4.03	3.15	3.00

Table 3. Comparison of average ACH50 (Two-sided exposure appear on the outside)

Content	A	B	C	D	E	F
Measuring at entrance door (BD)	1.49	3.14	2.46	3.49	3.88	2.75
Measuring at Window (KNS)	2.32	3.74	2.40	3.65	3.54	2.82

Table 4. Comparison of average ACH50 (Three-sided exposure appear on the outside)

CONCLUSION

This study analyzed the air-tightness performance of 36 houses in 6 apartment complexes built in Korea depending on the location of the air-tightness measuring equipment and the following conclusion was derived. The air-tightness performance of Korean apartments is on average ACH 50 3.01 h^{-1}, indicating that it is similar to the North European countries with relatively weak external environments.

Of the 7 apartment complexes, after carrying out air-tightness performance tests on one house each in the the low, middle and top floors - all vertically aligned – it could be seen that an increase in height led to a slight increase in ACH50 although the difference was insignificant. In addition, air-tightness analysis on houses with 2 sides exposed to the outside and hoses with 3 sides exposed to the outside showed that showed that houses with 2 sides exposed to the outside had higher ACH50 values to a certain degree compared to houses with 3 sides exposed to the outside.

Air-tightness performance measuring equipment was installed to the entrance door and kitchen window which are expected to have the largest influence on the infiltration or leakage of an apartment house and comparison experiments were carried out. Results show that apartments A, B and D have high measurement values when the airtightness performance equipment was installed on the windows. This is because the air-tightness performance of A, B and D's entrance doors were relatively lower. In the case of the other three complexes, there was no significant difference between the values from the Blower Door System or the KNS series. This is because the air-tightness performace of the front doors of these three complexes were high.

REFERENCES

[1] A Research on Standard of Ventilation and Air tightness Performance in Buildings, KICT, 2011

[2] ISO 9972, Thermal performance of buildings - Determination of air permeability of buildings - Fan pressurization method, 2006

[3] JIS A 2201, Test method for performance of building airtightness by fan pressurization, 2003

[4] Jo, Jae-Hun, Measurements of the Dwelling Unit Airtightness in High-rise Residential Buildings, Journal of the Architectural Institute of Korea, Vol. 26, No. 10, 2010

[5] Critical steps for wide scale implementation of building and ductwork airtightness, peter wouters, Air Infiltration and Ventilation Centre, 2011

[6] A Guide to Energy Efficient Ventrlation, Air Infiltration and Ventilation Centre, 2011

EXPERIMENTAL EVALUATION FOR THE DYNAMIC INSULATION APPLIED TO WINDOW FRAME

Shinsuke KATO[1], Sihwan LEE[*2], Kyosuke HIYAMA[1], Daisuke KAWAHARA[3]
Yoshikazu NOMURA[4], Yutaka OURA[4], Katsuhiko MORI[4], Satoshi SAWAKI[4]

1 Institute of Industrial Science, The University of Tokyo
4-6-1-Cw403, Komaba, Meguro-ku
153-8505, Tokyo, Japan

2 Department of Architecture, Tokyo University of Science
1-14-6-KS504, Kudankita, Chiyoda-ku
102-0073, Tokyo, Japan
**Corresponding author: shany@rs.tus.ac.jp*

3 Department of Architecture, The University of Tokyo
4-6-1-Cw403, Komaba, Meguro-ku
153-8505, Tokyo, Japan

4 Building Technology R&D, Sankyo Tateyama Al, Inc.
70, Hayakawa, Takaoka
933-8610, Toyama, Japan

ABSTRACT

An efficient thermal insulation of glazing or window frame is important because poor insulating performance usually cause the largest heat loss on any buildings. As one of the methods decreasing heat loss of buildings, we proposed a dynamic insulation system applied to window frame, and its energy saving performance and applicability for buliding had been confirmed using numerical simulation in previous study [1].

The aim of this study is to evaluate the thermal insulation efficiency of the proposed system by field test using experimental model houses among which we mainly focus on the dynamic insulation applied to window frame. First, the prototype of proposed system is designed to ventilate through the dynamic insulation in a window frame based on the fixed window which uses argon gas injected triple glazing. Then, we verify its thermal insulation efficiency in transient state, we constructed two experimental model houses located in Sapporo, Japan. The one of the house is installed a normal window frame with argon gas injected triple glazing coated Low-E film and the other one is installed in the proposed window frame. Finally, we evaluated the U-value of the proposed window frame by changing the ventilation rates through the dynamic insulation, after measuring the heat loss differences and the effective air leakage area to compare the airtightness and thermal insulation efficiency of two experimental model houses.

Although it could not measured the same U-value on the proposed window frame compare with the chamber test and field test unfortunately, the thermal insulation efficiency is increased by increasing ventilation rates through the dynamic insulation. The field test results show that the thermal insulation efficiency is increased approximately 89 % (U-value :1.80 $W/(m^2 \cdot K) \rightarrow 0.20\ W/(m^2 \cdot K)$) at the experimental model house installed in the proposed window frame by increasing ventilation rates ($0.00\ m^3/h \rightarrow 30.00\ m^3/h$). The ventilation volume led through the DI frame by pressure differences of inner/outer space heightens the thermal insulation performance. Therefore, DI window frame is an effective means of increasing the thermal insulation efficiency in any building.

KEYWORDS

Dynamic insulation, Window frame, Airtightness, Ventilation rate, Field test

INTRODUCTION

Approximately 24 % of global energy consumption goes on the heating, ventilating and air-conditioning of residential and commercial buildings. As one of the methods to reduce heat loss in buildings, we proposed a dynamic insulation system applied to window frames (the DI window frame), the energy saving performance of which was confirmed using a numerical simulation in a previous study [1].

OBJECTIVE

The purpose of this paper is to evaluate the insulation performance of the DI window frame via a field test using experimental model houses, among which we mainly focus on the DI window frame.

METHODS

1. Specification of the experimental model house

Figure 1(a) and Table 1 show experimental model houses and the specification of these models. The constructed model has insulation performance standardized by the society of heating, air-conditioning and sanitary engineers of Japan for cold districts. These models include timber framework and feature enhanced air-tightness by pasting adhesive tapes and sheets. The thermal resistances of the wall, ceiling and floor are 3.6, 6.6 and 3.9 $m^2 \cdot K/W$ respectively.
To keep variable pressure in the model, we achieve pull ventilation by installing an exhaust sirocco fan. The ventilation air quantity is varied using an inverter and dumper in the middle of the fan and vent cap in each model. The window, which is 2.3 m^2 in size, includes an argon gas-injected triple-glazing Low-E film coating, while the DI window frame has surface inlets (Fig. 1(b)). The normal window frame in the other model has the same internal shape without allowing the passage of air for ventilation and without porous material in the frame (Fig1(d),Fig(e)). We installed the window system on the north side to control disturbance due to sunlight. Walls were also erected adjacent to the models to control the amount of solar radiation heat by considering the solar radiation calculation shown in Fig. 1(c).

2. Primary experiment for the experimental model house

(1) Air leakage test for the experimental model houses
A precise relational expression between differential pressure and the ventilation air quantity is required for a simple air leakage tester based on JIS-Z-8762, comprising a venturi tube and fan, a wall with a connecting hole and a digital manometer (NAGANO KEIKI GC15). To approximate a condition equivalent to that of airtight experimental models, we adjusted the air vent of the model with a normal window frame.

(2) Thermal insulation performance of the experimental models
To evaluate the heat loss coefficient of these models, we measured the electrical differences amount per hour in each case under steady conditions. As the heat source, we used an 800 W radiant heater, the installation of which is shown in Figure 3(a). In the experiment, we set 90 mm insulation to decrease heat loss at the door. It was assumed that all electricity consumption per hour was heat loss. The measurement was performed in December from 20:00 and we measured the electricity consumption between 9:00 and 10:00 the following

day. We considered the interior air temperature distribution to be constant because an electric fan was used to circulate the inner room air. The method used to measure the sol-air temperature here was a SAT temperature measuring instrument installed in the middle of the window.

3. Evaluation of thermal insulation efficiency

(1) Methodology

The U-value of the window frame, U_{window_frame} determines the composition between the total heat transfer coefficient (α_{out}) and the thermal resistance of the window frame including the inner total heat transfer coefficient ($r_{window\_frame+inner}$) as shown in Eq. 1.

$$U_{window\_frame} = \frac{1}{r_{window\_frame+inner} + \frac{1}{\alpha_{out}}} \tag{1}$$

where, U_{window_frame} is the U-value of the window frame [$W/(m^2 \cdot K)$], α_{in} is the total indoor heat transfer coefficient [$W/(m^2 \cdot K)$], α_{out} is the total outdoor heat transfer coefficient [$W/(m^2 \cdot K)$] and r_{window_frame} is the thermal resistance of the window frame [$m^2 \cdot K/W$].

To determine r_{window_frame}, the measurement points on the frame surface are shown in Figs. 2(a) and 3(b). We obtained the window glass heat flux (q_{window_glass}) by using heat flux meter.The total outer heat transfer coefficient, α_{out} is obtained from the heat flux and differences between the surface temeperature of the heat flux meter ($T_{s\_heatflux\_meter}$) and outer sol-air temperatures ($T_{outer\_environment}$).

$$\alpha_{out} = \frac{q_{normal\_winodow\_frame}}{(T_{s\_heatflux\_meter} - T_{outer\_environment})} \tag{2}$$

In this experiment, the total outer heat transfer coefficient of the normal window frame is regarded as that of the window frame. The heat resistance, including the inner total heat transfer coefficient, $r_{window\_frame+inner}$, is calculated by measuring the outlet air temperature ($T_{outlet\_temperature}$), outdoor temperature ($T_{s\_out\_DI\_window\_frame}$) and sol-air temperature. In addition, we also measured the total interior heat transfer coefficient α_{in}.

$$r_{window\_frame+inner} = \frac{(T_{outlet\_temperature} - T_{s\_outer\_window\_frame})}{\alpha_{out} * (T_{s\_outer\_window\_frame} - T_{outer\_environment})} \tag{3}$$

(2) Experiment method

A radiant heater is set in the model as the heat source. The air pressure of the inner room was kept lower than that of the outside by introducing an exhaust fan to avoid disturbance from outside during the measurement. Measurement was carried out overnight with minimal temperature variation due to solar radiation in August. Preparation was performed for 12 hours before the experiment and the U-value was calculated based on the room temperature remaining steady between 3 and 4 am. An electric fan circulated the interior air to ensure uniform temperature distribution in the room. The outer environment temperature measured in the middle of the window was regarded as the representative temperature of each model and the surface temperature was determined by the weighted average temperature by the 4 parts shown in Fig. 3(b) and each part has 4 points in Fig3(c).

In this study, 4 cases with different ventilation rates (0, 5, 10, 20 and 30 m^3/h) were set in the experiment to evaluate the correlation between U-value and ventilation rates. To consider heat conduction alone, we closed the opening of the DI window frame for 0 m^3/h. All the cases were measured using digital manometer. It must also be noted that the case of 5 m^3/h was actually 6.5 m^3/h due to external disturbance.

RESULTS AND DISCUSSION

1. Results of the primary experiment for experimental model houses

(1) Results of the air leakage test
Because the clearance of the model house with the DI window frame installed with an opening in a closed condition was lower than 2.0 cm^2/m^2 as the standard for cold districts in Japan, we confirmed its effective airtight performance in Table 2. Such model clearance can control the effect of disturbance and enhance the experimental quality. Figures 4(a) and 4(b) show the result of the relationship between ventilation rate and pressure difference. Based on this expression, we controlled the ventilation rates to determine the U-value shown in Table 4 in which NO type means the frame without DI system and DI type means the frame with DI.

(2) Thermal insulation performance
Table 3 shows the result of thermal insulation performance. The heat loss coefficient does not include ventilation load. The heat loss of a normal window frame model is 20 W more than that of a model house with the DI window frame. This result shows how insulation performance varies depending on the construction.

2. Evaluation results of thermal insulation efficiency

Figure 5(a)-(d) show the surface temperature of window frames as taken by an infrared camera under 6 am. Compared with the considerable temperature difference between the case of 0 m3/h and the other case (6.5 m^3/h, 10.3 m^3/h, 20.0 m^3/h), the surface temperature of the DI window frame was much closer to the outdoor sol-air temperature, which confirmed far lower heat loss in qualitative terms in the DI window frame. And we confirm the temperature differences between upper parts and lower parts. The more quantity of ventilation, the closer temperature differences between the upper and lower part. In case of the 6.5 m^3/h, the difference is measured approximately 6 °C between the upper and the lower parts because of buoyancy. And estimation of the u-value of DI window frame in case of 30 m^3/h is impossible because outer environment temperature and temperature of outer DI window frame are almost equal. It means the measurement of the heat flux is not possible under this condition.

CONCLUSION

This paper described the insulation performance of the DI window frame using an experiment, when the frame model was applied to real buildings. Consequently, we confirmed that the difference in insulation performance between a model with a normal frame and with the DI window frame was approximately 20 W depending on the construction. In addition, we verified the efficiency of a DI window frame applied to an experimental model house by changing the ventilation rates.
The result shows that the U-value of the DI window frame decreases with increasing ventilation air rates and becomes approximately 0 $W/(m^2 \cdot K)$ in case of 30 m^3/h. Besides, the primary consideration in terms of the actual ventilation air rates through the DI window frame should be estimated based on the air leakage area, which is approximately half the measured

value. As for the future problems to be solved, we must verify the effect of the DI window frame when its air exchange rate is 0.5 1/h.

Part		Specification
Outside	Roof	Colored steel sheet, Asphalt roofing
		Structural plywood t = 9 [mm]
	Wall	Siding
	Window frame	DI window frame W1,690×H1,370
		Normal window frame W1,690×H1,370
	Door	W1,870×H730, U-value : 1.23 [W/(m^2·K)]
	Insulation	Ceiling: Wood fiber t=250 [mm]
		Wall: Wood fiber t=140 [mm]
		Floor: Wood fiber t=250 [mm]
	Vapour barrier	Ceiling: Airtight moisture-proof
		Wall: Airtight moisture-proof
		Floor: Airtight moisture-proof
Inside	Ceiling	PB t = 9.5 [mm]
	Wall	PB t = 9.5 [mm]
	Floor	Structural plywood t=24 [mm]

Table 1. Specification of experimental model house

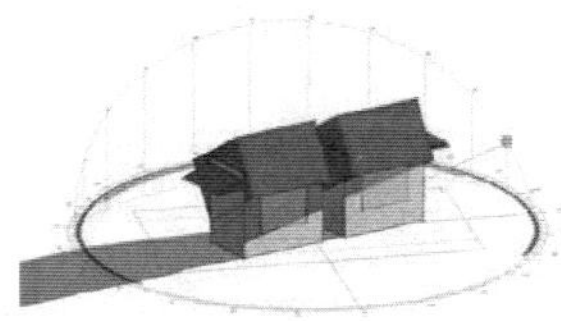

(a) Experimental model houses (b) Opening of DI window frame (c) Solar radiation calculation

Figure 1. Experimental model house and detail of the DI window frame

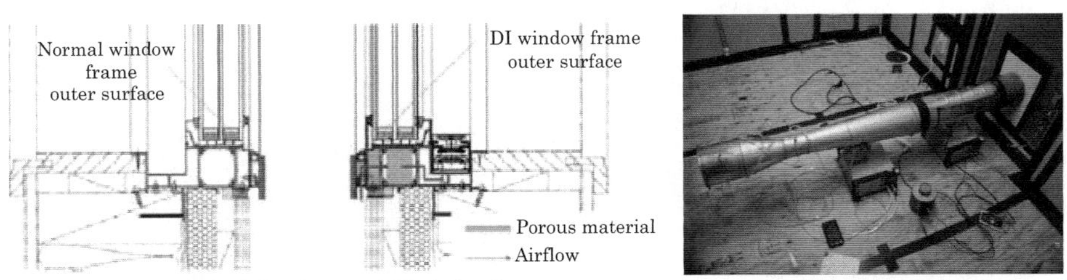

(a) Normal window frame and DI window frame (b) Air-leakage tester

Figure 2. Compare with normal window frame and DI window frame, and air-leakage tester

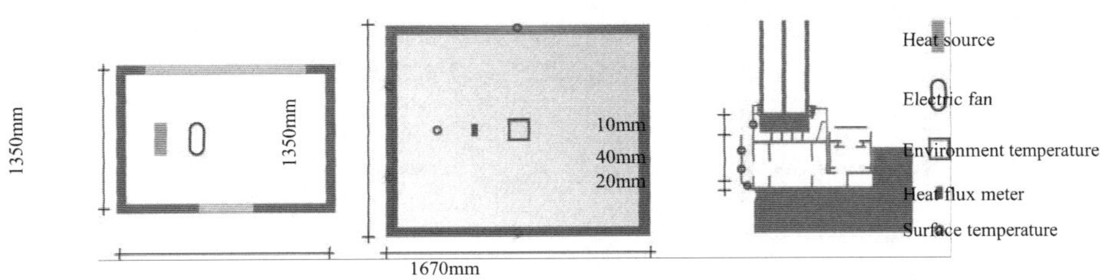

(a) Position of the heater (b) Measured parts (c) Measured points on window frame

Figure 3. Compare with normal window frame and DI window frame, and air-leakage test

	Ventilation rates (9.8Pa)	Air leakage area
Closed the opening of the DI winodw frame	5.24 [m^3/h]	3.6 [cm^2]
Opened the opening of the DI window frame	12.33 [m^3/h]	8.4 [cm^2]
Closed the opening by sheet	1.58 [m^3/h]	1.1 [cm^2]

Table 2. Result of air leakage test

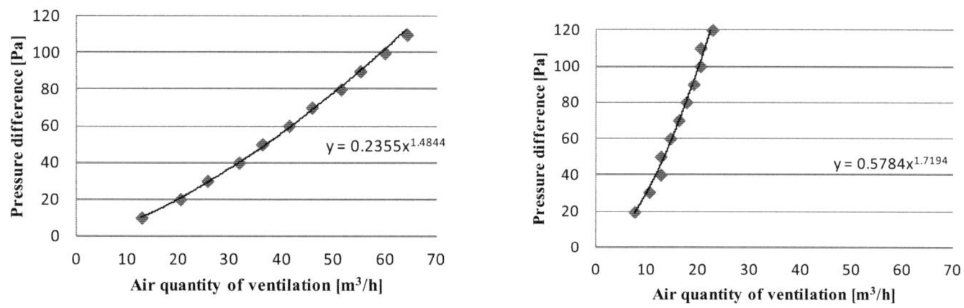

(a) Opened the opening of the DI window frame (b) Closed the opening of the DI window frame

Figure 4. Relation of ventilation rates and pressure difference

	Electrical consumption	Inner temperature	Sol-air temperature	Coefficient of heat loss without ventilation load
Model house installed normal window frame	670 [W]	46.0 [°C]	-1.28 [°C]	2.85 [$W/m^2 \cdot K$]
Model house installed the DI window frame	650 [W]	41.7 [°C]		3.04 [$W/m^2 \cdot K$]

Table 3 : Measurement results of thermal insulation performance

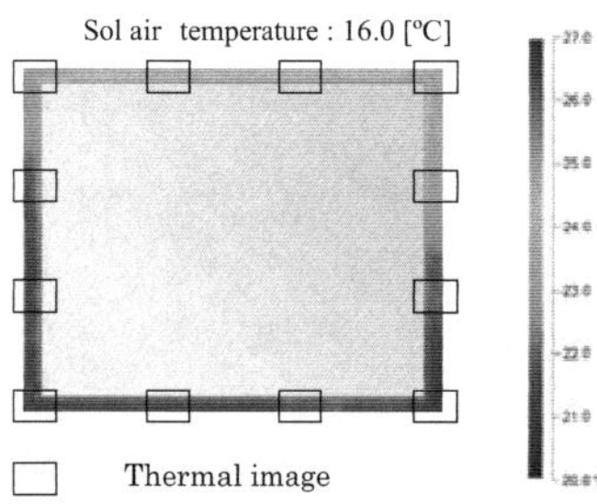

(a) Surface temperature of the window frame air quantity of ventilation is 0 m^3/h

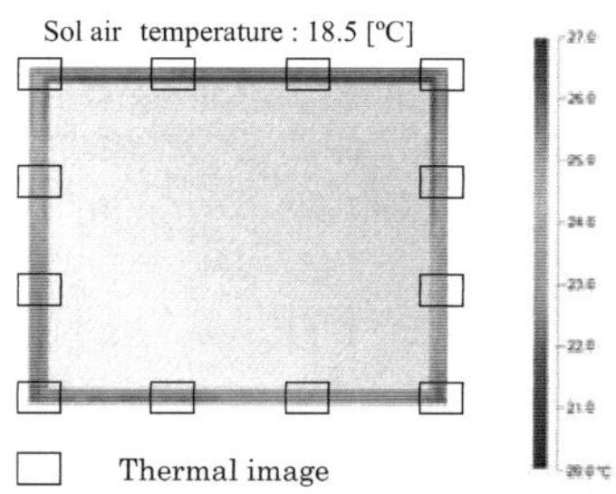

(b) Surface temperature of the window frame air quantity of ventilation is 6.5 m^3/h

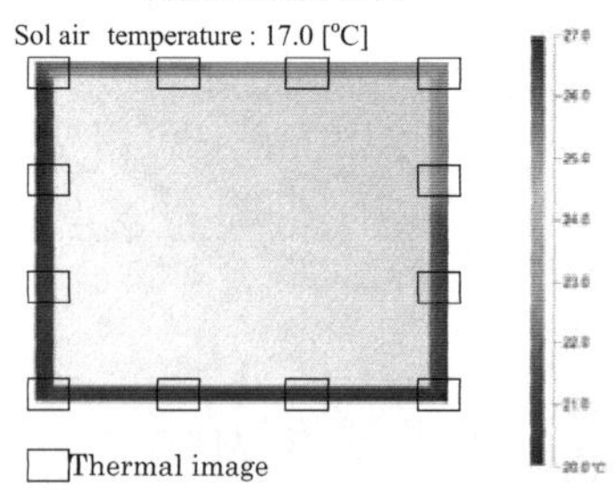

(c) Surface temperature of the window frame air quantity of ventilation is 10.3 m^3/h

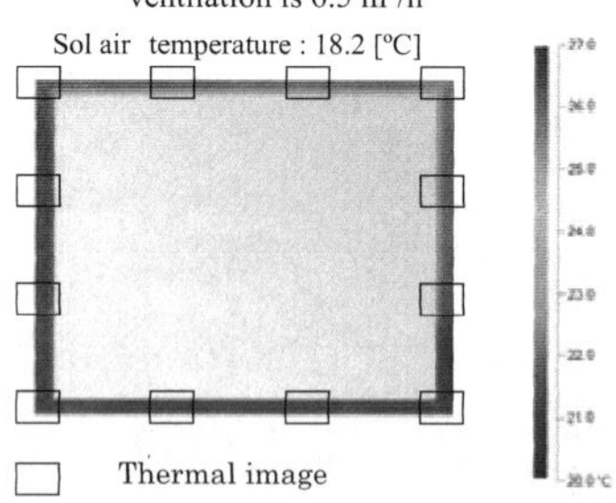

(d) Surface temperature of the window frame air quantity of ventilation is 20.0 m^3/h

Figure 5. Compare with surface temperature

Type	Ventilation rates	Total heat transfer coefficient		U-value of the window frame
		Inner surface	Outer surface	
NO	0 [m^3/h]	13.5 [W/(m^2·K)]	14.5 [W/(m^2·K)]	2.34 [W/(m^2·K)]
DI	0 [m^3/h]	12.4 [W/(m^2·K)]	15.5 [W/(m^2·K)]	1.80 [W/(m^2·K)]
DI	6.5 [m^3/h]	14.3 [W/(m^2·K)]	14.3 [W/(m^2·K)]	1.17 [W/(m^2·K)]
DI	10.3 [m^3/h]	14.8 [W/(m^2·K)]	18.8 [W/(m^2·K)]	0.96 [W/(m^2·K)]
DI	20.0 [m^3/h]	12.4 [W/(m^2·K)]	15.8 [W/(m^2·K)]	0.20 [W/(m^2·K)]
DI	30.0 [m^3/h]	4.8 [W/(m^2·K)]	21.4 [W/(m^2·K)]	--- [W/(m^2·K)]

Table 4. Result of thermal insulation performance

ACKNOWLEDGEMENTS

This study was made possible by the financial support received from the Ministry of the Environment, Japan. Moreover, we would like to thank the staff of Sankyo Tateyama Aluminium. Inc., J Architecture System. Inc., and Tsuchiya Hometopia. Inc. for their generous cooperation.

REFERENCES

[1] S. Lee, M. Tanaka, S. Kato. 2010. *The evaluation of the thermal insulation efficiency at the proposed window frames - Dynamic Insulation System applied to Window Frames (Part1)*, Air Infiltration and Ventilation Centre, AIVC2010, Seoul, Korea.

[2] M. Tanaka, S. Lee, S. Kato. 2010. *Energy saving effects of the proposed system on the residential buildings - Dynamic insulation system applied to window frames (Part 2)*, Air Infiltration and Ventilation Centre, AIVC2010, Seoul, Korea.

[3] A.Dimoudi, A.Androutsopoulos, S.Lykoudis. 2004. *Experimental work on a linked, dynamic and ventilated, wall component*, Energy and Buildings Vol.36, 443–453.

[4] M.S.Imbabi. 2006. *Modular breathing panels for energy efficient, healthy building construction*, Renewable Energy Vol.31, 729–738.

[5] Dalehaug Arvid, Fukushima Akira, Honma Yoshinori. 1993. *Dynamic insulation in a wall*, Report Collection of Architectural Institute of Japan, No.66, 261–264.

[6] J.M.Wong, F.P.Glasser, M.S.Imbabi. 2007. *Evaluation of thermal conductivity in air permeable concrete for dynamic breathing wall construction*, Cement & Concrete Composites Vol.29, 647–655.

[7] Guohui Gan. 2000. *Numerical evaluation of thermal comfort in rooms with dynamic insulation*, Building and Environment Vol.35, 445–453.

[8] S. Kato, K. Kuroki, S. Hagihara. 2007. *Method of in-Situ measurement of thermal insulation performance of building elements using infrared camera*, 6th International Conference on Indoor Air Quality, Ventilation & Energy Conservation in Buildings.

APPLICABILITY OF AIR SUPPLY TYPE AIRFLOW WINDOW SYSTEM APPLIED TO DOUBLE-PANE WINDOW

Sihwan LEE[*1], Shinsuke KATO[2], Kyosuke HIYAMA[2], Daisuke KAWAHARA[3]
Yoshikazu NOMURA[4], Yutaka OURA[4], Katsuhiko MORI[4], Satoshi SAWAKI[4]

1 Department of Architecture, Tokyo University of Science
1-14-6-KS504, Kudankita, Chiyoda-ku
102-0073, Tokyo, Japan
**Corresponding author: shany@rs.tus.ac.jp*

2 Institute of Industrial Science, The University of Tokyo
4-6-1-Cw403, Komaba, Meguro-ku
153-8505, Tokyo, Japan

3 Department of Architecture, The University of Tokyo
4-6-1-Cw403, Komaba, Meguro-ku
153-8505, Tokyo, Japan

4 Building Technology R&D, Sankyo Tateyama Al, Inc.
70, Hayakawa, Takaoka
933-8610, Toyama, Japan

ABSTRACT

It still remains heat loss and high risk of moisture condensation occurrence at glass of window because they have relatively poor insulating qualities and usually contribute the greatest heat loss by heat conduction in residential buildings. Although many attractive window systems are proposed to reduce heat loss such as double and triple glazing, low emissivity film coated glazing, argon gas injected glazing, vacuum insulated glazing, double-pane and triple-pane window etc., it has also demerits such as high initial cost and indoor air quality problem. To solve these problems, this paper propose air supply type airflow window system applied to double-pane window combine with a mechanical ventilator and a heat-recovery heat pump system.
The aim of this paper is to evaluate the thermal insulation efficiency and probability of moisture condensation in the air supply type airflow window system applied to double-pane window combine with 2 double glazing using numerical simualtion in order to confirm its feasibility and applicability in residential buildings. First, the proposed system is designed to ventilate through the air space of a double-pane window combine with 2 double glazing. Then, to verify its thermal insulation efficiency, the temperature distribution of the glass of window was evaluated using computer fluid dynamics (CFD) with a coated position of low emissivity film, after confirming calculation accuracy using double-pane window model, double glazing, and low emissivity film coated glazing. In addition, to verify the probability of moisture condensation, the dew-point temperature in the glass of window was calculated based on the various low emissivity film positions.
The calculated results show the thermal insulation efficiency of the proposed system is enhanced approximately 25.48% and 62.95% by airflow effect in comparison with the double-pane window and single-pane window with double glazing. Moreover the calculation results show that it is effective to use low emissive film in the proposed system to avoid moisture condensation occurrence. Therefore, it is confirmed technically feasible to reduce the energy consumption and to avoid moisture condensation occurrence the in residential buildings.

KEYWORDS

Air supply type airflow window system, Double-pane window, Moisture condensation, CFD

INTRODUCTION

Window glass remains prone to heat loss and considerable moisture condensation, due to its relatively poor insulating qualities and is usually the major source of heat loss via heat conduction in residential buildings. Although many attractive window systems have been proposed to reduce heat loss, such as double- and triple-glazing, glazing with low-emissivity film coating, argon gas-injected glazing, vacuum-insulated glazing, double- and triple-pane windows etc., these also have disadvantages such as the high initial cost and indoor air quality problem. To solve these, the author's previous paper proposed air supply via an airflow window system applied to the double-glazing combined with mechanical ventilation and heat-recovery heat pump systems. However, the problem of moisture condensation on the internal surface of the window remains.

OBJECTIVE

The purpose of this paper is to evaluate thermal insulation efficiency in the proposed new air supply via an airflow window system applied to dual double-glazed double-pane windows to confirm its feasibility and applicability in residential buildings. We also evaluate whether it produces excessive moisture condensation, depending on the position of the low emissivity film coating.

METHODS

CFD calculation is used to evaluate the proposed air supply type airflow window system. Although this method has been widely used to simulate air movement, heat transfer and mass transfer in indoor and outdoor environments, the experimental data or theoretical values must be validated. Accordingly, this paper evaluates the thermal insulation efficiency of the proposed system using CFD calculation, after confirming the calculation accuracy using a double-pane window model.

1. CFD validation on the double-pane window

In this section, we evaluate the CFD accuracy on the double-pane window with the position of the low-emissivity film coating. The model size is 765 × 1,870 × (6-12-6-12-6-12-6 double-glazing) mm and the calculation cases used are shown in Fig. 1. A low-Reynolds number k-epsilon turbulence model was used to compute the turbulent viscosity and diffusivity. The boundary conditions and material properties for the calculation are summarized in Tables 1 and 2. Outdoor/indoor air temperatures are assumed to be 00.00 ℃ and 22.00 ℃ respectively. The total heat transfer coefficient of the window surface is assumed to be 23.25 $W/(m^2 \cdot K)$ outdoors and 9.09 $W/(m^2 \cdot K)$ indoors for calculation. Moreover, the theoretical method makes reference to ISO10092 to calculate the theoretical U-value of the double-pane window.

2. Evaluation of thermal insulation efficiency on the air supply window system

To evaluate the thermal insulation efficiency of the proposed air supply window system, the temperature contribution of the window panes were calculated using 3-D steady-state CFD calculation, including detailed ray tracing-based radiation modeling. Figure 3(a) shows a plan of the building model used for the calculation, which is proposed as a model of a standard dwelling house in Japan by the Institute for Building Environment and Energy Conservation. In this study, a children's room on the second floor (2.40 (H) × 2.90 (W) × 3.50 (D) m = 24.36 m^3) is only used for calculations as shown in Fig. 3(b). Figure 3(c) shows the detail of the window (1.95 (H) × 1.65 (W) m) installed in the calculation model with Figs. 3(d) and 3(e). The air supply window system applied to the window glass allows fresh outdoor air into the room. The opening size was determined at 0.001 (H) × 0.765 (W) m and the applicable ventilation rate was applied 12.93 m^3/h to satisfy the requirement of the room, because the required indoor ventilation rate is at least one half air change per hour (0.5 ACH) over an entire 24 hour period in Japan. The turbulence model, material properties and boundary conditions for the CFD calculation used the same values as summarized in Tables 1 and 2, respectively. Unfortunately, to keep the calculation model simple in this study, the window frame is not considered.

RESULTS AND DISCUSSION

1. Result of the CFD validation on the double-pane window

The result of the CFD validation on the double-pane window with the position of the low-emissivity film coating is shown in Fig. 2. Figure 2(a) shows the surface temperature of the 4 plate glasses forming the double-pane window. Figure 2(b) shows a comparison of the theoretical and calculated U-values by CFD calculation. To calculate the U-value from the CFD results, the integrated value of the outdoor surface temperature of the plate glass, the outdoor air temperature and the total heat transfer coefficient of the outdoor window surface are used. As shown, the temperature difference peaks at the air space in contact with the glass coated with the low-emissivity film. Moreover, the indoor to outdoor heat loss fell when increasing the number of low-emissivity film coatings. Conversely, the outdoor surface temperature approached the outdoor air temperature of 0°C by increasing the number of low-emissivity film coatings. The calculation results of the double-pane window showed the U-value to be 1.20 $W/(m^2{\cdot}K)$ when using glazing with the low-emissive film coating. This represents an insulation performance enhancement of approximately 20.00 % based on the comparison 1.50 $W/(m^2{\cdot}K)$ with glazing without low-emissivity film applied. The error value of the CFD calculation was calculated to within approximately 5.62 % by comparison of the theoretical surface temperature and within approximately 2.04 % by comparison of the theoretical U-value. Therefore, the reliability of the CFD calculation was confirmed.

2. Result of calculation on the air supply window system

(1) U-value (thermal insulation efficiency)

The thermal insulation efficiency was predicted for the air supply type airflow window system, based on the position of the low-emissivity film coating. The result of the calculated U-value and the surface temperature of the glass are given in Table 3 and Fig. 4. Although the calculation results show that thermal insulation efficiency rose when increasing the number of low-emissivity film coatings, the glazing insulation performance is unchanged with the position of low-emissivity film coating. As shown, the thermal insulation efficiency increased approximately 25.48 ~ 38.43 % (U-value : 1.50 $W/(m^2{\cdot}K)$ → 1.12 $W/(m^2{\cdot}K)$ (Case 1), 0.90 $W/(m^2{\cdot}K)$ → 0.55 $W/(m^2{\cdot}K)$ (Case 9)) by applying the air supply window system in comparison with a double-pane window.

(2) Moisture condensation

The result of the calculated surface temperature of glass to confirm the occurrence of moisture condensation is also given in Table 3. The calculated results show that moisture condensation only occurred on the lower surface of glass where a low-emissivity film was applied close to indoors (Cases 6, 7), when the indoor air temperature and relative humidity exceeded 22.00 ºC and 50.00 %RH respectively (dew-point temperature : 11.10 ºC). Considering this, the outdoor air did not warm quickly because the low-emissivity film was applied close to indoors. It did not occur in the other cases (Cases 1 ~ 5) and the 2 low-emissivity film coated cases (Case 8, 9). The presence of moisture condensation depends not only on the outdoor temperature and humidity ratio but also the position of low-emissivity film coating and supply air flow rates.

CONCLUSION

This paper proposed an air supply type airflow window system to reduce energy consumption in residential buildings, the effectiveness of which was assessed with a feasibility study using CFD calculation. As a result, we found that the thermal insulation efficiency of the proposed system showed an insulation performance enhancement of approximately 25.48 % in comparison with the double-pane window. Moreover the calculation results showed that it is effective to use low-emissivity film in the proposed system to avoid moisture condensation. Therefore, it is confirmed as a technically feasible method to reduce energy consumption in residential buildings. A summary of the general findings of this study is as follows :

- The error in CFD calculation results was small and the reliability was confirmed by comparison of the theoretical U-value on the double- pane window.
- In the CFD accuracy simulation, the calculation results of the double-pane window showed the U-value to be 1.20 W/(m^2·K), where glazing was used with the low-emissivity film coating. It shows insulation performance enhancement of approximately 20.00 % in comparison to 1.50 W/(m^2·K) in the case of glazing with no low-emissive film applied.
- For the combined airflow glazing with no low-emissivity film applied, the U-value was calculated at 1.12 W/(m^2·K). This shows an insulation performance enhancement of approximately 25.48 % in comparison with a double-pane window using glazing without low-emissivity film applied. Moreover, it is confirmed that the proposed system can further boost thermal insulation efficiency by increasing the number of low-emissivity film coatings.
- Moisture condensation occurs only on the lower surface of glass where low-emissivity film is coated close to indoors when the indoor temperature and relative humidity exceed 22.00 ºC and 50.00 %RH respectively. Moisture condensation not only depends on the outdoor temperature and humidity ratio but also the position of low-emissivity film coating and the supply air flow rates.

FUTURE PERSPECTIVE

A future study is required to evaluate various design models such as triple glazing, argon gas-injected glazing, vacuum-insulated glazing and triple-pane window etc., because it is important to prevent moisture condensation on indoor glass surfaces. Moreover, thermal comfort should be calculated for a realizable room model by examining the effects of any cold drafts, thermal radiation effects and the energy-saving effects of the heat pumps included in the proposed system, which will be evaluated in future investigations. First, the following themes will be studied to avoid backflow and ventilate a stable air supply in the near future :

- The effect of indoor/outdoor pressure caused by outside wind conditions.
- The ventilation rates due to buoyancy caused by differences in indoor/outdoor temperature.

- In case of double-sided-openings (when installed in different directions).
- The effect of thermal environments caused by drafts or thermal radiation.

Item	Air	Water vapour	Glass pane	Low-E film
Specific heat (C_P)	1006.43 [J/(kg·K)]	2014 [J/(kg·K)]	753 [J/(kg·K)]	-
Conductivity (λ)	0.0242 [W/(m·K)]	0.0261 [W/(m·K)]	0.65 [W/(m·K)]	-
Viscosity (μ)	1.79×10^{-5} [kg/(m·s)]	1.34×10^{-5} [kg/(m·s)]	-	-
Moler weight (M)	28.97 [kg/kgmol]	18.02 [kg/kgmol]	-	-
Thickness (l)	-	-	6.00 [mm]	-
Emissivity (ε)	-	-	0.90 [-]	0.10 [-]

Table 1. Material properties.

Item		Contents
Model		765 x 1,870 x (6-12-6-12-6-12-6 double glazing) mm : 3-dimentional calculation
Turbulence model		Abe-Kondoh-Nagano low-Reynolds number k-epsilon model
Mesh		About 800,000 meshes (near wall y+ < 1)
Surface of glass		Velocity : No slip, $k\|_{wall}$: No slip, $\varepsilon\|_{wall} = 2\nu\left(\frac{\partial\sqrt{k}}{\partial y}\right)^2$
Heat transfer coefficient	Indoor	9.09 [W/(m²·K)]
	Outdoor	23.26 [W/(m²·K)]
Temperature condition	Indoor	22.00 [°C]
	Outdoor	0.00 [°C]

Table 2. Boundary condition.

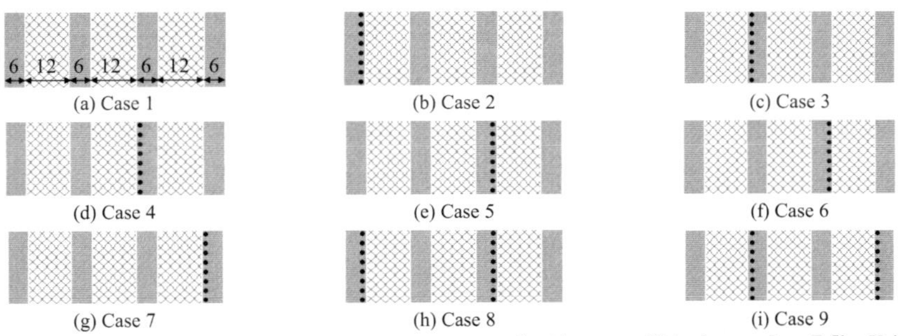

* ×:Air space, ■:Plate glass, · : Low-E film, Unit [mm]

Figure 1. Calculation cases by position of Low-E film.

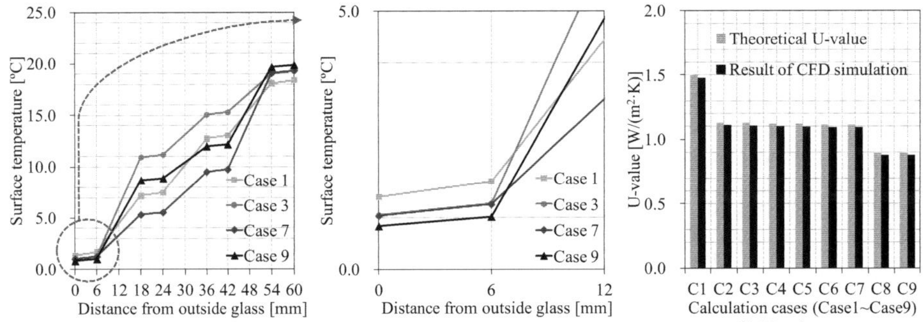

(a) Surface temperature of glass (b) U-value

Figure 2. CFD validation results on the double-pane window

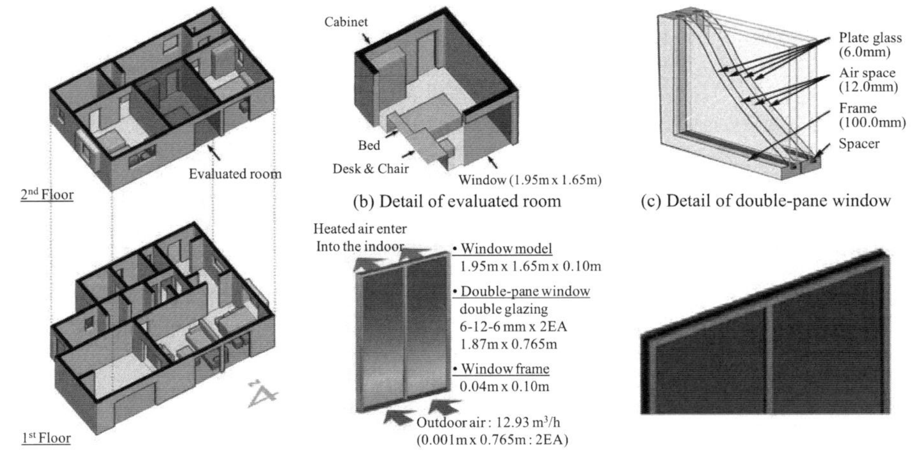

(a) Applied dwelling house (b) Detail of evaluated room (c) Detail of double-pane window (d) Detail of calculation model (e) Calculation mesh

Figure 3. Schematic of calculation model

Cases	U-value [W/(m²·K)]	T_{supply} [°C]	$T_{s\_outdoor}$ [°C]			T_{s_indoor} [°C]			Moisture condensation < 11.10°C*
			Max.	Min.	Avg.	Max.	Min.	Avg.	
Case 1	1.12	10.63	1.66	0.05	1.06	18.42	12.59	17.57	×
Case 2	0.77	13.01	1.21	0.03	0.74	19.12	13.78	18.01	×
Case 3	0.77	12.98	1.22	0.03	0.74	19.11	13.77	18.01	×
Case 4	0.76	11.06	1.31	0.02	0.72	19.20	13.07	18.37	×
Case 5	0.76	11.03	1.35	0.02	0.72	19.21	12.65	18.38	×
Case 6	0.78	8.33	1.67	0.04	0.74	19.27	11.04	18.79	○
Case 7	0.77	8.27	1.92	0.04	0.74	19.29	10.84	18.80	○
Case 8	0.55	10.28	1.45	0.02	0.52	19.63	12.60	19.01	×
Case 9	0.55	10.23	1.26	0.02	0.52	19.66	12.40	19.02	×

* Dew-point temperature is 11.10 ℃ when the indoor thermal environment controlled 22.00 ℃, 50.00 %RH.

Table 3. Results of calculation on the air supply window system.

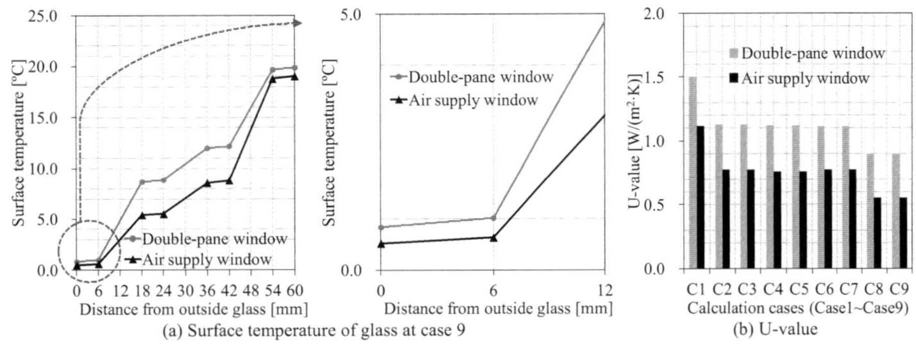

Figure 4. Comparison of the double-pane window and the air supply window system

ACKNOWLEDGEMENTS

This study was made possible by the financial support received from the Ministry of the Environment, Japan. Moreover, we would like to thank the staff of Sankyo Tateyama Aluminium. Inc., J Architecture System. Inc., and Tsuchiya Hometopia. Inc. for their generous cooperation.

REFERENCES

[1] S. Lee, M. Tanaka, S. Kato. 2010. *Dynamic Insulation System applied to Window Frames (Part1) The evaluation of the thermal insulation efficiency at the proposed window frames*, Air Infiltration and Ventilation Centre 2010, Korea.

[2] M. Tanaka, S. Lee, S. Kato. 2010. *Energy saving effects of the proposed system on the residential buildings - Dynamic insulation system applied to window frames (Part 2)*, Air Infiltration and Ventilation Centre, AIVC2010, Seoul, Korea.

[3] S. Kato, S. Lee, 2011. *The applicability of glazing system with dynamic insulation for residential buildings*, Air Infiltration and Ventilation Centre, AIVC2011, Brussels, Belgium.

[4] Abe, K., Kondoh, T., Nagano, Y. 1994. *A new turbulence for predicting fluid flow and heat transfer in separating and reattaching flows – I. Flow field calculations*, Heat Mass Transfer Vol.37, 139–151.

[5] Abe, K., Kondoh, T., Nagano, Y. 1995. *A new turbulence for predicting fluid flow and heat transfer in separating and reattaching flows – II. Thermal field calculations*, Heat Mass Transfer Vol.38, 1467–1481.

[6] Dimoudi, A., Androutsopoulos, A., Lykoudis, S. 2004. *Experimental work on a linked, dynamic and ventilated, wall component*, Energy and Buildings Vol.36, 443–453.

[7] Imbabi, M.S. 2006. *Modular breathing panels for energy efficient, healthy building construction*, Renewable Energy Vol.31, 729–738.

[8] Dalehaug A., Fukushima A., Honma Y. 1993. *Dynamic Insulation in a Wall*, Report Collection of Architectural Institute of Japan, No.66, 261–264.

[9] Wong, J.M., Glasser, F.P., Imbabi, M.S. 2007. *Evaluation of thermal conductivity in air permeable concrete for dynamic breathing wall construction*, Cement & Concrete Composites Vol.29, 647–655.

[10] Gan, G. 2000. *Numerical evaluation of thermal comfort in rooms with dynamic insulation*, Building and Environment Vol.35, 445–453.

[11] ISO 10292. 1994. *Glass in building – Calculation of steady-state U value (thermal transmittance) of multiple glazing.*
[12] ISO technical committee. 2006. *Thermal performance of windows, doors and shutters - Calculation of thermal transmittance* – ISO10077–1, –2.
[13] Siegel, R., Howell, J.R. 1981. *Thermal radiation heat transfer*, second edition, Hemispher Publishing Corporation, Washington.
[14] Tin-tai, C., Zhang, L., Kwong-fai, F., Lok-shun, C., Miao-miao, H. 2009. *Thermal performance of natural airflow window in subtropical and temperate climate zones – A comparative study*, Energy Conversion and Management, Vol.50, pp.1884–1890.